Contents of Volume 2

Gian Luigi Chierici

1 Principles of Petroleum Reservoir Engineering

Translated from the Italian by

Peter J. Westaway

With 220 Figures and 9 Tables

Springer-Verlag Berlin Heidelberg GmbH

Professor GIAN LUIGI CHIERICI
Chair of Petroleum Reservoir Engineering
Faculty of Engineering
Viale del Risorgimento, 2
I-40136 Bologna
Italy

Translator

PETER J. WESTAWAY
9 Nile Grove
Edinburgh EH10 4RE, Scotland
Great Britain

Title of the Original Italian edition:
Principi di ingegneria dei giacimenti petroliferi, Vol. 1

ISBN 978-3-662-02966-4   ISBN 978-3-662-02964-0 (eBook)
DOI 10.1007/978-3-662-02964-0

Library of Congress Cataloging-in-Publication Data. Chierici, Gian Luigi [Principi di ingegneria dei giacimenti petroliferi. English] Principles of petroleum reservoir engineering/Gian Luigi Chierici; translated from Italian by Peter J. Westaway. p. cm. Includes bibliographical references and index. ISBN 978-3-662-02966-4 1. Oil
reservoir engineering. 2. Oil fields. 3. Gas reservoirs.   I. Title. TN871.C494413 1994
622'.3382—dc20 93-26019

Typesetting: Macmillan India Ltd., Bangalore 25
SPIN: 10039899   32/3130/SPS – 5 4 3 2 1 0 – Printed on acid-free paper

Preface

Six years ago, at the end of my professional career in the oil industry, I left my management position within Agip S.p.A., a major multinational oil company whose headquarters are in Italy, to take up the chair in reservoir engineering at the University of Bologna, Italy. There, I decided to prepare what was initially intended to be a set of lecture notes for the students attending the course.

However, while preparing these notes, I became so absorbed in the subject matter that I soon found myself creating a substantial volume of text which could not only serve as a university course material, but also as a reference for wider professional applications. Thanks to the interest shown by the then president of Agip, Ing. Giuseppe Muscarella, this did indeed culminate in the publication of the first Italian edition of this book in 1989. The translation into English and publication of these volumes owes much to the encouragement of the current president of Agip, Ing. Guglielmo Moscato. My grateful thanks are due to both gentlemen.

And now – the English version, translated from the second Italian edition, and containing a number of revisions and much additional material. As well as providing a solid theoretical basis for the various topics, this work draws extensively on my 36 years of worldwide experience in the development and exploitation of oil and gas fields.

The first volume deals with the basics of reservoir engineering. After a look at the important aspects of reservoir geology, the thermodynamic properties of reservoir fluids are examined, and then the petrophysical and hydrodynamic properties of reservoir rock. This is followed by a description of the volumetric method for evaluating reserves, the interpretation of welltests in oil and gas wells, wellbore logging measurements to monitor the behaviour of the completion and reservoir under producing conditions, water influx calculations, and, finally, the use of the material balance equation to characterise (and in the case of gas, to predict) the reservoir behaviour.

The second volume covers more specialised aspects of reservoir exploitation: immiscible displacement of oil, water injection as a means of improving production rate and recovery factor, numerical techniques and their application in numerical models for reservoir simulation, short term prediction of well and reservoir behaviour through decline curve analysis and identified model technique, and, to conclude, enhanced oil recovery (EOR) processes to increase the recovery factor.

My principal objective in this first volume has been to explain clearly and concisely the physical concepts underlying the whole of petroleum reservoir engineering. To this end, I have wherever possible drawn on simplistic mathematical techniques which can be understood even by those with only a basic knowledge of differential equations. It is my firm belief that it is only by developing a solid appreciation of the fundamentals of reservoir engineering that one can hope to tackle the more complex problems encountered in professional life.

At the end of each chapter, the reader will find an extensive list of published technical references which can be consulted should further research on a particular topic be felt necessary.

Some of the material contained in these two volumes derives from the work of a number of people who collaborated with me in various petroleum engineering industry activities; other material is the fruit of countless discussions and exchanges of ideas over the years, with colleagues from Europe, U.S.A., the CIS (ex-USSR), Africa and the Arabian Gulf. My thanks to all concerned for their contribution of information.

A major part of the scientific work that I have been able to contribute to the field of reservoir engineering – much of which is to be found in these two volumes – originates from research work which was steered for many years by Ing. Giuseppe Faverzani, in his time executive vice-president, production, in Agip. It was he who taught me to avoid the "divine revelations" and "cookbook recipes" which proliferate in our profession, and to seek the truth through careful reasoning and logic. Thank you, boss, for the instruction you gave to me.

But my greatest and heartfelt thanks must go to the translator of this work, Mr. Peter J. Westaway. He has not only realised a perfect translation, but, having many years of reservoir engineering experience himself – particularly in the running and interpretation of wireline logs – has improved the text with a number of very valuable modifications and additions. His contributions are most significant in Chapters 6 and 7, and, especially, in Chapter 8, to the extent that I consider Mr. Westaway to be a co-author in these sections. Once again, many thanks, Peter.

I am also grateful to Prof. Guido Gottardi, associate professor at the Faculty of Engineering of the University of Bologna, Italy, who, while acting as co-lecturer for my own course, proposed a number of modifications and improvements in the drafting of the Italian edition of this book.

Finally, I would like to thank the two people who are responsible for the careful, patient and very professional preparation of the figures accompanying the text – Mrs. Irma Piacentini and Mr. Carlo Pedrazzini.

And so, once more, thank you everyone!

Bologna   GIAN LUIGI CHIERICI
February 1994

Contents

6   The Interpretation of Production Tests in Oil Wells

7   The Interpretation of Production Tests in Gas Wells

8 Downhole Measurements for the Monitoring of Reservoir Behaviour

9   The Influx of Water into the Reservoir

10   The Material Balance Equation

10.1 Material Balance Equation: Planning the Development
of a Dry Gas Reservoir to Maintain Steady Production for as
Long as Possible given Certain Operating Constraints, p. 397. –
10.2 Material Balance Equation: Estimation of Initial Gas
Reserves in Communication with an Aquifer, p. 400. –
10.3 Calculation of Drive Indices and Water Influx for an Oil
Reservoir with Gas Cap and Aquifer, p. 401. – 10.4 Calculation
of Oil Production for an Oil Reservoir not in Contact with
an Aquifer, p. 402.

List of Symbols

Letter symbol	Quantity	Dimensions
a	constant in the equation of state	$m\,L^5\,t^{-2}$
A	area, also:	L^2
	drainage area of a well	L^2
A_v	surface area per unit volume of rock	L^{-1}
A_w	areal extent of the aquifer	L^2
AOF	absolute open flow potential of a well	$L^3\,t^{-1}$
$^\circ$API	API gravity [Eq. (2.1)]	
b	covolume, also:	L^3
	slope of the linear section of the $m(p_{wf})$ vs. $\ln t$ plot, Eq. (7.46a)	$m\,L^{-1}\,t^{-3}$
b'	slope of the line defined by Eq. (7.51)	$m\,L^{-4}\,t^{-2}$
B_g	gas formation volume factor	
B_{gi}	gas formation volume factor at initial reservoir conditions	
B_o	oil formation volume factor	
B_{od}	oil formation volume factor resulting from the differential liberation of gas	
B_{of}	oil formation volume factor resulting from the flash liberation of gas	
B_{oi}	oil formation volume factor at initial reservoir conditions	
B_w	water (brine) formation volume factor	
c_b	rock bulk compressibility	$m^{-1}\,L\,t^2$
c_f	coefficient of porosity variation with pressure	$m^{-1}\,L\,t^2$
c_g	gas compressibility	$m^{-1}\,L\,t^2$
c_m	uniaxial compaction coefficient	$m^{-1}\,L\,t^2$
c_o	oil compressibility	$m^{-1}\,L\,t^2$
c_{oe}	oil equivalent compressibility [Eq. (10.68a)]	$m^{-1}\,L\,t^2$
c_p	pore volume compressibility	$m^{-1}\,L\,t^2$
c_r	rock grain compressibility	$m^{-1}\,L\,t^2$
c_t	total compressibility, pore volume + fluids	$m^{-1}\,L\,t^2$
c_w	water (brine) compressibility	$m^{-1}\,L\,t^2$
C	total solids content (TDS) in water, also:	$m\,L^{-3}$
	aquifer constant	$m^{-1}\,L^4\,t^2$
C_A	Dietz shape factor	
C_p	specific heat at constant pressure	$L^2\,t^{-2}\,T^{-1}$
CF	completion factor	
d	diameter	L
D	darcy, also:	L^2
	depth, also:	L
	non-Darcy flow skin coefficient, Eq. (7.19b)	$L^{-3}\,t$

Letter symbol	Quantity	Dimensions
e_w	water encroachment (or influx) rate	$L^3 t^{-1}$
$E_{R,g}$	gas recovery factor	
$E_{R,o}$	oil recovery factor	
f	friction factor	
$f_{i,L}$	fugacity of component "i" in the liquid phase	$m L^{-1} t^{-2}$
$f_{i,V}$	fugacity of component "i" in vapour phase	$m L^{-1} t^{-2}$
f_w	water fraction [Eq. (8.3b)]	
F	non-Darcy flow coefficient, Eq. (7.16), also:	$m L^{-7} t^{-1}$
	aquifer influence function, Sect. 9.5	$m L^{-4} t^{-1}$
F_s	well damage ratio	
g	acceleration of gravity	$L t^{-2}$
g_G	geothermal gradient	$L^{-1} T$
G	volume of gas initially in place (standard conditions)	L^3
G_p	cumulative volume of gas produced (standard conditions)	L^3
ΔG_p	incremental gas volume produced in a time step (standard conditions)	L^3
G_{pa}	gas recovery ultimate	L^3
GOR	gas/oil ratio	
GWC	gas–water contact	L
h	thickness	L
h_n	net pay thickness	L
h_t	total (or gross) pay thickness	L
H	specific enthalpy, per mass unit	$L^2 t^{-2}$
i	injection rate	$L^3 t^{-1}$
i_w	water injection rate	$L^3 t^{-1}$
I_H	hydrogen index	
J	productivity index	$m^{-1} L^4 t$
J_w	aquifer productivity index	$m^{-1} L^4 t$
$J(S_w)$	Leverett function	
k	permeability	L^2
$\bar{k}$	average permeability	L^2
$[k]$	permeability tensor	L^2
K	thermal diffusivity	$L^2 t^{-1}$
k_g	effective permeability to gas	L^2
k_h	horizontal permeability	L^2
k_l	cut-off permeability (layers with k less than k_l are considered non-pay)	L^2
k_L	permeability to liquid	L^2
k_o	effective permeability to oil	L^2
k_{rg}	relative permeability to gas	
k_{ro}	relative permeability to oil	
k_{rw}	relative permeability to water	
k_{rg}^*	normalized relative permeability to gas	
k_{ro}^*	normalized relative permeability to oil	
k_{rw}^*	normalized relative permeability to water	
$k_{rg,iw}$	relative permeability to gas, $S_g = 1 - S_{iw}$	

Letter symbol	Quantity	Dimensions
$k_{rg,or}$	relative permeability to gas, $S_g = 1 - S_{iw} - S_{or}$	
$k_{ro,gc}$	relative permeability to oil, $S_o = 1 - S_{iw} - S_{gc}$	
$k_{ro,iw}$	relative permeability to oil, $S_o = 1 - S_{iw}$	
$k_{rw,gr}$	relative permeability to water, $S_w = 1 - S_{gr}$	
$k_{rw,or}$	relative permeability to water, $S_w = 1 - S_{or}$	
k_s	permeability in the damaged (skin) zone	L^2
k_v	vertical permeability	L^2
k_w	effective permeability to water	L^2
K_i	equilibrium constant of component "i"	
L	length, also: number of moles in liquid phase	L
L_f	half-length of a fracture intersecting the well	L
$\mathscr{L}$	Lorenz heterogeneity coefficient	
m	ratio of initial reservoir free-gas volume to initial reservoir oil volume, also: slope of the linear section of the $p_{wf} = f(\ln t)$ plot, also:	$m\,L^{-1}t^{-2}$
	slope of the linear section of the $p_{ws} = f\left(\ln\dfrac{\Delta t}{t + \Delta t}\right)$ plot	$m\,L^{-1}t^{-2}$
m'	slope of the linear section of the $\dfrac{p_i - p_{wf}}{q_{sc}} = f(\ln t)$ plot, Eq. (6.26)	$m\,L^{-4}t^{-2}$
m''	slope of the linear section of the plot described by Eq. (6.34)	$m\,L^{-4}t^{-1}$
$m(p)$	real gas pseudo-pressure	$m\,L^{-1}t^{-3}$
$m(p_{1h})$	bottom-hole real gas pseudo-pressure 1 h after start of test	$m\,L^{-1}t^{-3}$
m_D	real gas pseudo-pressure, dimensionless	
$m_{D(MBH)}$	Matthews, Brons and Hazebroek function, expressed in terms of dimensionless real gas pseudo-pressure	
$m_{LIN}(p_{ws})$	(hypothetical) real gas pseudo-pressure on the extrapolation of the linear trend of the Horner plot	$m\,L^{-1}t^{-3}$
$m_{LIN}(p_{ws})(1\,h)$	value of $m_{LIN}(p_{ws})$ 1 h after start of buildup	$m\,L^{-1}t^{-3}$
M	molecular weight	m
M_{go}	gas/oil mobility ratio	
M_i	molecular weight of component "i"	m
M_{wg}	water/gas mobility ratio	
M_{wo}	water/oil mobility ratio	
n	number of moles, also: number of data points	
N	volume of stock-tank oil initially in place, also: total moles in a system, also: number of radioactive nuclei (Sect. 8.5)	L^3
N_p	oil produced, cumulative (stock tank conditions)	L^3
N_{pa}	oil recovery at abandonment, cumulative (stock tank conditions)	L^3
p	pressure	$m\,L^{-1}t^{-2}$
$\bar{p}$	average pressure in the drainage area	$m\,L^{-1}t^{-2}$
p^*	extrapolated average pressure, infinite drainage area	$m\,L^{-1}t^{-2}$
$\tilde{p}$	gas average pressure, Eq. (7.30a)	$m\,L^{-1}t^{-2}$

Letter symbol	Quantity	Dimensions
$\bar{p}_{aq}$	aquifer average pressure	$m\,L^{-1}t^{-2}$
p_b	pressure, bubble point (saturation) of an oil	$m\,L^{-1}t^{-2}$
p_c	critical pressure	$m\,L^{-1}t^{-2}$
$p_{Ch,i}$	parachor of component "i"	$m^{1/4}L^3t^{-1/2}$
p_{dp}	dew point pressure of a condensate gas	$m\,L^{-1}t^{-2}$
p_D	dimensionless pressure	
$p_{D(MBH)}$	Matthews, Brons and Hazebroek function, Eq. (6.18)	
p_e	pressure at the external boundary	$m\,L^{-1}t^{-2}$
p_f	fluid pressure	$m\,L^{-1}t^{-2}$
p_g	pressure in the gas phase	$m\,L^{-1}t^{-2}$
p_h	pressure in the hydrocarbon phase	$m\,L^{-1}t^{-2}$
p_i	initial (static) pressure	$m\,L^{-1}t^{-2}$
p_o	pressure in the oil phase	$m\,L^{-1}t^{-2}$
p_{ov}	overburden pressure	$m\,L^{-1}t^{-2}$
p_{pc}	pseudo-critical pressure	$m\,L^{-1}t^{-2}$
p_{pr}	pseudo-reduced pressure	
p_r	reduced pressure	
$\bar{p}_R$	average reservoir pressure	$m\,L^{-1}t^{-2}$
p_{sc}	pressure at standard conditions	$m\,L^{-1}t^{-2}$
p_{tf}	wellhead (tubing) pressure, flowing	$m\,L^{-1}t^{-2}$
p_{tr}	pressure at time "t", at a radial distance "r" from the well	$m\,L^{-1}t^{-2}$
p_w	bottom-hole pressure, also:	$m\,L^{-1}t^{-2}$
	pressure in the water phase	$m\,L^{-1}t^{-2}$
p_{wf}	bottom-hole pressure, flowing	$m\,L^{-1}t^{-2}$
p_{wh}	wellhead pressure, flowing	$m\,L^{-1}t^{-2}$
p_{ws}	bottom-hole static pressure	$m\,L^{-1}t^{-2}$
$p_{ws(LIN)}$	(hypothetical) bottom-hole static pressure on the extrapolation of the linear trend of the Horner plot	$m\,L^{-1}t^{-2}$
$p_{ws(LIN)}(1\,h)$	value of $p_{ws(LIN)}$ 1 h after the well was shut-in	$m\,L^{-1}t^{-2}$
p_{1h}	bottom-hole pressure 1 h after start of test	$m\,L^{-1}t^{-2}$
P_c	capillary pressure	$m\,L^{-1}t^{-2}$
$P_{c,go}$	capillary pressure between gas and oil phase	$m\,L^{-1}t^{-2}$
$P_{c,gw}$	capillary pressure between gas and water phase	$m\,L^{-1}t^{-2}$
$P_{c,ow}$	capillary pressure between oil and water phase	$m\,L^{-1}t^{-2}$
$(\Delta p)_{nD}$	additional pressure drop due to non- Darcy flow	$m\,L^{-1}t^{-2}$
Δp_s	additional pressure drop due to skin effect	$m\,L^{-1}t^{-2}$
PE	photoelectric absorption index	L^2
q	flow rate, volumetric, under reservoir conditions	L^3t^{-1}
$\boldsymbol{q}$	flow rate vector	L^3t^{-1}
q_D	dimensionless flow rate [Eq. (9.15)]	
q_g	gas flow rate, volumetric at reservoir conditions	L^3t^{-1}
$q_{g,sc}$	gas flow rate, volumetric, at standard conditions	L^3t^{-1}
q_L	liquid hydrocarbons (condensate) flow rate	L^3t^{-1}
q_m	mass flow rate	$m\,t^{-1}$
$q_{m,g}$	gas mass flow rate	$m\,t^{-1}$
q_{sc}	flow rate, volumetric, under standard conditions	L^3t^{-1}
Q_D	van Everdingen–Hurst function (Sect. 9.4)	
r	radius	L
r_D	dimensionless radius	
r_{De}	dimensionless external boundary radius	
r_e	external boundary radius	L

Letter symbol	Quantity	Dimensions
r_s	radius of damaged or stimulated zone (skin)	L
r_w	well radius	L
R	universal gas constant, per mole	$m\,L^2\,t^{-2}\,T^{-1}$
R_g	S_g normalization parameter, Eq. (3.58c), used in computing k_{ro} and k_{rg} values	
R_p	($=G_p/N_p$): cumulative gas/oil ratio (standard conditions)	
R_s	solution gas/oil ratio (gas solubility in oil)	
R_{sd}	solution gas/oil ratio by differential liberation	
R_{sf}	solution gas/oil ratio by flash liberation	
R_{si}	solution gas/oil ratio at initial reservoir conditions	
R_{sw}	gas solubility in water	
R_w	S_w normalization parameter, Eq. (3.59c), used in computing k_{ro} and k_{rw} values	
s	independent variable in the Boltzmann transform	
S	saturation, also: skin factor	
S'	($=S + Dq_{sc}$), skin + non-Darcy flow effect, Eq. (7.21)	
S_g	gas saturation	
S_g^*	normalized gas saturation	
S_{gc}	critical gas saturation	
S_{gr}	residual gas saturation	
S_{iw}	irreducible water saturation	
S_o	oil saturation	
S_{or}	residual oil saturation	
S_w	water saturation	
S_w^*	normalized water saturation	
$\bar{S}_w$	average water saturation	
S_{wi}	initial water saturation	
t	time	t
$\bar{t}$	($=N_p/q_n$), Eq. (6.55), conventional production time	t
t_D	dimensionless time [$=kt/(\phi\mu c_t r_w^2)$]	
t_{DA}	dimensionless time [$=kt/\phi\mu c_t A)$]	
t_{tr}	duration of the transient flow regime	t
T	temperature	T
T_c	critical temperature	T
$T_{c\prime}$	cricondentherm temperature	T
T_{pc}	pseudo-critical temperature	T
T_{pr}	pseudo-reduced temperature	
T_r	reduced temperature	
T_R	reservoir temperature	T
T_{sc}	temperature, standard conditions	T
Δt	closed-in time during a pressure buildup	t
u	Darcy velocity ($=q/A$)	$L\,t^{-1}$
$\boldsymbol{u}$	vector of Darcy velocity	$L\,t^{-1}$
u_g	Darcy velocity of the gas phase ($=q_g/A$)	$L\,t^{-1}$
u_o	Darcy velocity of the oil phase ($=q_o/A$)	$L\,t^{-1}$
u_w	Darcy velocity of the water phase ($=q_w/A$)	$L\,t^{-1}$
v	velocity	$L\,t^{-1}$
$\boldsymbol{v}$	velocity vector	$L\,t^{-1}$

Letter symbol	Quantity	Dimensions
V	volume, also:	L^3
	moles of vapour phase	
$\mathscr{V}$	permeability variation, as defined by Law	
V_{aq}	aquifer volume	L^3
V_b	rock bulk volume	L^3
V_g	gas volume, reservoir conditions	L^3
V_H	hydrocarbon volume, reservoir conditions	L^3
V_m	mole volume	L^3
V_o	oil volume, reservoir conditions	L^3
V_p	pore volume	L^3
V_r	grain volume in a rock sample	L^3
V_R	reservoir volume	L^3
V_w	water volume, reservoir conditions	L^3
W_e	cumulative volume of encroached water	L^3
W_{ei}	cumulative volume of water encroached into the reservoir when its pressure falls to zero	L^3
W_p	cumulative volume of produced water	L^3
WGC	water/gas contact	L
WOR	water/oil ratio	
x_i	mole fraction of component "i" in liquid phase	
y_i	mole fraction of component "i" in vapour phase	
z	vertical coordinate, also:	L
	gas compressibility factor, also:	
	atomic number, Chap. 8	
$\tilde{z}$	gas compressibility factor, average value in the pressure interval	
z_i	mole fraction of component "i" in the overall system	
z_{GOC}	depth, ssl, of gas/oil contact	L
z_{WOC}	depth, ssl, of water/oil contact	L
β	inertial resistance coefficient, Eq. (7.13a), also:	L^{-1}
	decay constant of a radioactive nucleus (Sect. 8.5)	t^{-1}
γ_g	gas gravity (air $= 1.0$)	
γ_o	weight of a unit volume of oil ($= g\rho_o$)	$m\,L^{-2}\,t^{-2}$
γ_w	weight of a unit volume of water ($= g\rho_w$)	$m\,L^{-2}\,t^{-2}$
δ_{ij}	interaction coefficient between components "i" and "j"	
ε_g	dielectric constant of gas at reservoir conditions	
ε_o	dielectric constant of oil at reservoir conditions	
ε_w	dielectric constant of water at reservoir conditions	
ε_x	strain in the "x" direction	
ε_y	strain in the "y" direction	
η	hydraulic diffusivity [$= k/(\phi\mu c_t)$]	$L^2\,t^{-1}$
Θ_c	water/rock contact angle	

Letter symbol	Quantity	Dimensions
λ	wavelength	L
λ_g	gas mobility $(=kk_{rg}/\mu_g)$	$m^{-1}L^3t$
$\lambda_{g,iw}$	gas mobility at $S_g = 1 - S_{iw}$	$m^{-1}L^3t$
$\lambda_{g,or}$	gas mobility at $S_g = 1 - S_{iw} - S_{or}$	$m^{-1}L^3t$
λ_o	oil mobility $(=kk_{ro}/\mu_o)$	$m^{-1}L^3t$
$\lambda_{o,iw}$	oil mobility at $S_o = 1 - S_{iw}$	$m^{-1}L^3t$
λ_w	water mobility $(=kk_{rw}/\mu_w)$	$m^{-1}L^3t$
$\lambda_{w,gr}$	water mobility at $S_w = 1 - S_{gr}$	$m^{-1}L^3t$
$\lambda_{w,or}$	water mobility at $S_w = 1 - S_{or}$	$m^{-1}L^3t$
μ	viscosity, also:	$mL^{-1}t^{-1}$
	Joule–Thomson coefficient, Eq. (8.4)	Lt^2T
μ_g	gas viscosity	$mL^{-1}t^{-1}$
$\tilde{\mu}_g$	gas viscosity, average value in the pressure interval	$mL^{-1}t^{-1}$
$\mu_{g,sc}$	gas viscosity, standard conditions	$mL^{-1}t^{-1}$
$(\mu_g)_w$	gas viscosity at wellbore	$mL^{-1}t^{-1}$
μ_o	oil viscosity	$mL^{-1}t^{-1}$
μ_w	water viscosity	$mL^{-1}t^{-1}$
ν	Poisson ratio, also:	
	frequency	t^{-1}
ρ	density	mL^{-3}
ρ_b	rock bulk density	mL^{-3}
ρ_f	fluid density	mL^{-3}
ρ_g	density of the gas phase	mL^{-3}
$\rho_{g,sc}$	gas density at standard conditions	mL^{-3}
ρ_L	density of liquid hydrocarbons (condensate)	mL^{-3}
ρ_o	oil density	mL^{-3}
$\rho_{o,sc}$	oil density at standard conditions	mL^{-3}
ρ_r	rock grain density	mL^{-3}
ρ_t	overburden rock density	mL^{-3}
ρ_w	water (brine) density	mL^{-3}
$\rho_{w,sc}$	water (brine) density at standard conditions	mL^{-3}
σ	surface, or interfacial, tension, also:	mt^{-2}
	standard deviation	
σ^2	variance	
$\bar{\sigma}$	hydrostatic component of the stress tensor	$mL^{-1}t^{-2}$
σ_{go}	interfacial tension between gas and oil phases	mt^{-2}
σ_{os}	oil/rock interfacial tension	mt^{-2}
σ_{ow}	interfacial tension between oil and water phases	mt^{-2}
σ_{xx}	component of the stress tensor in the "x" direction	$mL^{-1}t^{-2}$
σ_{yy}	component of the stress tensor in the "y" direction	$mL^{-1}t^{-2}$
σ_{ws}	water/rock interfacial tension	mt^{-2}
σ_{zz}	component of the stress tensor in the "z" direction	$mL^{-1}t^{-2}$
Σ_h	macroscopic capture cross section of the hydrocarbon phase	L^2
Σ_r	rock macroscopic capture cross section	L^2
Σ_w	water macroscopic capture cross section	L^2
τ	tortuosity, also:	
	decay time (thermal neutrons)	t

Letter symbol	Quantity	Dimensions
ϕ	porosity	
$\bar{\phi}$	average porosity	
ϕ_e	effective, or interconnected, porosity	
ϕ_l	cut-off porosity	
ϕ_o	porosity value measured at standard conditions	
ϕ_t	total porosity	
Φ	potential, per unit mass	$L^2 t^{-2}$
Φ^*	potential per unit mass, as defined in Eq. (3.27)	$L^2 t^{-2}$
Φ_g	gas potential, per unit mass	$L^2 t^{-2}$
Φ_o	oil potential, per unit mass	$L^2 t^{-2}$
Φ_w	water potential, per unit mass	$L^2 t^{-2}$
ψ	potential, per unit volume, also: fugacity coefficient	$m L^{-1} t^{-2}$
ω_i	acentric coefficient of component 'i", as defined by Pitzer	

Conversion Factors for the Main Parameters

Quantity	SI unit (1)	Practical unit (2)	Conversion factor, F (1) = F × (2)	
Space, time, speed				
Length	m	in	2.54*	E − 02
		ft	3.048*	E − 01
		mile (intl)	1.609344*	E + 03
Area	m²	sq in	6.4516*	E − 04
		sq ft	9.290304*	E − 02
		acre	4.046856	E + 03
		ha	1.0*	E + 04
Volume, capacity	m³	cu ft	2.831685	E − 02
		US gal	3.785412	E − 03
		bbl (42 US gal)	1.589873	E − 01
		acre-ft	1.233489	E + 03
Time	s	min	6.0*	E + 01
		h	3.6*	E + 03
		d	8.64*	E + 04
		yr	3.15576*	E + 07
Speed	m/s	ft/min	5.08*	E − 03
		ft/h	8.466666	E − 05
		ft/day	3.527777	E − 06
		mile/yr	5.099703	E − 05
		km/yr	3.168809	E − 05
Mass, density, concentration				
Mass	kg	lbm	4.535924	E − 01
		US ton (short)	9.071847	E + 02
		US cwt	4.535924	E + 01
Density	kg/m³	lbm/ft³	1.601846	E + 01
		lbm/US gal	1.198264	E + 02
		lbm/bbl	2.853010	E + 00
		°API	(1.415 E + 05)/ (131.5 + API)	
Concentration	kg/m³	lbm/bbl	2.853010	E + 00
	kg/kg	ppm (wt)	1.0*	E − 06
Pressure, compressibility, temperature				
Pressure	Pa	psi	6.894757	E + 03
		kg/cm²	9.80665*	E + 04
		atm	1.01325*	E + 05
		mmHg = torr	1.333224	E + 02

* Exact conversion factor

Quantity	SI unit (1)	Practical unit (2)	Conversion factor, F $(1) = F \times (2)$	
Pseudo-pressure	Pa/s	$(psi)^2/cP$	4.753767	$E + 10$
of real gases		$(kg/cm^2)^2/cP$	9.617038	$E + 12$
Pressure gradient	Pa/m	psi/ft	2.262059	$E + 04$
		$kg/cm^2 \times m$	9.80665*	$E + 04$
Compressibility	Pa^{-1}	psi^{-1}	1.450377	$E - 04$
		cm^2/kg	1.019716	$E - 05$
Temperature	K	°F	$(°F + 459.7)/1.8$	
		°C	$°C + 273.2$	
Temperature	K/m	°F/ft	1.822689	$E + 00$
gradient		°C/m	1.0*	$E + 00$

Flow rate, productivity index

Quantity	SI unit	Practical unit	Conversion factor, F	
Flow rate	m^3/s	ft^3/hr	7.865791	$E - 06$
(volume basis)		ft^3/d	3.277413	$E - 07$
		bbl/h	4.416314	$E - 05$
		bbl/d	1.840131	$E - 06$
		m^3/h	2.777778	$E - 04$
		m^3/d	1.157407	$E - 05$
Mass flow rate	kg/s	lbm/h	1.259979	$E - 04$
		lbm/d	5.249912	$E - 06$
		US ton/h	2.519958	$E - 01$
		US ton/d	1.049982	$E - 02$
Gas/oil ratio	m^3/m^3	cu ft/bbl	1.781076	$E - 01$
Productivity index	$m^3/(s\,Pa)$	bbl/(d psi)	2.668884	$E - 10$
		$m^3/(d\,kg\,cm^{-2})$	1.180227	$E - 10$

Transport properties

Quantity	SI unit	Practical unit	Conversion factor, F	
Permeability	m^2	md	9.869233	$E - 16$
		D	9.869233	$E - 13$
Viscosity	Pa s	μP	1.0*	$E - 07$
		cP	1.0*	$E - 03$
Mobility (kk_r/μ)	$m^2/Pa\,s$	md/cP	9.869233	$E - 13$
		D/cP	9.869233	$E - 10$
Diffusivity	m^2/s	ft^2/h	2.58064*	$E - 05$
		cm^2/s	1.0*	$E - 04$
Surface tension	N/m	dyn/cm	1.0*	$E - 03$

1 Hydrocarbon Reservoirs

1.1 Conditions for the Existence of an Oil or Gas Reservoir

A "reservoir" is defined as an accumulation of oil and/or gas in a porous and permeable rock – almost invariably of sedimentary origin. The conditions necessary (but not always sufficient) for the formation of such an accumulation are:

1. The presence in the sedimentary basin – in which the hydrocarbon accumulation is to form – of *source rock*;[5] namely, rocks (generally shales) highly charged with organic material of vegetable and animal origins.
2. The transformation of this organic material into hydrocarbons.[5] This occurs on a geological time scale of hundreds of thousands, or millions, of years, through the combined action of temperature, pressure, and the catalytic effects of certain components of the source rock, which cause the breaking down, or *cracking*, of the large organic molecules.
3. The migration of the hydrocarbons[2] out of the source rock, which is usually of extremely low permeability, into adjacent permeable strata (*primary migration*), and from here along permeable paths towards geological structures which might form suitable traps (*secondary migration*).
4. The presence, not too distant from the source rock, of a geological structure or rock configuration which is favourable to the entrapment and accumulation of the migrating hydrocarbons[1,4] (the *trap*).
5. A continuous impermeable stratum (*cap rock*) or other sealing structure, which prevents the further migration of the hydrocarbons into the overlying rock,[1,4] and their eventual seepage to surface.

We will now briefly examine the principal factors involved in the formation of an oil or gas reservoir.

1.2 Sedimentology

Sedimentary rocks are formed by a process of *deposition*, in an aqueous environment (sea, lake, river) or aeolian (desert). There are three main categories of sedimentary rock[3], based on their origins:

- clastic (or terrigenous)
- chemical
- organic

Clastic rocks are formed from fragments of pre-existing rocks, which have been broken down by chemical action or physical disintegration (erosion). There are a total of four stages involved in the process:

- fragmentation of the original rock (which may itself be of sedimentary or non-sedimentary origin)
- transport of the fragments by water or air
- deposition of the fragments
- transformation of the fragments into a rock mass (*diagenesis*)

During the transport phase, a process of sorting or gradation occurs – the largest fragments will be deposited first, followed by successively smaller ones. This means, for instance, that in a marine environment where material has been transported by a river, pebbles and gravel tend to be deposited closest to the shore, followed by coarse sands, then fine sands, and, furthest out, silts and clays.

The speed of the transporting medium (e.g. river current, wind) is of equal importance in determining the distance that the rock particles are transported before deposition. Thus we speak of *high energy* (fast) and *low energy* (slow) environments.

Referring to a marine depositional environment, successive cycles of *ingression* (where a rising sea level encroaches on once-dry land) and *transgression* (the opposite) explain the wide variations of grain size observed in a vertical section (gravels, sands, silts, clays) as shown in Fig. 1.1.

As deposition continues, earlier material becomes compressed beneath the weight of the more recent overlying sediment,[2] and various diagenetic changes may begin to occur. The principal change at this stage is *cementation* – the fragments and grains become cemented together by salts (calcium carbonate in particular) precipitated from the surrounding water.[3] Another important aspect of diagenesis is the partial dissolution of silica (from quartz grains), its removal and eventual reprecipitation elsewhere, where local chemical and physical conditions are suitable.[3]

Consequently, both calcium carbonate and silica are commonly found as the cementing medium in clastic rocks.

Rocks of chemical origin owe their formation to the precipitation of salts from aqueous solutions whose concentrations have reached saturation through evaporation or temperature variations.[3]

Among the commonest rocks of chemical origin are: *limestones* – travertine, beach rock, oolitic and pisolitic limestone – consisting of spheres of calcium carbonate; and *dolomites* – calcium magnesium carbonates formed from the reaction of magnesium chloride (in the sea water) with calcium carbonate.

Other examples are halite (rock-salt) and anhydrite (a calcium sulphate). These are too tight to be of interest as potential reservoir rock, but frequently form the cap rock to a reservoir.

Rocks of organic origin are formed by the deposition and cementation of animal and plant remains containing calcium or silicon minerals, extracted from the water in which they lived and fixed in their shells or skeletons.

Since the distribution of animal and plant populations can vary considerably over even small scales of distance and time, it is not surprising that reservoir rocks of organic origin are characterized by a high degree of heterogeneity[3].

Some important examples of this organic category are: *limestones* – nummulitic, fusulinid, bryozoan, ammonitic and lumachelle; *dolomites* – formed in atoll and

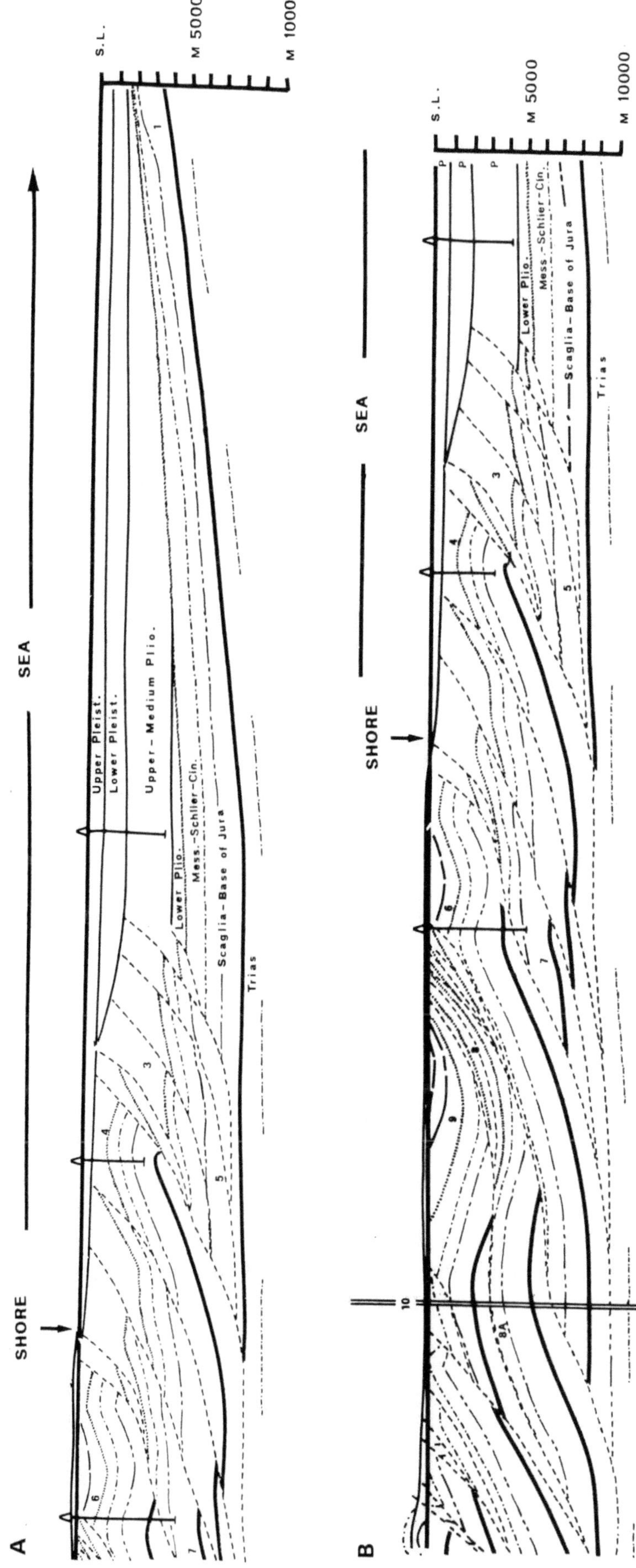

Fig. 1.1A, B. *Stratification caused by successive marine ingressions and transgressions*

barrier reef environments among coral remains; *jaspers* – formed from the oozes of siliceous sponges and radiolari; and *tripoli* or *diatomite marls*, consisting of the minute shells of marine or fresh-water diatoms.

Calcareous rocks of chemical or organic origin are grouped according to their texture into five distinct types.[3]

1. *Mudstone* (micrite): a matrix of calcareous mud with a grain content of less than 10%;
2. *Wackestone* (micritic limestone): a matrix of calcareous mud with a grain content of more than 10%;
3. *Packstone* (micritic calcarenite): a self-supporting calcitic grain structure, with intergranular spaces filled with micrite;
4. *Grainstone* (spathic calcarenite): a self-supporting calcitic grain structure, with intergranular spaces partly filled with crystalline calcite;
5. *Boundstone* (organogenic carbonates): consisting of structures and encrustations bound together while the organisms were alive (e.g. corals).

1.3 The Generation of Hydrocarbons, and Their Migration

To summarise so far, hydrocarbons originate from the thermal and/or microbic breakdown of dead animal and plant microorganisms (plankton) deposited along with clastic or organogenic sediments on the beds of seas, lagoons or lakes.

The maximum concentration of living organisms occurs in calm, shallow, warm waters found along continental shelves, not too far from the shore. The sediments laid down in these low energy conditions are predominantly clays or shales – and it is therefore these same shales that are potential hydrocarbon source rocks.

In fact, only a very small fraction of the dead plankton present in the water actually becomes part of the sediment; by far the major part is destroyed by bacterial action as it sinks to the bottom.[5]

However, bacterial activity continues within the bottom sediment, and it is this that can give rise to the generation of petroleum[5] from primary organic matter under the right conditions.

Ongoing sedimentation builds up the geostatic loading, causing compression and subsidence of the deeper sediments. Increasing pressure and, more importantly, temperature, helped by the catalytic action of some clay minerals, leads to the thermal break-down or cracking of the complex organic material into simpler hydrocarbon compounds. This process starts with bitumen and heavy oils, but, over geological time, leads to successively lighter components, and would culminate eventually in pure methane.[2, 5]

Figure 1.2 provides a qualitative illustration of how the evolution of hydrocarbons in the source-rock depends on *temperature* (which is a function of depth), and *time*. The older and deeper the source rock is, the greater the likelihood of light oil or gas being formed (the limiting case being pure methane): this is why, historically, deep exploration has found a preponderance of dry gas reservoirs.

Compaction of the deeper sediments beneath the overlying load causes the expulsion of some of the fluid (water and hydrocarbons) from the pore spaces of the source-rock. In this *primary migration*[2] phase, movement is towards more permeable strata in contact with the source rock. While it is fairly certain that the

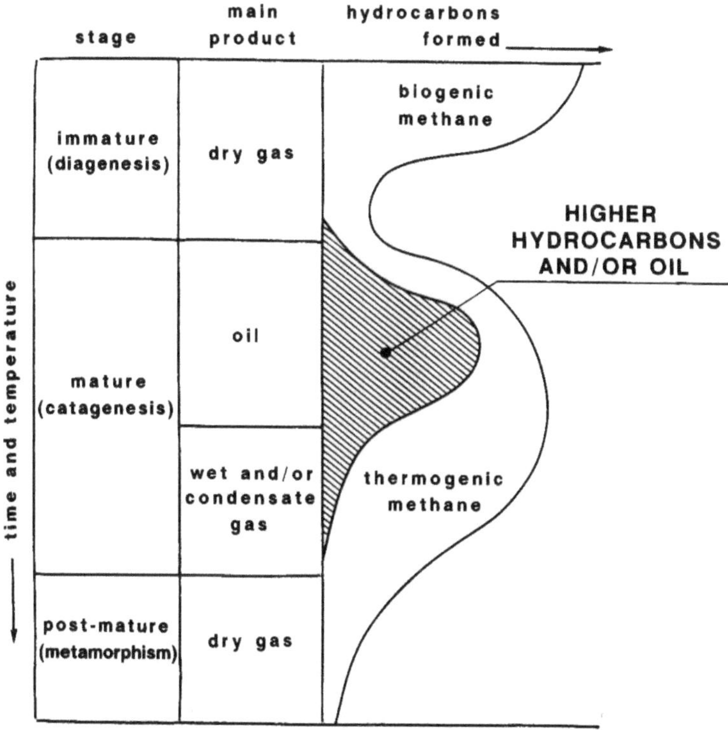

Fig. 1.2. *The effect of time and temperature (and therefore depth) on the generation of hydrocarbons*

gaseous hydrocarbons migrate in solution in the water, it is not clear whether the oil phase forms a micro-emulsion with the water, or is in solution.[5]

Given a suitably porous and permeable path through adjacent strata, the hydrocarbons and water will flow in the direction of the maximum hydrostatic potential gradient – this is the *secondary migration* phase. Since the hydrocarbons are of a lower density than water, gravitational forces cause them to segregate up-dip.

A *reservoir* is formed where the necessary conditions prevail to trap the oil or gas and prevent further migration (Fig. 1.3).

1.4 Hydrocarbon Traps

To the petroleum engineer, the term "trap" indicates a porous and permeable rock configuration capable of accumulating and retaining hydrocarbons.

However, an essential part of the trap is an overlying impermeable *cap rock* – often a tight layer of shale. An imperfect seal (non-continuity, or some small degree of vertical permeability), or total absence of suitable strata, are perhaps the major factors behind unsuccessful exploration drilling.

According to Levorsen's classification of hydrocarbon traps,[1,4] there are three major types: *structural, stratigraphic,* and *combination.*

Structural Traps: these are produced by the action of tectonic forces (Fig. 1.4). Rock strata are deformed by compression, upthrust or intrusion (particularly by salt

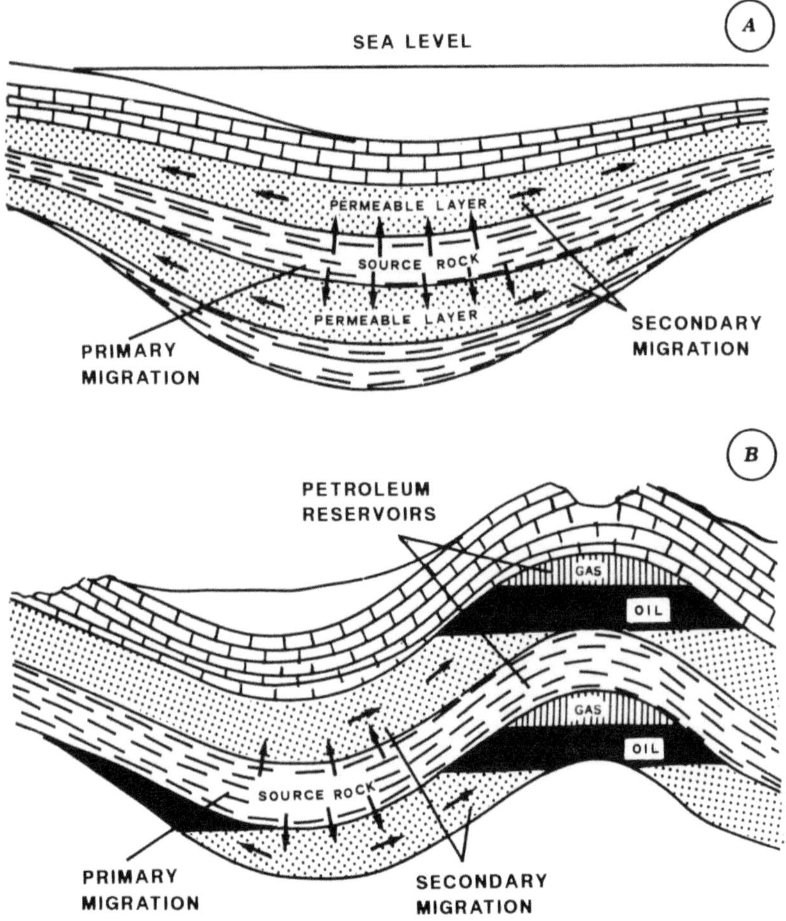

Fig. 1.3 A, B. *The formation of a hydrocarbon reservoir.* **A**: *initial stages of primary and secondary migration.* **B**: *advanced stage of hydrocarbon accumulation*

domes), leading in many instances to geometries suitable for the accumulation of hydrocarbons.

If the strata are brittle in nature, then intense tectonic activity may lead to a shear or *fault*. Relative displacement of each side of the break will produce a *normal* or *inverse* fault, or a more complex system consisting of both types (e.g. the horst and graben of Fig. 1.5). Structures like this are also potential hydrocarbon reservoirs.

Stratigraphic Traps: where local variations in the depositional process occur, the suspended material may be deposited unevenly, leading to the formation of porous, permeable structures (coarse-grained) bounded or completely confined by impermeable shales (fine-grained). Examples of this very common feature of clastic sediments are channel sands, shore-lines, dunes, sand bars, deltas and turbidite fans (Fig. 1.6).

Stratigraphic traps can also develop in organically produced carbonate sediments, as a result of irregular growth cycles in the living colony, and successive

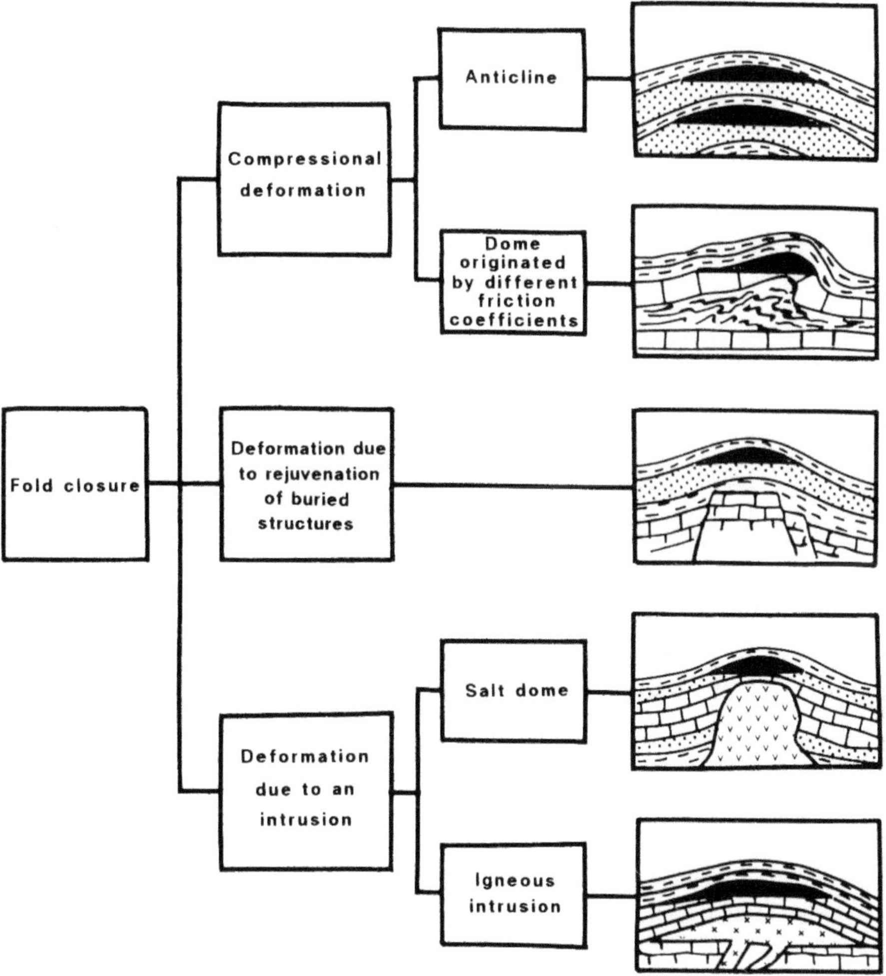

Fig. 1.4. *Structural traps caused by folding*

overlaying of clastic sediments. Examples include open platforms, reefs with banks, reef talus (Fig. 1.6).

Combination Traps: these are attributed to local variations in deposition combined with tectonic action and erosion occurring during or after the depositional phase. Some examples are shown in Fig. 1.7.

This has been a necessarily brief overview of the mechanisms – sedimentary, diagenetic and tectonic – which may produce geometrical configurations in porous, permeable rock strata suitable for the accumulation of hydrocarbons. Its purpose has been to provide the petroleum engineer with an appreciation of the extremely complex internal structure of a hydrocarbon reservoir.

Figure 1.8 portrays an idealised, uniform anticlinal structure containing an oil reservoir overlain by a gas cap, and with an aquifer beneath. This is what the man in the street mistakenly calls an "oil lake" or "pool". Although such a structure is among the easiest to distinguish from geophysical surveys, and perhaps the most

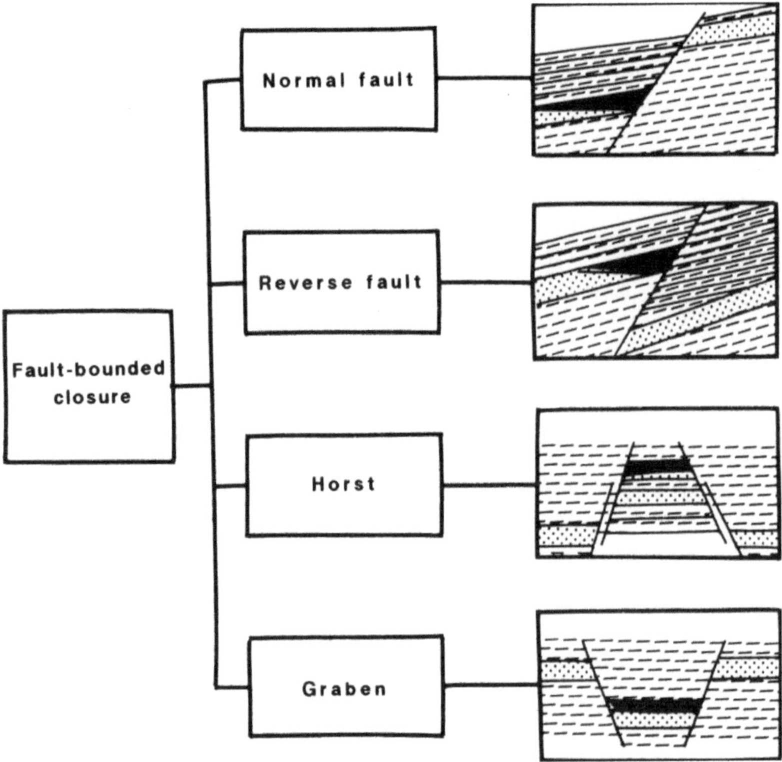

Fig. 1.5. *Structural traps caused by faulting*

straightforward to exploit, it is unfortunately not the only kind encountered in nature, as Figs. 1.4–1.7 show.

1.5 Temperature and Pressure in the Reservoir

It is well known that in the Earth's interior – particularly in the core – the predominant source of heat generation is the radioactive decay of unstable atomic nuclei, complemented to some extent by exothermic chemical reactions.

This heat flows outwards through the Earth's crust, and finally dissipates into the atmosphere. Such a heat flux produces a temperature gradient from the core to surface (the *geothermal* gradient,* g_G), which has an average value of 0.029 °C/m (1.6 °F/100 ft).

The local value of g_G depends not only on the thermal properties of the rock strata, but will be influenced by the possible presence of recent volcanic intrusions and lava flows (basalt, etc.) within the sedimentary basin or beneath it.

The local geothermal gradient can be deduced from wellbore temperature measurements, provided they are made under *stabilised* conditions – that is, after the transient cooling effects of drilling mud circulation have dissipated.

---

* This book uses the symbols recommended by the Society of Petroleum Engineers for all physical parameters. A list of symbols used is provided at the beginning of this volume. All parameters are expressed in SI units unless otherwise stated.

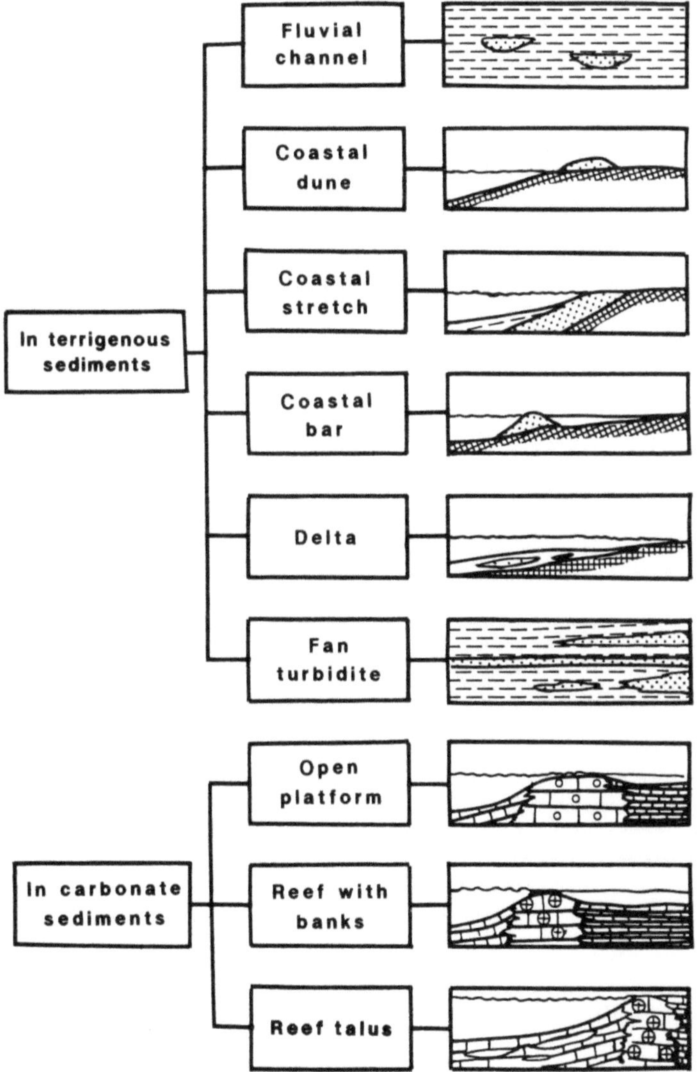

Fig. 1.6. *The principal types of stratigraphic traps*

In the absence of a direct measurement, the temperature T at a vertical depth D can be estimated using appropriate values for g_G and the surface temperature [conventionally taken as 15 °C (60 °F) or 288.2 K (519.4 °R)].

$$T(D) = 288.2 + g_G D \quad (\text{K}) . \tag{1.1}$$

As far as pressure is concerned, it must be appreciated that each rock stratum bears the weight of all the rock lying above it.

If we assume an average geostatic pressure gradient of 0.02262 MN/m^3 (1 psi/ft) for the sedimentary rocks of the crust, then the geostatic or overburden pressure p_{ov} at a vertical depth D will be:

$$p_{ov} = 0.02262 D \quad (\text{MPa}) . \tag{1.2}$$

This is balanced partly by the pressure of the fluid in the pore space of the rock, and partly by the rock grains themselves, which are under forces of compaction.

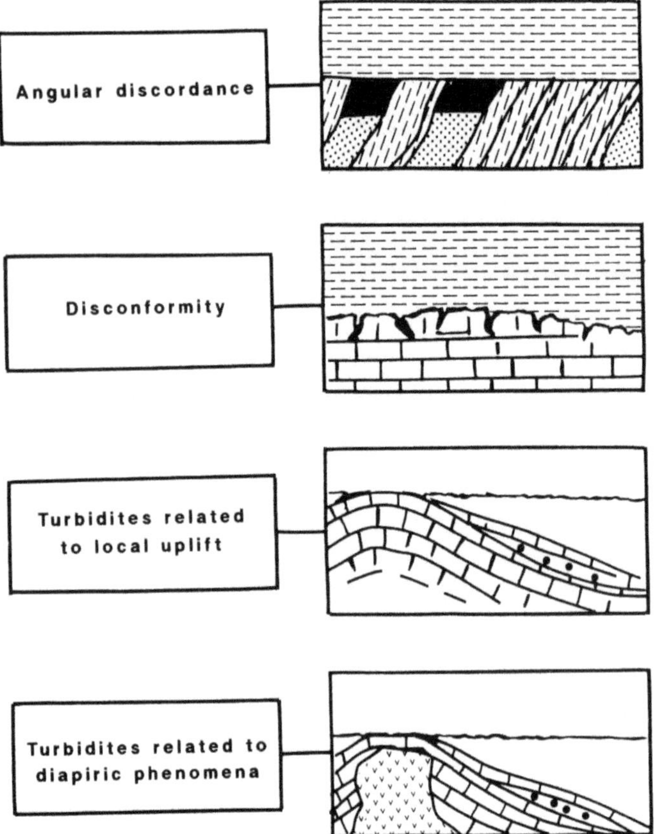

Fig. 1.7. *Examples of "combination" traps*

In the extreme case of an uncemented rock (loose sand) which is totally enclosed by an impermeable formation (shale), the *entire* overburden pressure is borne by the pore fluid. In this case, pore pressure p_f is equal to p_{ov}.

At the opposite extreme, where the stratum under consideration has been deposited in marine conditions, the continuity of the aqueous phase from surface to depth D means that the pore pressure will be equal to the hydrostatic:

$$p_f = g\rho_w D \ . \tag{1.3}$$

The value of ρ_w obviously depends on the water salinity and temperature profile between the surface and depth D. Typically, we assume a value of 0.01018 MN/m^3 (0.45 psi/ft) for $(g\rho_w)_{avg}$.

Intermediate cases can occur anywhere between these two extremes of hydrostatic and geostatic pore pressure. Most commonly, however, pore pressure will be found to be at or only slightly above hydrostatic.

Where the pore pressure is higher than the hydrostatic, the rock is said to be *overpressured*. It is essential during the drilling phase to use a sufficiently heavy mud to establish a wellbore pressure greater than pore pressure, so as to avoid "kicks" and a possible "blow-out". This is particularly important when drilling severely overpressured rock.

In hydrocarbon-bearing strata, the pressure gradients in the oil and gas phases are determined by the in-situ specific gravity (ρ_o and ρ_g) of each fluid.

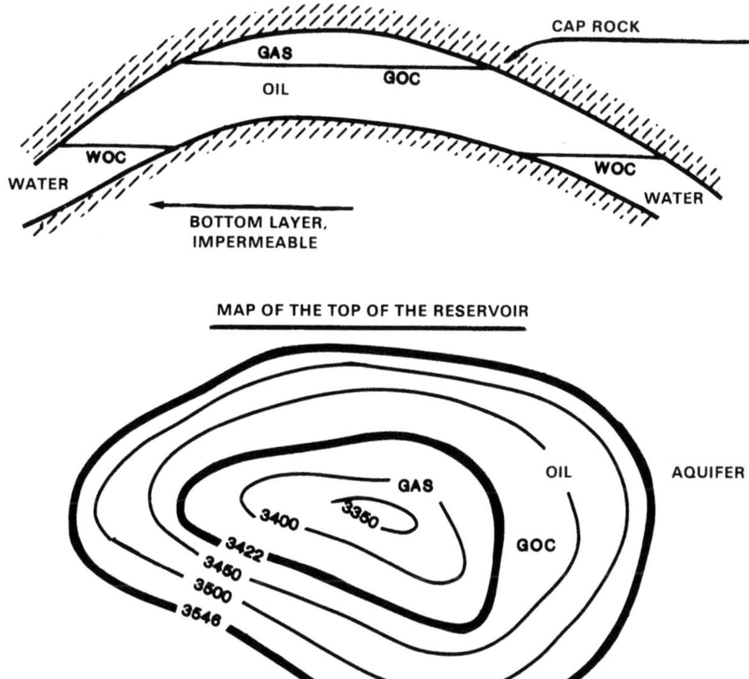

Fig. 1.8. *Schematic of a "classical" reservoir: an anticline containing a gas cap, oil zone, and edge aquifer*

Now water is always present to some degree in reservoir rock (the original depositional environment was water). The pressure p_h in the hydrocarbon phase, and p_w in the water phase, *if they could be measured at the same depth within the reservoir*, would be different.

Referring to Fig. 1.9, if p is the pressure in the water phase at the water/oil contact (the level at which traces of oil start to appear in the pore space), then the phase pressures at a vertical distance ΔD above the contact are:

$$p_h = p - g\rho_h\,\Delta D , \tag{1.4a}$$

$$p_w = p - g\rho_w\,\Delta D , \tag{1.4b}$$

from which:

$$p_h - p_w = g(\rho_w - \rho_h)\,\Delta D \quad (>0) . \tag{1.4c}$$

The positive pressure difference $(p_h - p_w)$ is referred to as the *capillary pressure*, P_c. As we shall see later, this capillary pressure determines the hydrocarbon saturation (the percentage of the pore volume) profile over the thickness of the reservoir.

In the absence of measured data, approximate calculations can be made using the following tentative values:

$$g\rho_o = 0.00792 \text{ MN/m}^3 \quad (0.35 \text{ psi/ft}) ,$$

$$g\rho_g = 0.00181 \text{ MN/m}^3 \quad (0.08 \text{ psi/ft}) .$$

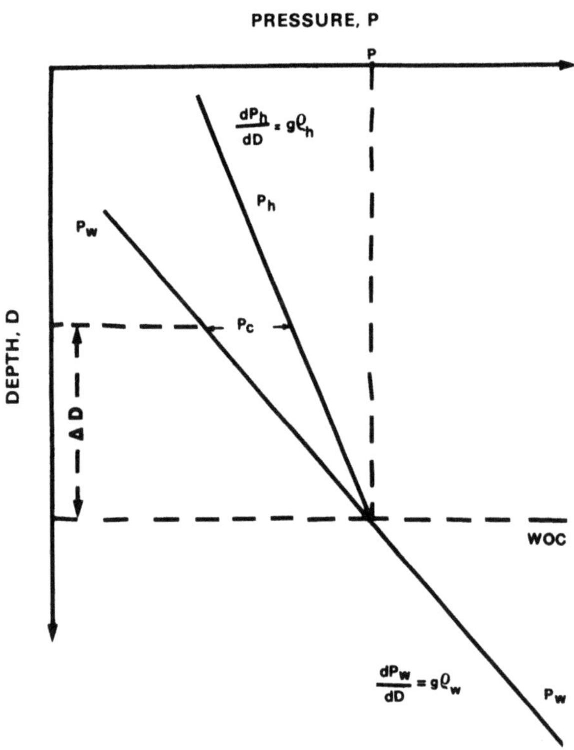

Fig. 1.9. *Variation of pressure with depth in the hydrocarbon and water phases, showing the significance of capillary pressure*

References

1. Dickey PA (1981) Petroleum development geology, 2nd edn. Pennwell, Tulsa
2. Magara K (1978) Compaction and fluid migration – practical petroleum geology. Elsevier, Amsterdam
3. Ricci-Lucchi F (1978) Sedimentologia – Parte II – Ambienti sedimentari e facies. CLUEB, Bologna
4. Selley RC (1985) Elements of petroleum geology, Freeman, New York
5. Tissot BP, Welte DH (1984) Petroleum formation and occurrence, 2nd edn. Springer, Berlin Heidelberg New York

EXERCISES

Exercise 1.1

Below is a list of repeat formation tester (RFT) measurements made in an uncompleted well:

Depth (m)	Pressure (kg/cm²)	(MPa)
2475	204.7	20.074
2500	205.3	20.133
2525	205.8	20.182
2550	207.4	20.339
2575	209.4	20.535
2600	211.3	20.721
2625	213.6	20.947
2650	216.2	21.202
2675	218.9	21.467

Depths are measured relative to the rotary table, which is 610 m above MSL.

Estimate:

- the density and the nature of the fluids present in the formation,
- the depths of the gas/oil and oil/water contacts,
- whether the aquifer is at hydrostatic pressure or not.

Solution

Plot the data on a pressure versus depth diagram (Fig. E1/1.1). Three fluid gradients (i.e. fluid densities) can be distinguished.

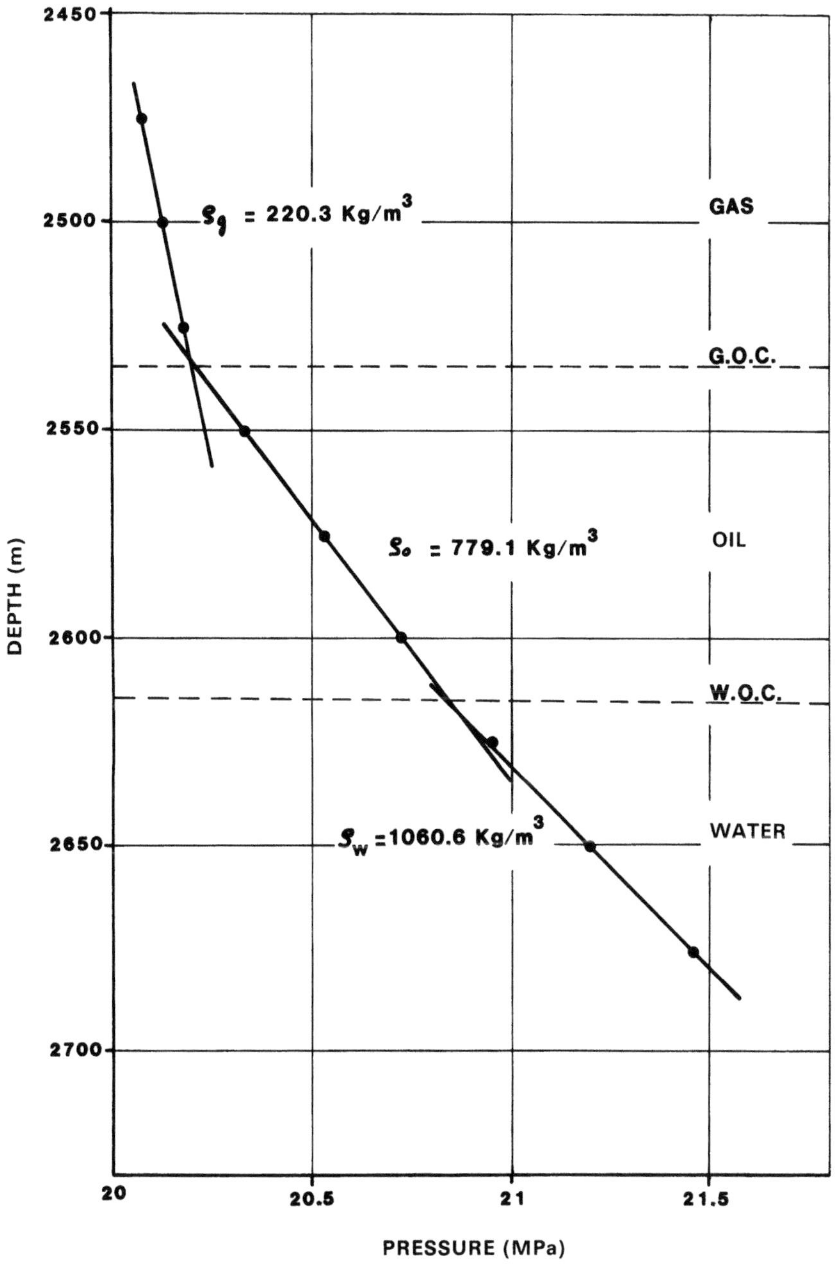

Fig. E1/1.1

Looking first at the intersections of the fluid gradient lines, the contact between the uppermost and middle fluids lies at approximately 2535 m, and between the middle and lowermost fluids at approximately 2610 m.

The density of the top fluid is:

$$\rho_1 = \frac{1}{g}\frac{\Delta p}{\Delta z} = \frac{1}{9.806}\frac{20.182 - 20.074}{2525 - 2475} \times 10^6 = 220.3 \text{ kg/m}^3 \ .$$

This is therefore *gas*.

The density of the middle fluid is:

$$\rho_2 = \frac{1}{9.806}\frac{20.721 - 20.339}{2600 - 2550} \times 10^6 = 779.1 \text{ kg/m}^3 \ .$$

and this is *oil*.

For the lowermost fluid:

$$\rho_3 = \frac{1}{9.806}\frac{21.467 - 20.947}{2675 - 2625} \times 10^6 = 1060.6 \text{ kg/m}^3 \ .$$

which corresponds to *water* with a salinity of about 100 kppm.

The pressure gradients are:

gas: $(9.806 \times 220.3) \text{ Pa/m} = 2.16 \times 10^{-3} \text{ MPa/m}$,

oil: $(9.806 \times 779.1) \text{ Pa/m} = 7.64 \times 10^{-3} \text{ MPa/m}$,

water: $(9.806 \times 1060.6) \text{ Pa/m} = 10.40 \times 10^{-3} \text{ MPa/m}$.

The gas/oil contact depth can be computed more precisely by finding the depth at which the pressure in the gas phase is equal to the pressure in the oil phase. We can start the calculations from the deepest gas data point, and the shallowest oil data point.

If z_{GOC} is the depth of the gas/oil contact, then:

$$20.182 + (z_{GOC} - 2525)\, 2.16 \times 10^{-3} = 20.339 - (2550 - z_{GOC})\, 7.64 \times 10^{-3} \ ,$$

from which we get:

$$z_{GOC} = 2531.2 \text{ m} \ .$$

By similar reasoning, if the oil/water contact depth is z_{WOC}, then:

$$20.721 + (z_{WOC} - 2600)\, 7.64 \times 10^{-3} = 20.947 - (2625 - z_{WOC})\, 10.40 \times 10^{-3} \ ,$$

so that:

$$z_{WOC} = 2612.3 \text{ m} \ .$$

The pressure at the deepest point in the aquifer is 21.467 MPa at 2675 m MSL. This corresponds to a hydrostatic pressure gradient measured from surface of:

$$\frac{\text{pressure at depth } z}{\text{depth of } z \text{ below MSL}} = \frac{21.467}{2675 - 610} = 1.039 \times 10^{-2} \text{ MPa/m}$$

$$= 0.106 \text{ kg/cm}^2/\text{m} \ .$$

The RFT indicates a local hydrostatic pressure gradient of 0.106 kg/cm^2/m in the water phase. Since this is the same as the gradient calculated from surface, we can conclude that the aquifer is in hydrostatic equilibrium (i.e. it is neither under- nor overpressured).

◇ ◇ ◇

Exercise 1.2

Five wells were drilled in a sedimentary basin. While penetrating the aquifer, the following pressure measurements were obtained in each well:

- Well	A	B	C	D	E
- Rotary table elevation (m above MSL)	650	400	825	1200	60
- Measurement depth (m BRT)	2970	4265	2510	5300	2860
- Pressure:					
(kg/cm^2)	298	398	174	422	288
(MPa)	29.2	39.0	17.1	41.4	28.2

Use this pressure data to investigate whether or not the wells are all in the same aquifer (or a system of communicating aquifers).

Solution

In a single aquifer, or where several are in good communication, the pressure gradient should be the same at any point. We need therefore to check whether:

$$\frac{\text{pressure at depth } z}{\text{depth of } z \text{ below MSL}} \text{ is the same in each well .}$$

First correct the measurement depths in each well to metres below MSL. Now calculate the pressure gradient indicated by the observed pressures:

- Well	A	B	C	D	E
- Pressure gradient $(kg/cm^2 \, m)$	0.1284	0.1030	0.1033	0.1029	0.1029

The conclusion is that all measurements pertain to the same aquifer or connected aquifer system, except the one at 2970 m BRT in well A, which is in an isolated section.

This is not obvious from a superficial examination of the data, especially in view of the apparent closeness of the measurements in wells A and E.

◇ ◇ ◇

Exercise 1.3

The following lithologies were encountered in an exploration well:

down to 1420 m:	sand with shale intercalations
1420–1630 m:	compacted shales
1630–2200 m:	fractured limestones, highly permeable
2200–2350 m:	compacted shales
2350–2500 m (BH):	sands and sandstones

No traces of oil or gas were observed, and in the interval 1000 m to 2500 m the pressure and temperature profiles shown in Fig. E1/3.1 were recorded.

The pressure profile indicates that only water is present, with a density of about 1050 kg/m³. Give a possible explanation for the temperature behaviour observed in the well.

Solution

The temperature in the interval 1600–2200 m is significantly higher than would be expected from the average geothermal gradient observed in the well (4 °C per 100 m, surface temperature 15 °C). This temperature anomaly covers the full thickness of the fractured limestone section, and extends for a few tens of metres into the overlying shale.

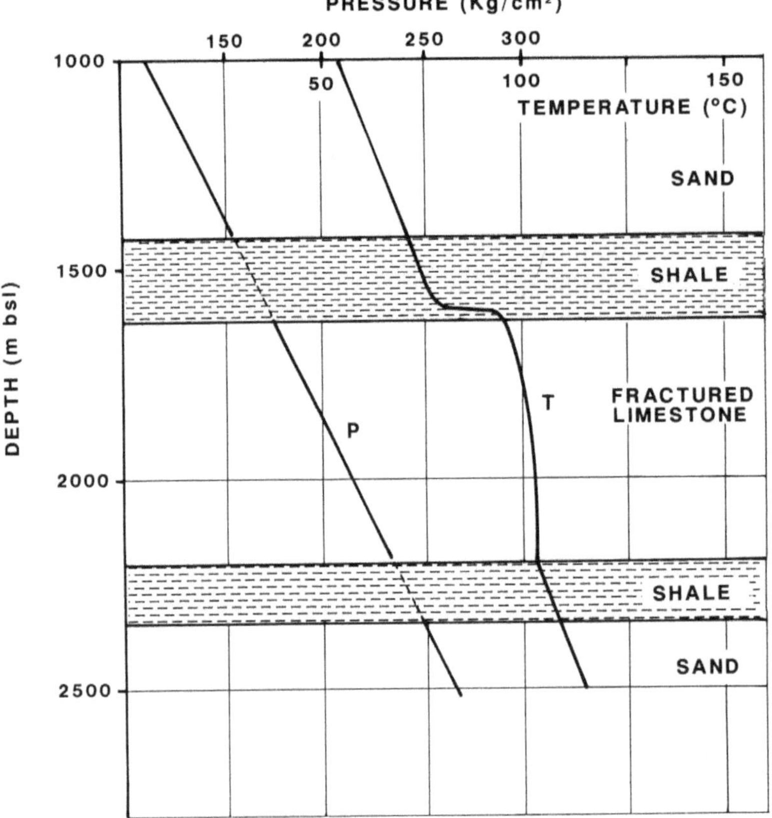

Fig. E1/3.1

The high vertical permeability of the limestone (resulting from the subvertical trends usually encountered in natural fracturing) suggests the possibility of *convection currents* in the water. Because the density of water is lower at higher temperature, the warmer water at the base of the limestone section will tend to rise towards the top, driven through the fracture system by buoyancy forces. As a result, the cooler water in the upper part of the section will be displaced downwards.

The rising warm water will lose heat to the cooler rock through which it passes, while the descending cool water will be gradually warmed up. A stable convection cell is eventually set up, which over geologic time puts the entire limestone layer at an almost uniform temperature – corresponding to that at its base.

Heat conduction into the overlying shale explains why the anomaly extends vertically beyond the limestone.

This type of phenomenon has been observed in Italy in the carbonate structures of Casaglia (Ferrara) and Sciacca (Sicily).

2 Reservoir Fluids

2.1 Composition

Reservoir fluids consist of sometimes highly complex mixtures of hydrocarbon molecules, from the lightest (C_1 methane) to naphthenes and polycyclics with molecular weights in excess of 1000. These molecules originated, as we have seen, from the thermal and bacterial breakdown of once-living organic material. They are almost invariably accompanied by non-hydrocarbon compounds of oxygen, sulphur or nitrogen, such as CO_2, H_2S and N_2.

Water is always present to a greater or lesser extent in the pore space. This is because the reservoir rock was originally laid down as sediment in an aqueous environment, and the subsequent influx of migrating hydrocarbons never achieves total displacement of this water, owing to surface tension forces, which will be described in the next chapter.

As far as production on the surface is concerned, petroleum reservoir fluids at atmospheric pressure and temperature are considered in commercial terminology to consist of *crude oil* (the liquid fraction, if any), and *gas* (the vapour fraction, consisting predominantly of C_1, with lesser amounts of C_2 and the heavier components of the reservoir fluid, plus any impurities).

Traditionally, crude oil is classified according to its *API gravity*, measured in °API.

$$°API = \frac{1.415 \cdot 10^5}{\rho_0} - 131.5 \tag{2.1}$$

where ρ_0 is the oil density in $kg\,m^{-3}$.

Crude oil is broadly categorised as:

light:	°API > 31.1	(ρ_0 < 870)
medium:	31.1 > °API > 22.3	(870 < ρ_0 < 920)
heavy:	22.3 > °API > 10.0	(920 < ρ_0 < 1000)

2.2 Phase Behaviour of Hydrocarbon Systems Under Reservoir Conditions

We will start by considering a "pure" hydrocarbon consisting of a single molecular type (for instance – propane or pentane). Its *phase diagram* (Fig. 2.1A) is typical of a single component system. Below the *critical point* C (critical temperature T_c and pressure p_c), the hydrocarbon will assume any one of three states – solid (region S), liquid (L), or gas (V), depending on the actual temperature and pressure.

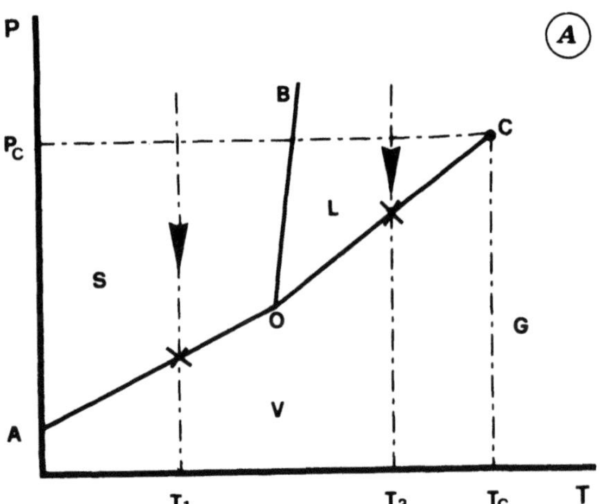

Fig. 2.1. A *Phase diagram for a single-component system in terms of T and p;* **B** *diagram of the volumetric and phase behaviour of a single-component system in terms of v and p (at different T)*

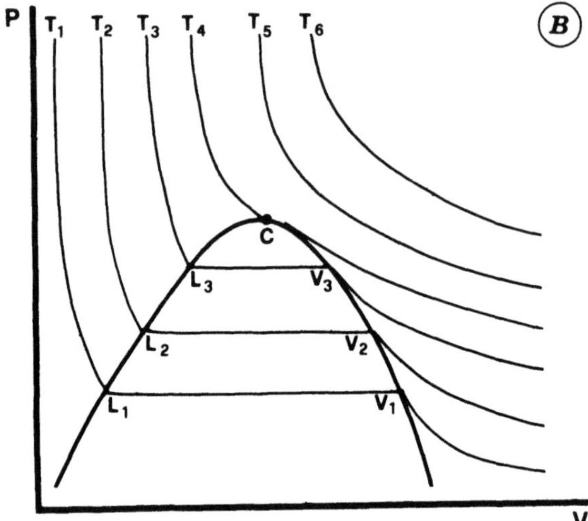

Above the critical temperature (region G), it can only exist in the gas phase, even at very high pressures.

The following phase rule applies:

degrees of freedom of the system = (no. of components) − (no. of phases) + 2

Our single component system is therefore *monovariant* (one degree of freedom) at every point on its phase diagram where two phases exist in equilibrium. This corresponds to the lines AO (solid + gas), OB (solid + liquid), and OC (liquid + gas). At any point on these lines, the value of either one of the thermodynamic variables p and T is determined by the other – they are not free to vary independently.

For example, if we reduce the pressure (by increasing the volume) along the isothermal T_2 (the vertical dashed line in Fig. 2.1A), then as we cross the line OC, the liquid will begin to vaporise. As we continue to increase the volume, there will be no further change in pressure while the liquid vaporises (Fig. 2.1B). The pressure remains constant at a value determined by the temperature T_2. Only when all the liquid has vaporised will the pressure start to decrease once more. The system is monovariant anywhere on the line OC, since the vaporising pressure is uniquely defined by the temperature, and vice versa.

Similarly, along isothermal T_1, the solid phase will sublimate at a constant pressure corresponding to the crossing point on line AO.

Finally, line OB defines the variation of the melting temperature with pressure.

Point O, at which all three phases can apparently coexist, is a characteristic of the system called the *triple point*. The phase rule indicates zero degrees of freedom – O is a unique pressure-temperature pair.

Everywhere else (i.e. away from the lines), the system is bivariant – temperature and pressure can be varied independently.

Real multi-component hydrocarbon systems, consisting of a mixture of components with a wide range of molecular weights, behave very differently from this.

Figure 2.2 represents the phase diagram of a certain multi-component petroleum fluid.[7, 17]

The curve AC describes the *bubble point* at which gas will start to appear in monophasic liquid (saturated oil) upon isothermal reduction of pressure. This line is defined by pairs of values (T, p).

Curve CB is the *dew-point* line – crossing this line from above as we reduce the pressure, droplets of liquid will start to appear from monophasic gas.

Within the envelope ACB, a bi-phasic mixture exists. The proportions of vapour and liquid vary with changing temperature and pressure as indicated by the liquid percentage lines in the figure.

The isotherm $T_{c'}$ is the *cricondentherm*. It is the vertical tangent to ACB and it represents the maximum temperature at which a liquid phase can exist. As such, it is the equivalent of the critical temperature of the single component system.

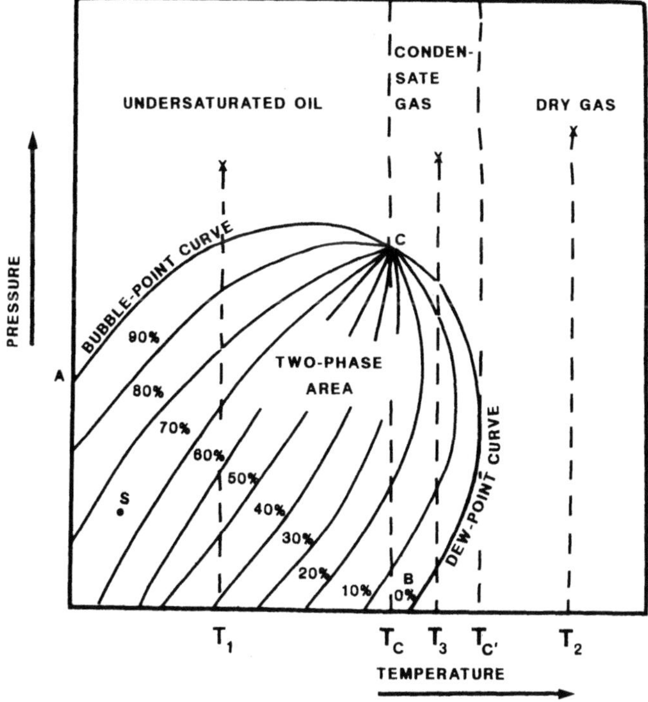

Fig. 2.2. *Phase diagram of a multi-component hydrocarbon mixture at reservoir conditions*

The critical temperature T_c of a multi-component system has a somewhat different significance. At temperature T_c, if the pressure is at or above its value at C the intrinsic properties of the phases (density, viscosity, refractive index, etc) are the same. In other words, it is no longer possible to distinguish liquid from vapour.

More importantly, since interfacial tension is zero, the gas and liquid phases are perfectly miscible. Several methods of enhanced oil recovery (EOR) using hydrocarbon gases or carbon dioxide are based on this principle.

We will now look at the significance of the phase diagram, such as Fig. 2.2, from the point of view of reservoir exploitation. The changes induced in the reservoir during production are isothermal (the temperature does not change significantly, except in the case of prolonged water or steam injection), and any reservoir pressure changes will follow a vertical path ($T = $ constant) on the phase diagram.

If the reservoir temperature (T_1) is lower than the critical temperature T_c then the reservoir is *oil bearing*. The oil is *undersaturated* if the initial reservoir pressure lies above the intersection of T_1 and the line AC (the bubble point). If the pressure lies within the phase envelope – below the bubble point – the presence of both phases constitutes a *saturated* oil with gas cap.

In all cases, the gradual reduction in pressure caused by production leads to evolution of gas from the reservoir oil once below the bubble point.

If the temperature (T_2) is higher than $T_{c'}$, the reservoir fluid is a *dry gas*. Since the isotherm T_2 is completely outside the phase envelope, a liquid phase will not appear at any pressure and there will only be single phase dry gas.

A more complex, and somewhat surprising situation occurs when the reservoir temperature (T_3) lies *between* T_c and $T_{c'}$. At the time of discovery, the reservoir pressure is normally above the intersection of the isotherm T_3 and the dew point curve CB. Here the reservoir fluid consists of a single gas phase.

As the gas is extracted, the reservoir pressure declines isothermally and will eventually cross CB. Here, at the *dew point*, the first droplets of liquid – formed from the heavier hydrocarbon components – will appear. As the pressure decreases further, more liquid will evolve, and as we shall see shortly, a small but commercially important liquid fraction is actually recovered at surface conditions.

This phenomenon of *retrograde condensation* – the appearance of liquid as pressure is *reduced* – is the exact opposite to the behaviour of a mono-component system, where liquefaction requires an *increase* of pressure. It is a result of molecular interactions (van der Waals forces) between the heavy molecules and methane present in the gas.

When producing these *condensate* reservoirs, special measures can be taken to prevent retrograde condensation (also called "liquid drop-out") occurring in the reservoir itself. This would otherwise result in loss of the valuable heavy components, which it is desirable to recover on surface. One approach is to *recycle* the produced gas fraction – by injecting it back into the reservoir – thereby maintaining a suitably high pressure.

Tracing a path on the phase diagram corresponding to the passage from the reservoir, through the production string, to the surface facilities, we might arrive at a point such as S in Fig. 2.2. Here, the pressure and temperature are considerably lower than in the reservoir, and S represents the oil/gas mixture that would be measured at the separator.

The liquid oil fraction at S depends, of course, on the reservoir fluid composition, but it also depends on the operating p and T of the separator – these are *independent of the reservoir pressure and temperature.* Consequently, liquid hydrocarbons will be recovered on surface regardless of the nature of the reservoir fluid – oil, oil/gas, condensate and, in some cases, gas.

Strictly speaking, the term *dry gas* introduced here applies to a gas reservoir where even the separator point S lies outside (i.e. to the right of) the phase envelope, so that a liquid phase *never* appears. A gas system where S falls *within* the envelope, so that there is a liquid fraction (albeit a small one) at surface conditions, is referred to as a *wet gas*. It is distinguished from a condensate by the fact that, like the dry gas, the reservoir temperature is higher than $T_{c'}$.

From this, it follows that the production of oil on surface does not mean that there is oil in the reservoir, any more than the production of gas means there must be gas in the reservoir. Certain of the more extreme cases can, of course, be identified with certainty – a heavy oil produced with a very small amount of gas must originate from an oil reservoir; gas production with minimal quantities of heavy components very probably indicates a dry gas reservoir, though the possibility of a wet gas or condensate cannot be excluded.

2.3 Thermodynamic Properties of Reservoir Fluids

The phase behaviour and physical properties of reservoir fluids are derived from laboratory PVT (pressure/volume/temperature) measurements. These are made on fluid samples either taken downhole (in the wellbore, or extracted directly from the reservoir), or recombined from separator samples of produced oil and gas.[2] These measurements are made at reservoir temperature and a range of pressures.

The equipment and methods used in the PVT laboratory are described in numerous references[1,7,17] and will not be detailed here.

Since the early 1980s, computerised numerical methods based on the *equation of state* (EOS) technique have been increasingly successful in predicting reservoir fluid properties, using as input a detailed molecular composition.[8,11,15,20]

The most commonly employed equations of state are those of Redlich and Kwong (RK):

$$p = \frac{RT}{V_{\mathrm{m}} - b} - \frac{a}{T^{0.5} V_{\mathrm{m}}(V_m + b)} \tag{2.2}$$

and Peng and Robinson (PR):

$$p = \frac{RT}{V_{\mathrm{m}} - b} - \frac{a(T)}{V_{\mathrm{m}}(V_{\mathrm{m}} + b) + b(V_{\mathrm{m}} - b)} \tag{2.3}$$

The coefficients a and b in the RK equation, and $a(T)$ and b in the PR equation, are calculated from the molar fractions. For the heavier components, there is considerable uncertainty in these terms – particularly the interaction coefficient. It is therefore common practice to calibrate them by comparing the EOS calculations against laboratory measurements.[8]

Please consult the reference list for more detail on this highly specialised subject.

2.3.1 Volumetric Behaviour and Viscosity of Gases and Condensates

2.3.1.1 Volumetric Behaviour

The equation of state for an ideal gas describes its change of volume in response to a change in pressure and/or temperature:

$$pV = nRT \,, \tag{2.4}$$

where n is the number of kg mol of gas, and R is the universal gas constant (8.312×10^{-3} if p is in MPa, T in K and V in m^3).

This equation can only be applied to an *ideal* gas, in which the volume of the molecules, and the van der Waals forces of attraction between them, are zero. *Real* gases only behave in this manner at low pressures (below atmospheric), and temperatures above atmospheric.

Van der Waals was one of the first to propose an equation of state for real gases:

$$\left(p + \frac{a}{V^2}\right)(V - b) = nRT \,. \tag{2.5}$$

The b term accounts for the finite volume of the molecules, and a for the forces of attraction between them. From Eq. (2.5), at high pressure, we have:

$$\lim_{p \to \infty} V = b \,. \tag{2.6}$$

Van der Waals' equation is not suited to reservoir engineering applications. The terms a and b are in fact functions of the gas composition, as well as the temperature. In addition, since V appears as V^3, it has three solutions at every value of p and T, only two of which have actual physical significance, at the points L (liquid) and V (gas) in Fig. 2.1B.

The preferred equation used in reservoir engineering is the empirical relationship:

$$pV = znRT \,. \tag{2.7}$$

Here, z is the *compressibility* or *deviation factor*, a dimensionless number which depends on the gas composition, pressure and temperature. The value of z is 1.0 for an ideal gas. T is the absolute temperature in K.

The general behaviour of z at different temperatures ($T_1 < T_2 < T_3$) is shown in Fig. 2.3. Note that at any $T < T_3$, the curve $z(p)$ decreases from an initial value of 1 with a negative slope ($\partial z/\partial p < 0$), passes through a minimum and then increases ($\partial z/\partial p > 0$). Boyle called this limiting temperature T_3 the *inversion temperature*. At high pressure, each curve crosses the horizontal line $z = 1.0$ – the ideal gas line AC.

Along any curve of the form ABC, the volume occupied by the gas is *less* than that of an ideal gas, owing to the forces of intermolecular attraction, which augment the compressive effect of the pressure. This is particularly evident along the part AB.

At pressures above C, the gas volume is now larger than the ideal gas. This is due to the finite volume of the molecules (the *covolume*), which in fact begins to influence the gas behaviour from point B on the curve.

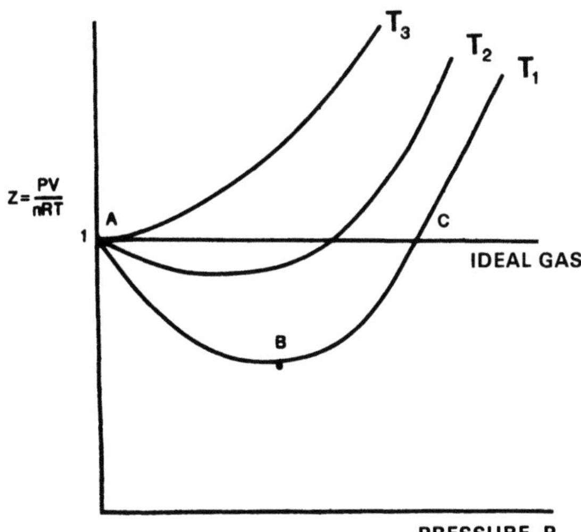

Fig. 2.3. *Schematic of the behaviour of the gas deviation factor (z) at different temperatures. $T_3 = $ Boyle inversion temperature*

The *isothermal compressibility* (c_g) of a real gas can be written in terms of Eq. (2.7) as:

$$c_g = -\frac{1}{V}\left(\frac{\partial V}{\partial p}\right)_T = \frac{1}{p} - \frac{1}{z}\left(\frac{\partial z}{\partial p}\right)_T . \tag{2.8a}$$

The *gas formation volume factor* (B_g) is defined as the volume occupied at pressure p and temperature T by $1\,\text{m}^3$ of the gas as measured at standard conditions of 0.1013 MPa and 15 °C. (Some countries define standard temperature as 0 °C. In oilfield units standard conditions are 14.65 psia and 60 °F.)

If n is the number of moles in $1\,\text{m}^3$ of gas at standard conditions, we can write Eq. (2.7) for 0.1013 MPa and 15 °C (288.2 K), and at reservoir p and T:

$$0.1013 \times 1 = nR \times 288.2,$$

$$pB_g = znRT .$$

From this we get:

$$B_g = 3.515 \times 10^{-4}\frac{zT}{p} . \tag{2.8b}$$

The gas density $\rho_g(p, T)$ at pressure p and temperature T can easily be calculated from the density at standard conditions, $\rho_{g,\,sc}$, because a given mass of gas will have $1/B_g$ times the volume on the surface that it had at p and T:

$$\rho_g(p, T) = \frac{\rho_{g,\,sc}}{B_g} . \tag{2.8c}$$

Sometimes, instead of the density $\rho_{g,\,sc}$, the molecular weight (M) or the specific gravity relative to air γ_g will be given. In such cases:

$$\rho_{g,\,sc} = \frac{M}{23.645}\ \text{kg}\,\text{m}^{-3} , \tag{2.9a}$$

$$\rho_{g,\,sc} = 1.225\,\gamma_g\ \text{kg}\,\text{m}^{-3} . \tag{2.9b}$$

The calculation of c_g, B_g and ρ_g requires a knowledge of the deviation factor z. This can be determined by laboratory PVT measurements,[7, 17] or estimated from the EOS.

In the following pages, we will employ an alternative and simpler semi-empirical approach to the estimation of the z-factor which is widely used in reservoir engineering. Firstly, the definition of a few terms:

For a single molecular type (for example: methane), there are two important dimensionless parameters:

$$p_r = \text{reduced pressure} = \frac{\text{absolute pressure}}{\text{critical pressure}} = \frac{p}{p_c}, \tag{2.10a}$$

$$T_r = \text{reduced temperature} = \frac{\text{absolute temperature}}{\text{critical temperature}} = \frac{T}{T_c}. \tag{2.10b}$$

The critical pressures and temperatures of the principal components of natural gas are listed in Table 2.1.

For a natural gas consisting of n components, where y_i is the mole fraction of component i ($\sum y_i = 1.0$), equivalent *pseudo* parameters have been defined. *They have no physical significance*, but are convenient for the purpose of calculating the mixture properties:

$$p_{pc} = \text{pseudo-critical pressure} = \sum_{i=1}^{n} y_i p_{c,i}, \tag{2.11a}$$

$$T_{pc} = \text{pseudo-critical temperature} = \sum_{i=1}^{n} y_i T_{c,i}, \tag{2.11b}$$

$$p_{pr} = \text{pseudo-reduced pressure} = \frac{p}{p_{pc}}, \tag{2.11c}$$

$$T_{pr} = \text{pseudo-reduced temperature} = \frac{T}{T_{pc}}. \tag{2.11d}$$

Van der Waal's law of corresponding states says that 1 mole of any mono-molecular gas at the same *reduced* temperature and pressure will occupy the same volume. This has been found to be rigorously true for all such gases.

This same principle has been applied empirically to the mixtures of alkanes (paraffinic hydrocarbons) constituting natural gas. Kay[17] performed an extensive series of experimental measurements of the z-factor. He found that all natural gases at the same *pseudo-reduced* temperature and pressure occupied approximately the same volume, and therefore had approximately the same z-factor.

Consequently, a single diagram of z as a function of p_{pr} and T_{pr}, Fig. 2.4, is enough to define the value of z for any natural gas of known molecular composition y_i.

When the molecular composition is not known, reasonably good approximations to p_{pc} and T_{pc} can be obtained from the gas specific gravity (relative to air), which should be available from a field measurement (Fig. 2.5).

The correlations shown in Fig. 2.4 can be easily programmed into a calculator as:[17]

$$z = A + \frac{1 - A}{\exp B} + C p_{pr}^{D}. \tag{2.12a}$$

Table 2.1. *Critical constants of the principal components of natural gas*

Molecular type			Molecular weight	Critical constants		Parachor
Symbol	Formula	Name		p_c (MPa)	T_c (K)	
C_1	CH_4	Methane	16.04	4.606	190.6	77.0
C_2	C_2H_6	Ethane	30.07	4.881	305.6	108.0
C_3	C_3H_8	Propane	44.10	4.247	370.0	150.3
$i\text{-}C_4$	$i\text{-}C_4H_{10}$	Isobutane	58.12	3.647	408.3	181.5
$n\text{-}C_4$	$n\text{-}C_4H_{10}$	Normal butane	58.12	3.799	425.0	190.0
$i\text{-}C_5$	$i\text{-}C_5H_{12}$	Isopentane	72.15	3.378	460.6	225.0
$n\text{-}C_5$	$n\text{-}C_5H_{12}$	Normal pentane	72.15	3.372	469.4	232.0
$n\text{-}C_6$	$n\text{-}C_6H_{14}$	Normal hexane	86.18	3.013	507.2	271.0
$n\text{-}C_7$	$n\text{-}C_7H_{16}$	Normal heptane	100.20	2.737	540.0	311.0
$n\text{-}C_8$	$n\text{-}C_8H_{18}$	Normal octane	114.23	2.489	568.9	352.0
$n\text{-}C_9$	$n\text{-}C_9H_{20}$	Normal nonane	128.26	2.289	594.4	392.0
$n\text{-}C_{10}$	$n\text{-}C_{10}H_{22}$	Normal decane	142.29	2.096	617.8	431.0
–	CO_2	Carbon dioxide	44.01	7.384	304.4	78.0
–	H_2S	Hydrogen sulphide	34.08	9.004	373.3	n.d.
–	N_2	Nitrogen	28.01	3.399	126.1	41.0

where:

$$A = 1.39(T_{pr} - 0.92)^{0.5} - 0.36\,T_{pr} - 0.101 \,, \tag{2.12b}$$

$$B = (0.62 - 0.23\,T_{pr})p_{pr} + \left[\frac{0.066}{T_{pr} - 0.86} - 0.037\right]p_{pr}^2$$

$$+ \frac{0.32}{\text{antilog}[9(T_{pr} - 1)]}\,p_{pr}^6 \,, \tag{2.12c}$$

$$C = (0.132 - 0.32\log T_{pr}) \,, \tag{2.12d}$$

and

$$D = \text{antilog}(0.3106 - 0.49\,T_{pr} + 0.1824\,T_{pr}^2) \,. \tag{2.12e}$$

These equations and diagrams are valid for dry and wet gases, and also describe the volumetric behaviour of condensates.

The phase behaviour of condensate systems, especially the variation of the gas phase composition in equilibrium with the heavy retrograde condensate fraction, has to be measured experimentally, or calculated (with less accuracy) from an equation of state such as RK or PR.

2.3.1.2 Viscosity

The viscosity of a natural gas is highly sensitive to the temperature and pressure, as well as the composition of the gas itself. This can be clearly seen in the example in Fig. 2.6.

It is not common practice to measure gas viscosity at *reservoir* conditions. It can be derived from empirical equations, and one of the simplest and most popular is the correlation of Lee, Gonzales and Eakin:[14]

$$\mu_g(\text{mPa}\,\text{s}) = K\,\exp\left[X\left(\frac{\rho_g}{1000}\right)^y\right]10^{-4} \,, \tag{2.13a}$$

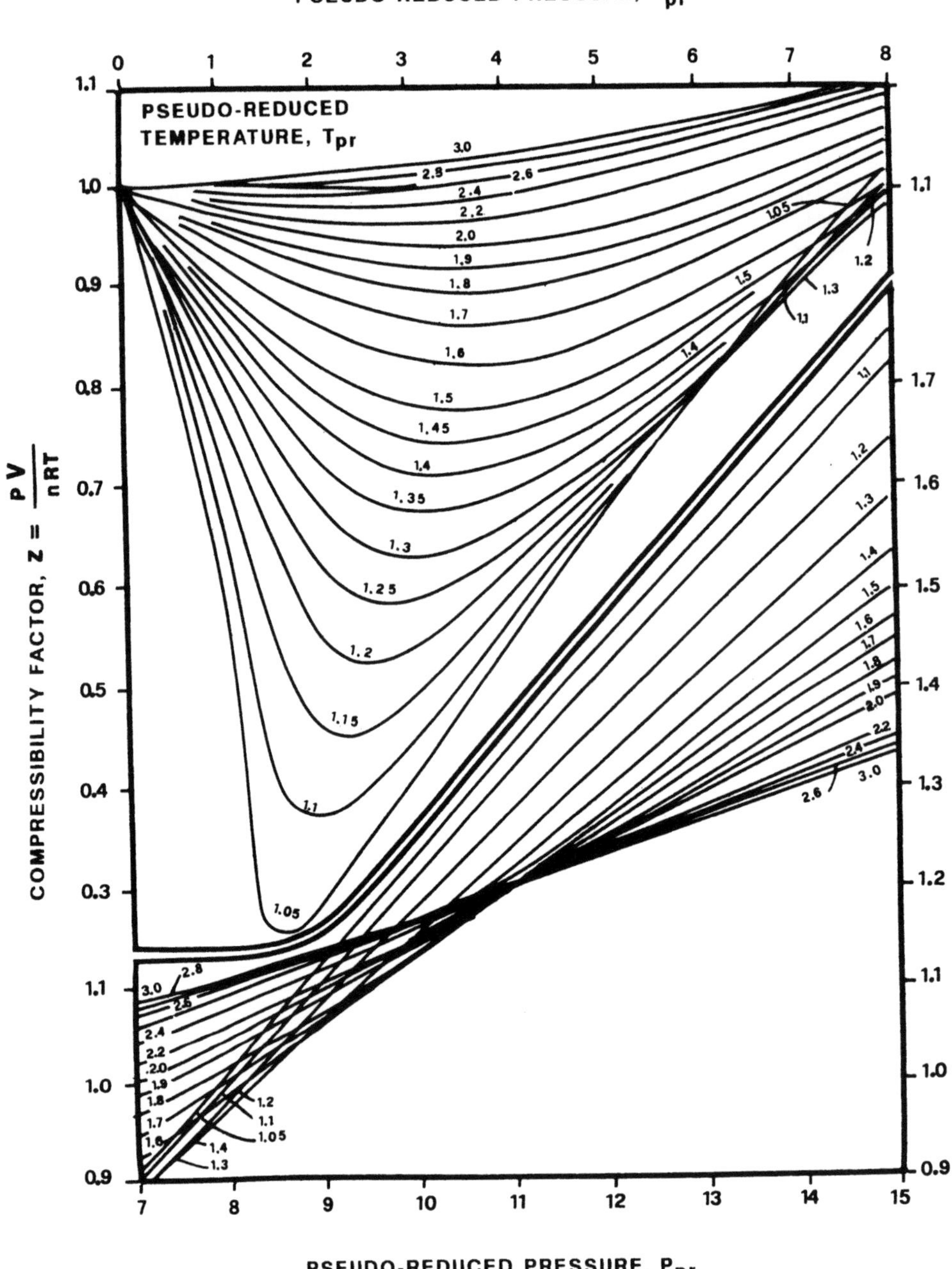

Fig. 2.4. *The z-factor as a function of* p_{pr} *and* T_{pr}. *From Ref. 17, 1942, Society of Petroleum Engineers of AIME. Reprinted with permission of the SPE*

where:

$$X = 3.5 + \frac{547.8}{T} + 0.01 M , \tag{2.13b}$$

$$y = 2.4 - 0.2X , \tag{2.13c}$$

$$K = \frac{(12.61 + 0.027M)\,T^{1.5}}{116.11 + 10.56M + T} . \tag{2.13d}$$

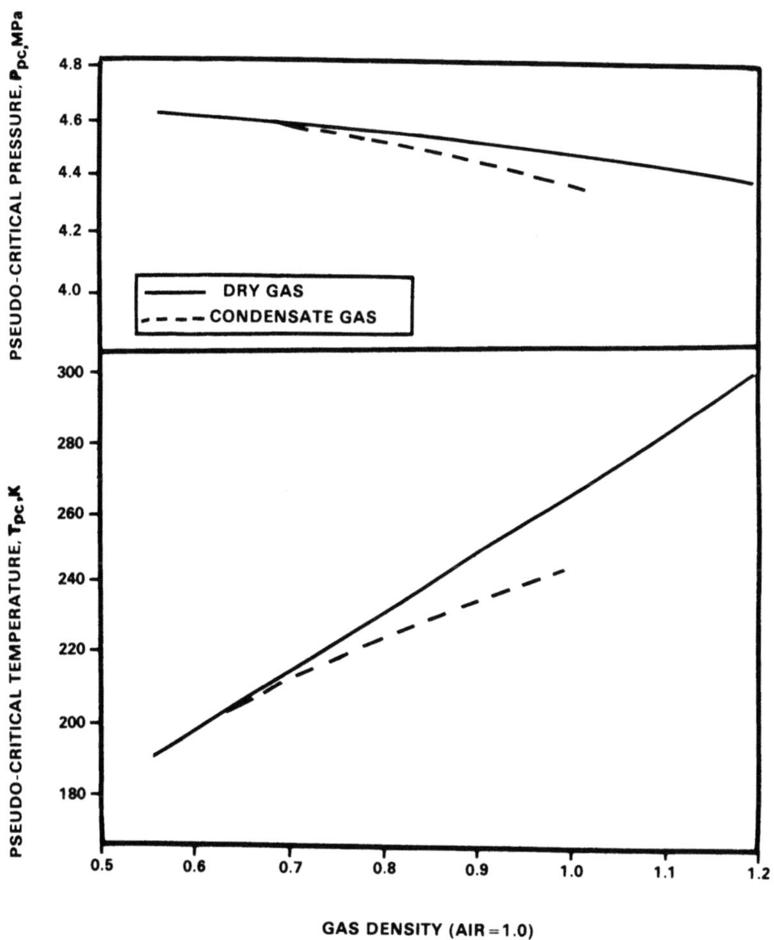

Fig. 2.5. p_{pc} and T_{pc} as a function of specific gravity for reservoir gases and condensates. From Ref. 17, 1942, Society of Petroleum Engineers of AIME. Reprinted with permission of the SPE

M is the molecular weight of the gas, and the absolute temperature T is in K. In the same units, the gas density ρ_g is:

$$\rho_g = \frac{M}{23.645 B_g} = 120.32 \frac{Mp}{zT} \tag{2.14}$$

with p expressed in MPa.

2.3.2 Volumetric Behaviour and Viscosity of Oil

2.3.2.1 Volumetric Behaviour

Reservoir oil almost always contains gas in solution (i.e. light components in liquid form). The liberation of this gas begins as soon as the process of production brings the pressure down below the bubble point (Fig. 2.2). Gas might therefore evolve in the reservoir itself.

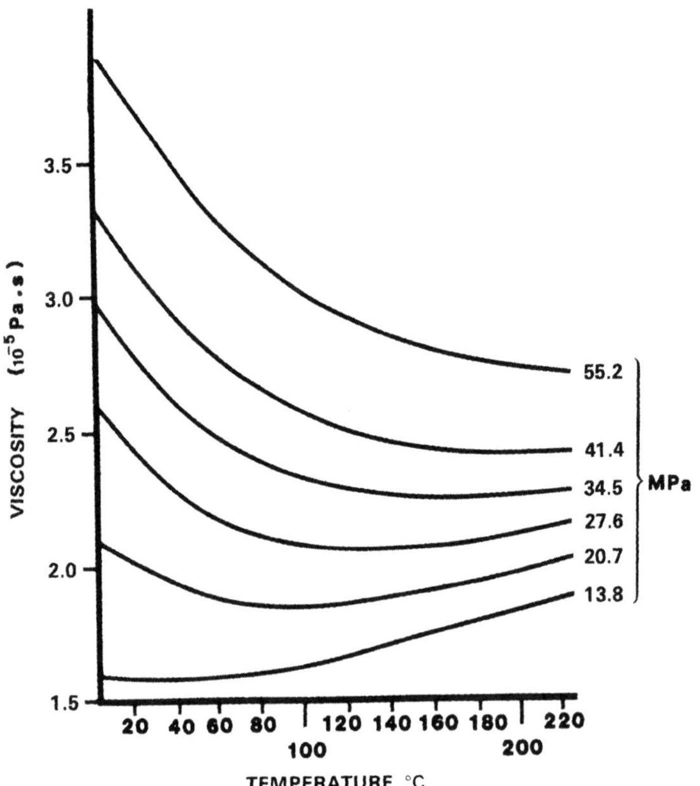

Fig. 2.6. *Viscosity of methane between 4 and 224 °C, at pressures from 13.8 to 55.2 MPa*

If not, it will almost certainly appear as the oil travels upwards towards the surface, or in the separator system, since the pressure along this path decreases continuously, finishing eventually at atmospheric in the stock tank.

These two processes of gas evolution will be considered separately.

In the reservoir, any decline in pressure as a result of the progressive removal of fluid from what is, to a first approximation, a *constant reservoir volume,* occurs at a *constant temperature* T_R. The gas liberation which may occur under these conditions is referred to as *differential liberation,* and the subscript *"d"* is appended to parameters measured during this process.

When defining these parameters, the reference volume V_o for the oil is taken as 1 m^3 of the *residual* oil resulting from the liberation process, measured at standard conditions of 1 atm (0.1013 MPa) and 15 °C (288.2 K).

As a consequence of the loss of the light components as gas, the liquid oil undergoes a *reduction* in volume as the pressure is decreased below bubble point. To describe this behaviour, we now define the following parameters:

R_{sd} = gas solubility in oil = $f(p)$
    = volume of gas in m^3 (measured at standard conditions) that were dissolved at p and T_R in a mass of oil that produced, through differential liberation, 1 m^3 of residual oil under standard conditions.

B_{od} = oil *formation volume factor* = $f(p)$
   = volume of oil in m³ at p and T_R, including dissolved gas, that produces, through differential liberation, 1 m³ of residual oil under standard conditions.

The dependence of R_{sd} and B_{od} on pressure is demonstrated qualitatively in Figs. 2.7 and 2.8. Note that above the bubble point, *since there is no more gas to dissolve* (it has all liquefied), R_{sd} = constant, yet B_{od} decreases with increasing pressure because of the small but finite compressibility c_o of the liquid oil (so-called *undersaturated*):

$$c_0 = -\frac{1}{V}\left(\frac{\partial V}{\partial p}\right)_T = -\frac{1}{B_o}\left(\frac{\partial B_o}{\partial p}\right)_T . \tag{2.15}$$

The curves $B_{od}(p)$ and $R_{sd}(p)$ can both be determined in the PVT laboratory, or calculated (albeit with some imprecision) from an equation of state.

As the reservoir fluids rise up the production string, flow through the surface piping to the separators, and eventually reach the stock tank, they experience

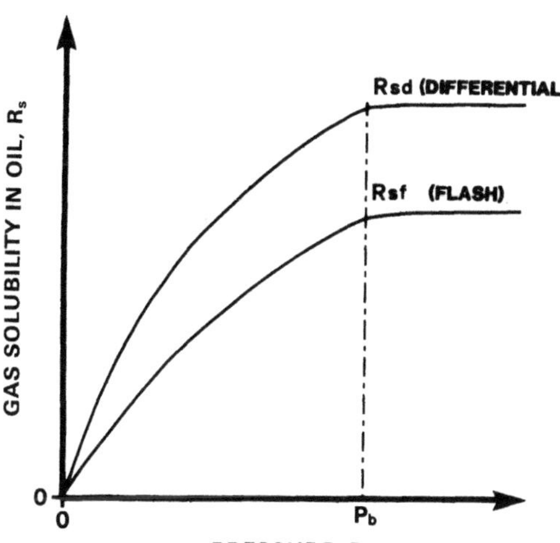

Fig. 2.7. *Schematic of the behaviour of the gas solubility (R_s) under "flash" and "differential" liberation*

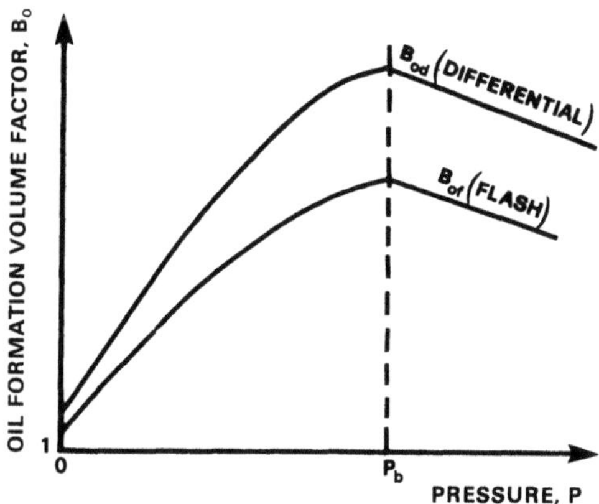

Fig. 2.8. *Schematic of the behaviour of the oil volume factor (B_o) under "flash" and "differential" liberation*

a reduction both in temperature (through heat loss to the surroundings), and pressure (through hydrostatic and dynamic losses).

By the time the fluid reaches the stock tank, it has undergone a series of separations, and it eventually finishes as degassed oil in the stock tank, fully separated from its quota of associated gas.

Each separation involves liberation of some of the dissolved gas, following which the oil proceeds to the next separation.

This process of *flash liberation* is distinguished from *differential* by the fact that it occurs over a wide range of temperature and with large increases in volume. Only the *total mass* of oil plus gas is conserved at every stage.

The solubility and volume factor coefficients for flash liberation are defined with reference to a unit volume of *stock tank* oil (again at standard conditions) resulting from the process. The subscript "f" is used.

$$R_{sf} = \frac{\text{volume of liberated gas in m}^3 \text{ (measured at standard conditions)}}{\text{volume of stock tank oil (measured under standard conditions)}},$$

$$B_{of} = \frac{\text{volume of oil at } p \text{ and } T_R, \text{ including dissolved gas}}{\text{volume of stock tank oil (measured under standard conditions)}}.$$

$R_{sf}(p)$ and $B_{of}(p)$ are compared with their differential counterparts in Figs. 2.7 and 2.8. Again, they may be measured directly in the PVT laboratory,[7,17] by controlled flash liberations in multi-stage separators, or estimated from an equation of state.

Experimentally derived correlations offer a third, much simpler, approach which can often provide a reasonable degree of accuracy. The following pages describe some of the correlations in common use.

2.3.2.2   Correlations for the Estimation of R_{sf}, B_{of}, and c_o

The principal correlations for R_{sf}, B_{of} and c_o will be presented in this section. All parameters are in SI units (in particular, *pressure is in* MPa and *temperature in* K) unless otherwise indicated.

Among the correlations that relate R_{sf}, p and T, and the produced oil and gas properties, perhaps the most widely used are those of Standing[17] and Lasater.[13]

STANDING[17]

$$R_{sf} = 6.931 \gamma_g (p + 0.176)^{1.205} \text{ antilog } (0.01506\,^\circ\text{API} - 0.001973\,T) \qquad (2.16)$$

where:

$$\gamma_g = \text{specific gravity of the gas (relative to air)}$$

and:

$$\text{antilog}(x) = 10^x$$

LASATER[13]

$$R_{sf} = 23.645 \frac{\rho_{o,sc}\, y_g}{M(1 - y_g)}. \qquad (2.17a)$$

where:

$$y_g = F(p^*), \quad \text{Fig. 2.9} \,,$$

$\rho_{o,\,sc}$ = density of the oil at stock tank conditions ,

M = corrected molecular weight of the stock tank oil $= f(\rho_{o,\,sc})$, Fig. 2.10 ,

and:

$$p^* = \frac{80.58\, p \gamma_g}{T} \, . \tag{2.17b}$$

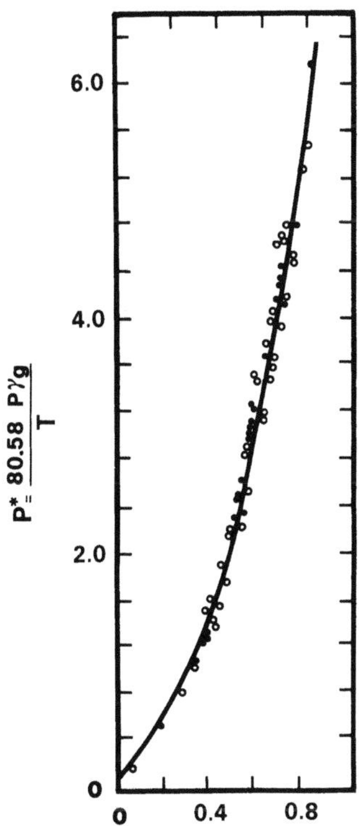

Fig. 2.9. *Experimental relationship between p* and* y_g *(see Eq. (2.17)). From Ref. 13, 1958, Society of Petroleum Engineers of AIME. Reprinted with permission of the SPE*

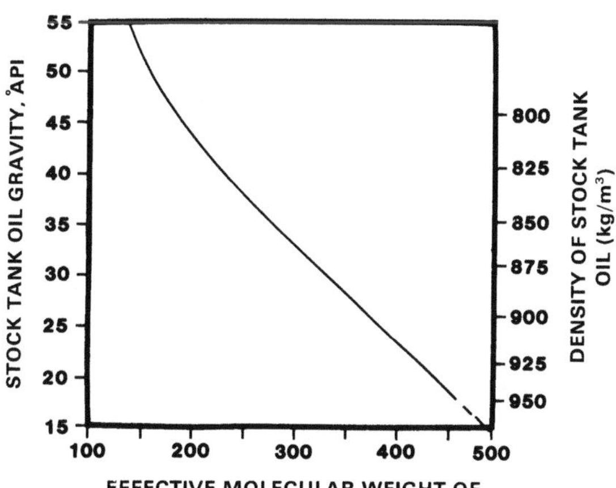

Fig. 2.10. *The relationship between the density (or API gravity) of a stock tank oil, and its molecular weight. From Ref. 13, 1958, Society of Petroleum Engineers of AIME. Reprinted with permission of the SPE*

Equations (2.16) and (2.17) give R_{sf} for an oil *saturated with gas at pressure p*. The inverse determination of the bubble point p_b of an oil saturated with gas is possible by rearranging the equations:

STANDING

$$p_b = 0.2 \left(\frac{R_{sf}}{\gamma_g}\right)^{0.83} \text{antilog}(0.001638\,T - 0.0125\,°\text{API}) - 0.176 \qquad (2.18)$$

LASATER

$$p_b = 0.01241 \frac{T}{\gamma_g} p^*, \qquad (2.19a)$$

where:

$$p^* = F^{-1}(y_g), \quad \text{Fig. 2.9},$$

$$y_g = \frac{R_{sf}}{R_{sf} + 23.645\,(\rho_{o,sc}/M)}, \qquad (2.19b)$$

$$M = f(\rho_{o,sc}), \quad \text{Fig. 2.10}.$$

Standing's correlation is generally found to give better results for heavier crudes with a gravity less than $15\,°\text{API}$, while Lasater's is more suited to crudes lighter than $15°$.

For the calculation of B_{of} from R_{sf}, T, and the produced oil and gas properties, the correlations of Standing[17] and Glasø[12] are probably the most suitable[18].

STANDING

$$B_{of} = 0.9759 + 0.06\left[R_{sf}\left(\frac{\gamma_g}{\rho_{o,sc}}\right)^{0.5} + 0.01267\,T - 3.237\right]^{1.2} \qquad (2.20)$$

GLASØ

$$B_{of} = 1.000 + \text{antilog}\,[-6.58511 + 2.91329\log B^*$$

$$- 0.27683(\log B^*)^2], \qquad (2.21a)$$

where the correlation parameter B^* is obtained from:

$$B^* = 212.5\left[R_{sf}\left(\frac{\gamma_g}{\rho_{o,sc}}\right)^{0.526} + 8.2\cdot 10^{-3}\,T - 2.094\right]. \qquad (2.21b)$$

From a knowledge of R_{sf}, B_{of}, ρ_o, γ_g, and T, the oil density $\rho_o(p, T)$ at p and T follows from the equation:

$$\rho_o(p, T) = \frac{\rho_{o,sc} + 1.225\,\gamma_g R_{sf}(p, T)}{B_{of}(p, T)}. \qquad (2.22)$$

The oil density $\rho_{o,sc}$ is measured at standard conditions. The 1.225 term corresponds to the density of air at standard conditions ($1.225\,\text{kg m}^{-3}$).

If the oil is undersaturated at the specified conditions, B_{of} calculated from Eq. (2.20) or (2.21) has to be corrected for the isothermal compressibility of the oil (c_o, Eq. (2.15)) above its bubble point.

Given R_{sf}, γ_g, $\rho_{o,sc}$, as well as p and T, p_b is first calculated from Eq. (2.18) or (2.19). If $p > p_b$, $B_{of}(p_b)$ is first computed at the bubble point pressure using

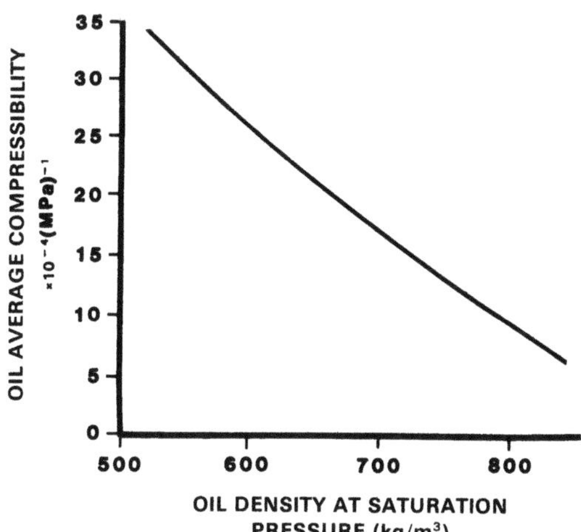

Fig. 2.11. *Isothermal compressibility (c_o) of oil with gas in solution, as a function of its density at the bubble point*

Eq. (2.20) or (2.21). $B_{of}(p)$ is then obtained from:

$$B_{of}(p) = B_{of}(p_b)\left[1 - c_o(p - p_b)\right] .$$ (2.23)

Several correlations are available to evaluate c_o. The simplest one to use, and the one giving best results in most cases, was proposed by Calhoun.[4] As shown in Fig. 2.11, c_o is a function of the oil density ρ_o at bubble point pressure, calculated from Eq. (2.22).

2.3.2.3 Viscosity of Oil with Dissolved Gas

The viscosity μ_o of oil at p and T_R is less than the value $\mu_{o,sc}$ for dead (i.e. degassed) oil at standard conditions, because of the dissolved gas content – expressed quantitatively by R_{sf} – and the increased temperature.

The variation of μ_o with p and T is a routine PVT laboratory measurement.[7, 17] Numerous empirical correlations also exist to calculate μ_o at specified p and T from R_{sf} and the oil API gravity.

This is done in two stages. Firstly, the *dead oil viscosity* μ_o^* is calculated at the temperature T. Glasø's correlation is:

$$\mu_o^*(m\text{Pa s}) = \frac{3.141 \times 10^{10}\,(\log{}^\circ\text{API})^{[10.313\,\log(1.8\,T - 459.76) - 36.447]}}{(1.8\,T - 459.76)^{3.444}} ,$$ (2.24)

where T is in K.

The oil viscosity at p can then be calculated from μ_o^* using the correlation of Beggs and Robinson:[3]

$$\mu_o(m\text{Pa s}) = A(\mu_o^*)^B ,$$ (2.25a)

where:

$$A = 4.406(R_{sf} + 17.809)^{-0.515} ,$$ (2.25b)

$$B = 3.036(R_{sf} + 26.714)^{-0.338} .$$ (2.25c)

R_{sf} is expressed in SI units. Equations (2.24) and (2.25) should be used to calculate the viscosity of a saturated oil (i.e. $p = p_b$).

If pressure is increased beyond the bubble point, the oil becomes under-saturated and its viscosity will increase as a consequence of its increased density.

The most straightforward equation, and one that is reasonably accurate, was proposed by Vasquez and Beggs:[19]

$$\mu_o(p) = \mu_o(p_b)\left(\frac{p}{p_b}\right)^m , \tag{2.26a}$$

where:

$$m = 956.4 \, p^{1.187} \, \text{antilog} \left[-5.656 \times 10^{-3} \, p - 5\right] . \tag{2.26b}$$

The pressure p is expressed in MPa.

2.3.2.4 Gas/Oil Interfacial Tension

The gas/oil interfacial tension σ_{go} plays a major role in all phenomena concerning the displacement of oil by gas, and its value must be known in order to estimate the height of the capillary transition zone.

A few specialised PVT laboratories are able to provide direct measurements of σ_{go}: in practice, however, correlations are generally used. The MacLeod-Sugden[10] correlation, which is based on quantum mechanics, is widely used:

$$\sigma_{go}^{0.25} = 10^{-3} \sum_{i=1}^{n} p_{\text{Ch},i}\left(x_i \frac{\rho_o}{M_o} - y_i \frac{\rho_g}{M_g}\right) , \tag{2.27}$$

where σ_{go} is in mN m^{-1} (dyne cm^{-1}), and $p_{\text{Ch},i}$ = parachor of component i, calculated according to Fanchi[9] (listed in Table 2.1), and:

x_i = mole fraction of component i in the oil phase ,

y_i = mole fraction of component i in the gas phase ,

M_o = molecular weight of the oil = $\sum_{i=1}^{n} x_i M_i$,

M_g = molecular weight of the gas = $\sum_{i=1}^{n} y_i M_i$,

ρ_o = oil density in kg m^{-3} under reservoir conditions (Eq. (2.22)) ,

ρ_g = gas density in kg m^{-3} under reservoir conditions (Eq. (2.14)) .

2.4 The Properties of Reservoir Water

Reservoir water is characterised by its salt content, or *salinity*, C, in kg m^{-3}. The dissolved salts are predominantly NaCl, with sometimes significant quantities of Ca and Mg chlorides and bicarbonates (especially in carbonate reservoirs), sulphates and traces of bromides and iodides.

In the same way as for oil, we define a gas solubility R_{sw}, a volume factor B_w, a compressibility c_w, a viscosity μ_w and a density ρ_w at reservoir conditions.

Figure 2.12 can be used to estimate R_{sw} as a function[16] of p, T, and C.

Figures 2.13 and 2.14 show how B_w varies with p and T for two different salinities $C = 0$ and 100 kg m^{-3}. These diagrams[6] are for water with no gas in

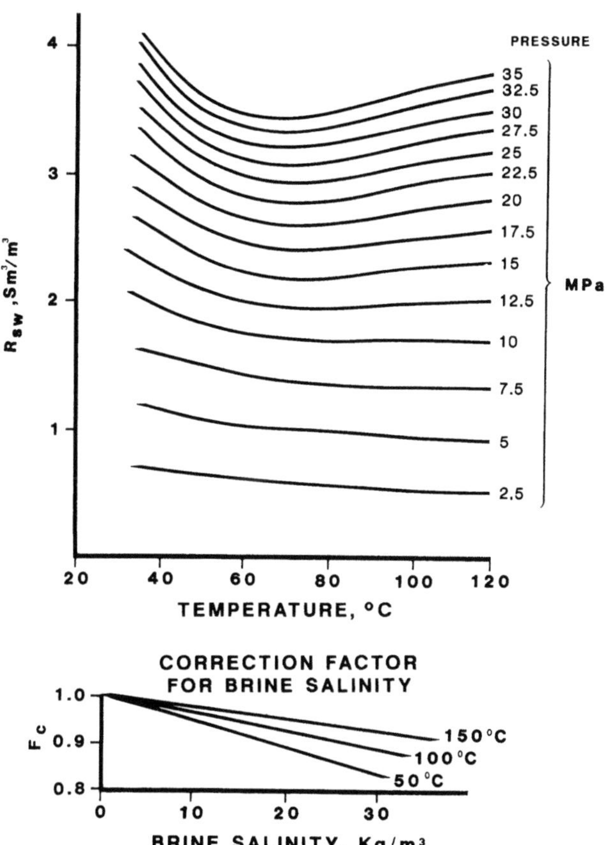

Fig. 2.12. *The solubility of methane in fresh and saline water at reservoir conditions. From Ref. 16, reprinted courtesy of Schlumberger*

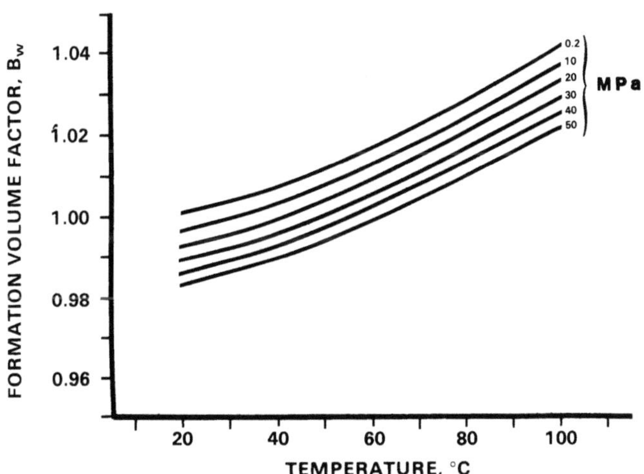

Fig. 2.13. *Volume factor (B_w) of fresh formation water at reservoir conditions. From Ref. 6, reprinted with permission of Petroleum Engineer International*

solution. A correction[6] for dissolved gas ($R_{sw} > 0$) can be made using Fig. 2.15 – this correction is strictly speaking only valid at zero salinity.

The isothermal compressibility $c_w(p, T)$ can be read[6] from Figs. 2.16 ($C = 0\,\text{kg m}^{-3}$) and 2.17 ($C = 100\,\text{kg m}^{-3}$).

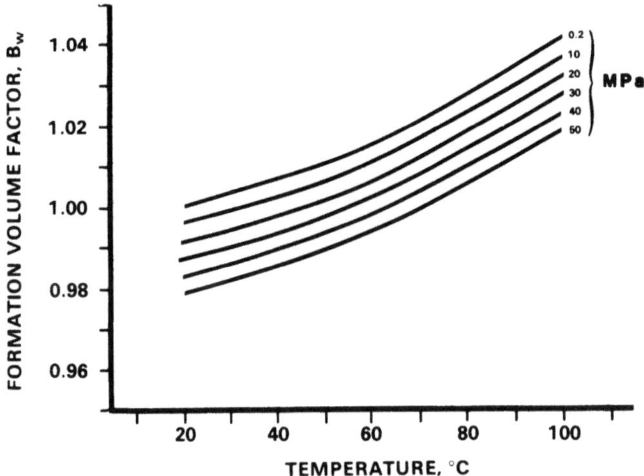

Fig. 2.14. *Volume factor* (B_w) *of formation water of 100* kg m^{-3} *NaCl salinity, at reservoir conditions. From Ref. 6, reprinted with permission of Petroleum Engineer International*

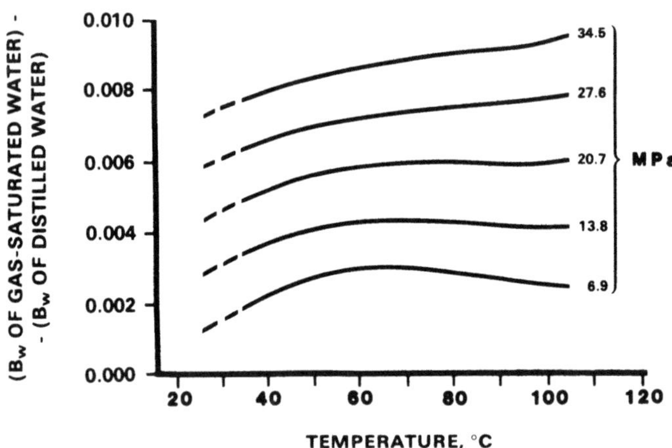

Fig. 2.15. *Dissolved gas correction to* B_w. *From Ref. 6, reprinted with permission of Petroleum Engineer International*

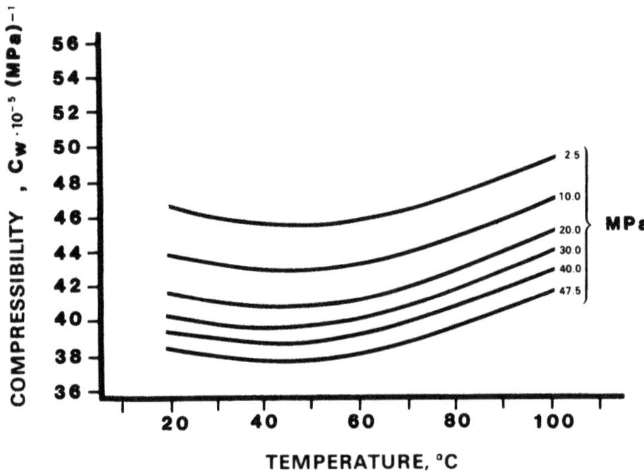

Fig. 2.16. *Isothermal compressibility of fresh formation water at reservoir conditions. From Ref. 6, reprinted with permission of Petroleum Engineer International*

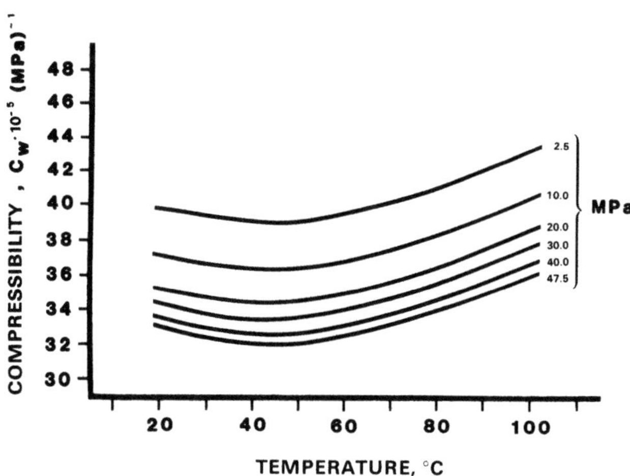

Fig. 2.17. *Isothermal compressibility of formation water of $100\ kg\,m^{-3}$ NaCl salinity at reservoir conditions. From Ref. 6, reprinted with permission of Petroleum Engineer International*

Chierici[5] derived the following correlation for the water density $\rho_w(p, T, C)$ from experimental data:

$$\rho_w = 730.6 + 2.025\,T - 3.8 \times 10^{-3}\,T^2$$
$$+ [2.362 - 1.197 \times 10^{-2}\,T + 1.835 \times 10^{-5}\,T^2]\,p$$
$$+ [2.374 - 1.024 \times 10^{-2}\,T + 1.49 \times 10^{-5}\,T^2 - 5.1 \times 10^{-4}\,p]\,C ,\qquad (2.28)$$

where p is in MPa, T in K and C in $kg\,m^{-3}$.

The range of validity of this relationship is:

salinity: $0-300\ \mathrm{kg\,m^{-3}}$,

pressure: $0-50\ \mathrm{MPa}$,

temperature: $293-373\ \mathrm{K}$.

At standard conditions (0.1013 MPa and 288.2 K):

$$\rho_{w,sc} = 998.6 + 0.660\,C, \quad \mathrm{kg\,m^{-3}} \qquad (2.29)$$

which is in good agreement with experimental data.

Note that from Eqs. (2.28) and (2.29) it is easy to deduce relationships for $B_w(p, T, C)$ and $c_w(p, T, C)$. We get:

$$B_w = \frac{\rho_{w,sc}}{\rho_w} , \qquad (2.30)$$

and:

$$c_w = -\frac{1}{B_w}\left(\frac{\partial B_w}{\partial p}\right)_{T,C} = -\frac{\rho_w}{\rho_{w,sc}}\left[\frac{\partial}{\partial p}\left(\frac{\rho_{w,sc}}{\rho_w}\right)\right]_{T,C} = \frac{1}{\rho_w}\left(\frac{\partial \rho_w}{\partial p}\right)_{T,C} , \qquad (2.31)$$

where, from Eq. (2.28), we can write:

$$\left(\frac{\partial \rho_w}{\partial p}\right)_{T,C} = 2.362 - 1.197 \times 10^{-2}\,T + 1.835 \times 10^{-5}\,T^2 - 5.1 \times 10^{-4}\,C . \qquad (2.32)$$

As far as the viscosity is concerned, $\mu_w(T, C, p)$ is best read from Fig. 2.18 (from the Shell Oil Co).[16]

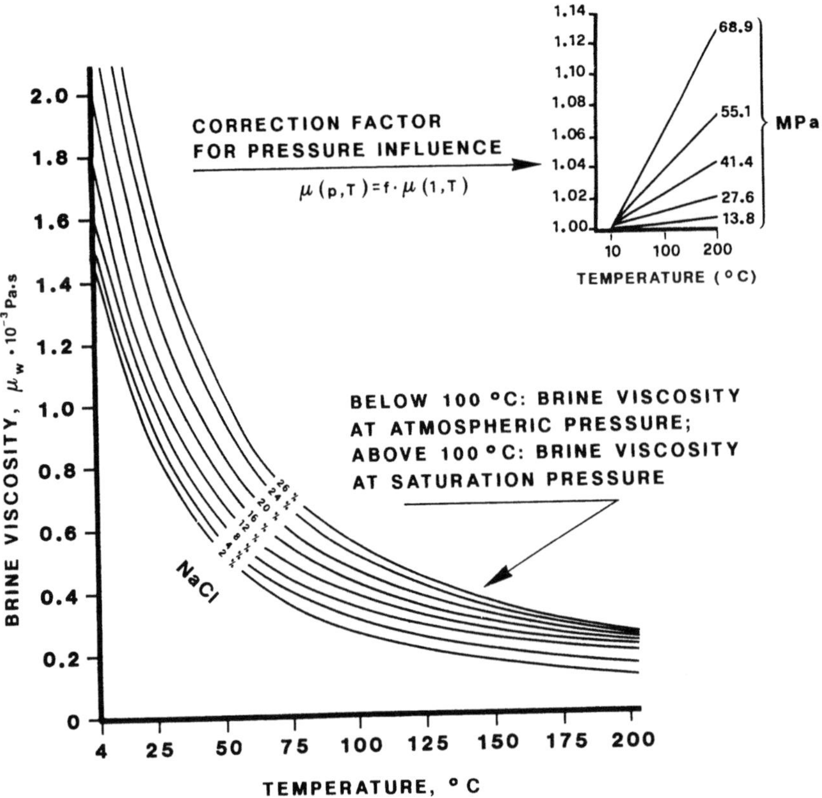

Fig. 2.18. *Viscosity of formation water as a function of salinity, temperature and pressure. Reprinted with permission of Shell Oil Company*

This diagram is closely approximated by the following equation:

$$\mu_w(T, C)(m\text{Pa s}) = (1 + 2.765 \times 10^{-3} C)\exp[11.897$$
$$- 5.943 \times 10^{-2} T + 6.422 \times 10^{-5} T^2] , \qquad (2.33)$$

with T in K.

The effect of pressure on the μ_w calculated from this equation should be obtained using the insert in Fig. 2.18.

References

1. Amyx JW, Bass DL Jr, Whiting RL (1960) Petroleum reservoir engineering-physical properties. McGraw-Hill, New York
2. API RP 44 (1966) Recommended practice for sampling petroleum reservoir fluids. American Petroleum Institute. New York
3. Beggs HD, Robinson JR (1975) Estimating the viscosity of crude oil systems. J Petrol Tech (Sept): 1140–1141
4. Calhoun JC Jr (1947) Fundamentals of reservoir engineering. University of Oklahoma Press, Norman
5. Chierici GL, Long G (1959) Compressibilité et masse spécifique des eaux de gisement dans les conditions du gisement. Application à quelques problèmes de reservoir engineering. Proc 5th World Petroleum Congr vol 2, New York, pp 187–210

6. Chierici GL, Long G (1961) Salt content changes compressibility of reservoir brines, Petrol Eng (July): B 25–B 31

7. Chierici GL (1962) Comportamento volumetrico e di fase degli idrocarburi nei giacimenti. Giuffrè Editore, Milano

8. Coats KH, Smart GT (1986) Application of a regression-based EOS PVT program to laboratory data. SPE Reserv Eng (May): 277–299

9. Fanchi JR (1985) Calculation of parachors for compositional simulation. J Petrol Tech (Nov): 2049–2050; Trans AIME 279

10. Frick TC (ed) (1962) Petroleum production handbook, vol II, Reservoir engineering. Society of Petroleum Engineers, Dallas

11. Fussell DD, Yanosik JL (1978) An iterative sequence for phase – equilibria calculation incorporating the Redlich–Kwong equation of state, Soc Petrol Eng (June): 173–182

12. Glasø Ø (1980) Generalized pressure-volume-temperature correlations. J Petrol Tech. (May): 785–795; Trans AIME 269

13. Lasater JA (1958) Bubble point pressure correlations. Trans AIME: 379–381

14. Lee AL, Gonzales MH, Eakin BE (1966) The viscosity of natural gases. Trans AIME: 997–1000

15. Mott RE (1980) Development and evaluation of a method for calculating the phase behaviour of multi-component hydrocarbon mixtures using an equation of state. AEE Winfrith, Rep 1331, Dorchester

16. Schlumberger Ltd (1974) Fluid conversion in production log interpretation. Schlumberger Ltd, Houston

17. Standing MB (1977) Volumetric and phase behavior of oil field hydrocarbon systems, 8th impr Society of Petroleum Engineers, Dallas

18. Sutton RP, Farshad FF (1984) Evaluation of empirically derived PVT properties for Gulf of Mexico crude oils. Pap SPE 13172 presented at the 59th Annual Technical Conference, Houston Sept 16–19

19. Vasquez M, Beggs HD (1980) Correlations for fluid physical property prediction. J Petrol Tech (June): 968–970; Trans AIME 269

20. Yarborough L (1979) Application of a generalized equation of state to petroleum reservoir fluids. Advances in Chemistry Series, vol 182. American Chemical Society, Washington, DC, pp 384–439

EXERCISES

Exercise 2.1

For a natural gas having the following composition:

Component	Symbol	Mole fraction (y_i)
Nitrogen	N_2	0.0080
Methane	C_1	0.8080
Ethane	C_2	0.1255
Propane	C_3	0.0380
i-Butane	$i\text{-}C_4$	0.0110
n-Butane	$n\text{-}C_4$	0.0065
Pentane	C_5	0.0030
		1.0000

calculate: the molecular weight and density (relative to air), the deviation factor z, the volume factor B_g, and the density and viscosity at the reservoir temperature of 78 °C, at initial reservoir pressure (14.5 MPa) and at abandonment pressure (2.0 MPa).

Solution

The molecular weight M of the gas is calculated as the average of the molecular weights M_i of the components, weighted according to their mole fractions y_i:

$$M = \sum M_i y_i$$

This gives us the following summation:

Component	M_i	y_i	$M_i y_i$
N_2	28.01	0.0080	0.2241
C_1	16.04	0.8080	12.9603
C_2	30.07	0.1255	3.7738
C_3	44.10	0.0380	1.6758
$i\text{-}C_4$	58.12	0.0110	0.6393
$n\text{-}C_4$	58.12	0.0065	0.3778
C_5	72.15	0.0030	0.2165
		1.0000	19.8676

The molecular weight of this gas is therefore 19.87.

Now, the molecular weight of air is 28.97. Since at the same conditions of temperature and pressure, one kg/mol of any gas occupies the same volume (*assuming ideal gas behaviour*), the density of the gas relative to air (i.e. the specific gravity) is given by

$$\gamma_g(\text{air} = 1) = \left(\frac{\text{weight of a given volume of gas}}{\text{weight of the same volume of air}} \right)_{p,\,T}$$

$$= \frac{19.87}{28.97} = 0.6859 \ .$$

So the specific gravity of the gas is 0.6859 relative to air.

Since 1 kg mol of any gas at standard conditions ($p = 0.1013$ MPa, T $= 288.2$ K) occupies a volume of 23.645 sm^3 (*standard* cubic metres), the density of the gas will be:

$$\rho_{g,\,sc} = \frac{19.87}{23.645} = 0.8403 \ \text{kg m}^{-3} \ \text{at standard conditions} \ .$$

For the gas deviation factor z, first of all we need to get the pseudo-critical pressure (p_{pc}) and temperature (T_{pc}) of the gas, using Eq. (2.11):

Component	y_i (fraz)	P_{ci} (MPa)	$y_i P_{ci}$ (MPa)	T_{ci} (K)	$y_i T_{ci}$ (K)
N_2	0.0080	3.339	0.0272	126.1	1.009
C_1	0.8080	4.606	3.7216	190.6	154.005
C_2	0.1255	4.881	0.6126	305.6	38.353
C_3	0.0380	4.247	0.1614	370.0	14.060
$i\text{-}C_4$	0.0110	3.647	0.0401	408.3	4.491
$n\text{-}C_4$	0.0065	3.799	0.0247	425.0	2.763
C_5	0.0030	3.378	0.0101	460.6	1.382
	1.0000		4.5977		216.063

This gives us:

$$p_{pc} = 4.60 \ \text{MPa}$$

$$T_{pc} = 216.1 \ \text{K}$$

and therefore:

$$T_{pr} = \frac{78 + 273.2}{216.1} = 1.625$$

At initial pressure:

$$p_{pr,i} = \frac{14.5}{4.6} = 3.152$$

and at abandonment pressure:

$$p_{pr,ab} = \frac{2.0}{4.6} = 0.435$$

We can now read values for the z-factor from Fig. 2.4:

$z(\text{initial}) = 0.83$

$z(\text{abandonment}) = 0.97$

Alternatively, if we use Eq. (2.12), we get:

$z(\text{initial}) = 0.829$

$z(\text{abandonment}) = 0.965$

Consequently:

at initial pressure: $\quad B_{gi} = 3.515 \times 10^{-4} \dfrac{0.829 \times 351.2}{14.5} = 7.058 \times 10^{-3}$

at abandonment pressure: $\quad B_{g,ab} = 3.515 \times 10^{-4} \dfrac{0.965 \times 351.2}{2} = 5.956 \times 10^{-2}$

This means that every $1\,\mathrm{m}^3$ of reservoir pore-space will contain:

at initial pressure: $\quad 1/B_{gi} = 141.7\,\mathrm{sm}^3$ of gas

at abandonment pressure: $\quad 1/B_{g,ab} = 16.8\,\mathrm{sm}^3$ of gas

For the gas density we have:

at initial pressure: $\quad \rho_{gi} = \dfrac{\rho_{g,sc}}{B_{gi}} = 0.8403 \times 141.7 = 119.1\,\mathrm{kg\,m}^{-3}$

at abandonment pressure: $\quad \rho_{g,ab} = \dfrac{\rho_{g,sc}}{B_{g,ab}} = 0.8403 \times 16.8 = 14.1\,\mathrm{kg\,m}^{-3}$

The gas viscosity can be calculated from Eq. (2.13), with:

$$X = 3.5 + \frac{547.8}{351.2} + (0.01 \times 19.87) = 5.258$$

$$y = 2.4 - (0.2 \times 5.258) = 1.348$$

$$K = \frac{(12.61 + 0.027 \times 19.87)351.2^{1.5}}{116.11 + 10.56 \times 19.87 + 351.2} = 127.78$$

Therefore:

at initial pressure: $\quad \mu_{gi} = 1.278 \times 10^{-2} \exp[5.258(0.119)^{1.348}]\,\mathrm{mPa\,s} = 17.2\,\mathrm{\mu Pa\,s}$

and at abandonment pressure:

$$\mu_{g,ab} = 1.278 \times 10^{-2} \exp[5.258(0.0141)^{1.348}]\,\mathrm{mPa\,s} = 13.0\,\mathrm{\mu Pa\,s}$$

$$\diamond\,\diamond\,\diamond$$

Exercise 2.2

PVT measurements were made on an oil sample at reservoir temperature (90 °C):

– reservoir pressure $p = 15.8$ MPa (161.1 kg cm^{-2})
– bubble-point pressure $= 9.5$ MPa (96.9 kg cm^{-2})

The differential liberation study gave the following results:

Pressure		Volume of oil in cell	Volume of gas removed	Specific gravity of gas removed
(MPa)	(kg cm^{-2})	(cm^3)	(s cm^3)	(air = 1)
15.8	161.1	138.861	—	—
9.5	96.9	140.184	—	—
8.0	81.6	137.251	728.2	1.374
6.0	61.2	134.603	934.5	0.927
4.0	40.8	130.480	873.9	1.180
2.0	20.4	126.839	788.9	0.950
0.0	zero	123.552	655.4	0.912
0.0 and 15 °C		121.370		

The density of the residual oil in the PVT cell was 933 kg m^{-3} at 15 °C

Calculate values for B_{od}, R_{sd}, and ρ_0 at reservoir pressure using these measurements.

Solution

Using the fact that the density of air at standard conditions is 1.225 kg m^{-3}, the volumes and masses in the PVT cell work out as follows:

Pressure	Volume of oil in cell	Total vol of dissolved gas	Mass of dissolved gas	Total mass in cell
(MPa)	(cm^3)	(s cm^3)	(g)	(g)
15.8	138.861	3980.9	5.200	118.438
9.5	140.184	3980.9	5.200	118.438
8.0	137.251	3252.7	3.974	117.212
6.0	134.603	2318.2	2.913	116.151
4.0	130.480	1444.3	1.650	114.888
2.0	126.839	655.4	0.732	113.970
0.0	123.552	zero	zero	113.238
0.0 and 15 °C	121.370 = 113.238 g			

The mass of gas evacuated from the cell at each pressure step was obtained by multiplying the volume of gas by its density ($\rho_g = 1.225\ \gamma_g$ kg m^{-3}).

B_{od} is calculated by dividing the volume of oil in the cell at a given pressure by the volume of residual oil at atmospheric pressure and 15 °C (121.370 cm^3 in the table).

Similarly, the value of R_{sd} is obtained by dividing the total volume of gas dissolved in the oil at a given pressure, by the volume of residual oil at 15 °C.

The density ρ_0 then follows by dividing the total mass (kg) of oil present in the cell at each pressure, by the volume occupied by the oil (m^3) at that pressure.

The next table lists the results, and Fig. E2/2.1 presents them graphically.

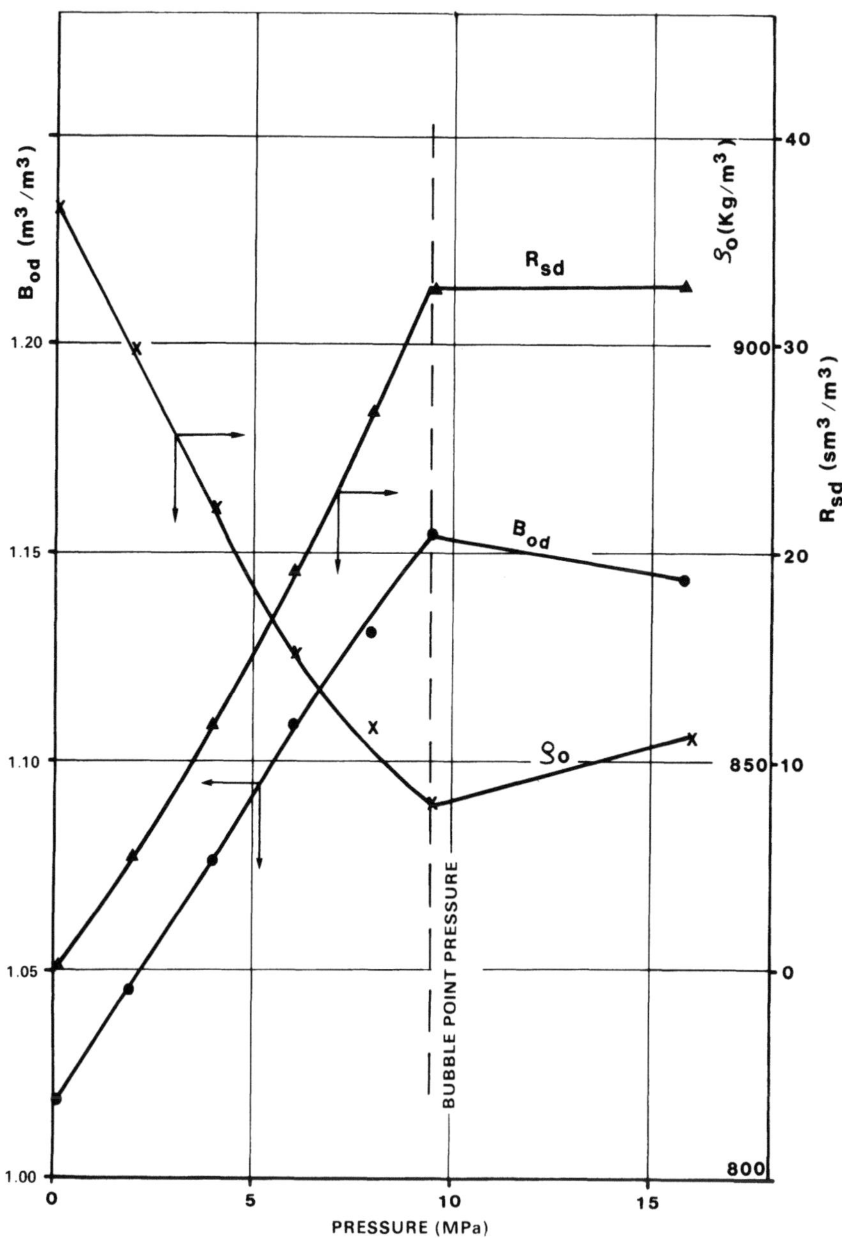

Fig. E2/2.1

Pressure (MPa)	(kg cm⁻²)	B_{od} (m³ m⁻³)	R_{sd} (s m³ m⁻³)	ρ_o (kg m⁻³)
15.8	161.1	1.144	32.80	853
9.5	96.9	1.155	32.80	845
8.0	81.6	1.131	26.80	854
6.0	61.2	1.109	19.10	863
4.0	40.8	1.075	11.89	881
2.0	20.4	1.045	5.40	898
0.0	0.0	1.018	zero	917

◇ ◇ ◇

Exercise 2.3

The following measurements were recorded during a production test on an oil well:

- depth to middle of pay-zone: 4500 m
- static reservoir pressure: 465 kg cm^{-2}
- reservoir temperature: 172 °C
- separator plus stock tank GOR: 210 sm^3 m^{-3} STO
- stock tank oil API gravity: 38 °
- separator gas gravity: 0.8 (air = 1.0)

Calculate the following using correlations:

- bubble-point pressure of the oil at reservoir temperature
- the oil formation volume factor at the bubble-point
- the oil formation volume factor at reservoir pressure
- oil viscosity at the bubble-point
- oil viscosity at reservoir pressure

Solution

The correlations provided in Sect. 2.3.2 are all in SI units, so a few units conversions will need to be performed on the data first of all.

- static reservoir pressure: $p\ = 465 \times 0.0980665 = 45.60$ MPa
- reservoir temperature: $T\ = 172 + 273.2 = 445.2$ K
- separator GOR: $R_{\mathrm{sf}} = 210$ m^3 m^{-3}
- stock tank oil API gravity: $= 38$ °
- stock tank oil density: $\rho_{\mathrm{o}}\ = \dfrac{141.5}{38 + 131.5}\,10^3 = 835$ kg m^{-3}
- separator gas gravity: $\gamma_{\mathrm{g}}\ = 0.8$

Since the stock tank oil gravity is higher than 15°, the Lasater correlation should be the more suitable for the calculation of the bubble-point.

Referring to Fig. 2.10, a 38 °API oil has a corrected molecular weight $M = 245$, and therefore, using Eq. (2.19b):

$$y_{\mathrm{g}} = \frac{210}{210 + 23.645\,(835/245)} = 0.727$$

With this value of y_{g}, we can read off $p^* = 4.7$ from Fig. 2.9. Equation (2.19a) for p_{b} from the Lasater correlation then becomes:

$$p_{\mathrm{b}} = 0.01241\,\frac{445.2}{0.8}\,4.7 = 32.46\ \mathrm{MPa} = 331\ \mathrm{kg\,cm}^{-2}$$

If we had used Standing's correlation (Eq. (2.18)) we would have got:

$$p_{\mathrm{b}} = 0.2\left(\frac{210}{0.8}\right)^{0.83}\ \mathrm{antilog}(0.001638 \times 445.2 - 0.0125 \times 38) - 0.176$$

$$= 36.40\ \mathrm{MPa} = 371\ \mathrm{kg\,cm}^{-2}$$

There is a difference of about 10% between the two estimates.

For the volume factor B_{of}, we can use the Glasø correlation (Eq. (2.21b)):

$$B^* = 212.5\left[210\left(\frac{0.8}{835}\right)^{0.526} + 8.2 \times 10^{-3} \times 445.2 - 2.094\right] = 1483.7$$

so that Eq. (2.21a) gives us:

$$B_{\mathrm{of}} = 1.000 + \mathrm{antilog}\left[-6.58511 + 2.91329\log 1483.7 - 0.27683\,(\log 1483.7)^2\right] = 1.741$$

For comparison, using Standing's correlation (Eq. 2.20):

$$B_{of} = 0.9759 + 0.06\left[210\left(\frac{0.8}{835}\right)^{0.5} + 0.01267 \times 445.2 - 3.237\right]^{1.2} = 1.803$$

The two values agree to within 4%.

Note that these values of B_{of} pertain the saturation or bubble-point pressure ($p_b = 32.46$ MPa). To convert to reservoir conditions ($p = 45.60$ MPa), we must allow for the compressibility of the liquid oil above its bubble point ($p > p_b$).

First, we calculate the oil density at $p = p_b$ using Eq. (2.22) with the Glasø estimate of B_{of}:

$$\rho_o(p_b, T) = \frac{835 + 1.225 \times 0.8 \times 210}{1.741} = 598 \text{ kg m}^{-3}$$

Entering Fig. 2.11 with this value, we obtain for the isothermal compressibility $c_o = 2.6 \times 10^{-3}$ MPa^{-1}.

The volume factor at reservoir pressure is therefore (Eq. (2.23)):

$$B_{of} = 1.741\left[1 - 2.6 \times 10^{-3}(45.60 - 32.46)\right] = 1.682$$

The Lasater estimate of p_b (32.46 MPa) has been used here.

The calculation of reservoir oil viscosity first requires the dead oil viscosity μ_o^* at reservoir temperature (Eq. (2.24)):

$$\mu_o^* = \frac{3.141 \times 10^{10}(\log 38)^{[10.313 \log(1.8 \times 445.2 - 459.76) - 36.447]}}{(1.8 \times 445.2 - 459.76)^{3.444}} = 0.528 \text{ mPa s (cP)}$$

Then, we determine the oil viscosity at bubble-point pressure (32.46 MPa) via Eq. (2.25), with:

$$A = 4.406(210 + 17.809)^{-0.515} = 0.269$$

$$B = 3.036(210 + 26.714)^{-0.338} = 0.478$$

so that:

$$\mu_o(p_b) = 0.269(0.528)^{0.478} = 0.198 \text{ mPa s (cP)}$$

Equation (2.26) then allows us to calculate μ_o at the reservoir pressure of 45.60 MPa:

$$m = 956.4 \times 45.6^{1.187} \text{ antilog}\left[-5.656 \times 10^{-3} \times 45.6 - 5\right] = 0.492$$

giving:

$$\mu_o(p = 45.60 \text{ MPa}) = 0.198\left(\frac{45.60}{32.46}\right)^{0.492} = 0.234 \text{ MPa s (cP)}.$$

◇ ◇ ◇

Exercise 2.4

For a formation water with an equivalent NaCl salinity of 120 kg m^{-3}, calculate:

- the density
- the formation volume factor
- the isothermal compressibility
- the viscosity

at the reservoir temperature 55 °C (328.2 K) and at pressures of 12 and 10 MPa.

Solution

We can use Eq. (2.28) to calculate $\rho_w(p, T)$. With $T = 328.2$ K we have:

$$\rho_w(p; 328.2 \text{ K}) = 985.887 + 0.410\,p + 120(0.6182 - 5.1 \times 10^{-4}p)$$

so that:

$$\rho_w(12\ \text{MPa},\ 328.2\ \text{K}) = 1064.257\ \text{kg m}^{-3}$$

and:

$$\rho_w(10\ \text{MPa},\ 328.2\ \text{K}) = 1063.559\ \text{kg m}^{-3}$$

Under standard conditions Eq. (2.29) gives:

$$\rho_{w,\,sc} = 998.6 + 0.66 \times 120 = 1077.8\ \text{kg m}^{-3}.$$

Consequently, from Eq. (2.30):

$$B_w(12\ \text{MPa},\ 328.2\ \text{K}) = \frac{1077.800}{1064.257} = 1.0127$$

$$B_w(10\ \text{MPa},\ 328.2\ \text{K}) = \frac{1077.800}{1063.559} = 1.0134$$

From Eq. (2.32):

$$\left(\frac{\partial \rho_w}{\partial p}\right)_{328.2\ \text{K};\ 120\ \text{kg m}^{-3}} = 0.3488$$

and therefore, from Eq. (2.31):

$$c_w(12\ \text{MPa},\ 328.2\ \text{K}) = \frac{0.3488}{1064.3} = 3.277 \times 10^{-4}\ (\text{MPa})^{-1}\,,$$

$$c_w(10\ \text{MPa},\ 328.2\ \text{K}) = \frac{0.3488}{1063.6} = 3.279 \times 10^{-4}\ (\text{MPa})^{-1}\,.$$

An average value for c_w between 10 and 12 MPa can also be calculated from the equation:

$$\bar{c}_w = \frac{1}{\bar{\rho}_w}\left(\frac{\Delta \rho_w}{\Delta p}\right)_{T,\,C} = \frac{1}{1063.908}\ \frac{0.698}{2} = 3.28 \times 10^{-4}\ (\text{MPa})^{-1}$$

which is in close agreement with the values of c_w calculated previously.

Since the effect of pressure is very slight as far as the viscosity is concerned (inset in Fig. 2.18), we can use Eq. (2.33) without further correction:

$$\mu_w(328.2\ \text{K};\ 120\ \text{kg m}^{-3}) = (1 + 2.765 \times 10^{-3} \times 120)\exp\left[11.897 - 5.943 \times 10^{-2} \times 328.2\right.$$
$$\left. + 6.422 \times 10^{-5} \times 328.2^2\right] = 0.67\ \text{mPa s}$$

A similar result can be obtained graphically using Fig. 2.18.

3 Reservoir Rocks

3.1 Introduction

Knowledge of the petrophysical and hydrodynamic properties of reservoir rock – properties which control both the distribution and movement of hydrocarbons – are obviously of fundamental importance to the petroleum engineer.

This sort of data can be acquired from two major sources:

1. Samples of rock (referred to as "cores") which are retrieved from the well during drilling,
2. Geophysical measurements (referred to as "logs") usually recorded after drilling, before the well has been completed with casing. In some cases logs are run after completion, even when the well is on production with tubing in place. Recent advances in equipment technology now also enable log data to be acquired *during* the drilling process.

We shall spend some time here to review the nature and quality of the information that can be deduced from cores and logs, and its usefulness to the reservoir engineer as a means of evaluating the reservoir rock.

Details about the methods used in the extraction and analysis of cores,[3, 27] and the basic theory, instrumentation, recording and interpretation of well logs[22, 23, 29–33] can be obtained from the references listed at the end of the chapter, and will not be covered here.

3.2 Cores and Logs: Their Characteristics and Relative Merits

3.2.1 Cores

Only "downhole cores" can provide valid indicators of reservoir rock properties. These are cut in the course of drilling the well by replacing the drill-bit with a *core barrel*. This is basically a hollow pipe about $2\frac{1}{2}''$–$5''$ (6.35–12.7 cm) in diameter and several tens of feet or metres long. It has a cutting tip which cuts out a continuous cylinder of rock which is retained within the core barrel and eventually withdrawn from the hole.

Coring while drilling is quite a costly procedure in terms of rig time: firstly it entails pulling the whole drill string out of the hole so as to replace the normal bit by the core barrel; the coring operation itself necessitates a reduced drilling rate in order to ensure a good recovery; then the barrel has to be brought back to surface. A single core is typically no more than about 30 ft (9 m) long, so this cycle must be repeated if further coring of the interval is required. There is also some increase in the risk of the drill string sticking during coring.

For completeness, it is worth mentioning "sidewall cores", which are an alternative source of rock sample. They are obtained after drilling has finished, but before the hole is cased, using a wireline-conveyed *core gun*. The gun is run in to a predetermined depth and a hollow cylindrical "bullet" is fired into the wall of the hole. The bullet is attached to the gun by a wire, so it (and the small cylindrical core it cuts out) can be pulled away from the hole wall and brought to surface. There are typically 20–30 such bullets on a gun, so a large number of cores can be taken from different depths. Sidewall cores have a diameter of about 1″ (2.5 cm) and are about 2″ (5.1 cm) long. Being rather small, and frequently contaminated with mud filtrate and mudcake, they are mainly of paleontological and sedimentological interest, and are of little value to the reservoir engineer.

Although downhole coring should obviously be programmed before the section of interest is drilled out, in an exploration well, it cannot always be planned with certainty, and in this situation sidewall coring may be a valuable after-the-event source of rock sample.

While the core is being cut, it becomes contaminated with drilling mud filtrate, which is overpressured with respect to the pore fluids. So even while it is still downhole, the core probably contains only a small fraction of the original reservoir fluids.

When it is brought to surface, the reduction in confining pressure from that of the reservoir to atmospheric leads to evolution of dissolved gas from any oil in the pores, and further expulsion of fluids.

Consequently, the fluid content of the core observed on surface has little or no bearing on the original situation at reservoir conditions, and can provide *no quantitative indications of the saturations of oil, gas and water, nor of their physical and chemical properties in the reservoir.*

It would, however, be safe to deduce from the presence of oil in the core that there was oil in the formation (provided of course that a water-based drilling mud had been used).

The complex geostatic stresses acting on the reservoir rock are absent once the core is on surface. These stresses influence some of the petrophysical properties of the rock – therefore, laboratory measurements should either be made at, or corrected to, reservoir conditions wherever possible.

A fundamental difference between log and core measurements is that the latter are *direct measurements of the desired parameter on an actual rock sample.*

Log measurements, on the other hand, are not direct; *none of the rock properties of interest can be measured directly with the logging sensors currently in use.* Instead, they must be *inferred* from the physical parameters which can be measured (such as the electrical resistivity, acoustic velocity, neutron absorption, gamma ray scatter, etc.) by means of theoretical and empirical correlations.

A direct measurement on a core is therefore considered preferable because it should be more precise and reliable than a log measurement.

This chapter will be restricted to the parameters of interest to the reservoir engineer that are measured from cores. We shall, therefore, ignore parameters such as the formation resistivity factor, the cementation factor, saturation exponent, etc., which are required for the interpretation of well logs.

The following two categories of rock property will be covered:

static: porosity
             compressibility
             wettability
             capillarity

dynamic: absolute permeability
             effective permeability

3.2.2 Logs

A "log" is a recording of the variations of certain physical rock properties with depth, over an interval of formation. The measurements are made using highly sophisticated logging tools which are lowered into the well by means of a metal-armoured cable whose core contains one or several (typically seven) insulated conductor wires. These wires transmit the measured signals to the surface recording system, as well as conveying power and control commands down to the tool electronics. Several different tools can be combined in a single run, and the various log measurements made simultaneously. As explained previously, the measurements made are rarely of *direct* use to the exploration geologist or reservoir engineer, and the rock properties have to be calculated from them.

Well logs can be grouped into three categories:

Lithological Logs:

- Spontaneous (or Self) Potential (SP): an electrical potential or e.m.f. is generated in the wellbore from the natural battery cell created between the mud filtrate in the flushed zone/the formation water in the uninvaded reservoir/the adjacent shale beds. This enables shale strata to be distinguished from more permeable rocks.
- Gamma Ray Log (GR): natural gamma ray emission varies with the nature of the sedimentary strata – in particular, clay minerals tend to be more radio-active than "clean" minerals such as silica and carbonates, so that shale beds (possibly including the reservoir cap rock) can be distinguished. The GR tool is almost *always* included in any logging run.
- Photo-electric Log (LDT): the photo-electric absorption index of the surrounding rock is measured from low energy gamma ray scatter. This index is a function of the mean atomic number of the rock, from which its mineral type can be deduced.
- Gamma Spectrometry (GST): gamma ray emission, induced in the formation by bombardment with high energy neutrons, is subjected to spectroscopic analysis, from which the presence or absence of certain elements (C, O, Fe, Ca, Si, S, H, Cl, etc) can be inferred, and an estimate of their relative abundance made.

Porosity Logs:

- Borehole Compensated Sonic Log (BHC): measures the velocity of elastic acoustic waves through the rock, which depends on the porosity.

- Formation Density Compensated Log (FDC, LDT): measures the scattering of gamma rays emitted from a source. This scattering is related to the bulk density of the rock.
- Compensated Neutron Log (CNL): measures the slowing down and absorption of neutrons emitted from a radio-active source. This is sensitive to the hydrogen present in water and oil.

The rock porosity is *calculated* from each of these log measurements. The term "compensated" implies that the tool design and measurement technique are designed to minimise the effects of the borehole (which represents a porosity of 100%). Some residual corrections for this and other environmental perturbations usually still need to be applied.

It should be mentioned that these log readings are all sensitive to lithology to some extent, and they are routinely used in combination (in "crossplot" techniques) to identify lithology.

Saturation Logs:

- Laterolog (LL): the electrical resistivity of the rock is measured using a focused current beam from an electrode array.
- Induction Log (IL): the electrical conductivity is measured using high frequency magnetic fields induced in the rock by a system of coils.
- Dielectric Log (EPT): this is a measurement of the dielectric permittivity of the rock using an electromagnetic wave generated by an antenna array.

The so-called "long spacing" configuration of these electrical logging tools is designed to read deeply into the formation – avoiding the flushed zone so as to respond to water saturation in the virgin reservoir – by means of large electrode or coil spacings (typically 1–2 m).

The "short" or "micro spacing" versions of these tools, by contrast, are designed to read the flushed zone. This shallow measurement is achieved by means of electrodes, coils or antenna spaced only a few centimetres apart, and as well as responding to flushed zone filtrate saturation, the fine depth resolution obtained is useful for delineating thin strata (e.g. Microlog (ML), Microlaterolog (MLL), Microproximity log (PL), Microspherically Focused log (MSFL), Dipmeters (HDT, SHDT), Dielectric log (EPT)).

- Pulsed Neutron Logs (PNL): the rate of absorption of thermalised neutrons (resulting from pulsed bombardment of the rock by high energy neutrons) is a function of the water saturation and its salinity. This is a fairly shallow reading. It is most commonly run in producing wells since the neutrons can penetrate tubing and casing, but it can be used in open hole.

The water saturation S_w is calculated from the resistivity or dielectric measurements, using the porosity estimated from other logs. In addition to porosity, the pulsed neutron methods require knowledge of other parameters related to the lithology.

The hydrocarbon saturation, which is simply $(1 - S_w)$, is thus derived at reservoir conditions.

As well as saturation and porosity, by appropriate choice of logging sensors, it is possible to obtain a quite detailed evaluation of the lithology and the mechanical properties of the rock, and to identify the presence of fractures in the survey interval.

3.2.3 Relative Merits of Core and Log Measurements

The advantages and disadvantages of core and log measurements as a means of evaluating reservoir rock properties are summarized below.

Advantages of Well Logs

1. Core-cutting must be programmed *before* the drilling bit traverses a zone of interest. There is always the possibility, particularly in an exploration well, that the opportunity to core will be missed by inadvertently drilling ahead. Logs, on the other hand, can be run at any time after the hole has been drilled. Although resistivity logs are limited to *open hole*, many of the radioactive type logs can be run even after casing is in place.
2. It is far less expensive to record and interpret well logs than to cut and analyse cores over the same interval.
3. Unlike cores, well logs are measurements made on essentially undisturbed rock at reservoir conditions. This is particularly significant as far as hydrocarbon saturation is concerned.

Advantages of Cores

Cores, in addition to providing the essential material for petrographic, sedimentological and paleontological study, offer several advantages over well logs:

1. Porosity cannot only be measured directly (ideally, with the geostatic confining pressure applied), but its *nature* and *distribution* can be observed – connected, non-connected, intergranular, vuggy, fissured, etc. This is more or less impossible from log data, although some indications – of fractures, for example – are sometimes discernible.
2. The capillary and wettability characteristics of the rock can be evaluated. These can be used to calibrate the saturation profile indicated by the well logs.
3. Dynamic rock properties which control fluid movement (absolute permeability, effective permeability to each phase at different saturations, directional permeability) can be measured. Although attempts have been made to correlate log readings quantitatively with permeability, logs tend to provide at best a qualitative indication of the permeable or impermeable nature of the strata.

3.3 Combined Use of Cores and Logs

The vertical and areal distributions of the petrophysical properties of the reservoir rock need to be evaluated in the most detailed and precise way possible. This is particularly relevant when, as is often the case, it is planned to model the behaviour of the producing reservoir using a numerical simulator. In order to achieve this in a cost-effective manner, it is common reservoir engineering practice to integrate core data with well logs.

Once a tentative development well programme has been set out for the new field, a number of key wells are selected. These are chosen according to their positions so as to provide a reasonable field-wide coverage of the trends that there may be in rock properties.

In these key wells, the entire reservoir section, and the aquifer where present, will be cored, and a complete suite of well logs will be run, including several porosity and lithology sensors, and both micro and deep resistivity tools for saturation.

As soon as the core reaches the laboratory, a continuous record of the natural gamma activity is made along its entire length. There then follows the "routine" measurement of porosity and permeability. In addition, "special" core analysis is performed on small plugs which are cut from the body of the main core at points of interest (for instance, in different lithologies). Grain density, acoustic velocity, cementation factor, saturation exponent, capillary pressures, etc., are investigated in this way.

The core-derived porosity data, corrected to reservoir pressure, is subjected to a numerical smoothing filter so as to replicate the same vertical resolution as the porosity logs [about 1 m (3 ft)]. The resulting porosity profile can then be used to calibrate the log-derived porosity. A simple linear regression analysis will provide the normalisation coefficients needed to match the log data to the core porosities.

These coefficients are then applied to all the porosity logs run in the field, perhaps interpolating values between the key wells if there are significant differences.

A correlation between core permeability and core porosity is determined from the experimental data (this will be covered in more detail later in the book). Using this relationship (permeability) = f(porosity), the permeability can be estimated from the porosity logs in the uncored wells.

From the "special" core analysis, the measurements of grain density are used to interpret the FDC log for porosity, while the cementation factor and saturation exponent are used with the resistivity logs to derive water saturations. The resulting saturation profile will then be compared with the capillary pressure data from the cores; this will be described more fully later in the chapter.

Once the logs from the key wells have been satisfactorily calibrated in this way, the calibration coefficients derived for each layer or zone will be applied to all the other wells in the field.

One factor which should not be overlooked, is that the depth attributed to the core can be in error, whether it was measured from the length of the drill string at coring depth, or from the length of cable run in hole during logging.

A discrepancy of only 0.1% between the two measurements (drill string and cable) at an actual depth of 3000 m, represents a difference of 3 m (almost 10 ft). Obviously, the consequences of calibrating log data against cores which in fact were 3 m higher up or lower down the hole can be disastrous, especially where complex lithologies or thin strata are concerned. Careful depth correction is therefore vital.

For this reason, one of the first measurements made on a core when it is brought into the laboratory is a natural GR profile. This will be correlated against the GR log to determine any depth shifts necessary.

An example of a comparison between GR curves measured from cores and from well logs is shown on the left hand side of Fig. 3.1.

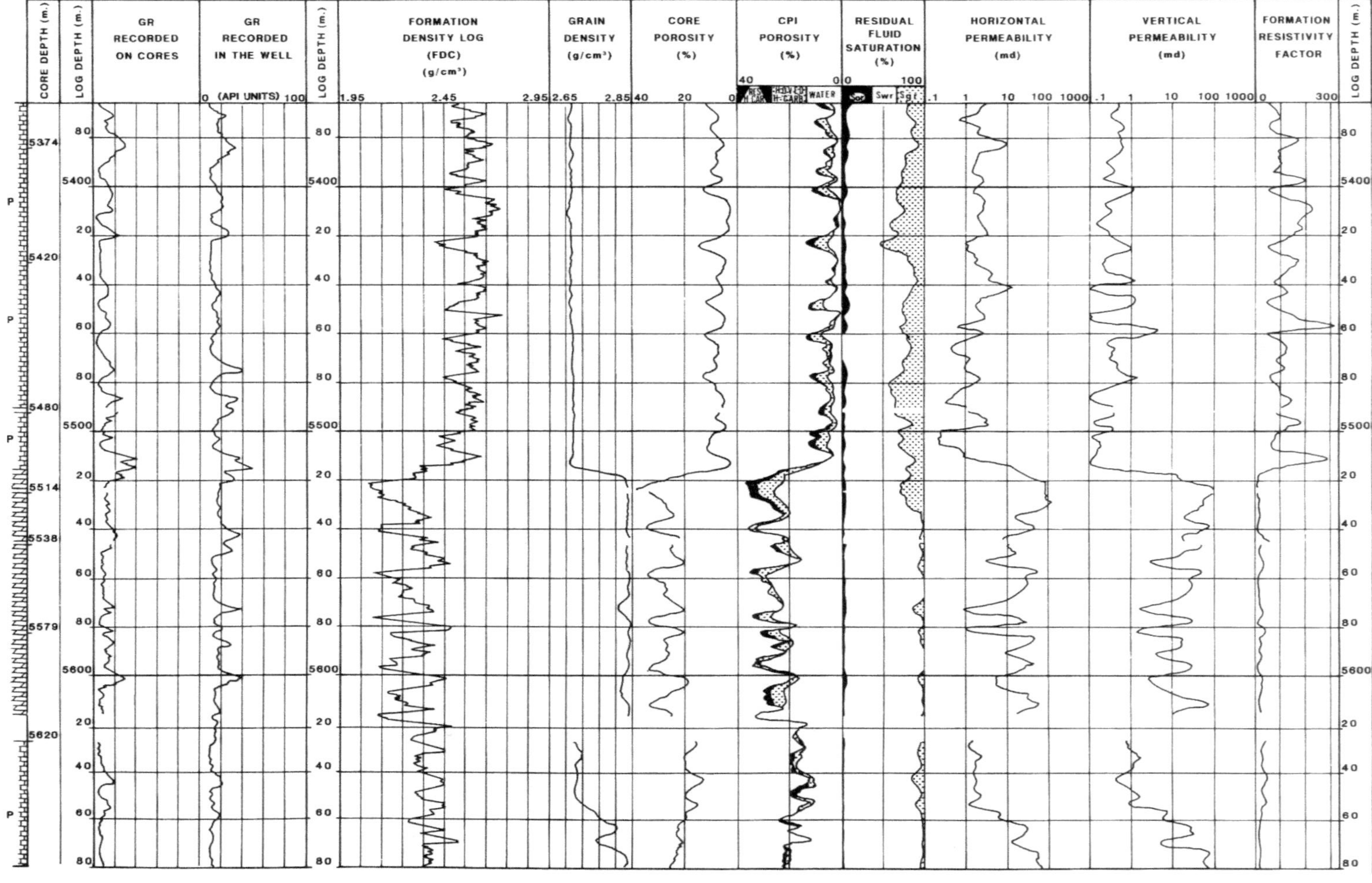

Fig. 3.1. *Variations in reservoir rock properties over a section of well (core graph)*

3.4 Scalar Properties Dependent on Position

3.4.1 Porosity

"Porosity", ϕ, is defined as the ratio

$$\phi = \frac{V_p}{V_b},\qquad (3.1)$$

where V_p is the total void volume in a bulk volume V_b of rock.

Porosity is a dimensionless scalar parameter $\phi(x, y, z)$; in theory it can have any value between 0 and 1, but in reality it rarely exceeds 0.45.

In practical usage by geologists and engineers, it is expressed as a percentage rather than a decimal fraction.

In clastic rocks (sandstones, shales), ϕ depends essentially on rock texture (grain size distribution), as well as the shape and orientation of the grains themselves, and the way in which they are arranged in the matrix lattice.

As an example, we will consider the theoretical case of an uncemented sand (Fig. 3.2), consisting of spherical grains of equal size. If they are arranged in a cubic lattice the porosity will be $\phi = 0.48$; if the packing mode were rhombohedral, ϕ would decrease to 0.26.

Other factors which influence the porosity of clastic rocks are the compression caused by geostatic stress, and the cementation of the grains by salts (particularly carbonates of Ca and Mg) from the action of circulating groundwater and diagenetic processes such as the dissolution and reprecipitation of SiO_2.

Carbonate rock porosity is mainly affected by diagenesis, such as dissolution in CO_2-rich waters, transformation of calcite into aragonite at high temperatures, and into dolomite by the action of water rich in Mg. Fractures and fissures caused by tectonic activity will also change the original porosity significantly.

As a general rule, in a sedimentary basin, ϕ decreases[1] with depth (Fig. 3.3) as a direct result of geostatic loading and diagenetic alteration.

The pores within a volume of rock may or may not be totally interconnected. Two types of porosity are therefore defined:

- *total porosity*, ϕ_t, which encompasses all the void space in the rock, and
- *effective porosity*, ϕ_e, which only includes the interconnected void space. ϕ_e is the useful pore space from which fluid can be extracted.

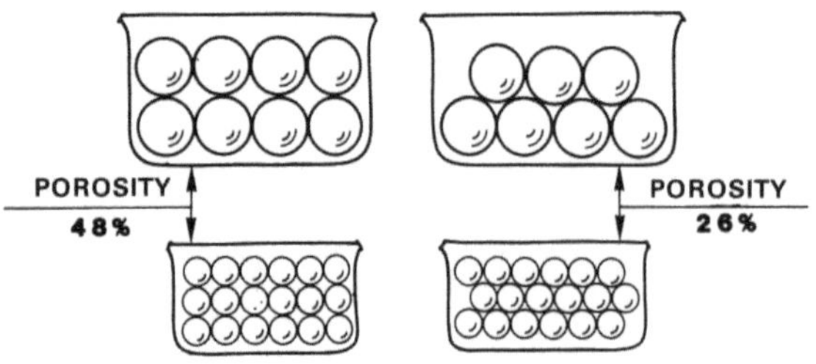

Fig. 3.2. *The effect of packing modes on porosity, using spheres of equal diameter*

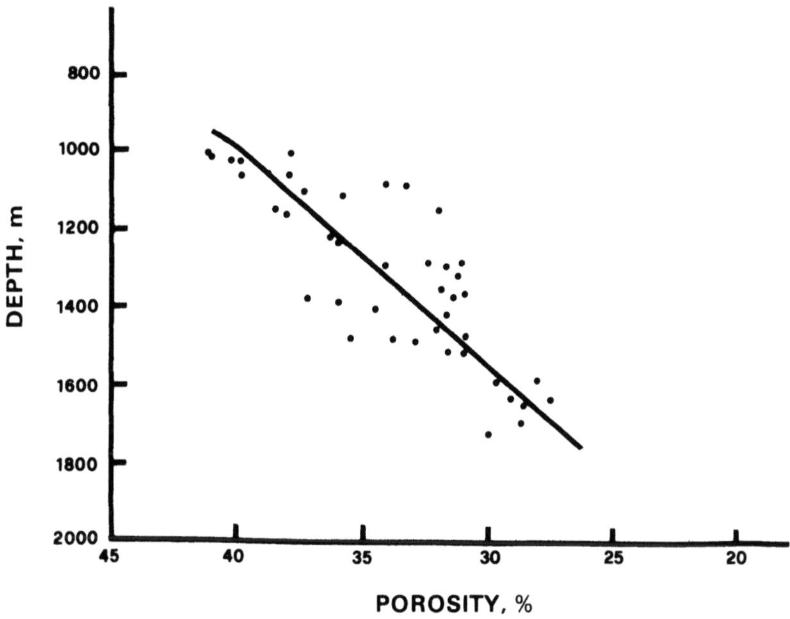

Fig. **3.3.** *Variation of average porosity with depth in a clean sand formation in the Po Valley*

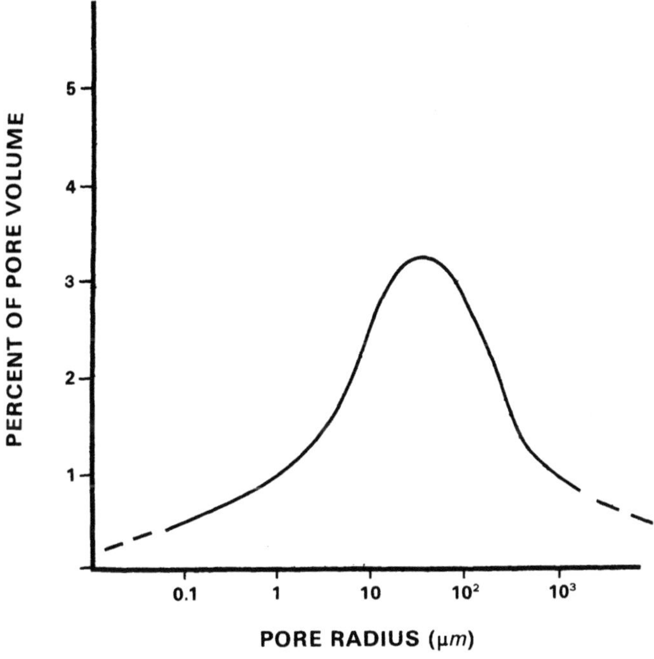

Fig. **3.4.** *Semilog plot of grain size distribution in a medium-fine grained sandstone*

Porosity is also classified in terms of the origin of the pore space:

- *primary porosity* was present at the time of deposition and is an original sedimentary feature;
- *secondary porosity* is void space created after deposition. This can be caused by *fracturing* (especially in carbonates), *solution* (wormholes, vugs) and *recrystallisation* (dolomitization).

Figure 3.4 is an example of pore-size distribution (in terms of pore radius) in a clastic rock containing only primary porosity.

In order to distinguish primary from secondary porosity, and to classify porosity by type, a core sample is essential. *Well logs tend to measure the bulk void space (i.e. ϕ_t) – although the acoustic type log (BHC) can be used to estimate fracture porosity.*

Since sedimentation and diagenesis are statistical in nature, we would not expect the porosity to be the same at all points in a given rock.

In a well-defined sedimentary rock, the distribution of ϕ will be found to be approximately Gaussian (Fig. 3.5); however, local variations in the depositional environment and/or diagenesis can give rise to asymmetrical distributions.

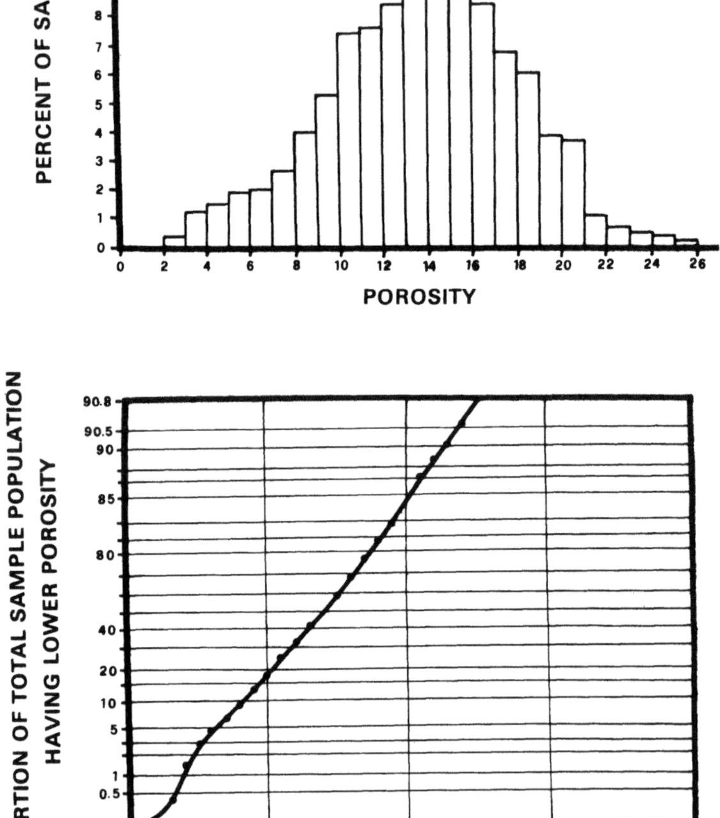

Fig. 3.5. *Histogram and cumulative frequency plots for the distribution of porosity in a sedimentary formation. Note the "normal" distribution*

3.4.2 Fluid Saturations

The "saturation" of a given fluid is the fraction of the pore volume which is occupied by that fluid. We define:

$$S_g = \text{gas saturation} \quad = \frac{V_g}{V_p}, \tag{3.2a}$$

$$S_o = \text{oil saturation} \quad = \frac{V_o}{V_p}, \tag{3.2b}$$

$$S_w = \text{water saturation} = \frac{V_w}{V_p}, \tag{3.2c}$$

It follows that

$$S_g + S_o + S_w = 1 . \tag{3.3}$$

The point has already been made that no useful measurement of reservoir fluid saturations can be obtained from cores. For this information we turn to the electrical logs, which provide a value for S_w from which the total hydrocarbon saturation can be estimated [from Eq. (3.3), $(S_o + S_g) = 1 - S_w$].

In theory, S_o and S_g can be determined using the FDC/CNL logs. In practice, the results are of qualitative value only.

The three equations [Eq. (3.2)] express a bulk or average value for the saturation of each fluid, and make no reference to the actual distribution of the fluids in the pore space. The distribution of each phase is, however, of fundamental importance in determining its dynamic properties – in particular, the ease (relative to any other fluids present) with which it will move through the rock.

This will be described later in the section on *relative permeability*.

3.4.3 Compressibility

It has already been mentioned several times that the reservoir rock is subjected to the pressure of the overlying strata (the overburden). At the same time, the fluid in the pore spaces exerts an opposing pressure.

The resulting forces are described[13, 17] by a scalar quantity p, the pore pressure, and by a second order tensor σ, whose components are the stresses on each face of an elementary cube with edges oriented according to the principal axes of the ellipsoid of forces.

If we define

$$\bar{\sigma} = \frac{\sigma_{xx} + \sigma_{yy} + \sigma_{zz}}{3}, \tag{3.4}$$

with σ positive in the direction of compression, the tensor σ can be subdivided into a hydrostatic component:

$$\begin{vmatrix} \bar{\sigma} & 0 & 0 \\ 0 & \bar{\sigma} & 0 \\ 0 & 0 & \bar{\sigma} \end{vmatrix}$$

and a deviatoric component

$$\begin{vmatrix} \sigma_{xx} - \bar{\sigma} & \sigma_{xy} & \sigma_{xz} \\ \sigma_{yx} & \sigma_{yy} - \bar{\sigma} & \sigma_{yz} \\ \sigma_{zx} & \sigma_{zy} & \sigma_{zz} - \bar{\sigma} \end{vmatrix} \; .$$

It has been shown[17] that it is only the hydrostatic component of the stress tensor which induces changes in rock volume. The deviatoric component causes distortion but without change of volume. As such, it will influence the properties of the porous medium to fluid flow, which means, in the current context, both absolute and relative permeability.

The z-axis of the ellipsoid of forces is obviously oriented vertically (the convention here is positive downwards), while the x- and y-axes are in the plane of sedimentation.

We therefore have

$$\sigma_{zz} = 10^{-6} g \int_0^D [\rho_w \phi + \rho_r(1 - \phi)] \, \mathrm{d}D \; , \tag{3.5}$$

where ρ_r is the grain density (kg/m^3), D is the depth (m) and σ_{zz} is expressed in MPa.

Now we will consider a rock sample which has the following characteristics:

V_b: total bulk volume of rock sample ,
V_p: total pore volume ($\phi = V_p/V_b$) ,
V_r: total rock grain volume ($V_b = V_p + V_r$).

The rock sample will now undergo a laboratory experiment in which the geostatic confining stress σ and the pore pressure p are increased simultaneously, but keeping the difference $(\bar{\sigma} - p)$ constant.

Since

$$\mathrm{d}(\bar{\sigma} - p) = 0 \; ,$$

we can write

$$\mathrm{d}\bar{\sigma} = \mathrm{d}p \; . \tag{3.6a}$$

The simultaneous variation of $\bar{\sigma}$ and p will provoke a change in the grain volume V_r, and therefore in V_b and V_p.

The *grain compressibility* c_r (a very small quantity) is related to these volume changes as follows:

$$\frac{\mathrm{d}V_r}{V_r} = \frac{\mathrm{d}V_b}{V_b} = \frac{\mathrm{d}V_p}{V_p} = -c_r \, \mathrm{d}\bar{\sigma} = -c_r \, \mathrm{d}p \; , \tag{3.6b}$$

Since V_b and V_p are functions of $\bar{\sigma}$ and p, Eq. (3.6b) can be expressed in terms of their partial derivatives as follows:

$$\frac{1}{V_b}\left(\frac{\partial V_b}{\partial \bar{\sigma}}\right)_p \mathrm{d}\bar{\sigma} + \frac{1}{V_b}\left(\frac{\partial V_b}{\partial p}\right)_{\bar{\sigma}} \mathrm{d}p = \frac{1}{V_p}\left(\frac{\partial V_p}{\partial \bar{\sigma}}\right)_p \mathrm{d}\bar{\sigma} + \frac{1}{V_p}\left(\frac{\partial V_p}{\partial p}\right)_{\bar{\sigma}} \mathrm{d}p$$

$$= -c_r \, \mathrm{d}\bar{\sigma} = -c_r \, \mathrm{d}p \; .$$

which, under the experimental conditions for which Eq. (3.6a) is valid, reduces to

$$\frac{1}{V_{\rm b}}\left(\frac{\partial V_{\rm b}}{\partial\bar\sigma}\right)_p + \frac{1}{V_{\rm b}}\cdot\left(\frac{\partial V_{\rm b}}{\partial p}\right)_{\bar\sigma} = \frac{1}{V_{\rm p}}\left(\frac{\partial V_{\rm p}}{\partial\bar\sigma}\right)_p + \frac{1}{V_{\rm p}}\left(\frac{\partial V_{\rm p}}{\partial p}\right)_{\bar\sigma} = -c_{\rm r}\,, \tag{3.6c}$$

Next we shall look at what happens when the two hydrostatic stresses $d\bar\sigma$ and dp are applied separately – dp on the total core, and $d\bar\sigma$ on the pore volume.

Betti's theorem states that, at equilibrium, the sum of the work done by each force on the deformation caused by the other must be zero. This is written as

$$d\bar\sigma\left(\frac{\partial V_{\rm b}}{\partial p}\right)_{\bar\sigma} dp + dp\left(\frac{\partial V_{\rm p}}{\partial\bar\sigma}\right) d\bar\sigma = 0\,, \tag{3.7a}$$

which simplifies to

$$\left(\frac{\partial V_{\rm b}}{\partial p}\right)_{\bar\sigma} = -\left(\frac{\partial V_{\rm p}}{\partial\sigma}\right)_p. \tag{3.7b}$$

We must now define the following parameters:

$$c_{\rm b} = -\frac{1}{V_{\rm b}}\left(\frac{\partial V_{\rm b}}{\partial\bar\sigma}\right)_p, \tag{3.8a}$$

$$c_{\rm p} = -\frac{1}{V_{\rm p}}\left(\frac{\partial V_{\rm p}}{\partial\bar\sigma}\right)_p, \tag{3.8b}$$

$$c_{\rm f} = \frac{1}{\phi}\left(\frac{\partial\phi}{\partial p}\right)_{\bar\sigma}, \tag{3.8c}$$

Substituting from Eq. (3.6c), we get

$$\frac{1}{V_{\rm b}}\left(\frac{\partial V_{\rm b}}{\partial p}\right)_{\bar\sigma} = c_{\rm b} - c_{\rm r}\,, \tag{3.9a}$$

$$\frac{1}{V_{\rm p}}\left(\frac{\partial V_{\rm p}}{\partial p}\right)_{\bar\sigma} = c_{\rm p} - c_{\rm r}. \tag{3.9b}$$

Finally, from Eqs. (3.7b), (3.8b) and (3.9a) we can write

$$\frac{1}{V_{\rm b}}\left(\frac{\partial V_{\rm b}}{\partial p}\right)_{\bar\sigma} = -\frac{1}{V_{\rm b}}\left(\frac{\partial V_{\rm p}}{\partial\bar\sigma}\right)_p = -\frac{\phi}{V_{\rm p}}\left(\frac{\partial V_{\rm p}}{\partial\bar\sigma}\right)_p$$

from which we conclude that

$$c_{\rm b} - c_{\rm r} = \phi c_{\rm p}\,.$$

Rearranging the terms:

$$c_{\rm p} = \frac{c_{\rm b} - c_{\rm r}}{\phi} \cong \frac{c_{\rm b}}{\phi} \tag{3.9c}$$

since $c_{\rm r}$ is negligible relative to $c_{\rm b}$.

We now take Eq. (3.8c), replace ϕ by $V_\mathrm{p}/V_\mathrm{b}$, and expand as partial derivatives:

$$c_\mathrm{f} = \frac{1}{\phi}\left(\frac{\partial \phi}{\partial p}\right)_{\bar{\sigma}} = \frac{1}{V_\mathrm{p}/V_\mathrm{b}}\left[\frac{\partial}{\partial p}\left(\frac{V_\mathrm{p}}{V_\mathrm{b}}\right)\right]_{\bar{\sigma}}$$

$$= \frac{1}{V_\mathrm{p}}\left(\frac{\partial V_\mathrm{p}}{\partial p}\right)_{\bar{\sigma}} - \frac{1}{V_\mathrm{b}}\left[\frac{\partial V_\mathrm{b}}{\partial p}\right]_{\bar{\sigma}}$$

$$= (c_\mathrm{p} - c_\mathrm{r}) - (c_\mathrm{b} - c_\mathrm{r})$$

and therefore:

$$c_\mathrm{f} = c_\mathrm{p} - c_\mathrm{b} = \frac{1}{\phi}\left[c_\mathrm{b}(1 - \phi) - c_\mathrm{r}\right] . \tag{3.9d}$$

Equations. (3.9a) and (3.9b) provide the relationship between the partial derivatives with respect to pore pressure and those with respect to the geostatic stress $\bar{\sigma}$ defined in (3.8a) and (3.8b). Equations (3.9c) and (3.9d) link c_b (the parameter which lends itself most easily to laboratory measurement from a rock sample) with c_p and c_f.

Note that the value of c_r is almost always assumed to be negligible in the literature on this topic.

A *uniaxial compaction coefficient*,[17] is also in common use:

$$c_\mathrm{m} = \frac{1}{h}\left(\frac{\partial h}{\partial p}\right)_{\bar{\sigma};\, \varepsilon_x = \varepsilon_y = 0} , \tag{3.10}$$

h is the thickness of the rock sample (or stratum), and the measurement is made such that the rock is deformed in the vertical (z) direction only, so that $\varepsilon_x = \varepsilon_y = 0$. This represents the downhole environmental conditions more closely.

c_m is related[17] to the parameters we defined previously as follows:

$$c_\mathrm{m} = \frac{1}{3}\frac{1 + v}{1 - v}(c_\mathrm{b} - c_\mathrm{r}) , \tag{3.9e}$$

where v = Poisson's ratio (0.25–0.30 in clastic rocks).

The results of laboratory measurements of c_m made by Shell[17] on sandstones from about 3000 m depth are shown in Fig. 3.6 as a function of porosity and degree of consolidation.

c_m is an important parameter when considering the compaction of the reservoir rock and supporting aquifer which results from the reduction in pore pressure following extraction of hydrocarbons. It is used to predict[35] the degree of surface subsidence – sometimes quite appreciable – which is the direct consequence of this subterranean compaction.[28, 36]

The *physical* properties of the rock (particularly the porosity) at reservoir conditions *are uniquely dependent on the effective geostatic pressure* $(\bar{\sigma} - p)$. Where lateral deformation is impeded ($\varepsilon_x = \varepsilon_y = 0$), which is usually the case, we have[13]

$$\bar{\sigma} - p = \frac{1}{3}\frac{1 + v}{1 - v}(\sigma_{zz} - p) - \frac{2}{3}\frac{1 - 2v}{1 - v}\frac{c_\mathrm{r}}{c_\mathrm{b}}p , \tag{3.11a}$$

where σ_{zz} is defined in Eq. (3.5).

Since porosity measurements are commonly made on core samples at $(\bar{\sigma} - p) = 0$, they have to be corrected to reservoir conditions before they can be used in any engineering calculations.

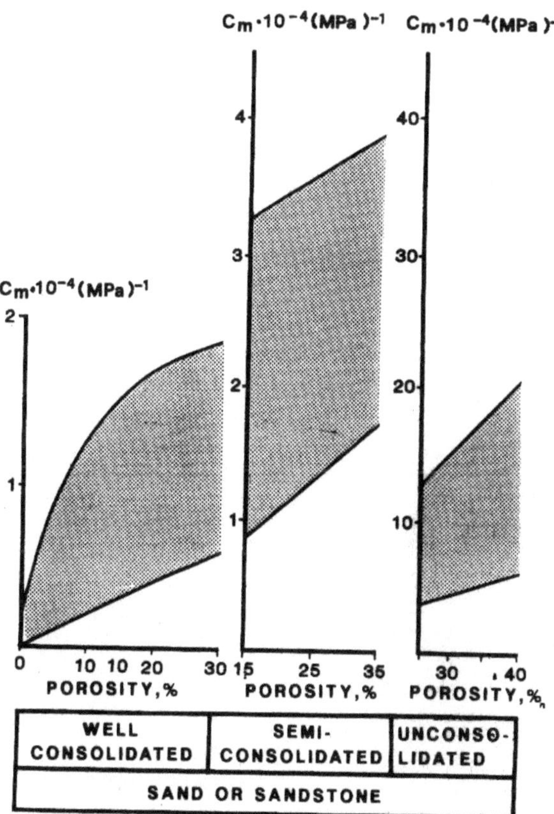

Fig. 3.6. *Values of the uniaxial compaction coefficient c_m for sandstones from around 3000 m depth, showing different degrees of consolidation. From Ref. 17, 1973, Society of Petroleum Engineers of AIME. Reprinted with permission of the SPE*

Chierici[13] has shown, and it can be deduced from Eqs. (3.8c) and (3.9d), that (ignoring c_r):

$$\phi = 1 - (1 - \phi_0)\exp[c_b\,(\bar{\sigma} - p)]\,, \tag{3.11b}$$

where ϕ_0 is the porosity as measured in the laboratory.

3.4.4 Wettability and Capillarity

3.4.4.1 Basic Concepts of Wettability

Suppose we have a glass beaker (Fig. 3.7A) containing a medium grade distillate such as oil of vaseline. Using a pipette, we now deposit a drop of water at the bottom of the beaker.

We will observe the following (Fig. 3.7B): the oil/water interface (the meniscus) will assume a shape corresponding to minimum energy, such that the contact angle Θ_c at static equilibrium is given by:

$$\sigma_{ow} \cos \Theta_c = \sigma_{os} - \sigma_{ws}\,, \tag{3.12}$$

where σ_{os} and σ_{ws} are the interfacial tensions between oil/solid and water/solid respectively. σ_{ow} is the interfacial tension between oil/water.

Θ_c is the "angle of contact" between the oil/water interface and the solid (in this case the bottom of the beaker).

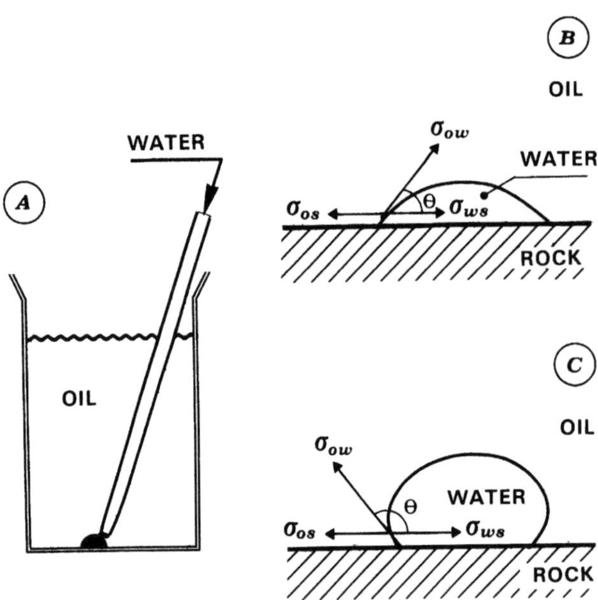

Fig. 3.7. *A basic experiment on wettability. A drop of water is placed at the bottom of an oil-filled beaker* **A.** *If the droplet assumes the shape shown in* **B** *($\Theta_c < \pi/2$), the internal surface of the beaker is water-wet; in the alternative case shown in* **C** *($\Theta_c > \pi/2$), it is oil-wet*

Note that Eq. (3.12) is only valid in *static* conditions. In the experiment shown in Fig. 3.8, where the droplet is situated between two *moving* surfaces, the contact angle is altered to compensate for additional viscous forces between fluid and surface.

The solid surface is said[2] to be "preferentially wet to water" (or "water-wet" for short) when $0° \leq \Theta_c \leq 75°$, Fig. 3.7B.

Conversely, the solid surface is "preferentially wet to oil" (or "oil-wet" for short) when $105° \leq \Theta_c \leq 180°$, Fig. 3.7C.

For $75° < \Theta_c < 105°$, the surface is considered to be of "neutral" wettability.

On the physical level, wettability is determined by a film of fluid, a single molecule in thickness, which is held to the solid surface by electrostatic forces (van der Waals bonding). All solids have local deficiencies in the electron population on their surfaces. The resulting areas of net positive charge attract negatively polarized molecules from the adjacent pore fluid. By covering the surface of the solid, this layer of fluid molecules controls the wettability.

We have seen that all reservoir rock was originally deposited in an aqueous environment: the H_2O molecule is strongly polarized (Fig. 3.9A) and is therefore promptly adsorbed onto the grain surfaces during sedimentation. Consequently, all reservoir rocks start out preferentially water-wet.

During and after the formation of a hydrocarbon accumulation, it may happen that other strongly polarized molecules [especially those of heavy hydrocarbons (Fig. 3.9B) and their oxidised derivatives] displace some of the water molecules from the surface film. Depending on whether the water molecules are partly or totally replaced in this way, the rock may acquire partial wettability to oil, or become totally oil-wet (100% of the surface covered by a film of oil molecules).

It is widely accepted[2] that many rocks exhibit "mixed" or "dalmatian" wettability, whereby isolated patches of the grain surface are oil-wet, while the rest of the

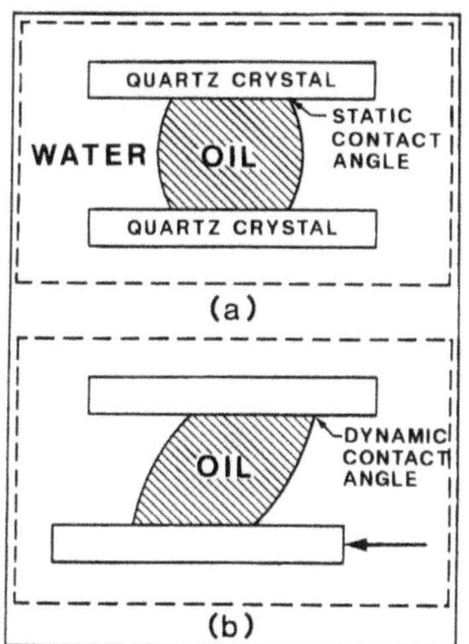

Fig. 3.8. *The oil/rock contact angle under static (top) and dynamic (bottom) conditions. The diagram represents a droplet of oil moving through the pore channels of the rock. From Ref. 2, 1986, Society of Petroleum Engineers of AIME. Reprinted by permission of the SPE*

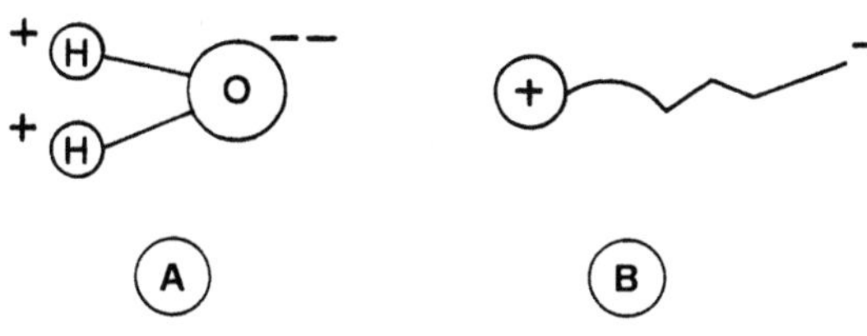

Fig. 3.9A, B. *Disposition of the electrical charges **A** in a molecule of water and **B** in strongly polarized hydrocarbon molecule*

surface remains water-wet. (This is reminiscent of the black spots on the otherwise white coat of a dalmatian dog.)

The means of making quantitative measurements of the preferential wettability of a reservoir rock are covered next.

3.4.4.2 Basic Concepts of Capillarity

We start with a glass beaker partly filled with water, with the same medium grade oil as in the previous section on top of the water (Fig. 3.10).

We immerse a short length of capillary tube of uniform internal diameter completely in the liquids, so that it crosses the oil/water interface as shown in the Fig. 3.10.

The water level will rise inside the tube. If the oil/water interface in the tube is now a height h above the interface in the beaker, the following relationship is simple to derive by considering the balance of hydrostatic and interfacial tension forces:

$$p_A - p_D = g\rho_w(z_A - z_B) + g\rho_o(z_B - z_D)$$

$$= g\rho_w(z_A - z_C) + g\rho_o((z_C - z_D) - \frac{2\pi r\,\sigma_{ow}\cos\Theta_c}{\pi r^2}$$

from which we obtain

$$g\rho_w(z_B - z_C) - g\rho_o(z_B - z_C) = \frac{2\sigma_{ow}\cos\Theta_c}{r}$$

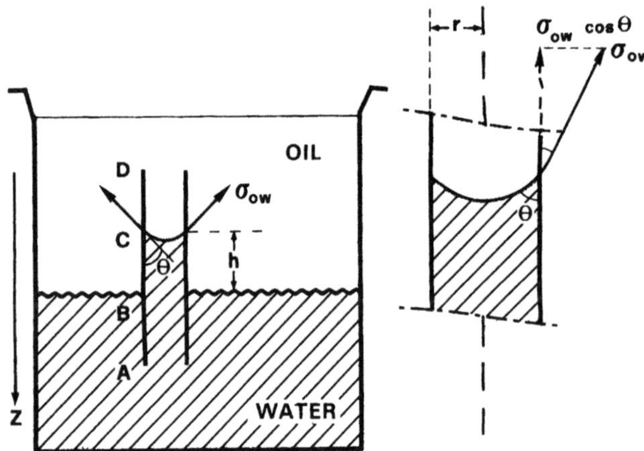

Fig. 3.10. *Water rises up a capillary tube whose internal walls are preferentially water-wet. The forces acting at the water/oil to solid contact are shown on the* right

or

$$g(\rho_w - \rho_o)h = \frac{2\sigma_{ow}\cos\Theta_c}{r} = P_{c,ow} , \qquad (3.13)$$

The term on the right is the *oil/water capillary pressure* $P_{c,ow,}$. It can also be regarded as the force per unit area of tube generated at the contact between the oil/water interface and the wall of the tube. This force is balanced by the water column of height *h*.

$$P_{c,ow} = \frac{2\sigma_{ow}\cos\Theta_c}{r} = p_o - p_w . \qquad (3.14)$$

It follows that:

$h > 0$, so that $P_{c,ow} > 0$, if $\cos\Theta_c > 0$ (i.e. $0° \leq \Theta_c < 90°$). In this case the inside wall of the tube is preferentially water-wet.

$h < 0$, so that $P_{c,ow} < 0$, if $\cos\Theta_c < 0$ (i.e. $90° < \Theta_c \leq 180°$). In this case the inside wall of the tube is preferentially oil-wet.

$h = 0$, so that $P_{c,ow} = 0$, if $\cos\Theta_c = 0$ (i.e. $\Theta_c = 90°$).

3.4.4.3 *Drainage and Imbibition*

Suppose now we had a pair of capillary tubes of different internal diameters, connected in parallel as shown in Fig. 3.11A. They are filled with water, their walls are preferentially water-wet, and they are immersed in a water-filled container.

We now wish to inject oil into the capillaries from the top, and displace the water. This could also be achieved with gas, which is also a non-wetting fluid.

The displacement of the wetting fluid by a non-wetting fluid is referred to as *drainage*.

According to Eq. (3.14), since $0° \leq \Theta_c \leq 90°$, we would need to apply a pressure behind the oil to move it against the opposing capillary force (we will ignore any hydrostatic pressure difference which may exist between the two phases for the moment).

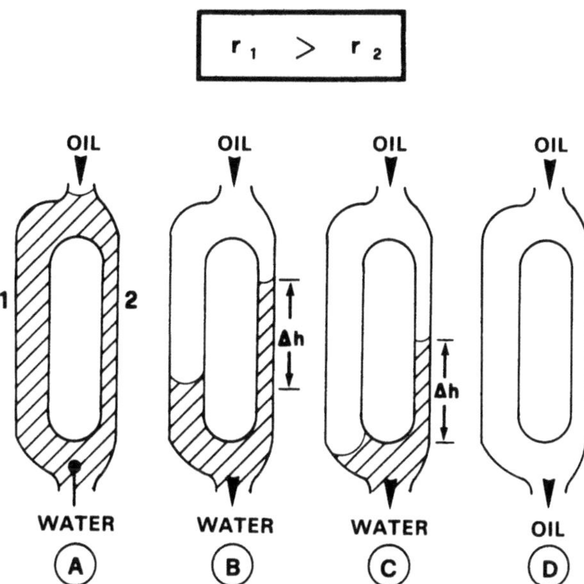

Fig. 3.11A–D. *The drainage process in a pair of connected capillary tubes of different internal diameter. The irreducible water saturation in* **D** *is represented by the film of water remaining on the inside walls of the tubes. This is an idealisation of a 'pore doublet' found in real rock*

Obviously, the oil will move into the larger capillary first [the larger r means a smaller $P_{c,ow}$ in Eq. (3.14)]. If we increase the pressure of injection it will eventually enter the narrower tube as well.

If this displacement is performed very slowly, so that viscous forces do not need to be considered, the difference in height Δh between the two oil/water interfaces will be such that

$$g(\rho_w - \rho_o)\Delta h = 2\sigma_{ow}\cos\Theta_c\left(\frac{1}{r_2} - \frac{1}{r_1}\right).\tag{3.15}$$

After a certain time, all the water in the larger capillary will be displaced by oil, while the narrower one will contain water to a height Δh. From this moment on, oil will flow from the bottom end of the system. The oil/water interface in the narrower capillary will continue to move downwards, until it, too, contains no more water (Fig. 3.11D).

The *residual* (or *irreducible*) *saturation* S_{iw} of the wetting fluid in the drainage process consists of the molecule-thick layer which remains on the inner walls of the capillaries, plus whatever is contained in non-connected pores and dead end (or blind) pore channels.

One very important aspect of the drainage mechanism which follows from the preceding discussion[15] is the displacement of oil by gas in the presence of an irreducible water saturation S_{iw}, under gravity-stabilized conditions.

In this case the gas is the non-wetting phase with respect to the oil. If the displacement progresses very slowly (as happens in the reservoir), *all* the oil present in the interconnected pore channels can be displaced by the gas.[15] Consequently, there will be no residual oil saturation S_{or} in the system.

In Fig. 3.12 we see the capillary doublet at the end of the drainage cycle. The internal walls are still water-wet (because of the layer of water molecules), but the capillaries are now full of oil, and are immersed in oil. (The same considerations would apply if the system was filled with, and immersed in, gas.) We now wish to displace the oil with water injected at the bottom (Fig. 3.12A).

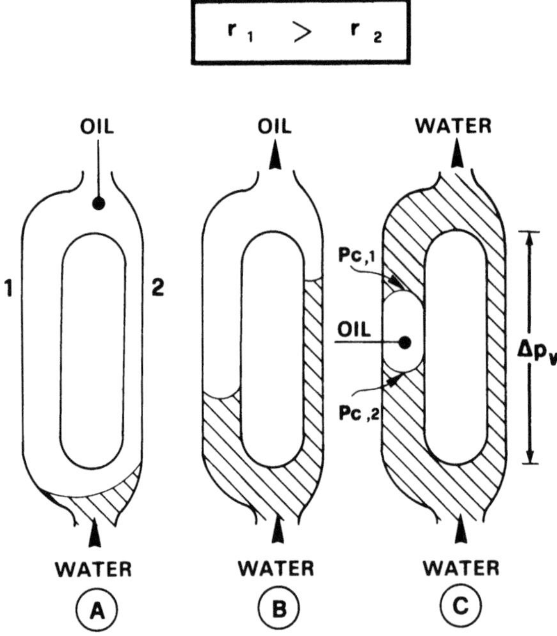

Fig. 3.12A–C. *The imbibition process in a pair of connected capillary tubes of different internal diameter. Note the presence of a trapped oil droplet (or ganglion) at the end of the process* (**C**)

The displacement of a non-wetting fluid by a wetting fluid is referred to as *imbibition.*

In this case, capillary forces tend to draw the water into the capillaries. From Eq. (3.14), the value of $P_{c,ow}$ will be larger in the smaller diameter tube.

If the displacement of the oil is allowed to proceed very slowly, so that viscous forces can be ignored, the water will be drawn more readily up the narrower capillary, and will reach the juncture of the two tubes before all the oil has been displaced from the wider tube (Fig. 3.12C).

Thus, a certain quantity of oil is left behind by the imbibition process. There is no longer any continuity between this oil droplet and the rest of the oil phase, and this renders it immobile. It therefore becomes part[1] of the *residual oil saturation,* S_{or} (or, in the case of a gas/water system,[12] S_{gr}).

Note that once water flows from the narrower capillary, there exists a pressure difference Δp_w between the inlet and outlet points of the doublet system.

The surfaces of each side of the residual oil droplet assume shapes such that the respective capillary pressures $P_{c,1,ow}$ and $P_{c,2,ow}$ (Fig. 3.12C) satisfy the equation

$$\Delta p_w = P_{c,1,ow} - P_{c,2,ow} \tag{3.16}$$

and the droplet remains immobile.

In order to move the droplet, thereby recovering the residual oil, it would be necessary to eliminate the interfacial tension σ_{ow} (and therefore $P_{c,ow}$). This can be achieved in principle by replacing the water with a fluid which is miscible with oil ($\sigma_{ow} = 0$; $P_{c,ow} = 0$): this is the principle of "enhanced recovery" using oil-miscible fluids.

3.4.4.4 Capillary Pressure Curves

The sedimentary rocks which may later become hydrocarbon reservoirs, are laid down in an aqueous environment, and so are initially saturated with water ($S_w = 1$).

With the influx of hydrocarbons during the migration phase, the height of the column of oil (or gas) in the reservoir structure increases, and with it the local value of P_c ($P_{c,ow}$ or $P_{c,gw}$) wherever the hydrocarbon has encroached. As we have seen in the previous section, this is a progressive process of *drainage*.

Referring to Eq. (3.14), the influx of hydrocarbon into the rock will begin ($S_w < 1$) once P_c attains a sufficient value (the *threshold pressure*) for the hydrocarbon to enter the largest pore channels.

As P_c increases, the oil or gas will be able to penetrate pore channels of smaller and smaller diameters, displacing water from them and increasing $S_o = 1 - S_w$. But even at extremely large values of P_c (corresponding to a very thick hydrocarbon column), the molecular layer of water will always remain on the grains, and, along with water contained in unconnected pores and blind channels, this constitutes the irreducible water saturation S_{iw}.

At the culmination of the migration process, the hydrocarbon accumulation will have a saturation profile (S_w) with respect to height as shown in Fig. 3.13 (right-hand curve). The height h is measured from the water table (usually referred to in this context as the *free water level*) where $P_c = 0$.

This profile $S_w = f(h)$, or $S_w = f(P_c)$, is called the *capillary pressure curve*.

Starting from the free water level, $S_w = 1.0$ up to a height at which P_c is equal to the threshold pressure for the rock. As h increases beyond this, S_w starts to decrease up through the *transition zone* until it reaches the irreducible water saturation S_{iw} (provided the rock is sufficiently thick).

The y-axis of the capillary pressure diagram can be expressed in terms of h or P_c since these two parameters are linked by Eq. (3.13):

$$P_{c,ow} = g(\rho_w - \rho_o)h , \tag{3.17a}$$

$$P_{c,gw} = g(\rho_w - \rho_g)h . \tag{3.17b}$$

In the reservoir interval over which $S_{iw} < S_w < 1$, *the water phase is mobile*. But so also is the hydrocarbon phase, and both water and oil (or gas) will be produced from the transition zone. A later section headed "relative permeability" will describe the laws governing the simultaneous flow of two or three phases.

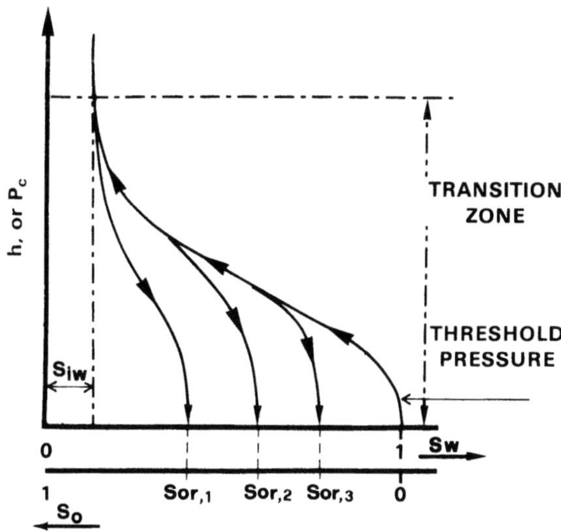

Fig. 3.13. *Capillary pressure curves for drainage and imbibition*

Note that for a given rock type, the thickness of the gas/water transition zone in a gas reservoir is always less than the oil/water transition in an oil reservoir, because, in Eq. (3.17), it is always the case that $(\rho_w - \rho_g) > (\rho_w - \rho_o)$.

In a reservoir where a gas cap exists above the oil, similar considerations apply to the gas/oil system, not forgetting that they are in the presence of S_{iw}. The gas is now the non-wetting phase relative to the oil, and the capillary curve will exhibit a transition zone where both oil and gas are mobile.

In the case of a reservoir which has natural water drive, or where water is injected for water flooding and pressure maintenance, the water level will rise as hydrocarbons are withdrawn. This is the onset of an *imbibition* cycle if the rock is water-wet, and P_c will begin to decrease progressively throughout the reservoir as the height h of any point is reduced by the rising free water level [Eq. (3.17)].

For the same rock type, the imbibition curve $P_c(S_w)$ (S_w increasing) is very different from the drainage curve (S_w decreasing). In particular, the oil or gas saturation is not restored to zero at the end of the cycle (i.e. we do not return to $S_w = 1$). It will not decrease beyond a non-zero value corresponding to the residual oil saturation S_{or} (or, for gas, S_{gr}) as described in Section 3.4.4.3.

The value of the residual saturation, and the shape of the capillary pressure curve, depends on the value of S_o at which imbibition began, and therefore on h (Fig. 3.13).

This amounts to hysteresis of the *capillary pressure curve*, since it does not return to its initial form. This must be taken into account in all cases where reservoir depletion involves the imbibition process.

3.4.4.5 Conversion of Laboratory Capillary Pressure Curves to Reservoir Conditions

Capillary pressure curves are usually established from core samples under laboratory conditions.[18,26] Only for highly specialised applications would they be measured at reservoir p and T, using reservoir fluids.

The assumption that the cutting and extraction of the core (with its incumbent contamination by mud filtrate), and the subsequent preparation in the laboratory, do not alter the preferential wettability of the grain surfaces, is debatable.

The core sample is initially saturated with water ($S_w = 1$). Using oil free of any polarized molecules (Soltrol), or air, as the displacing fluid, the drainage curve is defined by stepwise increasing $P_{c,ow}$ (or $P_{c,gw}$) and measuring S_w, which decreases as far as S_{iw}.

For the imbibition curve, the oil or gas is subsequently displaced by water. P_c is allowed to decrease progressively to zero or, if desired, to negative values ($p_w > p_o$).

There are two principal laboratory methods in use: the semi-permeable membrane, and the centrifuge. Detailed descriptions can be found in the list of references[1,27] at the end of this chapter.

The laboratory curve $(P_c)_L$ is converted to $(P_c)_R$ at reservoir conditions using an equation derived from Eq. (3.14):

$$(P_c)_R = (P_c)_L \frac{\sigma_R \cos \Theta_R}{\sigma_L \cos \Theta_L}, \tag{3.18}$$

where the subscripts L and R indicate laboratory and reservoir conditions respectively.

The variation $S(h)$ in the saturation above the level where $S = 1$ can then be calculated from an equation derived from Eq. (3.17):

$$h(S) = \frac{[P_c(S)]_R}{g(\rho_2 - \rho_1)}, \tag{3.19a}$$

where $[P_c(S)]_R$ is the capillary pressure at reservoir conditions at fluid saturation S, and ρ_1 and ρ_2 are the densities of the two fluid phases in contact.

For example, in the case of the oil/water capillary transition zone, we have

$$h(S_w) = \frac{[P_c(S_w)]_R}{g(\rho_w - \rho_o)}, \tag{3.19b}$$

3.4.4.6 Porosity as a Function of Pore Radius

The size of the pore channels in a sedimentary rock can vary between wide limits. The porosity distribution as a function of pore radius can be calculated from the capillary pressure curve for the drainage cycle, using any two non-miscible fluids. Since a high σ will produce a large P_c for a given r, the combination air (wetting phase)/mercury (non-wetting phase) is often used. From Eq. (3.14) and the relationship:

$$\frac{dP_c}{dS} = \frac{dP_c}{dr}\frac{dr}{dS}, \tag{3.20}$$

we can write

$$\frac{dS}{dr} = -\frac{2\sigma\cos\Theta_c}{r^2}\frac{1}{(dP_c/dS)_r}. \tag{3.21a}$$

dS represents the incremental change in saturation attributed to filling (or emptying) pores which have a radius between $(r - dr/2)$ and $(r + dr/2)$.

We therefore have the following equation:

$$\left(\frac{\Delta\phi}{\Delta r}\right)_{\bar{r}} = \frac{2\sigma\cos\Theta_c}{\bar{r}^2}\frac{1}{-(dP_c/dS)_{\bar{r}}}, \tag{3.21b}$$

with which we can calculate the porosity distribution as a function of pore radius. Figure 3.4 is an example of this.

Equations (3.21) are based on the assumption that the porous medium consists of a bundle of capillary tubes of different diameters, not interconnected. Since the real nature of porous rock is very different from this idealised model, the results derived from the equation should be considered a first approximation only.

3.4.4.7 Calculation of the Average Capillary Pressure Curve for a Reservoir – the Leverett J-Function

The form of the capillary pressure curve $P_c(S)$ for a porous rock depends basically on the distribution and mean value of the pore radii.

Very fine-grained rocks will exhibit an extended transition zone like curve 1 in Fig. 3.14; while coarse-grained rocks tend to have relatively short transitions (curve 3 in Fig. 3.14).

In reservoirs consisting of a number of sedimentary strata, each with a widely different grain size, it is quite common to find what at first appears to be an abnormal saturation profile, with values of S_w which increase suddenly where a fine-grained stratum overlies a coarse one. This situation is illustrated in Fig. 3.14. Although this appears to contradict the intuitive notion that we should not find water above oil (or even water saturations increasing upwards), an understanding of the mechanism of capillarity will explain the changes in saturation. (Remember that each of the strata is dipping downwards into the aquifer, and each behaves as a distinct capillary system.)

But even within a given sedimentary layer, formed from statistically uniform material deposited in statistically constant conditions, there will be local variations in ϕ and the permeability k which will have an effect on the behaviour of the P_c curve from point to point. (The parameter k will be dealt with in a later section.)

Suppose now we have a certain number of $P_c(S)$ curves, determined from cores taken from the same sedimentary unit (one curve per core). We wish to establish an average capillary pressure curve which can be applied anywhere in the unit, particularly in uncored sections.

Leverett[26] defined a function $J(S)$:

$$J(S) = \frac{P_c(S)}{\sigma \cos \Theta_c} \sqrt{\frac{k}{\phi}} . \tag{3.22}$$

This dimensionless function, which is characteristic of a given sedimentary unit, represents a *normalised* capillary pressure curve.

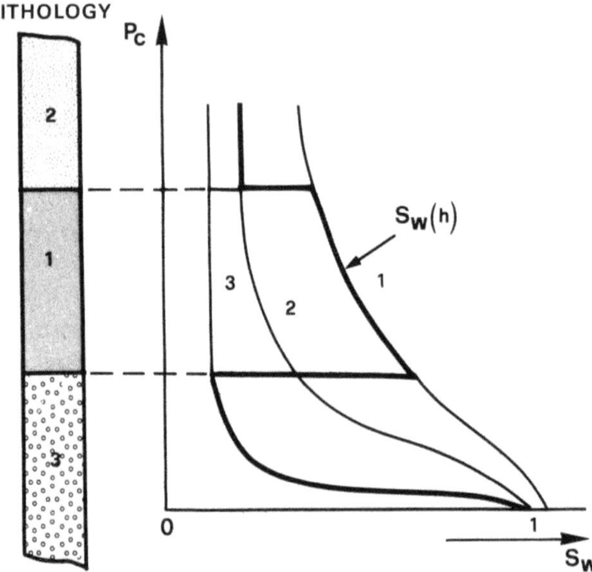

Fig. 3.14. *Water saturation profile (*heavy solid line*) across the pay zone in a reservoir consisting of three sedimentary units (*$k_1 \ll k_2, k_3$*)*

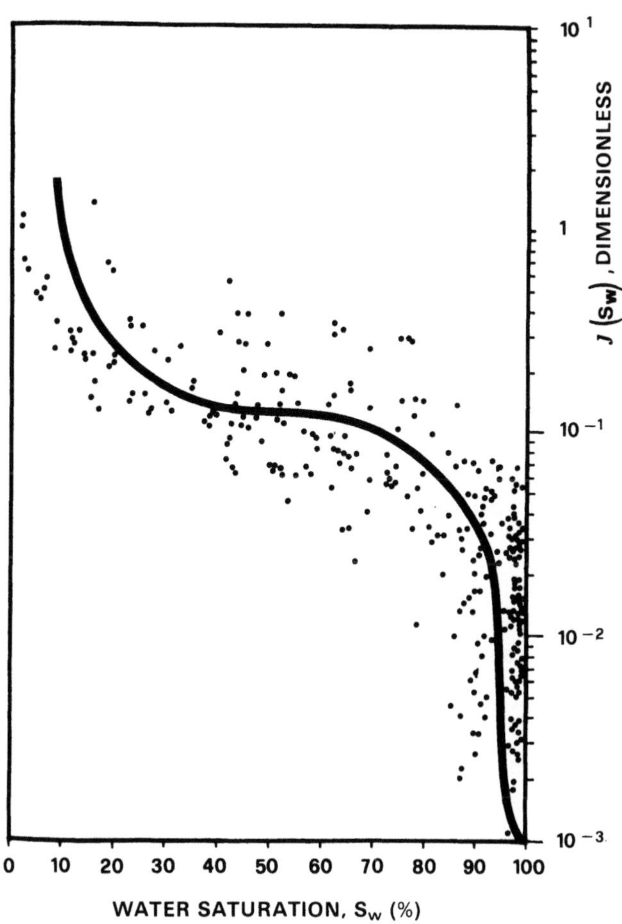

Fig. 3.15. *The Leverett J-function for a sedimentary unit. Data are obtained from individual core measurements*

$J(S)$ is obtained by fitting a curve through the points calculated from the available $P_c(S)$ curves normalised with values of k, ϕ and $\sigma \cos \Theta_c$ measured in each core (Fig. 3.15).

Having determined the J-function for the unit, the capillary pressure $P_c(S)$ corresponding to any saturation S can be predicted using a unit-wide average porosity ϕ and permeability k, by inversion of Eq. (3.22):

$$\bar{P}_c(S) = \sigma \cos \Theta_c \, J(S) \sqrt{\frac{\bar{\phi}}{\bar{k}}} \,. \tag{3.23}$$

3.4.4.8 Determination of Preferential Wettability

Even though the contact angle Θ_c is not one of the parameters used in calculations concerning reservoir behaviour, a knowledge of the nature of the preferential wettability is essential, especially when considering improved oil recovery by water-flooding.

Many methods have been proposed to measure the wettability of rock. The one most widely used[2] was devised by the United States Bureau of Mines (USBM).

Their method is based on the fact that the energy expended in forcing a non-wetting fluid into the pores is always greater than the energy needed to inject the wetting fluid. The work, A, performed is

$$A = \int_0^1 P_c(S)\,dS ,\tag{3.24}$$

which is in fact the area between the $P_c(S)$ curve and the S-axis.

The core, initially saturated with water ($S_w = 1$), is immersed in oil and subjected to a process of drainage by centrifuging. The $P_c(S_w)$ curve is measured up to a maximum capillary pressure $P_c = 0.0687$ MPa (10 psia). We will call A_1 the area beneath the drainage curve.

The core is then immersed in oil and subjected to imbibition, again using a centrifuge. The curve $P_c(S_w)$ is measured down to a minimum value of $P_c = -0.0687$ (-10 psia). We will call A_2 the area under the imbibition curve.

Referring to Fig. 3.16, the USBM wettability index I is defined as[2]:

$$I = \log \frac{A_1}{A_2} .\tag{3.25}$$

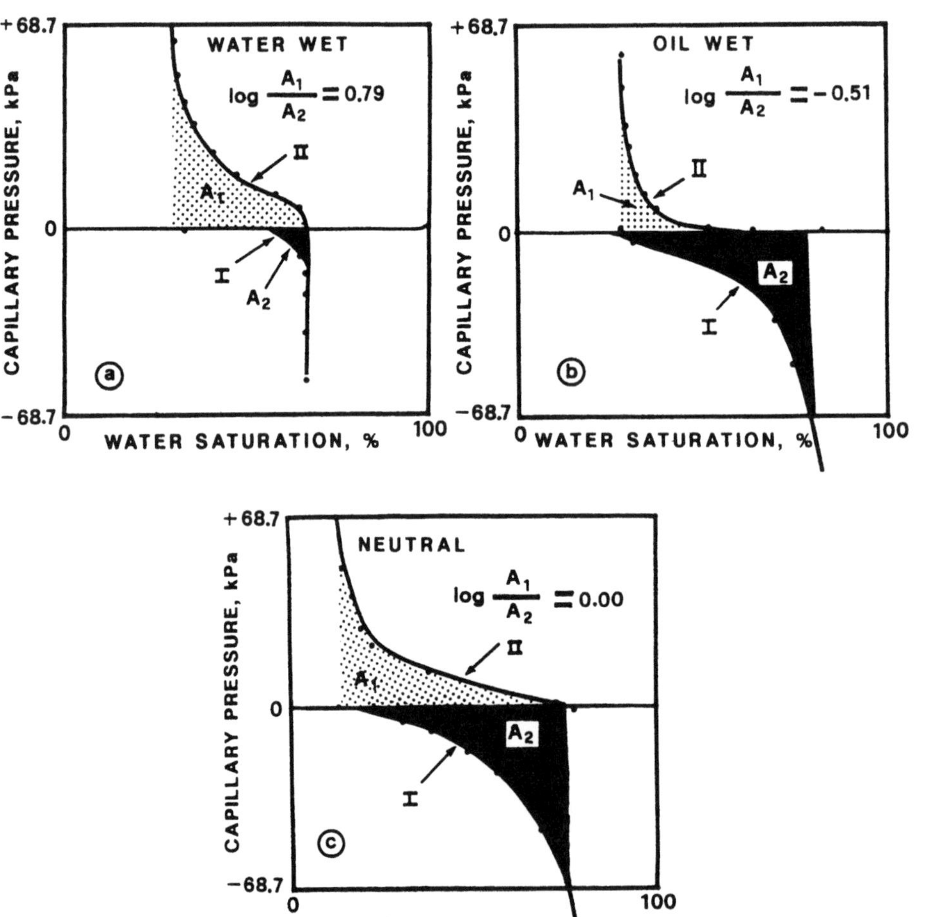

Fig. 3.16a–c. *Schematic of the USBM method for determining preferential wettability. Three cases are shown:* **a** *water-wet,* **b** *oil-wet,* **c** *neutral. From Ref. 2, 1986, Society of Petroleum Engineers of AIME. Reprinted by permission of the SPE*

It follows that

> if $I > 0$ the rock is preferentially water-wet,
> if $I < 0$ the rock is preferentially oil-wet,
> if $I = 0$ the wettability is neutral.

3.5 Dynamic Properties

3.5.1 Permeability

3.5.1.1 Some Basics on the Concept of Potential

In hydrodynamics, Bernoulli's theorem states that when a perfect fluid ($\mu = 0$) is moving at a constant velocity, with no internal energy dissipation or exchange of energy with its surroundings, the total energy *per unit weight of fluid* as defined by the equation

$$E_\mathrm{w} = \frac{p}{g\rho} - z + \frac{v^2}{2g} \tag{3.26}$$

is constant along any flow tube. The z-direction here is positive downwards. E_w has the dimension of length, and its three components are, in the order they appear in Eq. (3.26), piezometric height, geometric height, and dynamic height.

From this we now define the following parameter:

$$\Phi^* = \int_0^P \frac{\mathrm{d}p}{\rho} - gz + \frac{v^2}{2}, \tag{3.27}$$

Φ^* is called the *fluid potential*. It is work per unit mass of fluid, and has the dimensions $[\mathrm{L}^2\,\mathrm{t}^{-2}]$.

All work done by the fluid within the system, and any exchange of energy with the surroundings, will appear as a change $\Delta\Phi^*$ in its potential.

The velocity at which fluid moves through porous media is very slow, and changes in velocity (i.e. accelerations) are usually negligible. Consequently, we can assume $v = \mathrm{const.}$ ($\mathrm{d}v = 0$), and this allows us to simplify Eq. (3.27) as follows:

$$\Phi = \text{potential per unit mass} = \int_0^P \frac{\mathrm{d}p}{\rho} - gz. \tag{3.28a}$$

In addition, we define a second parameter

$$\Psi = \rho\Phi = \text{potential per unit volume} = p - \rho gz. \tag{3.28b}$$

Ψ has the dimensions of pressure $[\mathrm{m}\,\mathrm{L}^{-1}\,\mathrm{t}^{-2}]$, and is often referred to as the *datum pressure.*

3.5.1.2 The Definition of Permeability

In a porous medium which is totally saturated ($S = 1$) with a fluid which does not interact chemically or physically with the solid material, the incremental work per unit mass required to drive the fluid past the grains of the rock (and dispersed as heat) is given by

$$\mathrm{d}W = \mathrm{d}\Phi = -\frac{\mu}{k}\frac{u}{\rho}\,\mathrm{d}L\,. \tag{3.29}$$

Here, u is the fluid (or flux) velocity, defined as q/A, *the volumetric flowrate q per unit cross-sectional area A of the porous medium* across which it is flowing (it is also called the Darcy velocity) μ is the fluid viscosity.

The negative sign in Eq. (3.29) accounts for the fact that the velocity u is positive in the direction of decreasing potential ($\mathrm{d}W < 0$).

k is a property of the porous medium which is *independent of the fluid type* (provided there is no interaction between fluid and solid). It is called the *permeability* – sometimes referred to as the "absolute permeability".

$[k]$ is a second order tensor. Two of its principal axes (x,y) lie in the plane of deposition of the sedimentary stratum, while the third (z) is oriented perpendicular to it.

We speak of a *vertical permeability*, k_v, in the z-direction; and a *horizontal permeability*, k_h, which is the average value of the permeability in the x- and y-directions.

k has the dimensions of area $[L^2]$ and, in the SI units system, is expressed in m^2. Since this is in fact a very large unit, it is more convenient to work in μm^2 ($=10^{-12}$ m^2).

An alternative, empirical, unit which has very wide usage in reservoir engineering, is the "Darcy" (D) and, more particularly, the "millidarcy" (md) (1 md $= 10^{-3}$ D). They are related to their SI counterparts by

$$1\,\mathrm{D} = 0.9869233\,\mu\mathrm{m}^2 \cong 1\,\mu\mathrm{m}^2\,, \tag{3.30a}$$

$$1\,\mathrm{md} = 0.9869233 \times 10^{-3}\,\mu\mathrm{m}^2 \cong 10^{-3}\,\mu\mathrm{m}^2 \tag{3.30b}$$

Equation (3.29) can be rewritten in a more general form

$$\boldsymbol{u} = -\frac{[k]\rho}{\mu}\,\mathrm{grad}\,\Phi\,, \tag{3.31a}$$

and, in terms of volumetric flow rate:

$$\boldsymbol{q} = -\frac{[k]\rho A}{\mu}\,\mathrm{grad}\,\Phi\,, \tag{3.31b}$$

which is the well-known *Darcy's law*, named after the French fluid mechanics engineer who derived it from experimental observations in 1856. It was not until a century later, in 1956, that King Hubbert[21] established the theoretical justification for Darcy's empirical law, and it took its rightful place as a fundamental theorem of fluid mechanics.

3.5.1.3 Sign Conventions

In the case of fluid flow in a linear system, if we define the axis l as positive in the direction of motion, the potential gradient $\mathrm{d}\Phi/\mathrm{d}l$ in this direction will be negative, since the fluid will move towards a decreasing potential. Darcy's law will therefore be written as

$$u = -\frac{k\rho}{\mu}\frac{d\Phi}{dl} \, , \tag{3.31c}$$

In a *radial system* (which is the most common regime for fluid flow towards a well in a horizontal layer), the sign convention is that the flow should be positive into the wellbore. The radial coordinate r has the axis of the well as its origin, and increases away from the well, so q (and therefore u) is positive in the direction of *decreasing* r. (It follows that q and u are positive for a producing well, negative for an injector.)

Since $d\Phi/dr$ is positive (Φ decreases in the negative direction of the r-axis), the Darcy equation will not include a minus sign:

$$u = \frac{k\rho}{\mu}\frac{d\Phi}{dr} \, , \tag{3.31d}$$

Equations (3.31c) and (3.31d) demonstrate the sign conventions to be respected in the use of Darcy's law.

3.5.1.4 Horizontal Flow of an Incompressible Fluid

This section describes linear and radial flow of an almost incompressible fluid ($\rho \approx$ const.) in a horizontal porous medium ($dz = 0$). From Eq. (3.28a), which defined the potential Φ per unit mass, we can derive

$$d\Phi = \frac{1}{\rho}dp \quad (z = \text{const.}) \, . \tag{3.28c}$$

For *linear* flow in a porous medium of length L and cross-sectional area A, we can substitute Eq. (3.28c) into (3.31c) and get

$$q = Au = -\frac{k\rho A}{\mu}\frac{1}{\rho}\frac{dp}{dl} \, .$$

If we integrate this between the inlet pressure p_{in} and outlet pressure p_{out} (in MPa), we obtain

$$q = \frac{kA}{\mu}\frac{P_{In} - P_{out}}{L} \times 10^6 \, . \tag{3.32a}$$

For *radial* flow, we first consider a cylindrical annulus of the porous medium, of internal radius r and thickness dr, centred on the axis of the well. The layer thickness is h, the wellbore radius r_w, and the external radius of the porous medium (also assumed to be cylindrical) is r_e.

The fluid therefore flows through a surface area A given by

$$A = 2\pi rh \, .$$

From Eqs. (3.31d) and (3.28c), since $q = uA$, we get

$$q = \frac{k\rho}{\mu}2\pi rh\frac{1}{\rho}\frac{dp}{dr} \, ,$$

which, if we integrate between r_w and r_e, becomes

$$q = \frac{2\pi\,kh}{\mu}\,\frac{p_e - p_w}{\ln(r_e/r_w)} \times 10^6 \;. \tag{3.32b}$$

p_e and p_w are the pressures at the outer and inner boundaries respectively, and are expressed in MPa. Equations (3.32a) and (3.32b) are only valid for *constant rate* horizontal flow of a single phase fluid of negligible compressibility.

3.5.1.5 Steady State Horizontal Flow of Gas

The volumetric flow rate of gas, $q_g(l)$ or $q_g(r)$, does *not* remain constant across a porous medium, because it has a non-negligible compressibility. The volume factor B_g is pressure-dependent [Eq. (2.8b)], causing the volumetric flowrate to increase towards decreasing pressure. However, since at the same time the gas *density* is inversely proportional to B_g, the *mass flow rate* $q_{m,g}$ does remain constant:

$$q_{m,g} = q_g\,\rho_g \;.$$

The compressibility of the gas will also cause the mass flow rate to pass through an initial *transient phase* during which $p = p(x,t)$ or $p = p(r,t)$ (i.e. pressure and rate are time-dependent). They will only stabilise (steady state) after a certain period of time, beyond which $p = p(x)$ or $p = p(r)$ and $q_{m,g} = $ const.

This section will only describe the steady state behaviour; the transient period will be covered in Chaps. 5 and 7.

In a *linear system* we have

$$q_{m,g} = \text{const.} = \rho_g q_g = -\rho_g\,\frac{kA\rho_g}{\mu_g}\,\frac{1}{\rho_g}\,\frac{dp}{dl} \;.$$

Substituting for the gas density from Eq. (2.14), integrating between the inlet and outlet sections of the porous medium, and expressing all parameters in SI units, with p in MPa, we have

$$q_{m,g} = 1.2032 \times 10^8\,\frac{M}{\bar{z}T}\,\frac{kA}{\mu_g}\,\frac{p_{in}^2 - p_{out}^2}{2L} \;. \tag{3.33a}$$

Equation (3.33a) is open to several interpretations. Firstly, if we define $q_{g,sc}$ as the gas volumetric flow rate at standard conditions (0.1013 MPa, 288.2 K), and $\rho_{g,sc}$ as its density, we have

$$q_{m,g} = q_{g,sc}\,\rho_{g,sc}$$

and from Eq. (2.9):

$$q_{g,sc} = 2.845 \times 10^9\,\frac{1}{\bar{z}T}\,\frac{kA}{\mu_g}\,\frac{p_{in}^2 - p_{out}^2}{2L} \;. \tag{3.33b}$$

If core permeabilities are measured at laboratory conditions where $p_{out} = 1$ atm $= 0.1013$ MPa and $T = 288.2$ K, such that $z = 1$) and if the other parameters are expressed in the following units:

$q_{g,sc}$	cm^3/s
A	cm^2
L	cm
μ_g	cP
$p_{in},\, p_{out}$	atm

then Eq. (3.33b) simplifies to:

$$\dot{q}_{\mathrm{g,sc}} = \frac{kA}{\mu_{\mathrm{g}}} \frac{p_{\mathrm{in}}^2 - p_{\mathrm{out}}^2}{2L}, \tag{3.33c}$$

with k *in Darcy units*. Equation (3.33c) is the classical equation for the laboratory measurement of core permeability using air (or some other gas).

If we define $\bar{q}_{\mathrm{g}}$ as the flow rate at mean pressure $(p_{\mathrm{in}} + p_{\mathrm{out}})/2$ in the porous medium, and $\bar{\rho}_{\mathrm{g}}$ the corresponding gas density at this pressure, Eq. (3.33) can be developed as

$$q_{\mathrm{m,g}} = \bar{q}_{\mathrm{g}}\,\bar{\rho}_{\mathrm{g}} = \bar{q}_{\mathrm{g}} \times 120.32 \frac{M}{\bar{z}T} \frac{p_{\mathrm{in}} + p_{\mathrm{out}}}{2}$$

$$= 1.2032 \times 10^8 \frac{M}{\bar{z}T} \frac{kA}{\mu_{\mathrm{g}}} \frac{p_{\mathrm{in}}^2 - p_{\mathrm{out}}^2}{2L},$$

from which:

$$\bar{q}_{\mathrm{g}} = \frac{kA}{\mu_{\mathrm{g}}} \frac{p_{\mathrm{in}} - p_{\mathrm{out}}}{L} \times 10^6 . \tag{3.33d}$$

This equation is identical in form to Eq. (3.32a), which was valid for incompressible fluids.

The conclusion from this is that in linear flow, Darcy's equation for incompressible liquid can be used for gas, provided the gas flowrate is measured at the average pressure $(p_{\mathrm{in}} + p_{\mathrm{out}})/2$ of the porous medium.

Following the same steps for the *radial* flow of gas, we have

$$q_{\mathrm{m,g}} = \mathrm{const.} = \rho_{\mathrm{g}}q_{\mathrm{g}} = 120.32 \frac{Mp}{zT} \frac{k\rho_{\mathrm{g}}}{\mu_{\mathrm{g}}} 2\pi rh \frac{1}{\rho_{\mathrm{g}}} \frac{\mathrm{d}p}{\mathrm{d}r}$$

and, by integrating:

$$q_{\mathrm{m,g}} = 120.32 \frac{M}{\bar{z}T} \frac{\pi kh}{\mu_{\mathrm{g}}} \frac{p_{\mathrm{e}}^2 - p_{\mathrm{w}}^2}{\ln(r_{\mathrm{e}}/r_{\mathrm{w}})} \times 10^6 \tag{3.33e}$$

with pressure in MPa.

Therefore:

$$q_{\mathrm{g,sc}} = 2.845 \times 10^9 \frac{1}{\bar{z}T} \frac{\pi kh}{\mu_{\mathrm{g}}} \frac{p_{\mathrm{e}}^2 - p_{\mathrm{w}}^2}{\ln(r_{\mathrm{e}}/r_{\mathrm{w}})} . \tag{3.33f}$$

From this it is straightforward to show that, in radial flow as well, the Darcy equation for gas reduces to that for incompressible liquid, if the gas flow rate $\bar{q}_{\mathrm{g}}$ is measured at a pressure $(p_{\mathrm{e}} + p_{\mathrm{w}})/2$.

The radial system differs significantly from the linear system in that, by virtue of its radial geometry (as we shall see later), the *average pressure* is not the simple arithmetic mean of the outer and inner boundary pressures of the porous medium.

3.5.1.6 The Klinkenberg Effect

It has already been mentioned that core permeabilities are measured using air (or another gas) as the fluid, with the outlet pressure at atmospheric, and the flow rate measured at outlet conditions.

The value of k is calculated in *Darcy* units using Eq. (3.33c) (for linear flow):

$$k = \frac{2L}{A} \frac{q_{sc}\,\mu_g}{p_{in}^2 - p_{out}^2}\,.$$

(3.34)

The parameters here are in laboratory units [q_{sc} in cm^3 s^{-1}; L in cm; A in cm^2; μ_g in cP ($=$mPa s); p in atm abs].

In laminar flow theory, which is the basis of Darcy's law, the boundary condition at the fluid/solid interface is that the fluid molecules are static. The potential gradient is caused by drag forces between these molecules and the adjacent "layer" of moving molecules, as they slide past.

At low pressures, the mean free path of a gas molecule may be larger than the size of the pores – this is especially true in very fine-grained porous media (low k). In this case the fluid/solid boundary molecules cannot be assumed static. This "slippage" increases the flow rate over and above that predicted by Darcy's law.

Consequently, $q_{sc} > q_{sc}$ [Eq. (3.33c)], and Eq. (3.34) overestimates k. This is the well-known[1] *Klinkenberg effect*.

Since the mean free path of the gas molecules decreases with increasing pressure, the Klinkenberg effect becomes less significant at higher average core pressures, $p_{av} = (p_{in} + p_{out})/2$, Fig. 3.17.

Consequently, we have

$$k_g = k_L\left(1 + \frac{b}{p_{av}}\right),$$

(3.35a)

where k_g is the gas permeability measured at average pressure p_{av}, k_L is the liquid permeability (where the Klinkenberg effect is non-existent), and b is a constant[24] which depends on the liquid permeability (Fig. 3.18) as well as, to a lesser degree, the nature of the gas used.

It follows that

$$\lim_{p_{av} \to \infty} k_g = k_L\,.$$

(3.35b)

The Klinkenberg effect is corrected for by measuring k_g at different pressures p_{av}. The data are plotted against $1/p_{av}$ as shown in Fig. 3.17, and the straight line trend is extrapolated back to $1/p_{av} = 0$ (corresponding to $p_{av} = \infty$) to obtain the true value of k.

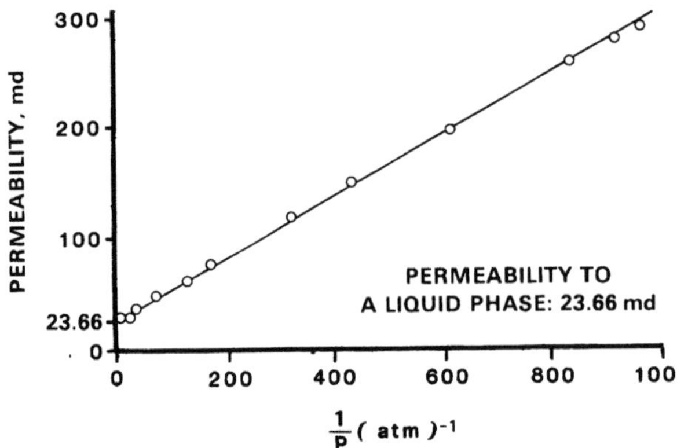

Fig. 3.17. *Plot of air permeability versus reciprocal pressure – the Klinkenberg effect. From Ref. 24, 1959, McGraw-Hill Inc. Reprinted with permission of McGraw-Hill Inc*

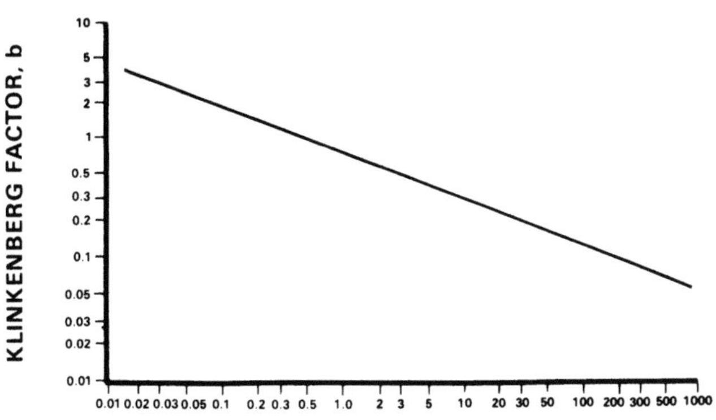

Fig. 3.18. *The dependence of the coefficient b in Eq. (3.35a) on k_L (p_{av} in atm abs). From Ref. 24, 1959, McGraw-Hill Inc. Reprinted with permission of McGraw-Hill Inc*

The reservoir engineer must ascertain that measurements of k provided by the core laboratory have been corrected for the Klinkenberg effect (as well as reservoir confining pressure), especially when dealing with low permeability rock.

3.5.1.7 *Heterogeneity and Average Permeability in a Sedimentary Unit*

The term "sedimentary unit" or "depositional unit" refers to a thickness of rock (a "layer" or "stratum") within which the conditions of deposition (the energy and local direction of the paleocurrent transporting the solid material) were *statistically* uniform during the period of sedimentation.

Consequently, the petrophysical properties of the rock in a sedimentary unit will vary statistically about a mean value ($\bar{\phi}, \bar{k}$).

An initial indication of the degree of heterogeneity in a unit can be obtained by calculating the *Lorentz coefficient*.

If we had n measurements of permeability k_j ($j = 1, 2, \ldots n$) from n cores, each of which represents a thickness Δh_j of the sedimentary unit of total thickness h, then it follows that:

$$h = \sum_{j=1}^{n} \Delta h_j, \tag{3.36a}$$

$$\bar{k} = \frac{\sum_{j=1}^{n} k_j \, \Delta h_j}{h}, \tag{3.36b}$$

where $\bar{k}$ is the thickness-averaged (arithmetic mean) permeability of the unit.

Suppose the k values are arranged in increasing order $k_1 < k_2 < \cdots k_j < \cdots k_n$, and we calculate the following two parameters for the ith sample:

$$A_i = \frac{\sum_{j=1}^{i} \Delta h_j}{h}, \tag{3.37a}$$

$$B_i = \frac{\sum_{j=1}^{i} k_j \, \Delta h_j}{h\bar{k}}. \tag{3.37b}$$

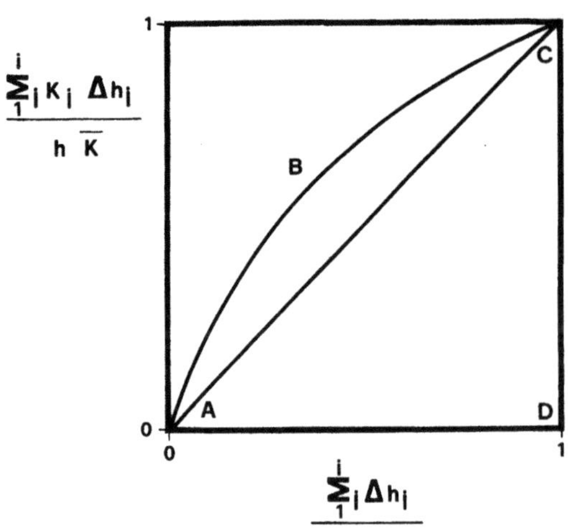

Fig. 3.19. *Diagram for the calculation of the Lorenz coefficient of heterogeneity*

The *n* points (A_i, B_i) are plotted on cartesian axes to produce the *Lorenz diagram*, shown schematically in Fig. 3.19.

Obviously, in a perfectly homogeneous unit, where the *n* values of *k* would all be equal, the points would fall along the diagonal AC.

The ratio

$$\mathscr{L} = \frac{\text{area ABCA}}{\text{area ACDA}} \tag{3.38}$$

is called[1] the *Lorenz coefficient of heterogeneity*.

$\mathscr{L}$ is a number between zero (perfect homogeneity) and 1 (maximum heterogeneity), and it is used as a quantitative indicator of the degree of heterogeneity within a sedimentary unit.

Note that $\mathscr{L}$ does not define the permeability distribution uniquely: in fact, a number of different curve shapes ABC will produce the same area ABCA.

A more effective description of heterogeneity can be obtained by applying a few basic principals of statistical theory to the population of *k* samples.

According to the central limit theorem, if the number of samples (*k* measurements) is large enough, they should be normally distributed about the mean value. Their probability density should then be Gaussian, centred on the mean *k*.

Experimentally, we find that it is not *k* but $y = \log k$ which assumes the normal distribution, with probability density

$$f(y) = \frac{1}{\sigma\sqrt{2\pi}} \exp\left[-\frac{(y - \bar{y})^2}{2\sigma^2}\right] \tag{3.39}$$

and the *distribution function*

$$F(y) = \frac{1}{\sigma\sqrt{2\pi}} \int_{-\infty}^{y} \exp\left[-\frac{(y - \bar{y})^2}{2\sigma^2}\right] dy \tag{3.40}$$

where

$$\bar{y} = \log \bar{k} = \frac{1}{n} \sum_{j=1}^{n} \log k_j$$

= mean, or average value of $\log k$, $\tag{3.41}$

and

$$\sigma = \text{standard deviation} . \tag{3.42}$$

In a normal distribution, the mean, the median, and the mode all coincide and represent the most probable value of the variable under consideration.

From Eq. (3.41) we can write straight away:

$$\bar{k} = \left(\prod_{j=1}^{n} k_j \right)^{1/n} . \tag{3.43}$$

In other words, as long as a large enough number of measurements has been made of k across the sedimentary unit, the average permeability of the unit is the geometric mean of these measurements.

The standard deviation σ describes the degree of heterogeneity in the samples.

σ is determined graphically – the n values of k are arranged in decreasing order, and for each one we note the percentage $F(k_j)$ of samples having $k \geq k_j$.

$F(k_j)$ is then plotted against $\log k_j$ on a log–probability diagram (Fig. 3.20). The n points should be scattered about a straight line, which can be estimated by eye or calculated from a least-squares fit.

The mean permeability $\bar{k}$ corresponds to $F = 50\%$.

In a normal distribution, 34.14% of the samples lie within a distance σ from the mean value – i.e. between $F = 50\%$ and $F = 84.14\%$.

σ is therefore the difference between the $\log k$ values at $F = 50$ and $F = 84.14$, that is

$$\sigma = \log k_{50} - \log k_{84.14} = \log \frac{k_{50}}{k_{84.14}} = - \log \frac{k_{84.14}}{k_{50}} . \tag{3.44}$$

σ can be considered as an indicator of the variation in permeability within the unit, and therefore of its heterogeneity.

Law[25] prefers to determine the permeability variation as

$$\mathcal{V} = \frac{k_{50} - k_{84.14}}{k_{50}} = 1 - \frac{k_{84.14}}{k_{50}} . \tag{3.45a}$$

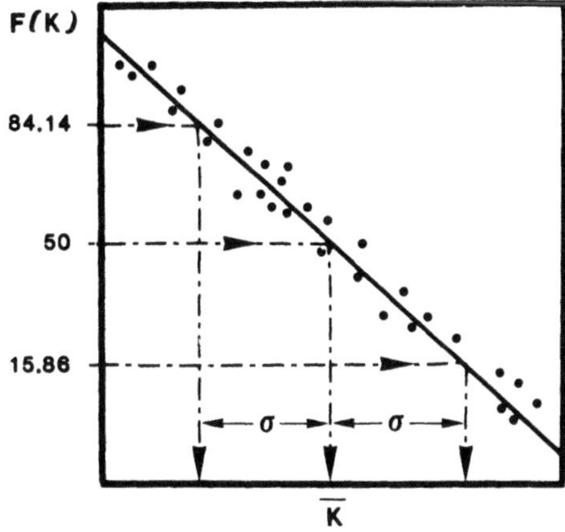

Fig. 3.20. *Log–probability diagram of the distribution function for k measurements in a sedimentary unit. The mean value* $\bar{k}$, *and the standard deviation* σ *are shown*

The relationship between σ and $\mathscr{V}$ is quite simply

$$\sigma = -\log(1 - \mathscr{V}). \qquad (3.45b)$$

In a homogeneous unit, $\sigma = \mathscr{V} = 0$. Both σ and $\mathscr{V}$ increase with the degree of heterogeneity.

3.5.1.8 Permeability-Porosity Correlations

Many researchers have proposed equations for the prediction of k for any given ϕ. Probably the best known from among these is the Carman–Kozeny[6] equation

$$k = \frac{\phi^3}{C_0 A_v^2 \tau}. \qquad (3.46)$$

A_v is the specific surface area (per unit volume) of the porous medium, τ the tortuosity of the network of capillaries making up the medium, and C_0 a numerical coefficient.

In fact, none of the equations gives universally satisfactory results. Only experimental correlations adapted to local conditions offer any solution to the problem, which can be expressed as follows:

> "given a well-defined sedimentary unit with an adequate number of (ϕ, k) measurements from cores taken from key wells in the field, calculate k from values of ϕ obtained from well logs in the same unit anywhere else in the field."

It has been found that there is almost always a linear correlation between $\log k$ and ϕ (Fig. 3.21A); sometimes there is a better correlation between the logarithm of both parameters.

Consequently, it should be relatively easy, after a few attempts, to find a correlation which satisfies all the core data from a single sedimentary unit. k can then be calculated from ϕ in parts of the unit where core data are not available.

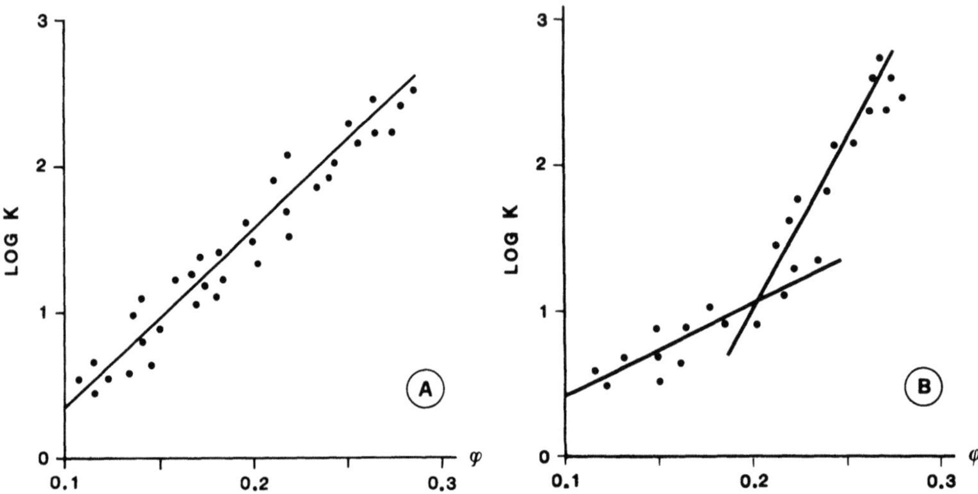

Fig. 3.21A, B. *Permeability-porosity correlations of the form $\log_{10} k = f(\phi)$ for a reservoir consisting of* **A** *a single depositional unit, and* **B** *two depositional units*

In some cases we may find that in deriving a suitable correlation from the core data, two or more straight line trends are apparent (Fig. 3.21B). This is a clear indication that we are not dealing with a single depositional unit, and the change in correlation between k and ϕ can be used to localise the cut-off between successive units.

This is one of the techniques[16] employed in reservoir "zonation", where the objective is to identify layering. Each layer has internal petrophysical properties which are statistically homogeneous, but differ from those of the adjacent layers.

3.5.2 Effective and Relative Permeability

3.5.2.1 The Concept of Potential - a Further Recap

For each of the fluid phases (oil, gas, water) present in the reservoir we define a "potential"

$$\Phi_g = \int_0^{p_g} \frac{dp_g}{\rho_g} - gz \; , \tag{3.47a}$$

$$\Phi_o = \int_0^{p_o} \frac{dp_o}{\rho_o} - gz \; , \tag{3.47b}$$

$$\Phi_w = \int_0^{p_w} \frac{dp_w}{\rho_w} - gz \; , \tag{3.47c}$$

The pressures within each phase at any point are different; $p_g \neq p_o \neq p_w$ because:

$$p_g = p_o + P_{c,go} \; , \tag{3.48a}$$

$$p_w = p_o - P_{c,ow} \; , \tag{3.48b}$$

Equations (3.48) allow us to express the three potentials Φ_g, Φ_o and Φ_w in terms of the pressure in any one of the three phases, by adding or subtracting the appropriate capillary pressure, which is, in turn, a function of the saturation and process of drainage or imbibition.

3.5.2.2 Effective Permeability

Referring to Eq. (3.31), we can extend Darcy's law to cover multiphase flow by substituting from Eqs. (3.47):

$$\boldsymbol{u}_g = -\frac{[k_g]\,\rho_g}{\mu_g}\,\text{grad}\,\Phi_g \; , \tag{3.49a}$$

$$\boldsymbol{u}_o = -\frac{[k_o]\,\rho_o}{\mu_o}\,\text{grad}\,\Phi_o \; , \tag{3.49b}$$

$$\boldsymbol{u}_w = -\frac{[k_w]\,\rho_w}{\mu_w}\,\text{grad}\,\Phi_w \; , \tag{3.49c}$$

where k_g, k_o and k_w are, like k, second order tensors, whose components are now not only functions of the porous medium, but also of the quantity and distribution of the relevant fluid phase in the pore-space, and the interaction between the phases.

Each of these parameters is referred to as an *effective permeability* – to gas, oil and water respectively.

3.5.2.3 Relative Permeability

For the purposes of reservoir engineering work, the following parameters have been defined:

$$k_{rg} = \frac{k_g}{k} = \text{relative permeability to gas ,} \tag{3.50a}$$

$$k_{ro} = \frac{k_o}{k} = \text{relative permeability to oil ,} \tag{3.50b}$$

$$k_{rw} = \frac{k_w}{k} = \text{relative permeability to water ,} \tag{3.50c}$$

Note that both the permeability and the effective permeabilities have been written as scalar quantities, rather than the tensors that they really are.

This is equivalent to saying that the ratio (effective permeability)/(permeability) is independent of the direction of flow, an assumption which is debatable.[5]

We are on even less certain ground when we write, for two-phase flow:

$$k_{rg} = f(S_g) , \tag{3.51a}$$

$$k_{ro} = f(S_o) , \tag{3.51b}$$

$$k_{rw} = f(S_w) , \tag{3.51c}$$

because here we are assuming that the permeability depends solely on the saturation of the fluid phase, and not on its manner of distribution in the pore-space. This in turn is a function[6] of relative magnitudes of the local gravity, capillary and viscous forces.

The group of Eqs. (3.51) must be taken to refer to the particular process of drainage or imbition that prevails at the time.

In the case of the displacement of oil or gas by water, the relative permeabilities apply to the imbition process, while for the displacement of oil by gas, it would be drainage (gas being the non-wetting phase relative to oil).

The validity or otherwise of the concept of relative permeability will be discussed in the next section.

3.5.2.4 Relative Permeability Curves for Oil/Water and Gas/Water (Imbibition)

The laboratory measurement of relative permeability curves from cores is fully described in the references listed at the end of this chapter.

The general shape of these curves, for the imbition cycle (S_w increasing), is similar for oil/water and gas/water systems (Figs. 3.22 and 3.23).

As S_w increases, k_{ro} (or k_{rg}) declines steadily, eventually reaching zero at the residual saturation (S_{or} for oil, S_{gr} for gas). The actual values of S_{or} and S_{gr} depend on the texture and grain size of the porous medium, but as a general rule[12] S_{gr} *is less than* S_{or}.

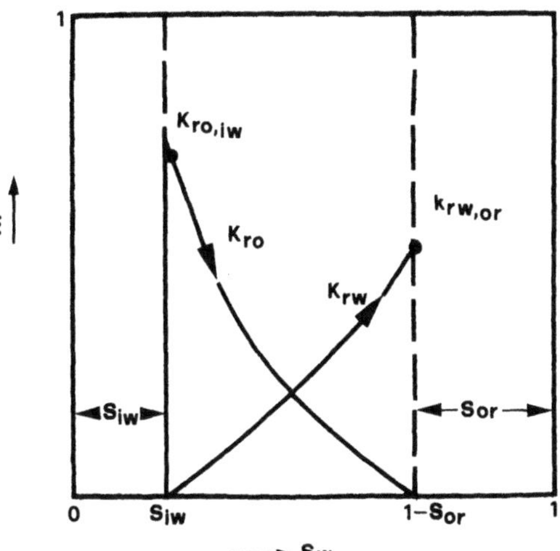

Fig. 3.22. *Schematic of relative permeability curves for imbibition in an oil/water system*

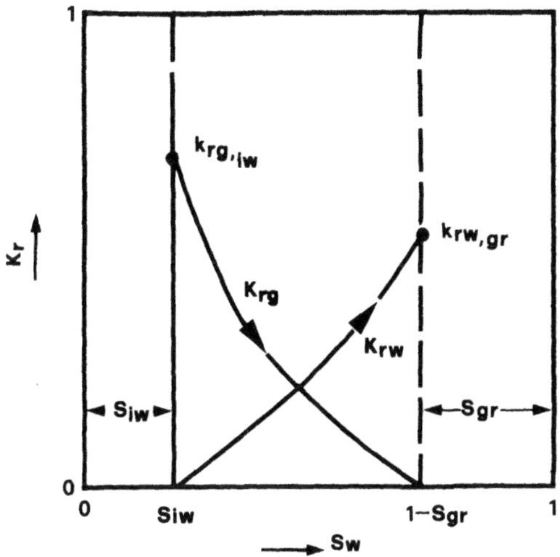

Fig. 3.23. *Schematic of relative permeability curves for imbibition in a gas/water system*

k_{rw} assumes a finite value as soon as $S_w > S_{iw}$, and reaches a maximum at $S_w = 1 - S_{or}$.

So the relative permeability curves for imbibition in oil/water or gas/water are non-zero for S_w between S_{iw} and $1 - S_{or}$ (or $1 - S_{gr}$); k_r varies between $k_{ro,iw}$ and 0 for oil (or $k_{rg,iw}$ and 0 for gas), and between 0 and $k_{rw,or}$ (or $k_{rw,gr}$) for water. The maximum value attained by each curve is called the *end point relative permeability*.

Two new parameters are now defined:

$$M_{wo} = \frac{\lambda_{w,or}}{\lambda_{o,iw}} = \frac{k_{rw,or}/\mu_w}{k_{ro,iw}/\mu_o} , \tag{3.52a}$$

$$M_{wg} = \frac{\lambda_{w,gr}}{\lambda_{g,iw}} = \frac{k_{rw,gr}/\mu_w}{k_{rg,iw}/\mu_g} , \tag{3.52b}$$

M is the *mobility ratio*, expressed as the ratio of the *maximum* mobility of the displacing fluid to the *maximum* mobility of the displaced fluid.

The mobility itself is defined as the ratio of the maximum effective permeability (end point) to the fluid viscosity.

3.5.2.5   Relative Permeability Curves for Gas/Oil (Drainage)

The relative permeability curve for drainage in a gas/oil system has the general shape shown in Fig. 3.24.

The most significant feature is that k_{rg} is zero until S_g reaches a minimum value S_{gc}, the critical gas saturation; then it starts to increase. The gas phase does not become continuous in the pore channels until $S_g = S_{gc}$, and only above S_{gc} is it possible for the gas to actually flow through the porous medium.

Note how k_{ro} declines in the range $0 < S_g < S_{gc}$, as an increasing fraction of the pore space is taken up by gas, which is, however, not yet mobile.

As S_g increases, k_{ro} falls progressively until it reaches zero at a value of $S_g = 1 - S_{iw} - S_{or}$. S_{or} is once again called the *residual oil saturation*.

So the relative permeability curves for drainage in a gas/oil system are non-zero for $0 \leq S_g \leq (1 - S_{or} - S_{iw})$, and k_r varies between zero and the end points $k_{ro,iw}$ and $k_{rg,or}$.

The mobility ratio is:

$$M_{go} = \frac{\lambda_{g,or}}{\lambda_{o,iw}} = \frac{k_{rg,or}/\mu_g}{k_{ro,iw}/\mu_o}, \tag{3.52c}$$

Since $\mu_o \gg \mu_g$, M_{go} is always very large.

3.5.3   Comments on the Concept of Relative Permeability

Although the idea of a relative permeability is in itself perfectly valid in physical terms, it is an over-simplification to represent it as a scalar parameter, dependent only on the saturation.

The saturation (Sect. 3.4.2) is merely the fraction of the pore-space filled with a particular fluid: it is quite correct to express this as a scalar quantity.

But the effective permeabilities from which k_r is derived depend on the spatial distribution of each fluid phase within the pores, as well as the local interaction

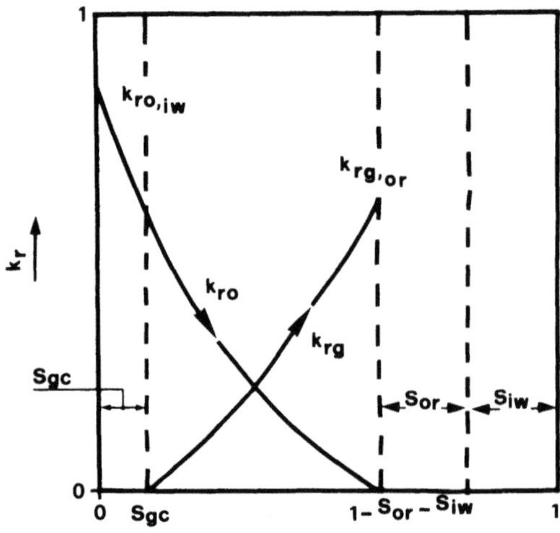

Fig. 3.24. *Schematic of relative permeability curves for drainage in a gas/oil system*

between the phases. For these reasons, k_r is most definitely not a scalar, but a tensor – the ratio between two tensors $\mathbf{k_f}$ (f = fluid) and $\mathbf{k}$.

Fluid saturations at any point in the reservoir depend on local conditions; especially the local relationship between viscous forces, gravity and capillarity, and the relative direction of flow. This topic is discussed in detail by Chierici[9] and Houpeurt.[20]

Laboratory measurement of k_r is effected on horizontal cores which are kept rotating slowly to prevent gravitational segregation of the fluids. A constant saturation is established along the length of the core to eliminate capillary effects, and both test fluids flow in the same direction.

The first question that comes to mind[20] is: how relevant are curves for $k_{ro}(S_o)$ and $k_{rg}(S_g)$ measured in this manner to the movement of oil and gas in a reservoir producing by solution gas drive?

Here, as the gas comes out of solution it tends to migrate upwards, while the oil segregates downwards, because of gravitational forces. This constitutes a *counter-current* flow of gas and oil in the reservoir, while the $k_{ro}(S_o)$ and $k_{rg}(S_g)$ curves were measured for *unidirectional* flow in the cores.

Secondly: local gravitational, capillary and viscous forces have different values in the various parts of the reservoir. Two dimensionless groups have been defined:

$$N_{vg} = \frac{u\mu_o}{kg(\rho_w - \rho_o)} = \frac{\text{viscous force}}{\text{gravitational force}} , \tag{3.53a}$$

$$N_{vc} = \frac{u\mu_o}{\sigma \cos \Theta_c} = \frac{\text{viscous force}}{\text{capillary force}} , \tag{3.53b}$$

where u is the local flow velocity (Darcy velocity).

In the immediate vicinity of the wellbore, the flow is radial horizontal and, because of the relatively high fluid velocity, it is dominated by viscous forces. The core-derived relative permeability curves are well suited to describe the fluid behaviour in these conditions.

At increasing distance from the well, as fluid velocity becomes smaller, viscous forces become less significant and the gravitational and capillary forces dominate [the values of N_{vg} and N_{vc} in Eqs. (3.53) decrease]. This is very different from the laboratory conditions under which the relative permeability curves were derived, and they become meaningless[5, 9, 20] as far as describing local multiphase flow is concerned.

A simple experiment[11] will amply demonstrate the problem.

A core saturated with oil in the presence of a water saturation S_{iw}, is held vertically and gas is injected *at very low velocity* at the top. To eliminate capillary effects at the lower face of the core, it is placed in contact with a semipermeable membrane saturated with oil. This set-up ensures that essentially only gravitational forces are acting. The capillary pressure $(p_g - p_o)$ applied to the core is now increased until no more oil flows from the lower face.

Now, we will find that the residual oil saturation at the end of this experiment is zero (Fig. 3.25, above).

We then take the same core, resaturate it with oil (still in the presence of water at S_{iw}), and then perform a conventional laboratory relative permeability analysis.[19] Now only viscous forces act – the capillary and gravitational forces are nil.

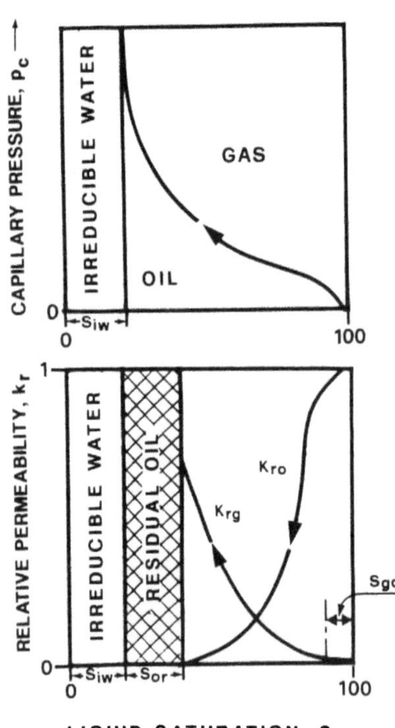

Fig. 3.25. *Capillary pressure (upper) and relative permeability (lower) curves, both measured for drainage in a gas/oil system, in the presence of S_{iw}. Note the absence of any residual oil saturation in the capillary pressure curve*

The curves $k_{ro}(S_o)$ and $k_{rg}(S_g)$ that we would obtain are shown in Fig. 3.25 (lower). Note that the residual oil saturation in this case is not only finite but quite large.

A wealth of experimental data from various researchers shows clearly that the value of S_{or} depends on the capillary number,[4] N_{vc}, (Fig. 3.26). S_{or} will be zero if N_{vc} is sufficiently large.

The preceding comments should be sufficient to raise a few serious doubts about the concept of relative permeability, as presented in any number of authoritative books on reservoir engineering. It follows that relative permeability curves should be applied with caution to real reservoir problems.

There is one final point worth considering.

Physically, relative permeability curves describe the phenomenon of dispersion of fluids as they flow through the tortuous system of pore channels;[6] the shape of the curves depends on heterogeneity within the porous medium on a *microscopic* scale.

But reservoir rocks are also heterogeneous on a scale of metres (*macroscopic* heterogeneity) and even kilometres (*megascopic* heterogeneity).

Multiphase flow through a reservoir is obviously going to be influenced significantly by these larger scale heterogeneities. This is why it is nearly always necessary to modify the laboratory relative permeability curves used in a numerical reservoir simulator in order to obtain a satisfactory history match to past reservoir behaviour. (Reservoir simulation is covered in Chap. 13.)

Relative permeability curves represent, in fact, a convenient means of providing a "rough and ready" description of multiphase fluid flow through porous media. Although they have been with us since the 1930s, they have not been replaced by anything better.

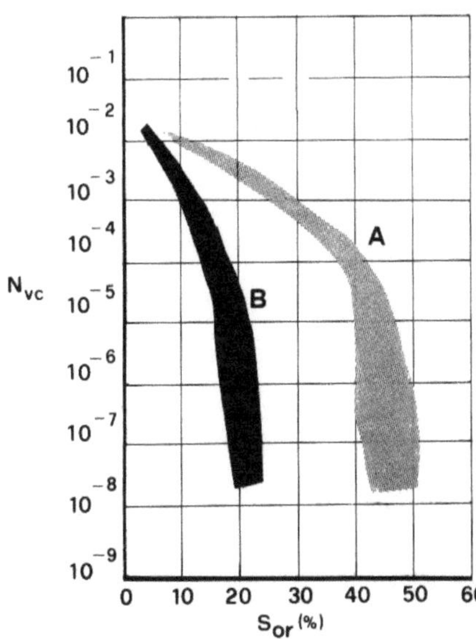

Fig. 3.26. *Residual oil saturation as a function of the capillary number, N_{vc}, during displacement. A: after Foster; B: after du Prey. 1973, Society of Petroleum Engineers of AIME. Reprinted with permission of the SPE*

However, it is foreseeable[11] that the notion of relative permeability might be superseded by the concept of *multiphase fluid dispersion* on a macroscopic scale; with behaviour on the microscopic (pore) scale characterised by just S_{iw}, S_{or}, S_{gr}, S_{gc}, $k_{rw,or}$, $k_{ro,iw}$, and $k_{rg,or}$, expressed as a function of the local gravity number N_{vg}, and capillary number N_{vc}.

3.5.4 Calculation of the Average Relative Permeability Curve in a Sedimentary Unit

Relative permeability curves can be normalised by appropriate functioning of the variables:[7,8]

If we define:

for water/oil relative permeability curves:

$$S_w^* = \frac{S_w - S_{iw}}{1 - S_{iw} - S_{or}}, \tag{3.54a}$$

$$k_{ro}^* = \frac{k_{ro}}{k_{ro,iw}}, \tag{3.54b}$$

$$k_{rw}^* = \frac{k_{rw}}{k_{rw,or}}, \tag{3.54c}$$

for gas/oil relative permeability curves:

$$S_g^* = \frac{S_g - S_{gc}}{1 - S_{iw} - S_{gc} - S_{or}}, \tag{3.55a}$$

$$k_{ro}^* = \frac{k_{ro}}{k_{ro,gc}}, \tag{3.55b}$$

$$k_{rg}^* = \frac{k_{rg}}{k_{rg,or}}, \tag{3.55c}$$

we find that the *normalised* parameters S_w^*, k_{ro}^*, k_{rw}^*, and S_g^*, k_{ro}^*, k_{rg}^* all vary between 0 and 1.0, these limits corresponding to the end points.

The most important feature is that the *normalised curves* for cores from the same sedimentary unit are, to a good approximation, identical (Fig. 3.27).

Furthermore, the parameters appearing in the normalising functions – viz:

for water/oil relative permeability:

$$S_{iw}, \quad (1 - S_{iw} - S_{or}), \quad k_{ro,iw}, \quad k_{rw,or}$$

for gas/oil relative permeability:

$$S_{gc}, \quad (1 - S_{iw} - S_{gc} - S_{or}), \quad k_{ro,gc}, \quad k_{rg,or}$$

are, also to a good approximation, a unique function of the *tortuosity* of the porous medium on which the measurements were made (Figs. 3.28 and 3.29). The tortuosity is expressed as $(k/\phi)^{1/2}$.

It is therefore a straightforward matter to derive an average relative permeability curve from all the different curves measured on cores from the same sedimentary unit. For this purpose, all we need are the global average (or most probable) values $\bar{k}$ and $\bar{\phi}$ of the permeability and porosity of the unit.

The following example is for gas/oil; the same procedure is applied for water/oil.

Firstly, we plot the normalisation diagrams (like Figs. 3.27, 3.28 and 3.29), using the relative permeability curves $k_{ro}(S_g)$ and $k_{rg}(S_g)$ from each core.

Then, having computed the average k and ϕ for the unit, we calculate the average tortuosity $(\bar{k}/\bar{\phi})^{1/2}$, and read off the corresponding values of S_{gc}, $(1 - S_{iw} - S_{or} - S_{gc})$, $k_{ro,gc}$ and $k_{rg,or}$ on the diagrams equivalent to Figs. 3.28 and 3.29.

Finally, we read off a series of points (S_g^*, k_{ro}^*) *and* (S_g^*, k_{rg}^*) from the curve fitted on the normalisation diagram like Fig. 3.27 (a sufficient number of points to define

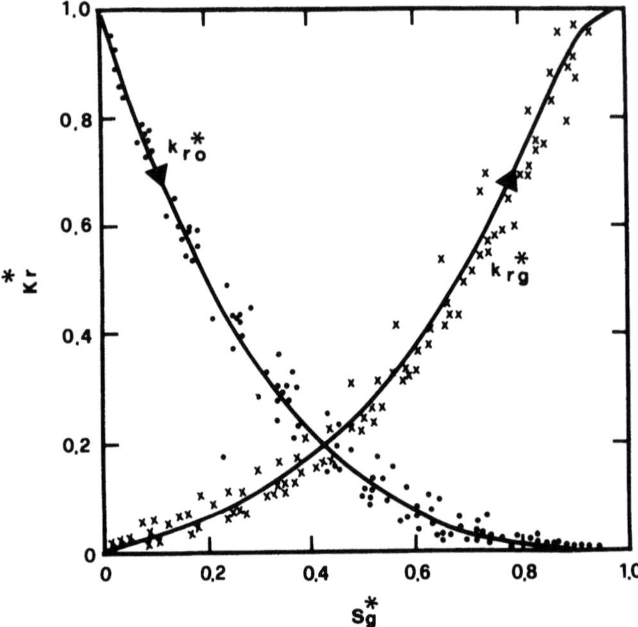

Fig. 3.27. *Normalised drainage relative permeability curves for gas/oil in a sedimentary unit*

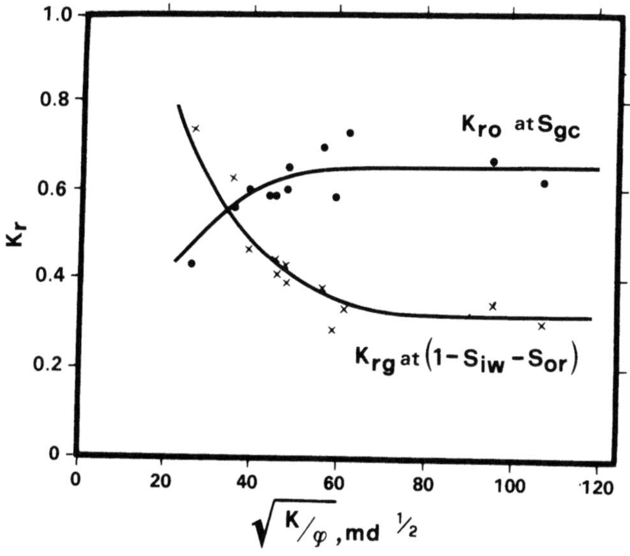

Fig. 3.28. *End-point relative permeabilities as a function of tortuosity $(k/\phi)^{1/2}$, from the same sedimentary unit as Fig. 3.27*

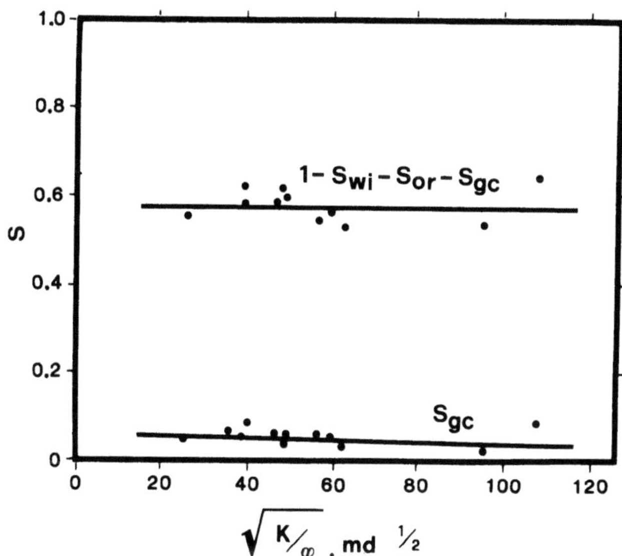

Fig. 3.29. *Critical and residual saturations as a function of tortuosity $(k/\phi)^{1/2}$, from the same sedimentary unit as Figs. 3.27 and 3.28*

the shape of the curve). We can then use Eqs. (3.55) to calculate the corresponding real points $[S_g, k_{ro}(S_g)$ and $(S_g, k_{rg}(S_g)]$. With these we plot the average gas/oil relative permeability diagram for the unit.

3.5.5 Equations for the Calculation of Relative Permeability Curves

Wherever core data are not available, it is common practice to resort to empirical correlations for a first estimate of relative permeability curves.

The most widely used correlations are listed below:

gas/oil drainage cycle

$$k_{ro} = \left(\frac{S_o - S_{or}}{1 - S_{iw} - S_{or}}\right)^4 , \tag{3.56a}$$

$$k_{rg} = (S_g - S_{gc})^3 \frac{2(1 - S_{iw} - S_{or}) - (S_g - S_{gc})}{(1 - S_{iw} - S_{or})^4} , \qquad (3.56b)$$

$[k_{rg} = 0 \text{ for } S_g \leq S_{gc}]$

water/oil imbibition cycle:

$$k_{rw} = \left(\frac{S_w - S_{iw}}{1 - S_{iw}}\right)^4 , \qquad (3.57a)$$

$$k_{ro} = (S^*_{OF})^3 (2 - S^*_{OF}) , \qquad (3.57b)$$

where

$$S^*_{OF} = \frac{1}{2} \left\{ \frac{S_o - S_{or}}{1 - S_{iw}} + \frac{(S_o^2 + 2 S_o S_{or} - 3 S_{or}^2)^{1/2}}{1 - S_{iw}} \right\} . \qquad (3.57c)$$

The values of S_{iw}, S_{gc} and S_{or} which appear in Eqs. (3.56) and (3.57) must be drawn from local field experience, taking into account, of course, whether they refer to drainage or imbibition.

A different type of empirical correlation, based on numerical analysis of a large number of relative permeability curves, has been proposed by Chierici:[10]

gas/oil drainage cycle

$$k_{ro} = \exp(-A R_g^L) , \qquad (3.58a)$$

$$k_{rg} = \exp(-B R_g^{-M}) , \qquad (3.58b)$$

where

$$R_g = \frac{S_g - S_{gc}}{1 - S_{iw} - S_g} , \qquad (3.58c)$$

with the condition

$$S_g - S_{gc} = 0 \text{ for } S_g \leqslant S_{gc} , \qquad (3.58d)$$

water/oil imbibition cycle

$$k^*_{ro} = \exp(-C R_w^N) , \qquad (3.59a)$$

$$k^*_{rw} = \exp(-D R_w^{-P}) , \qquad (3.59b)$$

with k^*_{ro} and k^*_{rw} defined as in Eqs. (3.54b) and (3.54c) and

$$R_w = \frac{S_w - S_{iw}}{1 - S_{or} - S_w} . \qquad (3.59c)$$

Equations (3.58), for the calculation of the relative permeability curves for gas/oil, contain two physical constants (S_{iw} and S_{gc}) and four numerical coefficients (A, B, L, M). Note that S_{or} does not appear; following what was said in Sect. 3.5.3, its value is zero in this case.

Equations (3.59), for the calculation of relative permeability curves for water/oil, contain four physical constants (S_{iw}, S_{or}, $k_{ro,iw}$, and $k_{rw,or}$) and four numerical constants (C, D, N, P).

As far as the physical constants are concerned, their values are a matter of local experience, as stated previously.

The suggested values of the numerical coefficients are as follows:

$A = 4.0–4.5$
$B = 0.8–1.3$
$C = 2.0$
$D = 1.5–3.0$
$L = 0.5–0.8$
$M = 1.0–2.5$
$N = 0.7–1.2$
$P = 0.6–1.5$

Equations (3.58) and (3.59) have the advantage over their counterparts (3.56) and (3.57) of modelling the relative permeability curve more closely at lower values of k_r [the initial part of the curves $k_{rg}(S_g)$ and $k_{rw}(S_w)$, and the end part of $k_{ro}(S_o)$]. This improvement is especially significant in the case of oil reservoirs producing by gas drive (internal or gas cap).

3.6 Reservoir Zonation

The term "zonation", in a reservoir engineering context, means the vertical subdivision of a sedimentary formation (particularly reservoir rock or "pay") into a series of strata or layers, in vertical communication, each of which can be considered as a *statistically homogeneous unit* for the purpose of reservoir engineering calculations. Some or all of the mean dynamic and petrophysical properties of each "zone" thus defined are significantly different from those of the adjacent zones.

This concept of zones almost invariably coincides with the geological concept of the "sedimentary (depositional) unit", although this is not a mandatory condition.

A sedimentary unit is in fact characterised by the uniformity, within certain limits, of the characteristics of its sedimentary material, and of the depositional environment (paleogeography, direction and energy of the transporting current). In such conditions, the petrophysical properties of the sediment at the time of deposition are statistically constant over the entire unit.

If successive diagenetic and tectonic phenomena have not altered the rock in an irregular manner, the "zone" is the same as the "unit". Therefore, a sedimentological study of the reservoir rock provides the first valid indication of the presence of or absence of zones within the pay, and of the cut-off points between zones.

3.6.1 Quick Methods of Zonation

Within a zone (as defined in the previous section), the petrophysical properties are distributed in a normal or log-normal manner about the mean values, usually with quite small variation on either side. This is one of the first criteria to be applied when deciding if an interval consists of one or several zones.

Suppose we had the two permeability distributions shown in Fig. 3.30A and B, from a large number of measurements in two different reservoirs.

The nature of the distribution in Fig. 3.30A – unimodal, with a small deviation about the mean – clearly indicates that the data come from a single zone.

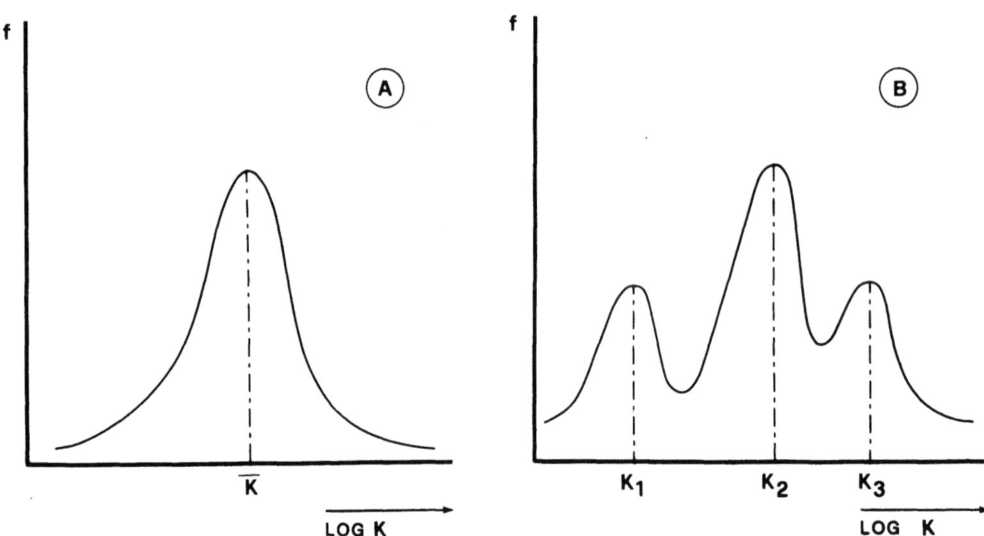

Fig. 3.30A, B. *Two different core permeability distributions:* **A** *unimodal, representing a single zone;* **B** *trimodal, where the three different rock types suggest three distinct zones*

Figure 3.30B, on the other hand, suggests that there are three different rock types – three zones – characterised by mean permeabilities k_1, k_2, and k_3.

Another simplistic approach to reservoir zonation is to plot log k against ϕ for all the cores examined.[16] This type of plot was already introduced in Sect. 3.5.1.8: if the data fall on a single straight line trend (Fig. 3.21A) we have a single zone; if several distinct lines can be fitted (Fig. 3.21B), each segment corresponds to a zone of different lithology.

In the case of a very thick pay section, analysis of the vast amount of data which may be available from cores and well logs will necessitate the use of a calculator or computer. The next section introduces some of the methods of statistical analysis used to distinguish zone characteristics.

3.6.2   The Use of Statistical Analysis in Reservoir Zonation

The most commonly used method used for zoning a single well was devised by Testerman.[34]

In identifying discreet zones, and the parameter cut-offs between them, the following rules are applied to the parameter under test:

– the sum of the variances of each zone is a minimum,
– the variance of the mean values of the parameter is a maximum over the zones.

Testerman's method is usually applied to the permeability. With:

N: total number of measurements of k in the interval considered,
$\bar{k}$: overall arithmetic mean of the N measurements of k,
L: number of zones postulated,
n_i: number of k measurements in zone i,
$\bar{k}_i$: arithmetic mean k in zone i,
k_{ij}: jth measurement of k in zone i.

he defined the following parameters:

W is the average of the variances of all the L zones:

$$W = \frac{1}{N - L} \sum_{i=1}^{N} \sum_{j=1}^{n_i} (k_{ij} - \bar{k}_i)^2 \, . \tag{3.60a}$$

B is the variance of the mean $\bar{k}_i$ of each zone, about the overall mean $\bar{k}$:

$$B = \frac{1}{L - 1} \sum_{i=1}^{L} (\bar{k}_i - \bar{k})^2 \, . \tag{3.60b}$$

Testerman's quality index R for the postulated zonation is then defined as:

$$R = \frac{B - W}{B} = 1 - \frac{W}{B} \, . \tag{3.60c}$$

In the extreme situation where the variance of k is zero in all L zones (i.e. all zones are perfectly homogeneous), we get $W = 0$ and therefore $R = 1$.

This would indicate that the zone limits have been correctly defined in terms of L homogeneous zones.

The further R is from 1 ($0 \leq R \leq 1$), the poorer the quality of the zonation.

At the other extreme, if we attempt to define zones within what is in fact a single homogeneous layer, B and W will equal 0, so R is undefined.

In practice, the following procedure is used. Given N measurements of k arranged in order of increasing depth, we try $(N - 1)$ different subdivisions of the interval into two zones, with the boundary placed successively below the 1st, 2nd, 3rd, ... $(N - 1)$th data point. For each zonation, calculate R.

The maximum value of R is where the interval should be subdivided into two zones.

Keeping this zone boundary, we now try a three-zone scenario by placing a second boundary in each of the $(N - 3)$ possible places, and calculating R each time. As before, the maximum value of R will indicate where the boundary should be, and we will have the optimum zonation into three layers.

As this procedure is continued, the value of R will increase provided the addition of a new zone is justified. When further subdivision does not increase R by a significant amount (arbitrarily, $\Delta R \leq 0.06$), the zonation is considered to be complete.

This method gives us the mean value of k and its variance in each zone, as well as the zone boundaries.

3.6.3 Zonation Using Cluster Analysis

The analysis of a population of samples by "cluster analysis" is one of the most powerful statistical techniques available where the similarities or dissimilarities between samples (live, or inanimate objects) is involved.[14]

The use of cluster analysis in reservoir engineering is a relatively recent advance. It offers the important advantage over Testerman's method of being able to handle the zonation of a reservoir interval *simultaneously in several wells*, and to indicate whether a particular zone is continuous from well to well.

The input can be supplied from core analysis (typically k and ϕ), and from those well logs that respond to lithology (SP, GR, CNL, FDC/LDT, BHC).

Resistivity logs are excluded because they are strongly influenced by the nature of the pore fluids and their saturation.

More parameters are incorporated in the analysis than with the Testerman method (e.g. k, ϕ, GR, BHC), the only proviso being that the same measurements have been made in all the wells under examination.

Detailed descriptions of the inner workings of cluster analysis can be found in many text books on statistical analysis,[14] and will not be covered in the present work.

Very briefly, we start with a population of N samples (data sets measured at points in the pay section of each well). Each sample consists of measured values of the *same* group of r parameters (e.g. k, ϕ, GR, BHC: $r = 4$). These values must firstly be *normalised* to eliminate the influence of different units of measurement on the cluster process (some values are small, some are large).

Taking each of the r parameters, we calculate its mean and variance. If k represents any parameter in the sample group ($1 \leq k \leq r$), we have, summing over all N samples:

$$\bar{x}_k = \frac{1}{N} \sum_{i=1}^{N} x_{ik} \, . \tag{3.61a}$$

$$\sigma_k^2 = \frac{1}{N-1} \sum_{i=1}^{N} (x_{ik} - \bar{x}_k)^2 \, . \tag{3.61b}$$

If we call X_{ik} the normalised value of x_{ik} we can write

$$X_{ik} = \frac{x_{ik} - \bar{x}_k}{\sigma_k} \, . \tag{3.62}$$

In accordance with the central limit theorem, for each of the r parameters considered over all N data sets, the normalised values will exhibit a standardised normal distribution with mean $= 0$, and variance $= 1$.

Among the N samples, there are $N!/2(N-2)!$ possible pairs of normalised values (i, j) of a particular parameter. For each pair, we calculate the Euclidean distance:

$$d_{ij} = \left[\frac{\sum_{k=1}^{r} (X_{ik} - X_{jk})^2}{r} \right]^{1/2} \, . \tag{3.63}$$

Where the physical (lithological) characteristics at two different measurement points are very similar (with respect to all r parameters within each sample data set), d_{ij} will be very small. The converse also applies.

The N points are now subdivided into two or more groups, each having a Euclidean distance less than a certain arbitrary value, which means they have a certain degree of similarity within each group.

A second, separate, pass of cluster analysis is then applied to *each* of these groups of points.

In this way a number of subgroups is distinguished, in each of which the similarity between the samples they comprise is greater than in the first pass.

Further passes are made and a "dendrogram" is constructed (Fig. 3.31), with each branch splitting into two or more sub-branches at each successive stage of the analysis. The physical similarity between samples associated with a particular branch increases from pass to pass.

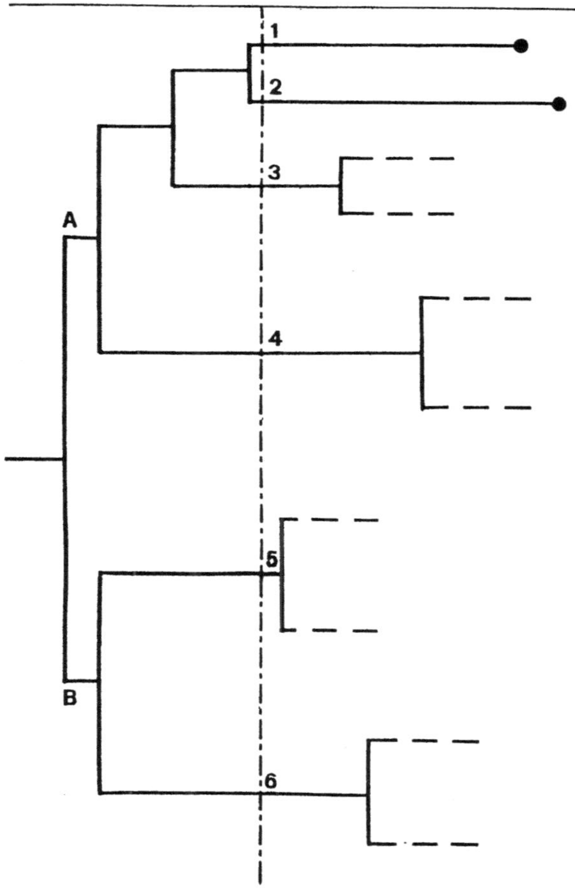

	LITHOLOGY	DYNAMIC CHARACTERISTICS	DEPOSITIONAL CHARACTERISTICS
1	MEDIUM GRAINED SANDSTONE WITH CROSS STRATIFICATIONS	GOOD 300-2000 md	CHANNEL
2	PARALLEL-LAMINATED FINE SAND AND SILT	MODERATE TO GOOD 100-700 md	CHANNEL (UPPER REACHES)
3	CROSS-LAMINATED SILT AND ARGILLACEOUS SILT	POOR 2-100 md	FOOT
4	MEDIUM TO COARSE GRAIN SANDSTONE	VERY GOOD 1 TO 5 DARCY	CHANNEL
5	VARICOLORED CLAYSTONE	IMPERMEABLE	MEANDER PLANE
6	VARICOLORED CLAYSTONE	IMPERMEABLE	MEANDER PLANE

Fig. 3.31. *Dendrogram obtained by cluster analysis, with lithological interpretation*

The process requires validation against available core data to ascertain not only the lithological significance of each of the branches of the diagram, but also at what stage further subdivision would be meaningless in physical terms.

Every lithology constitutes a different zone of reservoir rock. Cluster analysis provides a means of examining the lateral continuity of zones across the reservoir, and possible variations of facies across a given zone.

References

1. Amyx JW, Bass DM Jr, Whiting RL (1960) Petroleum reservoir engineering – physical properties. McGraw-Hill, New York
2. Anderson WG (1986) Wettability literature survey. Part 1: Rock/oil/brine interactions and the effects of core handling on wettability. J Petrol Tech (Oct): 1125–1144; Part 2: Wettability Measurement. J Petrol Tech (Nov): 1246–1262
3. API (1960) American Petroleum Institute: Recommended practice for core analysis procedure – API RP 40, 1st edn. American Petroleum Institute, Washington, DC
4. Chatzis I, Morrow NR (1984) Correlation of capillary number relationship for sandstone. Soc Petrol Eng J (Oct): 555–562; Trans AIME 277
5. Chierici GL (1971) Meccanismo microscopico del flusso polifasico nei mezzi porosi – Revisione critica del concetto di permeabilità relativa. Boll Assoc Miner Subalp (Sept–Dec): 503–540
6. Chierici GL (1972/73) Modelli di mezzo poroso per il calcolo curve di permeabilità relativa – Parte I. Boll Assoc Miner Subalp (Sept–Dec): 500–538; Parte II. Boll Assoc Miner Subalp (March–June): 82–104
7. Chierici GL (1975) Curve normalizzate di permeabilità relativa per lo studio del flusso bifase in giacimenti petroliferi statisticamente omogenei. Boll Assoc Miner Subalp (Dec): 401–422
8. Chierici GL (1976a) On the normalization of relative permeability curves. "Unsolicited paper" to the Society of Petroleum Engineers (SPE), Pap SPE 5962 (March)
9. Chierici GL (1976b) Modelli per la simulazione del comportamento di giacimenti petroliferi. Una revisione critica. Boll Assoc Miner Subalp (Sept): 237–251
10. Chierici GL (1984) Novel relations for drainage and imbibition relative permeabilities. Soc Petrol Eng J (June): 275–276; Trans AIME 277, 56th Annual Conf of the Society of Petroleum Engineers, San Antonio, TX. Oct 5–7, 1981, Pap 10165
11. Chierici GL (1985) Petroleum reservoir engineering in the year 2000. Energy Explor Exploit 3(3): 173–193
12. Chierici GL, Ciucci GM, Long G (1963) Experimental research on gas saturation behind the water front in gas reservoirs subjected to water drive. Proc 6th World Pet Congr. Frankfurt, Vol 2, Verein zur Förderung des 6. Welt Erdöl Kongress, Hamburg, pp 483–498
13. Chierici GL, Ciucci GM, Eva F, Long G (1967) Effect of the overburden pressure on some petrophysical parameters of reservoir rock. Proc 7th World Petrol Congr Mexico City. vol 2. Elsevier, Barking, pp 309–338
14. Davis JC (1973) Statistics and data analysis in geology. Wiley, New York
15. Dumoré JM, Schols RS (1974) Drainage capillary pressure functions and the influence of connate water. Soc Petrol Eng J (Oct): 437–444
16. Dupuy M, Pottier J (1963) Application de méthodes statistiques à l'étude détaillée des réservoirs. Rev Inst Fr Pétrole (Numéro hors série): 258–283
17. Geertsma J (1973) Land subsidence above compacting oil and gas reservoirs. J Petrol Tech (June): 734–744: Trans AIME 255
18. Hassler GL, Brunner E (1945) Measurement of capillary pressure in small core samples. Trans AIME: 114–123
19. Honarpour MM, Koederitz F, Harvey AH (1986) Relative permeability of petroleum reservoirs. CRC Press, Boca Raton
20. Houpeurt A (1974) Mécanique des fluides dans les milieux poreux. Critiques et recherches. Editions Technip, Paris
21. Hubbert MK (1956) Darcy's law and the field equations of the flow of underground fluids. Trans AIME: 222–239

22. Jorden JR, Campbell FL (1985) Well logging I – rock properties, borehole environment, mud, and temperature logging. Society of Petroleum Engineers, Dallas
23. Jorden JR, Campbell FL (1986) Well logging II – electric and acoustic logging. Society of Petroleum Engineers, Dallas
24. Katz DL, Cornell D, Kobayashi R, Poettmann FH, Vary JA, Elenbaas JR, Weinaug CF (1959) Handbook of natural gas engineering. McGraw-Hill, New York
25. Law J (1944) A statistical approach to the interstitial heterogeneity of sand reservoirs. Trans AIME 155: 202–222
26. Leverett MC (1941) Capillary behavior in porous solids. Trans AIME: 142: 152–169
27. Monicard R (1965) Cours de production – Tome I: charactéristiques des roches réservoirs – analyse des carottes. Editions Technip, Paris
28. Phillips Petroleum Co (1985/87) Ekofisk 2 and 1
29. Schlumberger Ltd (1972) Log interpretation – vol 1 – principles. Schumberger Ltd, New York
30. Schlumberger Ltd (1974) Log interpretation – vol II – applications. Schlumberger Ltd, New York
31. Schlumberger Well Services (1985) Log interpretation charts. Schlumberger Well Services, Houston
32. Serra O (1979) Diagraphies différées: bases de l'interprétation – Tome 1: acquisition des données diagraphiques. Bulletin des Centres de Recherche d'Exploration Production Elf-Aquitaine, Mémoire 1, Pau
33. Serra O (1985) Diagraphies différées: bases de l'interprétation – Tome 2: interprétation des données diagraphiques. Bulletin des centres de Recherches d'Exploration Production Elf-Aquitaine, Mémoire 7, Pau
34. Testerman JO (1962) A statistical reservoir – zonation technique. J Petrol Tech (Aug): 889–893; Trans AIME 225
35. Waal de JA, Smits RMM (1985) Prediction of reservoir compaction and surface subsidence: field application of a new model. Paper SPE 14214. presented at the 60th Annu Conf of the Society of Petroleum Engineers, Las Vegas, 22–25 Sept 1985
36. Wiborg R, Jewhurst J (1986) Ekofisk subsidence detailed and solutions assessed. Oil & Gas J (Feb 17): 47–51

EXERCISES

Exercise 3.1

The following measurements were made on a cylindrical core-plug of diameter $1''$ (2.54 cm) and length approximately $2''$ (5.1 cm):

dry weight of core-plug:	$p_{dry} = 54.821$ g
weight of water-saturated core-plug:	$p_{sat} = 60.837$ g
density of water used:	$\rho_w = 1.018$ g/cm^3
weight of saturated plug immersed in water ($\rho_w = 1.018$ g/cm^3):	$p_{imm} = 33.239$ g

A nearby core-plug was dried and crushed to a fine powder so as to obtain grains free of any pores. The rock matrix density was then measured directly:

matrix density: $\rho_{ma} = 2.640$ g/cm^3

Determine the total porosity ϕ_t, and the effective (interconnected) porosity ϕ_e of the core-plug under examination.

Solution

The weight of the saturated core-plug immersed in water is equal to the weight of the water-saturated plug minus the effect of buoyancy. This in turn is equal to the weight of the volume of water displaced by the plug (Archimedes' principle).

Now:

volume of plug $= (p_{\text{sat}} - p_{\text{imm}})/\rho_{\text{w}} = (60.837 - 33.239)/1.018 = 27.110\,\text{cm}^3 = V_{\text{b}}$.

The water can only saturate the interconnected pores. Therefore,

volume of interconnected pores

$$= (p_{\text{sat}} - p_{\text{dry}})/\rho_{\text{w}} = (60.837 - 54.821)/1.018 = 5.910\,\text{cm}^3 = V_{\text{e}} .$$

By definition we have

effective porosity $= V_{\text{e}}/V_{\text{b}} = 5.910/27.110 = 0.218 = 21.8\% = \phi_{\text{e}}$.

The total volume of the grain content of the core-plug (i.e. the rock matrix) can be calculated from the dry weight of the plug divided by the matrix density ρ_{ma}:

matrix volume $= p_{\text{dry}}/\rho_{\text{ma}} = 54.821/2.640 = 20.766\,\text{cm}^3 = V_{\text{ma}}$

So the total void volume (interconnected and non-interconnected pores) within the plug is:

total void volume $= V_{\text{b}} - V_{\text{ma}} = 27.110 - 20.766 = 6.344\,\text{cm}^3 = V_{\text{p}}$

and, by definition, the total porosity is

$\phi_{\text{t}} = V_{\text{p}}/V_{\text{b}} = 6.344/27.110 = 0.234 = 23.4\%$.

Exercise 3.2

Laboratory measurements were made on a number of cores extracted from the same reservoir interval to obtain estimates of the uniaxial compressibility c_{m}, Poisson's ratio v, and the porosity ϕ.

The average values obtained were as follows:

$c_{\text{m}} = 2.2 \times 10^{-4}\,\text{MPa}^{-1}$,

$v = 0.28$,

$\phi = 0.262$.

The compressibility of the aquifer water in contact with the reservoir is

$c_{\text{w}} = 3.5 \times 10^{-4}\,\text{MPa}^{-1}$.

If the reservoir is at 2600 m depth, has an average thickness of 100 m, and an area of 10 km^2, determine:

- the volume of water that will be expelled from 1 km^3 of aquifer if its pressure decreases by 1 MPa.
- the reduction in reservoir thickness after the pressure has declined from its initial value of 25.5 MPa to the abandonment pressure of 2 MPa.
- the maximum surface subsidence that can be expected over the centre of the reservoir as a result of producing to abandonment pressure.

Use an average density of 2220 kg/m^3 for the overlying sediments.

Solution

The vertical component σ_{zz} of the geostatic pressure can be calculated from the depth (TVD) as

$$\sigma_{zz} = g\rho_{\text{t}}D \times 10^{-6} = 9.807 \times 2220 \times 2600 \times 10^{-6} = 56.61\,\text{MPa}$$

The initial pressure of the virgin reservoir was:

$p = 25.5\,\text{MPa}$

and, if we consider c_r to be negligibly small, we can use Eq. (3.11a) to get

$$\bar{\sigma} - p = \frac{1}{3}\frac{1 + 0.28}{1 - 0.28}(56.61 - 25.50) = 18.43\ \text{MPa} .$$

If we also ignore c_r in Eq. (3.9a) we have

$$c_b = 3\frac{1 - v}{1 + v}c_m = 3\frac{1 - 0.28}{1 + 0.28}\times 2.2\times 10^{-4} = 3.71\times 10^{-4}\ \text{MPa}^{-1} .$$

From the laboratory-measured porosity $\phi_0 = 0.262$ (at $\bar{\sigma} - p = 0$) we can calculate the porosity at reservoir conditions ($\bar{\sigma} - p = 18.43\ \text{MPa}$) using Eq. (3.11b):

$$\phi = 1 - (1 - 0.262)\exp(3.71\times 10^{-4}\times 18.43) = 0.2569 = 25.7\% .$$

Again ignoring c_r, the pore compressibility c_p is given by [Eq. (3.9b)]:

$$\frac{1}{V_p}\left(\frac{\partial V_p}{\partial p}\right)_\sigma = c_p - c_r = \frac{c_f + \phi c_r}{1 - \phi} \cong \frac{c_b}{\phi} = \frac{3.71}{0.2569}\times 10^{-4} = 1.443\cdot 10^{-3}\ \text{MPa}^{-1} .$$

A reduction of 1 MPa in the average aquifer pressure will have the following consequences:

- each unit of pore volume will be reduced by an amount equal to c_p,
- each unit volume of water in the pores will expand by an amount equal to c_w.

The combined effect of these two volume changes will be the net expulsion of a volume of water equal to

$$c_t = c_p + c_w$$

per unit volume of water. Therefore if V_w is the total volume of aquifer water initially, and Δp the total reduction in the pressure, we have

$$\Delta V_w = c_t\, V_w(\Delta p)_{\bar{\sigma}} ,$$

where c_t is the "*effective compressibility*" of the aquifer.

Since 1 km³ of aquifer contains $\phi \times 10^9$ m³ of water, the volume of water expelled from 1 km³ of aquifer following a 1 MPa decrease in its average pressure will be

$$\Delta V_w = \phi\, c_t \times 10^9 = 0.2569\,(1.44\times 10^{-3} + 3.5\times 10^{-4})\times 10^9 = 460\,000\ \text{m}^3 .$$

The reduction in reservoir thickness caused by the decrease in pressure from $p_i = 25.5\ \text{MPa}$ to $p_a = 2\ \text{MPa}$ will be

$$\Delta h = c_m h \Delta p = 2.2\times 10^{-4}\times 100(25.5 - 2.0) = 0.517\ \text{m}.$$

A first approximation to the surface subsidence can be made by considering an equivalent circular reservoir of radius r_e, with the same surface area as the actual reservoir so that

$$\pi r_e^2 = 10\ \text{km}^2 .$$

Therefore,

$$r_e = 1784\ \text{m} .$$

The surface subsidence $\Delta z(0, 0)$ over the centre of the reservoir can now be approximated using the following equation:

$$\Delta z(0, 0) = 2(1 - v)\left(1 - \frac{\eta}{\sqrt{1 + \eta^2}}\right)\Delta h .$$

This equation comes from work on poroelasticity by Geertsma.[17] The term η is defined as:

$$\eta = \frac{\text{reservoir depth}}{\text{equivalent radius of reservoir}} .$$

In our case

$$\eta = \frac{2600}{1784} = 1.457 .$$

Therefore,

$$A = 2(1 - v)\left(1 - \frac{\eta}{\sqrt{1 + \eta^2}}\right) = 2(1 - 0.28)\left(1 - \frac{1.457}{\sqrt{1 + 1.457^2}}\right) = 0.253 \; .$$

and

$$\Delta z(0, 0) = A \, \Delta h = 0.253 \times 0.517 = 0.13 \text{ m}$$

Although the reservoir itself loses about 50 cm of thickness, the maximum surface subsidence is only 13 cm. The difference is "absorbed" by the elasticity of the intervening strata.

 ◇ ◇ ◇

Exercise 3.3

The average Leverett J-function $J(S_w)$ [Eq. (3.22)] was determined from a set of capillary pressure curves measured on a number of cores from a reservoir interval. Some values are tabulated below:

S_w (fraction)	$J(S_w)$ (dimensionless)
1.000	0.00
0.950	0.22
0.900	0.31
0.750	0.55
0.600	1.02
0.450	1.66
0.300	2.84
0.250	3.80
0.235	4.23
0.235	5.29

Note that $J(S_w)$ is dimensionless, because in Eq. (3.22) P_c is in Pa, σ in N/m, k in m^2 and ϕ is a fraction.

Calculate the thickness of the capillary transition zones between water and oil, and between oil and gas, using the following additional data on the reservoir:

average porosity	$\bar{\phi} = 0.20$
average permeability	$\bar{k} = 200$ md
water/oil interfacial tension	$\sigma_{wo} = 30$ dyn/cm
water/oil angle of contact	$\Theta_{wo} = 35°$
gas/oil interfacial tension	$\sigma_{go} = 5$ dyn/cm
gas/oil angle of contact	$\Theta_{go} = 10°$
oil density	$\rho_o = 850$ kg/m^3
water density	$\rho_w = 1050$ kg/m^3
gas density	$\rho_g = 120$ kg/m^3

If the reservoir contained only gas directly in contact with the aquifer, what would the thickness of the gas/water capillary transition be, given that $\sigma_{wg} = 45$ dyn/cm and $\Theta_{wg} = 0°$?

Solution

Looking at the J-function data, we can see that the irreducible water saturation $S_{iw} = 0.235$ is reached at $J(S_w) = 4.23$.

The average reservoir permeability $\bar{k}$ can be converted to SI units using the relationship:

$$1 \text{ md} = 9.869 \times 10^{-16} \text{ m}^2$$

giving:

$$\bar{k} = 200 \times 9.869 \times 10^{-16} = 1.974 \times 10^{-13} \, \text{m}^2 \, .$$

For the water/oil transition we have:

$$\sigma_{wo} = 30 \, \text{dyn/cm} = 30 \times 10^{-3} \, \text{N/m}$$

and

$$\cos \Theta_{wo} = \cos 35° = 0.819$$

so that

$$P_c(S_w) = J(S_w) \, \sigma \cos \Theta \sqrt{\frac{\phi}{k}}$$

$$= J(S_w) \times 30 \times 10^{-3} \times 0.819 \sqrt{\frac{0.2}{1.974 \times 10^{-13}}} \, \text{Pa}$$

$$= 2.473 \times 10^4 \, J(S_w) \, \text{Pa} \, .$$

The capillary pressure at the top of the transition zone will therefore be

$$P_c(S_{iw}) = 2.473 \times 10^4 \times 4.23 = 1.046 \times 10^5 \, \text{Pa}$$

From Eq. (1.4c):

$$P_c(S_{iw}) = g(\rho_w - \rho_o) \Delta D_{wo} \, ,$$

so that the transition zone thickness ΔD_{wo} will be:

$$\Delta D_{wo} = \frac{P_c(S_{iw})}{g(\rho_w - \rho_o)} = \frac{1.046 \times 10^5}{9.807(1050 - 850)} = 53.3 \, \text{m} \, .$$

Above the water/oil transition zone, $S_{iw} = \text{const.} = 0.235$, so that the following must be borne in mind when considering the gas/oil transition:

just below the gas/oil contact we have

$$S_o = 1 - S_{iw} = 0.765 \, ,$$

above the transition zone

$$S_g = 1 - S_{iw} = 0.765$$

and therefore

$$S_o = 0 \, .$$

We now follow the same procedure to determine the thickness of the gas/oil transition

$$\sigma_{go} = 5 \, \text{dyn/cm} = 5 \times 10^{-3} \, \text{N/m}$$

and

$$\cos \Theta_{go} = \cos 10° = 0.985 \, ,$$

so that

$$P_c(S_o) = J(S_w) \times 5 \times 10^{-3} \times 0.985 \sqrt{\frac{0.2}{1.974 \times 10^{-13}}} \, \text{Pa}$$

$$= 4.957 \times 10^3 \, J(S_w) \, \text{Pa} \, .$$

The capillary pressure at the top of the gas/oil transition zone will therefore be

$$P_c(S_o = 0) = 4.957 \times 10^3 \times 4.23$$

$$= 2.097 \times 10^4 \, \text{Pa}$$

which corresponds to a transition zone thickness of

$$\Delta D_{go} = \frac{P_c(S_o = 0)}{g(\rho_o - \rho_g)} = \frac{2.097 \times 10^4}{9.807(850 - 120)} = 2.9 \text{ m} .$$

For the final case where the reservoir contains only gas in contact with water, the procedure is the same:

$$P_c(S_w) = J(S_w) \times 45 \times 10^{-3} \times 1.00 \sqrt{\frac{0.2}{1.974 \times 10^{-13}}}$$

$$= 4.529 \times 10^4 \, J(S_w) \text{ Pa}$$

Therefore:

$$P_c(S_{iw}) = 4.529 \times 10^4 \times 4.23$$

$$= 1.916 \times 10^5 \text{ Pa}$$

so that:

$$\Delta D_{gw} = \frac{P_c(S_{iw})}{g(\rho_w - \rho_g)} = \frac{1.916 \times 10^5}{9.807(1050 - 120)} = 21 \text{ m} .$$

For any given set of reservoir rock characteristics, we always observe this sort of relationship between transition zone thicknesses: the water/oil transition always has the maximum thickness, followed by gas/water, with gas/oil the thinnest.

In effect, the transition zone will be thinner the smaller the interfacial tension, and the greater the difference between the densities of the two fluids.

Exercise 3.4

Show that permeability has the dimensions of length-squared, and calculate the conversion coefficient from millidarcy (md) to m^2.

Solution

Firstly, we should state Darcy's law:

$$\mathbf{u} = -\frac{[k]\rho}{\mu} \text{grad} \, \Phi \tag{3.31a}$$

and the potential per unit mass:

$$\Phi = \int_o^p \frac{dp}{\rho} - gz \tag{3.28a}$$

The dimensions of pressure are

$$p = \frac{\text{force}}{\text{area}} = \frac{m \, L \, t^{-2}}{L^2} = m \, L^{-1} t^{-2}$$

and those of density are

$$\rho = \frac{\text{mass}}{\text{volume}} = m \, L^{-3} .$$

Consequently, the dimensions of the potential per unit mass are

$$\Phi = \frac{m \, L^{-1} t^{-2}}{m \, L^{-3}} = L^2 t^{-2}$$

and therefore

$$\text{grad} \, \Phi = \frac{L^2 t^{-2}}{L} = L t^{-2} .$$

Now, viscosity μ has dimensions

$$\mu = m\,L^{-1}t^{-1}$$

and the Darcy fluid velocity u

$$u = \frac{\text{volumetric rate}}{\text{area}} = \frac{L^3 t^{-1}}{L^2} = L\,t^{-1}$$

so Eq. (3.31a) will give us:

$$L\,t^{-1} = [k]\frac{m\,L^{-3}}{m\,L^{-1}t^{-1}}L\,t^{-2}$$

$$= [k]\,L^{-1}t^{-1}$$

from which we establish the required proof:

$$k = L^2\,.$$

In the empirical Darcy system, a porous medium has a permeability of 1 Darcy (1 D) when an incompressible fluid of viscosity 1 cP flows through it at a velocity $u = 1$ cm/s under a pressure gradient of 1 atm/cm.

For an incompressible fluid, the density ρ is constant, and from Eqs. (3.28a) and (3.31a) we obtain

$$u = -\frac{k}{\mu}\mathrm{grad}(p - \rho gz)$$

where:

Parameter	SI units	Darcy units
k	m^2	D
u	m/s	cm/s
$p - \rho gz$	Pa	atm
μ	Pa s	cP

We will need the following conversion factors:

$$\text{m/s} = 10^{-2}\,\text{(cm/s)}\,,$$
$$\text{atm} = 1.033\,\text{(kg/cm}^2)\,,$$
$$\text{kg/cm}^2 = 9.80665 \times 10^4\,\text{(Pa)}\,,$$
$$\text{cP} = 10^{-3}\,\text{(Pa s)}\,.$$

In order to determine the conversion factor F from Darcy to m^2, we first write all the parameters in Darcy units, converting each in turn to SI units as follows:

$$u(\text{cm/s}) \times 10^{-2} = \frac{k(\text{darcy})\,F}{\mu(\text{cP}) \times 10^{-3}}\,\frac{p(\text{atm}) \times 1.033 \times 98066.5}{1\,(\text{cm}) \times 10^{-2}}$$

from which we get:

$$F = 9.869 \times 10^{-13}\,.$$

Therefore

$$1\ \text{Darcy} = 9.869 \times 10^{-13}\,m^2\,.$$

$$1\ \text{millidarcy} = 9.869 \times 10^{-16}\,m^2$$

You will appreciate that the m^2 is an impractically large unit to use for permeability. It is more convenient to use a submultiple – the micrometre^2 (μm^2):

$$1\,\mu m^2 = 1\,(\mu m)^2 = 10^{-12}\,m^2$$

We then have

$$1 \text{ Darcy} = 0.9869 \, \mu m^2$$
$$\cong 1 \, \mu m^2$$

$$1 \text{ millidarcy} \cong 10^{-3} \, \mu m^2 \; .$$

Exercise 3.5

A reservoir interval consists of 5 strata (zones) with the following characteristics:

Zone	Thickness (m)	Permeability (md)
1	10	100
2	1	3
3	15	1500
4	10	50
5	20	200

Calculate the average permeability to flow parallel to the strata (i.e. the horizontal permeability k_h), and in the perpendicular direction (i.e. the vertical permeability k_v). What is the average coefficient of anisotropy k_v/k_h in this reservoir?

Solution

The first step is to assemble the equations necessary for the calculation.

For the case of horizontal flow along to the strata, the total flow is the sum of the flow in each stratum, since they are in parallel.

Referring to Fig. E3/5.1, we have

k_i = the permeability of the ith zone,

h_i = the thickness of the ith zone,

$\Delta\Phi$ = the potential differential applied across a block of reservoir rock (between the inflow and outflow faces),

L = the length of this block (linear flow case),

W = the width of this block (linear flow case),

r_e = external radius of the block (radial flow case),

r_w = internal radius of the block (radial flow case),

μ = fluid viscosity,

ρ = fluid density,

q_i = flow in the ith zone,

Q = total flow.

Linear flow equation:

$$Q = \sum_{i=1}^{n} q_i = \sum_{i=1}^{n} \left[\frac{\rho \, k_i \, h_i \, W}{\mu} \frac{\Delta\Phi}{L} \right] = \frac{\rho \, W \, \Delta\Phi}{\mu L} \sum_{i=1}^{n} k_i \, h_i \; .$$

If $\bar{k}_h$ is the average horizontal permeability of the reservoir rock, then

$$Q = \rho \, \bar{k}_h \left(\sum_{i=1}^{n} h_i \right) \frac{W \, \Delta\Phi}{\mu L}$$

and therefore

$$\bar{k}_h = \frac{\displaystyle\sum_{i=1}^{n} k_i \, h_i}{\displaystyle\sum_{i=1}^{n} h_i} \; .$$

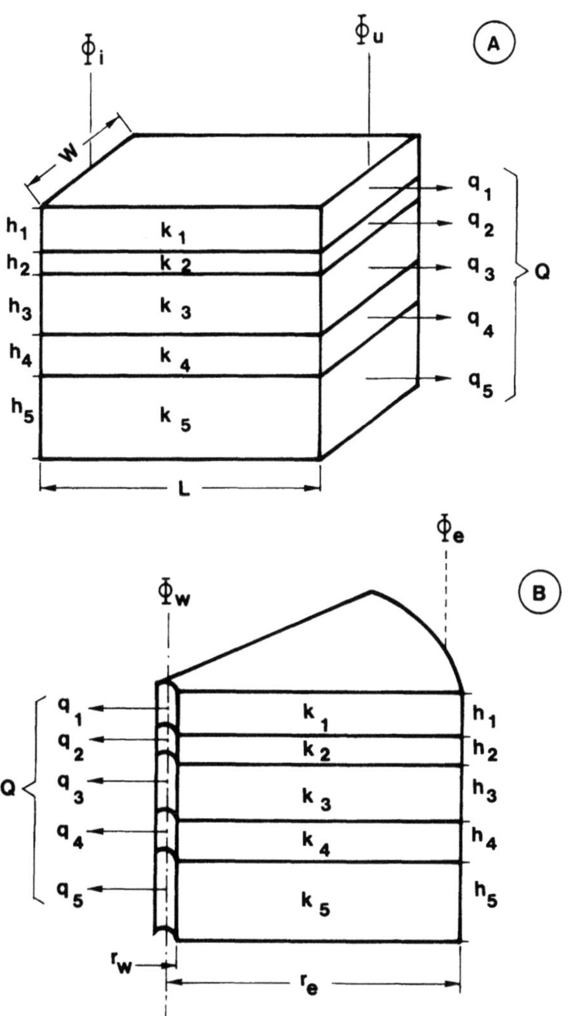

Radial flow equation:

$$Q = \sum_{i=1}^{n} q_i = 2\pi \sum_{i=1}^{n} \left[\frac{\rho\, k_i\, h_i}{\mu} \; \frac{\Delta\Phi}{\ln\,[r_e/r_w]} \right] = \frac{2\pi\,\rho\,\Delta\Phi}{\mu\,\ln\,[r_e/r_w]} \sum_{i=1}^{n} k_i\, h_i$$

and

$$Q = 2\pi\rho\,\bar{k}_h \left(\sum_{i=1}^{n} h_i \right) \frac{\Delta\Phi}{\mu\,\ln\,[r_e/r_w]}\,,$$

from which

$$\bar{k}_h = \frac{\displaystyle\sum_{i=1}^{n} k_i\, h_i}{\displaystyle\sum_{i=1}^{n} h_i}\,,$$

which is the same result as for linear flow.

The conclusion from this is that *in the case of flow parallel to the strata, the average permeability is equal to the thickness-weighted arithmetic mean of the permeabilities of the strata, regardless of whether the flow is linear or radial.*

In the perpendicular direction, the flow rate will be the same in each stratum, and the total potential differential across the reservoir will be the sum of the drops in potential in each individual stratum. The flow geometry can only be linear (assuming the well is not horizontal or highly deviated).

Referring to Fig. E3/5.2, we have:

$$\Delta\Phi = \sum_{i=1}^{n} \Delta\Phi_i = \sum_{i=1}^{n} \frac{q\,\mu}{\rho\,A}\,\frac{h_i}{k_i} = \frac{q\,\mu}{\rho\,A}\sum_{i=1}^{n}\frac{h_i}{k_i}\,,$$

where

q: the flow rate across all the strata (which are in series in this case),
A: the cross-sectional area to vertical flow (the same in all strata).

If $\bar{k}_v$ is the average permeability to vertical flow in the reservoir rock, then

$$\Delta\Phi = \frac{q\,\mu}{\rho\,A}\,\frac{\sum_{i=1}^{n} h_i}{\bar{k}_v}$$

and therefore:

$$\bar{k}_v = \frac{\sum_{i=1}^{n} h_i}{\sum_{i=1}^{n} (h_i/k_i)}\,.$$

Therefore, *in the case of strata in series, the average permeability is equal to the thickness-weighted harmonic mean of the permeabilities of the strata.*

For the particular reservoir in question, we have

$$\bar{k}_h = \frac{10\times100 + 1\times3 + 15\times1500 + 10\times50 + 20\times200}{10 + 1 + 15 + 10 + 20} = 500\text{ md}$$

$$\bar{k}_v = \frac{10 + 1 + 15 + 10 + 20}{\dfrac{10}{100} + \dfrac{1}{3} + \dfrac{15}{1500} + \dfrac{10}{50} + \dfrac{20}{200}} = 75.3\text{ md}$$

and therefore

$$\frac{\bar{k}_v}{\bar{k}_h} = 0.15.$$

This high degree of anisotropy ($\bar{k}_v \ll \bar{k}_h$) is due largely to the presence of zone 2, which is thin (1 m) but has a very low permeability (3 md) relative to the others.

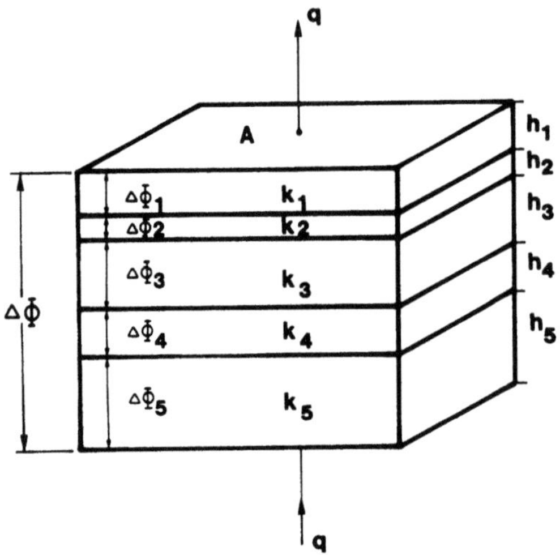

Fig. E3/5.2

◇ ◇ ◇

Exercise 3.6

A reservoir has been developed with a uniform well spacing of 40 acres. Evaluate the improvement in productivity that would be obtained by completing and perforating each well with $12\frac{3}{8}''$, $16''$, or $24''$ casing, compared to the standard $7\frac{5}{8}''$ casing.

Solution

The term "spacing" represents the drainage area of each well.
    Since

$$1 \text{ acre} = 4046.856 \text{ m}^2$$

a spacing of 40 acres corresponds to a drainage area of

$$A = 40 \times 4046.856 = 161\,874.24\,\text{m}^2 \ .$$

The distance d between wells (Fig. E3/6.1) must therefore be

$$d = \sqrt{A} = 402.35 \text{ m} \ .$$

For the purpose of the calculation, we assume an equivalent circular drainage area of radius r_e for each well, such that

$$r_e = \sqrt{\frac{A}{\pi}} = 227 \text{ m} \ .$$

Under steady state flow conditions we can write:

$$q = \frac{2\pi k h}{\mu}\,\rho_o \frac{\Delta\Phi}{\ln\left(r_e/r_w\right)} \ .$$

Since $\Delta\Phi$, k, h, μ and ρ_o are constant, it follows that

$$q\ln\frac{r_e}{r_w} = \frac{2\pi k h}{\mu}\,\rho_o\,\Delta\Phi = C \ .$$

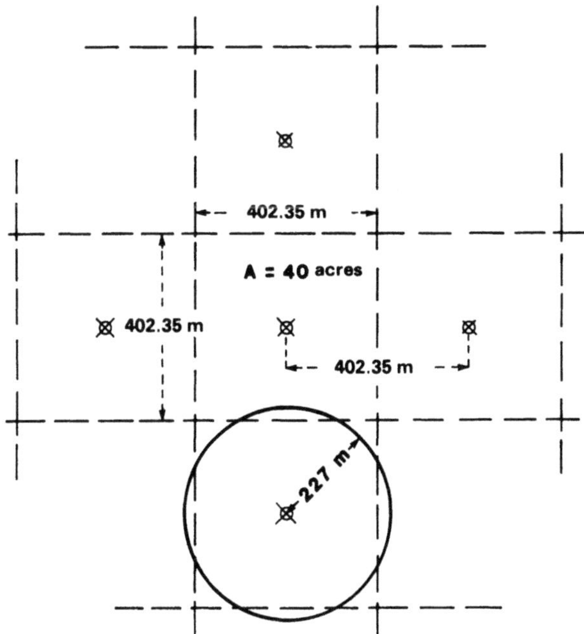

Fig. E3/6.1

Taking each casing in turn, with the conversion $1'' = 2.54$ cm:

$$r_w(7\tfrac{5}{8}'') \;=\; 9.684 \text{ cm} ,$$

$$r_w(12\tfrac{3}{8}'') = 15.716 \text{ cm} ,$$

$$r_w(16'') \;= 20.320 \text{ cm} ,$$

$$r_w(24'') \;= 30.480 \text{ cm} ,$$

so that

$$q(12\tfrac{3}{8}'')\ln\frac{227}{15.716\times 10^{-2}} = q(7\tfrac{5}{8}'')\ln\frac{227}{9.684\times 10^{-2}} ,$$

from which we get that

$$q(12\tfrac{3}{8}'') = 1.07\times q(7\tfrac{5}{8}'')$$

and, similarly

$$q(16'') = 1.11\times q(7\tfrac{5}{8}'')$$

$$q(24'') = 1.17\times q(7\tfrac{5}{8}'') .$$

Interestingly, the gain in productivity resulting from completing the wells with large diameter casing is marginal. Even at 24″ diameter (practically a surface casing), the increase is less than 20%.

$$\diamond\,\diamond\,\diamond$$

Exercise 3.7

A well completed with a $7\tfrac{5}{8}''$ casing in a reservoir with an effective permeability to oil $kk_{ro} = 12$ md is stimulated with a matrix acid treatment (flushing of the near wellbore rock with an acid and surfactant combination).

Calculations predict that the local permeability to oil should have been increased to $kk_{ro} = 250$ md out to a radius of 10 m from the axis of the well.

If the well has a drainage radius $r_e = 227$ m, estimate the increase in productivity that can be expected from this stimulation (assume it is the same well as in Ex. 3.6).

Solution

Fig E3/7.1 represents the situation after stimulation.
Between the radii,

$$r_w = \tfrac{1}{2}(7\tfrac{5}{8}'') = 9.684 \text{ cm}$$

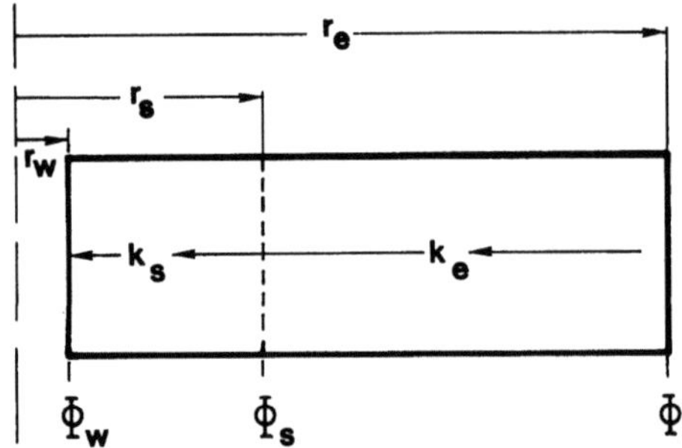

Fig. E3/7.1

and

$$r_s = 10 \text{ m} = 1000 \text{ cm}$$

the permeability to oil is 250 md.

Between the radii,

$$r_s = 10 \text{ m}$$

and

$$r_e = 227 \text{ m} ,$$

the permeability to oil is 12 md.

Therefore, the fluid will traverse an annulus of rock where $kk_{ro} = 12$ md, and then a concentric annulus where $kk_{ro} = 250$ md, as it moves from the extremity of the drainage area to the wellbore.

In order to simplify the calculation, we can assume that the potential is constant with time at the outer boundary, so that we have steady state flow. We will see later in Chap. 5 that a rigorous treatment would require using the flow equation for pseudo-steady state flow.

Since the two annuli are in series as far as the oil flow is concerned, we have (Fig. E3/7.1):

$$\Phi_e - \Phi_w = (\Phi_e - \Phi_s) + (\Phi_s - \Phi_w)$$

$$= \frac{q_s B_o \mu_o}{2\pi h \rho_o} \frac{1}{k_e k_{ro}} \ln \frac{r_e}{r_s} + \frac{q_s B_o \mu_o}{2\pi h \rho_o} \frac{1}{k_s k_{ro}} \ln \frac{r_s}{r_w} , \tag{1}$$

where

q_s = the oil production rate after simulation,
h = pay thickness,
k_e = the permeability of the reservoir rock (unstimulated),
k_s = the permeability of the rock in the acid-stimulated zone,

If the oil rate before stimulation was q_o, then for the same potential drop, or drawdown, $(\Phi_e - \Phi_w)$, we have

$$\Phi_e - \Phi_w = \frac{q_o B_o \mu_o}{2\pi h \rho_o} \frac{1}{k_e k_{ro}} \ln \frac{r_e}{r_w} . \tag{2}$$

Equating the right-hand terms of Eqs. (1) and (2), and simplifying, we obtain

$$q_s \left(\frac{1}{k_e} \ln \frac{r_e}{r_s} + \frac{1}{k_s} \ln \frac{r_s}{r_w} \right) = q_o \frac{1}{k_e} \ln \frac{r_e}{r_w}$$

or:

$$\frac{q_s}{q_o} = \frac{\ln \dfrac{r_e}{r_w}}{\ln \dfrac{r_e}{r_s} + \dfrac{k_e}{k_s} \ln \dfrac{r_s}{r_w}} .$$

In the present case, this works out as

$$\frac{q_s}{q_o} = \frac{\ln \dfrac{227}{9.684 \times 10^{-2}}}{\ln \dfrac{227}{10} + \dfrac{12}{250} \ln \dfrac{10}{9.684 \times 10^{-2}}} = 2.32 .$$

In other words, in this example the acid stimulation has (theoretically) more than doubled the well productivity – an increase of 132% to be precise.

$$\diamond \diamond \diamond$$

Exercise 3.8

An oil reservoir situated at an average depth of 2650 m below sea level, is in communication with an aquifer which outcrops 1500 m *above* sea level and is fed by meteoric water.

The aquifer has an average thickness of 100 m and a permeability of 150 md. Its lateral extent is limited by two parallel faults 1.5 km apart.

There is a distance of 10 km (measured along the aquifer) between the reservoir and the point at which water feeds into the aquifer.

The reservoir is on production, and its current pressure is 30 MPa.

Estimate the rate of influx of water into the reservoir from the aquifer.

Solution

The aquifer geometry is approximately linear, with the following dimensions:

width: $\quad W = 1.5\,\text{km} = 1.5 \times 10^3\,\text{m}$
thickness: $h = 100\,\text{m}$,

so that

cross-sectional area: $A = 1.5 \times 10^5\,\text{m}^2$
length: $\qquad\quad L = 10\,\text{km} = 10^4\,\text{m}$
permeability: $\qquad k = 150\,\text{md} = 1.48 \times 10^{-13}\,\text{m}^2$.

The aquifer water is fresh. If we ignore any geothermal temperature effects we can assume

density of water: $\rho_\text{w} = 1000\,\text{kg/m}^3$

We will work with a coordinate system where the sea level corresponds to zero depth and the z-axis increases vertically downwards:

elevation of aquifer feed point: $D_\text{w} = -1500\,\text{m}$,
elevation of reservoir: $\qquad\qquad D_\text{res} = +2650\,\text{m}$.

The pressure at the feed point of the aquifer is atmospheric (101 325 Pa), so that with the assumption of a constant water density, the potential Φ_1 at this point is

$$\Phi_1 = \frac{p_\text{a}}{\rho_\text{w}} - g\,D_\text{w} = \frac{101\,325}{1000} + 9.807 \times 1500 = 14\,812\,\text{m}^2/\text{s}^2$$

and therefore

$$\rho_\text{w}\,\Phi_1 = 1000 \times 14\,812\,\text{Pa} = 14.812\,\text{MPa}$$

At the level of the reservoir, the potential Φ_2 in the aquifer is

$$\Phi_2 = \frac{p_\text{res}}{\rho_\text{w}} - g\,D_\text{res} = \frac{30 \times 10^6}{1000} - 9.807 \times 2650 = 4011.5\,\text{m}^2/\text{s}^2$$

and therefore

$$\rho_\text{w}\,\Phi_2 = 1000 \times 4011.5\,\text{Pa} = 4.0115\,\text{MPa},$$

so that

$$\rho_\text{w}(\Phi_1 - \Phi_2) = 14.812 - 4.0115 = 10.8005\,\text{MPa} = 110.13\,\text{kg/cm}^2.$$

The water flow rate can now be calculated using the linear flow equation:

$$q_\text{w} = \frac{k}{\mu_\text{w}}\,A\,\frac{\rho_\text{w}(\Phi_1 - \Phi_2)}{L}$$

$$= \frac{1.48 \times 10^{-13}}{10^{-3}}\,1.5 \times 10^5\,\frac{10.8 \times 10^6}{10^4} = 0.024\,\text{m}^3/\text{s}$$

$$= 756\,000\,\text{m}^3/\text{yr}.$$

The corresponding "Darcy velocity" of the water in the aquifer is

$$u_\text{w} = q_\text{w}/A = 5\,\text{m/yr}.$$

From this we can conclude that at the current reservoir pressure, the aquifer can supply a drive energy equivalent to the influx of $756\,000$ m^3/yr of water.

$$\Diamond \Diamond \Diamond$$

Exercise 3.9

Use the correlations provided in Sect. 3.5.5, to calculate imbibition relative permeability curves $k_{ro}(S_w)$ and $k_{rw}(S_w)$, for a core containing primary porosity with $S_{iw} = 0.15$ and $S_{or} = 0.20$.

If oil and water are made to flow simultaneously through the core, such that the ratio $q_w/q_o = 10$, determine:

1. The equilibrium water saturation at this flow ratio,
2. The volumetric fraction of oil recovered when $q_w/q_o = 10$, if the viscosity ratio $\mu_o/\mu_w = 1$, 10 and 100.

Solution

$k_{rw}(S_w)$ can be calculated with Eq. (3.57a) and $k_{ro}(S_w)$ with Eqs. (3.57b) and (3.57c), with $S_w = 0.15$ and $S_{or} = 0.20$.

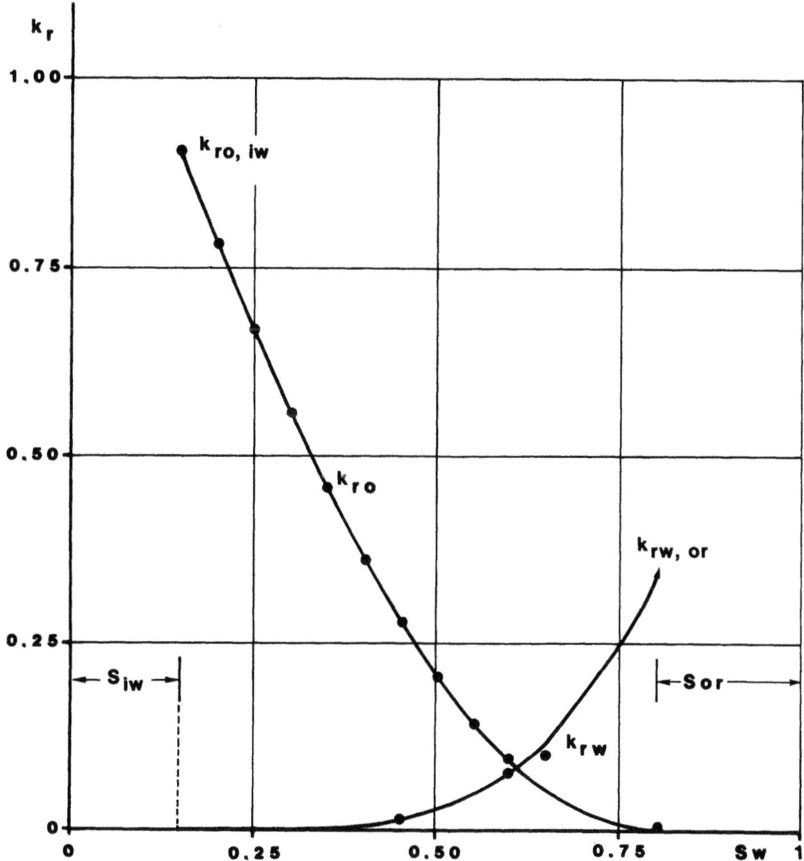

Fig. E3/9.1

We will get the following values:

S_w	S_o	S_{oF}^*	k_{rw}	k_{ro}	k_{ro}/k_{rw}
0.15	0.85	0.95343	zero	0.9070	∞
0.20	0.80	0.89207	1.20×10^{-5}	0.7865	6.57×10^4
0.25	0.75	0.83040	1.92×10^{-4}	0.6697	3.50×10^3
0.30	0.70	0.76837	9.70×10^{-4}	0.5587	576
0.35	0.65	0.70588	3.07×10^{-3}	0.4552	149
0.40	0.60	0.64284	7.48×10^{-3}	0.3605	48.2
0.45	0.55	0.57908	0.0155	0.2759	17.8
0.50	0.50	0.51439	0.0287	0.2022	7.03
0.55	0.45	0.44844	0.0490	0.1399	2.85
0.60	0.40	0.38071	0.0786	0.0894	1.14
0.65	0.35	0.31029	0.1197	0.0505	0.422
0.70	0.30	0.23529	0.1753	0.0230	0.131
0.75	0.25	0.15068	0.2483	0.0063	0.0255
0.80	0.20	zero	0.3420	zero	zero

The curves are plotted in Fig. E3/9.1.

Assumig the core to be horizontal for the experiment ($z = $ const.), and ignoring capillary effects ($P_c = p_o - p_w = 0$), we can write

$$q_w = - \frac{k\,k_{rw}}{\mu_w} A \frac{dp}{dx} ,$$

$$q_o = - \frac{k\,k_{ro}}{\mu_w} A \frac{dp}{dx} ,$$

and therefore, dividing one equation into the other and rearranging

$$\frac{k_{ro}}{k_{rw}} = \frac{q_o}{q_w} \cdot \frac{\mu_o}{\mu_w} .$$

In our case, $q_o/q_w = 0.1$, so we can calculate

$\dfrac{\mu_o}{\mu_w}$	$\dfrac{k_{ro}}{k_{rw}}$	S_w	E_D
1	0.1	0.71	0.659
10	1	0.61	0.541
100	10	0.47	0.376

where the values of S_w were determined by interpolation of the curve $k_{ro}/k_{rw} = f(S_w)$ in Fig. E3/9.2, and E_D was calculated using the equation:

$$E_D = \frac{\text{volume of oil produced}}{\text{initial volume of oil}} = \frac{S_w - S_{iw}}{1 - S_{iw}} .$$

This means that for a given water/oil ratio (WOR) – 10 in this case – the fraction of oil recovered is reduced markedly by an increase in the oil/water viscosity ratio μ_o/μ_w.

Note that in this example, we are assuming that the oil and water flow simultaneously through the core with a fixed ratio q_w/q_o.

The displacement of oil by water will be dealt with in Chap. 11.

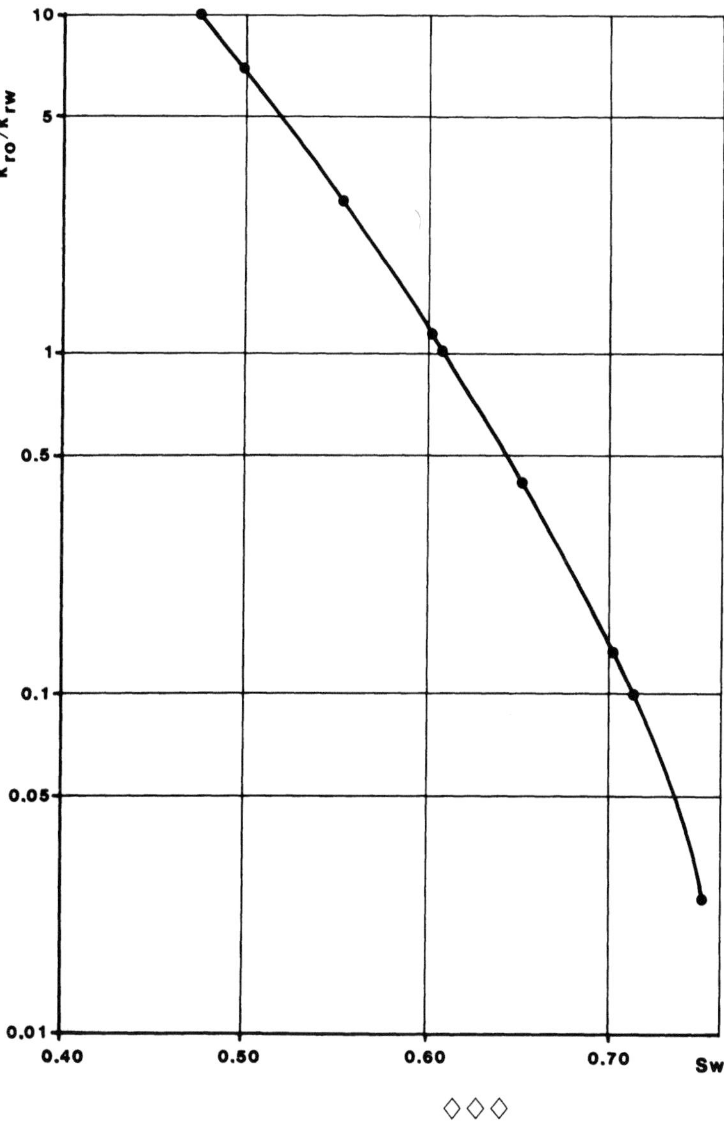

Fig. E3/9.2

◇ ◇ ◇

Exercise 3.10

Cores were taken from the same zone (or depositional unit) in each of 6 wells, and permeability measurements were made on about 1000 core plugs. The log-probability distribution shown in Fig. E3/10.1 was observed.

As described in Sect. 3.5.1.7, determine the mean permeability $\bar{k}$ of the unit, its standard deviation σ and the permeability variation as defined by Law.

Solution

Referring to Sect. 3.5.1.7, the mean permeability for a sample population corresponds to a cumulative probability $F = 50\%$.

From Fig. E3/10.1, for $F = 50\%$:

$$\bar{k} = k_{50} = 56.2 \text{ md} .$$

At $F = 84.14\%$:

$$k_{84.14} = 7.2 \text{ md} .$$

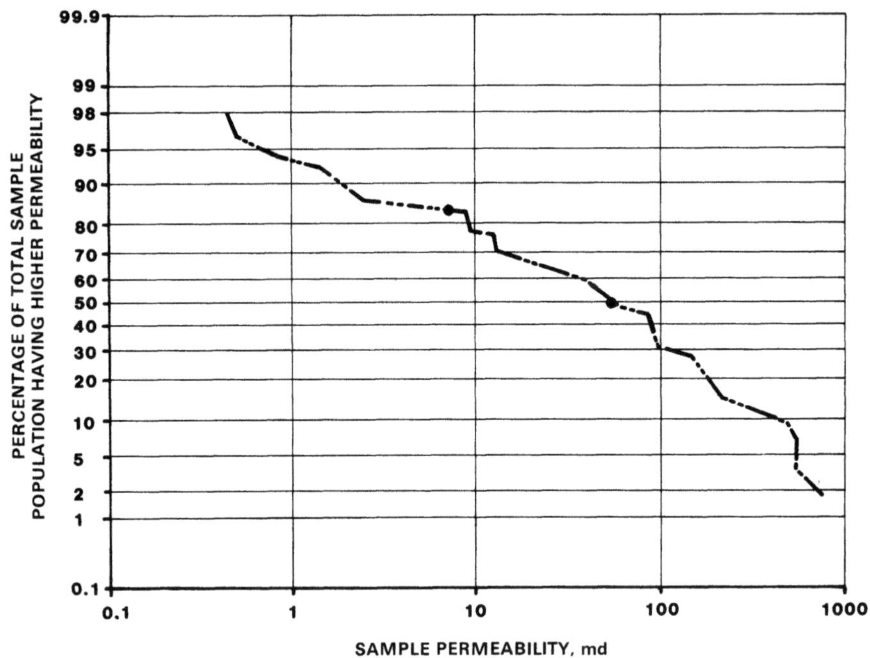

Fig. E3/10.1

Therefore,

$$\sigma = \log \frac{k_{50}}{k_{84.14}} = \log \frac{56.2}{7.2} = 0.892 \ .$$

So 68.3% of the samples $[= 2 \times (84.14 - 50)]$ have permeabilities in the range

$$(\log k_{50} - \sigma) \leq \log k \leq (\log k_{50} + \sigma) \ ,$$

that is

$$7.2 \ \text{md} \leq k \leq 438 \ \text{md} \ .$$

The permeability variation is then [Eq. (3.45)]:

$$\mathscr{V} = \frac{k_{50} - k_{84.14}}{k_{50}} = \frac{56.2 - 7.2}{56.2} = 0.872 \ .$$

The high value of σ (and consequently of $\mathscr{V}$) classifies this depositional unit as "strongly heterogeneous".

4 The Evaluation of Oil and Gas Reserves

4.1 Introduction

An essential factor in planning the development and production of a reservoir is a reliable estimate of the volume of hydrocarbons in place (the *resources*), and how much can be recovered.

At that early stage, since we obviously do not yet have any data on the behaviour of the reservoir under producing conditions, the only approach that can be used is based on *volumetrics*.

We will see shortly that later in the life of the field, this initial reserves evaluation made by the volumetric method can be compared and refined by *material balance* calculations based on the production history. In general, the longer the period covered by this production data, the better the estimate will be.

Before describing the volumetric method, we will list the definitions of some important terms. These are based on the definitions[7] approved by the committee of the Society of Petroleum Engineers on 27th February 1987, and accepted by most of the petroleum industry worldwide.

4.2 Reserves

Reserves are defined as the estimated quantities of crude oil, natural gas, gas condensate, liquids recovered from natural gas, and associated substances (such as sulphur from hydrocarbons containing H_2S), that it is considered commercially viable to recover from a given accumulation, *beginning at a certain future date, under specified economic conditions, using current technology, and subject to present-day legal restrictions.*

The estimation of reserves is based on the interpretation of geological and/or engineering data available at the time.

In general, the reserves estimate is revised during the course of production, as progressively more geological and other data on the reservoir becomes available, or when a change of economic conditions warrants a review (as per the SPE definition).

The reserves figure does not include the volume of oil, natural gas, condensate or liquids from natural gas that may be held in field storage tanks or underground. Financial and other specialised reports often include an estimate of the expected reduction in reserves attributable to supplying the field with fuel and/or spillage and other losses during exploitation.

Reserves which can be produced by the *natural drive mechanisms* of the reservoir are distinguished from those that require *improved recovery* methods. The term "improved recovery" encompasses all methods of maintaining or supplementing the natural energy of the reservoir in order to increase the total hydrocarbon recovery. These methods include: (1) pressure maintenance, (2) gas recycling, (3) water flooding, (4) thermal processes (5) surfactant and polymer injection, (6) displacement of oil by miscible or immiscible fluids.

Any reserves estimate carries a degree of uncertainty, which depends on the quality and availability of data needed for the computation, as well as the way in which these data have been processed, and the nature of the computation itself.

This uncertainty is presented in terms of two categories of reserves: *"proven reserves"* and *"non-proven reserves"*. The probability of successfully producing non-proven reserves is less than for proven reserves.

The non-proven category is further subdivided into two levels of probability: *"probable reserves"* and *"possible reserves"*.

4.2.1 Proven Reserves

Proven reserves are those for which recovery can be predicted with a reasonable degree of certainty under specified economic conditions. These economic considerations include, among others, the sale price and cost of production of the hydrocarbons. Proven reserves may already be under development, or still to be developed.

In general, reserves will be classified as "proven" if production or formation tests show the economic feasibility of producing them. *The term "proven" refers to the estimated volume of the total reserves, and is not confined to the productivity of a single well or even of a single reservoir unit.* In certain cases, reserves might be classified as proven solely on the basis of well (or other) logs, or core analysis, which indicate that the reservoir contains hydrocarbons, and can be considered similar to nearby reservoirs which are either producing, or have been tested as potential producers.

The lateral extent of a proven reserve covers: (1) the area delimited by the wells drilled and by the water/oil (or water/gas) and gas/oil contacts if they exist, and (2) the area not yet drilled that might reasonably be considered economically productive on the basis of geological and other data available. In the absence of data about the water/oil or water/gas contacts, the extent of the proven reserves is limited to the deepest level in the structure at which commercially producible hydrocarbon has been identified. (The expressions "oil down to . . . " or "gas down to . . ." are used in this context.) This assumes, of course, that no unambiguous alternative, or even contradictory, evidence is provided by other data.

In order for reserves to be classified as proven, all the necessary facilities for producing the hydrocarbons and transporting them to the point of sale must either be in place, or there must be a reasonable likelihood that such facilities will be installed in the future, backed up by a formal commitment to do so.

In general, *non-developed proven reserves* are reserves contained in as-yet undrilled areas which meet the following conditions: (1) they lie within one well spacing of wells producing from the same formation, (2) they are with reasonable certainty within the limits of the proven productive area of the formation, (3) the planned well locations satisfy local regulations, and (4) it is reasonably certain that wells will be drilled in these locations. The key factor in the definition of a non-developed proven reserve is that the interpretation of data from existing wells drilled into the formation in question (i.e. in the proven reserve) suggests that there is lateral continuity and producible hydrocarbons beyond one well spacing from these wells.

Reserves which can be produced by means of established improved recovery methods are classified as proven when: (1) pilot production tests, or the pressure response to a programme of injection in the same reservoir (or in an area immediately adjacent with similar rock and fluid characteristics), provide support for the engineering analysis upon which the project is based; and (2) it is reasonably certain that such a project will go ahead.

Reserves intended for improved recovery by methods which have *not* yet established commercially successful track records may only be classified as proven if: (1) a representative pilot test, or an existing injection programme in the reservoir in question, have yielded favourable results, on which the feasibility analysis for the whole reservoir will be based, and (2) it is reasonably certain that the project will go ahead.

4.2.2 Non-Proven Reserves

Although the evaluation of both non-proven and proven reserves is based on similar kinds of geological and engineering data, a greater degree of uncertainty inherent in the technical, contractual, legislative and commercial aspects of a given field would warrant the classification "non-proven". Such reserves can be estimated assuming future economic conditions which differ significantly from those prevailing at the time of the evaluation.

Non-proven reserve estimates can be made for long-term planning purposes, or for special projects, but might not be routinely compiled.

They must *under no circumstances* be added to the proven reserves, owing to the different levels of uncertainty involved in each estimate.

There are two categories of non-proven reserves: *probable* and *possible.*

4.2.2.1 Probable Reserves

The likelihood of being able to produce effectively from probable reserves is less than that for proven reserves; the evaluation will however show that successful exploitation is more probable than unsuccessful exploitation.

In general, probable reserves include: (1) reserves where currently available geological evidence is not sufficient by itself to justify classification as proven, but where normal development of the field will yield production data which can be expected to indicate proven reserves; (2) reserves in formations which appear to be productive on the evidence of well logs, but for which there is no core or test data, and which cannot be compared with similar reservoirs in the area which are proven or are already on production; (3) additional reserves that could be recovered by in-fill drilling, which although technically classifiable as proven, cannot be so classified because, at the time of evaluation, approval for the reduced well spacing had not been granted by the regulatory body; (4) reserves found in part of a formation which has been certified as proven in another area of the reservoir, which appear to be isolated from this proven area by faulting, and seem from geological evidence to be structurally higher; (5) reserves whose recovery depends on the success of workover, stimulation or restimulation of wells, or modification of the well completion or other facilities, where these procedures have not produced the desired results in comparable wells drilled in similar reservoirs; and (6) additional reserves identified in a proven and productive reservoir, where an alternative interpretation of production or volumetric data indicates the existence of significantly more reserves than were originally classified as proven.

4.2.2.2 *Possible Reserves*

Possible reserves have a greater degree of uncertainty associated with their evaluation and, as a consequence, they cannot be determined as productive or not.

In general, possible reserves can include: (1) reserves whose presence is suggested by a structural or stratigraphic extrapolation (based on geological and geophysical data) outside an area which has been classified as *probable*; (2) reserves in formations which appear to be hydrocarbon-bearing according to log and core data, but which may not be productive on a commercial scale; (3) additional reserves which could be realised by means of in-fill drilling, but there are technical uncertainties concerning this aspect of the operation; (4) reserves which could be retrieved by improved recovery methods, where a pilot or main project is planned but has not yet been started, and the characteristics of the reservoir, rock and fluids give rise to reasonable doubt as to the commercial viability of such an attempt at improved recovery; and (5) reserves situated in an area which is part of a formation classed as productive elsewhere in the reservoir, but which appears to be separated from those proven reserves by faults, and which is indicated by geological evidence to be structurally lower than the productive area.

4.2.3 Classification of Reserves According to Production Status

The classification of reserves according to production status depends on the current phase of development of the wells and/or the reservoir.

4.2.3.1 *Developed Reserves*

Developed reserves are those which are expected to be produced from existing wells (including wells which have still to be perforated and are, therefore, not as yet on production). Reserves which can be extracted by improved recovery methods are only considered as "developed" when all the necessary equipment has been installed, or when the costs of installing such equipment are negligible.

There are two categories of developed reserves: *producing reserves* and *non-producing reserves*.

Producing Reserves: this category represents the reserves still to be recovered by continued production from a field which is already producing at the time of the evaluation. Reserves dependent on improved recovery are only considered to be "on production" when the recovery project has started.

Non-producing Reserves: these are reserves pertaining to shut-in wells and wells awaiting workover (behind-pipe reserves).

In the case of shut-in wells, production of these reserves from intervals already perforated had not yet started at the time of evaluation, or had been shut down for commercial or mechanical reasons, or was awaiting connection of surface piping; and no future date has been set to initiate commercial production.

In the case of a well awaiting workover, these reserves, contained in hydrocarbon bearing formations around the well, will eventually be produced but require further work on the completion, or a recompletion, before being ready to produce.

4.2.3.2 Non-Developed Reserves

Non-developed reserves are expected to be produced: (1) from new wells to be drilled in an area of the reservoir where no other wells have been drilled; (2) from wells which will be drilled to a greater depth so as to penetrate a separate, deeper reservoir unit, or (3) where costs are relatively high for (a) the recompletion of an existing well, or (b) the installation of production or transportation equipment for primary or improved recovery.

4.3 Basic Data for the Volumetric Calculation of Reserves

4.3.1 Equations Used

The following equations are used in the volumetric calculation of reserves:

For oil reservoirs:

$$N_{pa} = \iint_A \frac{h_n \phi (1 - S_w)}{B_o} E_{R,o} \, dx \, dy = \frac{A \bar{h}_n \bar{\phi} (1 - \bar{S}_w)}{\bar{B}_o} \bar{E}_{R,o} \, , \tag{4.1a}$$

For gas reservoirs:

$$G_{pa} = \iint_A \frac{h_n \phi (1 - S_w)}{B_g} E_{R,g} \, dx \, dy = \frac{A \bar{h}_n \bar{\phi} (1 - \bar{S}_w)}{\bar{B}_g} \bar{E}_{R,g} \, , \tag{4.1b}$$

where N_{pa} and G_{pa} are the oil or gas reserves (the estimated cumulative produced hydrocarbons at the time of abandonment of the reservoir, when the reserves are exhausted), measured at stock tank (i.e. standard) conditions of 288 K and 0.1013 MPa.

h_n, ϕ and S_w are, respectively, the *net* thickness of the hydrocarbon-bearing formation (pay), the effective porosity and the water saturation: all these parameters may vary spatially, and are therefore functions of the coordinates (x, y) within the area A of the reservoir.

B_o and B_g are the volume factors of the oil ($B_o \equiv B_{of}$, Sect. 2.3.2.1) and gas (Sect. 2.3.1.1) and E_R is the recovery factor. These too are a function of position. The bar over each term in the equations indicates the *average* value.

A is the area of the reservoir. It will correspond to the proven, probable or possible area depending on the category of reserves to be evaluated.

We shall now examine in some detail how to determine the values of each of the terms on the right-hand side of Eqs. (4.1a), and (4.1b).

4.3.2 Reservoir Area, A

The reservoir area, A, is evaluated separately for each non-connected productive interval (pool) present in the reservoir, and sometimes even for each depositional unit (zone) present within each pool.

Obviously, the area is defined according to the reserves to which it pertains – proven, probable or possible – as described in Sect. 4.2.

To construct a map of the hydrocarbon-bearing area as a function of depth we need to ascertain the shallowest and deepest point at which hydrocarbons have been encountered in each well. Well logs and cores are the main source of this information. For regions where no wells have been drilled, seismic maps are referred to (these must be migrated to *depth* – use of the raw transit time data can lead to serious errors).

Figure 4.1 is a typical map derived in this way. A planimeter can then be used to construct a diagram such as Fig. 4.2, which provides a first estimation of the vertical distribution of reservoir volume.

4.3.3 Net Pay Thickness

Within the reservoir interval, there are almost always intercalations of shale or other rock, which, owing to their low ϕ and k or high S_w, do not contain recoverable reserves.

We should therefore subtract the cumulative thickness of these non-productive strata from the gross thickness h_t of the reservoir to obtain the net pay thickness h_n.

The estimation of h_n, and the *net to gross ratio* h_n/h_t, is a critical stage in the evaluation, as it will have major implications for the volume of reserves.

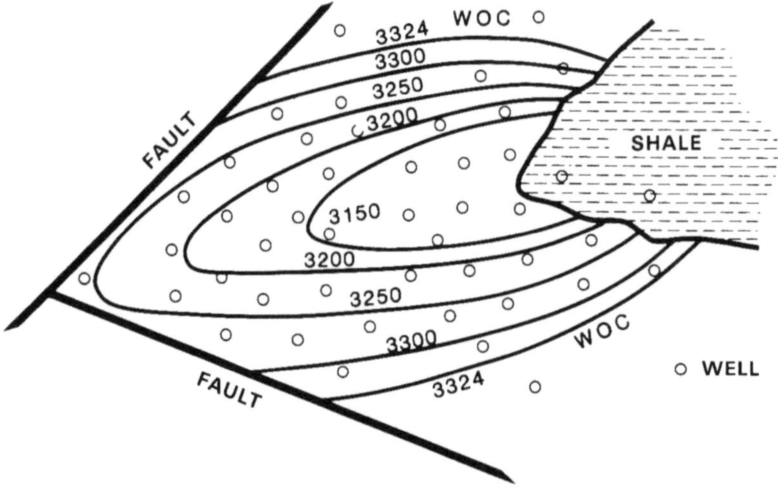

Fig. 4.1. *Isobath contour map of the top of a hydrocarbon-bearing pool*

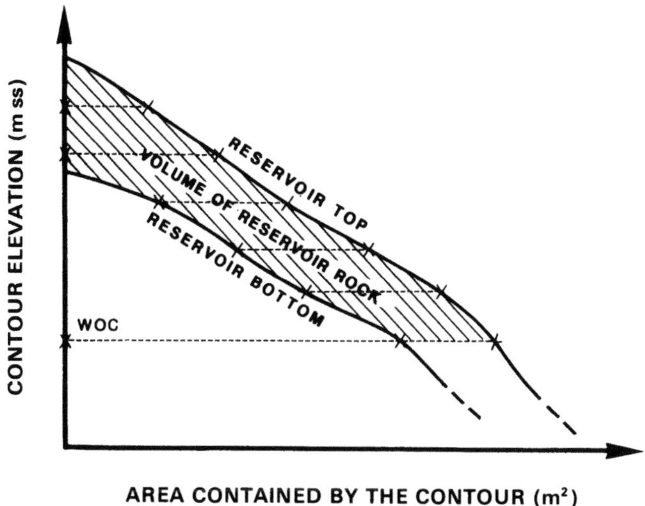

Fig. 4.2. *The hydrocarbon-bearing volume plotted against depth. From Ref. 4, reprinted with permission of Kluwer Academic Publishers and of Professors Archer and Wall*

If a cluster analysis is available (Sect. 3.6.3), it is a straightforward matter to distinguish pay and non-pay zones, and their distribution through the reservoir interval.

Otherwise, the non-pay footage will have to be identified from cores, using porosity (usually very low), and from log characteristics (for example, the SP, GR, ML, PL and MLL to pick out the shales).

A logical criterion for the definition of non-producing intervals would be to fix a lower limit to the permeability – the *permeability cut-off*, k_ℓ. Wherever $k < k_\ell$, the rock would be non-pay. However, we have seen, logs cannot provide reliable quantitative estimates of permeability. Some form of correlation $k = f(\phi)$ is therefore needed (Sect. 3.5.1.8), so that from k_ℓ a porosity cut-off ϕ_ℓ can be defined.

Using the computed porosity curve from a Computer Processed Interpretation (CPI) of the well logs, the net pay thickness h_n can then be calculated immediately by summing the intervals where $\phi > \phi_\ell$.

This method, however, does discriminate against certain rocks which, although of very low k, contain movable hydrocarbons.

Although this type of interval does not produce freely under normal drawdown conditions because of its low permeability, it can contribute to the producible reserves through imbibition processes which occur in the reservoir at large, as described in Sect. 3.4.4.3

Use of a cut-off based on log data may therefore underestimate h_n.

The isopay map shown in Fig. 4.3 can now be drawn by plotting the values of h_n from each well onto a map of the reservoir. The net volume of hydrocarbon-bearing rock will be

$$V_R = \iint_A h_n \, dx \, dy \, . \tag{4.2}$$

Several methods are employed to perform this calculation.

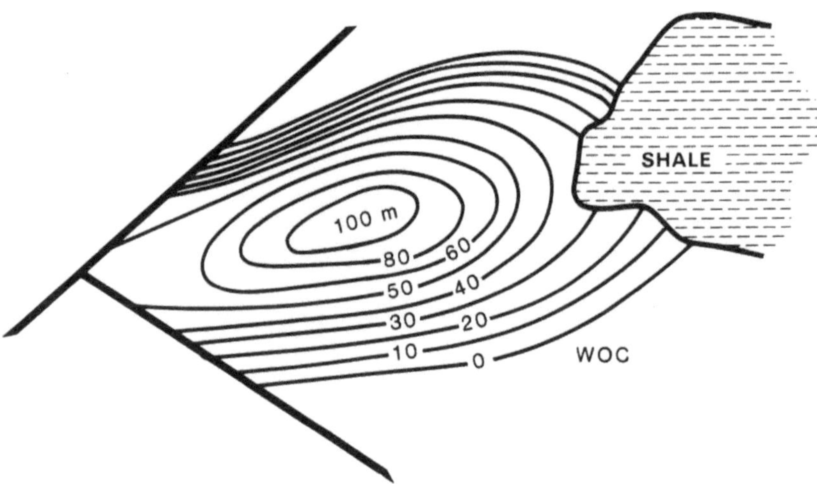

Fig. 4.3. *Isopay map for a hydrocarbon-bearing pool*

The most rigorous approach is a by-product of the initialisation phase of numerical simulators used in reservoir modelling. The hydrocarbon-bearing volume is subdivided into parallelepiped blocks using a fine grid in orthogonal coordinates (x, y). The volume of each block (base area $\times \bar{h}_n$) is calculated, then summed over all blocks.

A second procedure is to planimeter the area contained within each isopay contour, and to plot each area against the corresponding pay thickness (Fig. 4.4). The total net pay volume V_R is then obtained by integrating the resulting curve.

Yet another approach involves dividing the isopay map into a number of overlying trapezoids. The volume of the trapezoid between the isopay contours $h_{n,i}$ and $h_{n,i+1}$ is

$$V_{i,i+1} = \frac{A_{i+1} + A_i}{2} \left(h_{n,i+1} - h_{n,i} \right),$$

(4.3a)

i.e. the average area multiplied by the incremental thickness.

The topmost trapezoid represents a special case and is treated as a pyramid:

$$V_{top} = A_{min} \frac{h_{max} - h_{i,max}}{3},$$

(4.3b)

A_{min}, its base area, being of course the smallest isopay area on the map.

4.3.4   Porosity ϕ, and Average Porosity $\bar{\phi}$

Using the (CPI) log interpretations calibrated against cores, the values of ϕ are read off *across the intervals which constitute net pay* in each well in the field. The average porosity, ϕ_w, in each well is calculated as the thickness-weighted mean of the log porosities:

$$\phi_w = \frac{\sum_{k=1}^{m} \phi_k h_{n,k}}{h_n}.$$

(4.4)

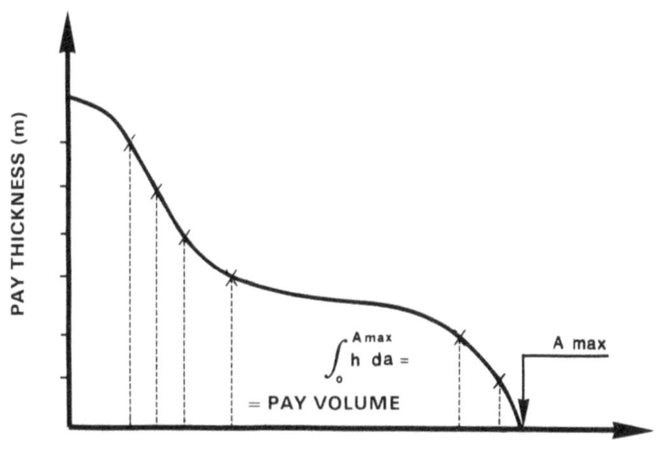

Fig. 4.4. *Graphical method for the calculation of the net pay volume. From Ref. 4, reprinted with permission of Kluwer Academic Publishers and of Professors Archer and Wall*

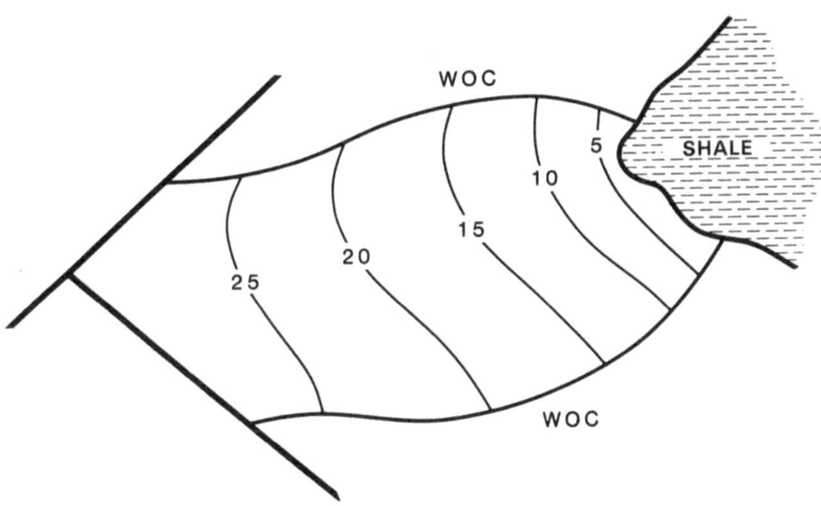

Fig. 4.5. *Isoporosity map for a hydrocarbon-bearing pool*

ϕ_k being the log-derived porosity over a small section, of thickness $h_{n,k}$, within the net pay, thickness h_n.

The values of ϕ_w are plotted well by well to produce the isoporosity map shown in Fig. 4.5.

The average porosity of the reservoir, $\bar{\phi}$, is then computed as the volume weighted mean of the well porosities ϕ_w:

$$\bar{\phi} = \frac{\iint_A \phi_w h_n \, dx \, dy}{V_R} \, . \tag{4.5}$$

This requires preparation of a contour map of constant $\phi_w h_n$ (iso-porosity thickness). The numerical integration over the field can then be performed in a similar way to the calculation of V_R described earlier.

4.3.5 Calculation of the Water Saturation, S_w, and Mean $\bar{S}_w$

As we have seen, in a given lithology the water saturation S_w is dependent on the height above the free water level. This fact must be taken into account in the large proportion of reservoirs which have water in contact with oil or gas.

Firstly, the average reservoir saturation curve $S_w = S_w(h)$ must be established versus height h. This can be obtained by interpolation of log (CPI)-derived values of S_w noted at various depths in each well; or, where cores are available, by the normalisation process (Leverett J-function) described in Sects. 3.4.4.5 and 3.4.4.7.

The average curve is then used to correct or eliminate any values present on the CPI of each well which may be anomalous with respect to their height above the free water level.

The next step is to calculate the average water saturation $S_{w,w}$ in each well, taking the volume-weighted mean *across only those intervals classed as pay*:

$$S_{w,w} = \frac{\sum\limits_{k=1}^{m} S_{w,k}\phi_k h_{n,k}}{\phi_w h_n}. \tag{4.6}$$

These values of $S_{w,w}$ can now be plotted on the reservoir map and contours of constant S_w ("isosaturation") constructed (Fig. 4.6).

The average reservoir saturation $\bar{S}_w$ is computed as the mean of these saturations weighted by total pore volume, over the area of the reservoir:

$$\bar{S}_w = \frac{\iint_A S_{w,w}\phi_w h_n \, dx\, dy}{\bar{\phi} V_R} \tag{4.7}$$

This requires preparation of a contour map of constant "water thickness" $h_n\phi_w S_{w,w}$, before the integral can be evaluated. The numerical integration follows the same procedure as described in Sect. 4.3.3.

4.3.6 Calculation of the Volume Factors of Oil (B_o) and Gas (B_g)

The volume factor of oil, B_o, and especially gas, B_g, do not in general vary much within a reservoir. An experimental value is therefore usually adequate, failing which one of the correlations described in Sect. 2.3.2.2 (B_o) or 2.3.1 (B_g) can be used.

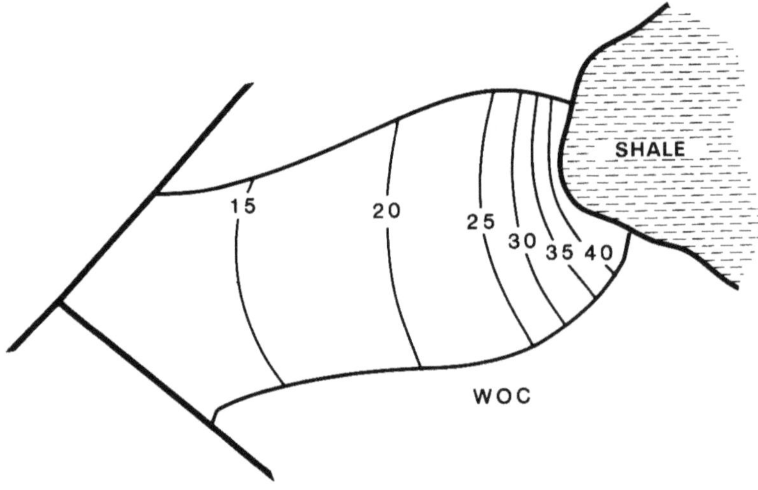

Fig. 4.6. *Isosaturation (S_w) map for a hydrocarbon-bearing pool*

In very thick oil reservoirs, where gravity segregation can be significant, the oil will tend to be heavier towards the bottom.

In this case it would be advantageous to have B_o measured or calculated at different depths. The average $\bar{B}_o$ can then be estimated as the mean weighted by the hydrocarbon volume $[\phi(1 - S_w)V_R]$ at each depth.

4.3.7 Recovery Factor E_R

The calculation of the oil recovery factor $E_{R,o}$ is perhaps the most delicate, and controversial, part of reserves evaluation by the volumetric method.

$E_{R,o}$ depends on a number of interrelated factors: reservoir drive mechanism (water drive, gas cap, solution gas, etc.), mobility ratio [displacing fluid/oil – Eq. (3.52)], heterogeneity of reservoir rock, number of wells and their distribution, production schedule of each well, possible implementation of improved recovery).

An accurate prediction of the recovery factor is only accessible through the use of a numerical simulator to model the reservoir behaviour. This is rarely feasible at such an early stage because of the scarcity of data during the discovery phase.

Evaluation of $E_{R,o}$ therefore usually has to depend on final recovery figures achieved elsewhere in a comparable reservoir rock – preferably in the same sedimentary basin – where the oil, and the drive mechanism, are of the same type.

Using final recoveries achieved in reservoirs in which the oil properties and petrophysical characteristics are well known, a number of correlations have been published. The most commonly used of these are supplied by the American Petroleum Institute:[1]

For sandstone and carbonate reservoirs with solution gas drive:

$$\bar{E}_{R,o} = 0.41815 \left[\frac{\bar{\phi}(1 - \bar{S}_w)}{B_{ob}}\right]^{0.1611} \left(\frac{\bar{k}}{\mu_{ob}}\right)^{0.0979} (\bar{S}_w)^{0.3722} \left(\frac{p_b}{p_a}\right)^{0.1741} \qquad (4.8a)$$

For sandstone reservoirs with water drive:

$$\bar{E}_{R,o} = 0.54898 \left[\frac{\bar{\phi}(1 - \bar{S}_w)}{B_{oi}}\right]^{0.0422} \left(\frac{\bar{k}\mu_{wi}}{\mu_{oi}}\right)^{0.0770} (\bar{S}_w)^{-0.1903} \left(\frac{p_i}{p_a}\right)^{-0.2159} \qquad (4.8b)$$

In both these equations, ϕ and S_w are decimal fractions, k is in Darcy, and μ in cP. The subscript b indicates "value at bubble point", i is "initial value", and p_a is the abandonment pressure (depleted reservoir). The two formulae are derived from a statistical analysis of data from, respectively, 80 and 70 reservoirs.

The distribution of E_R, S_{or} (for water drive) and S_{gr} (for solution gas drive) are shown in Figs. 4.7 and 4.8 for the reservoirs in the survey.

It should be mentioned that in a subsequent publication,[3] the API expressed doubts about the accuracy of these correlations, and recommended they be used with caution!

The recovery factor for gas production, $E_{R,g}$, is a far simpler quantity to predict.

For expansion drive reservoirs (without water drive), recovery is controlled essentially by the level set for the abandonment pressure p_a. This in turn depends on the minimum acceptable wellhead pressure.

$$\bar{E}_{R,g} = 1 - \frac{p_a/z_a}{p_i/z_i}. \qquad (4.9a)$$

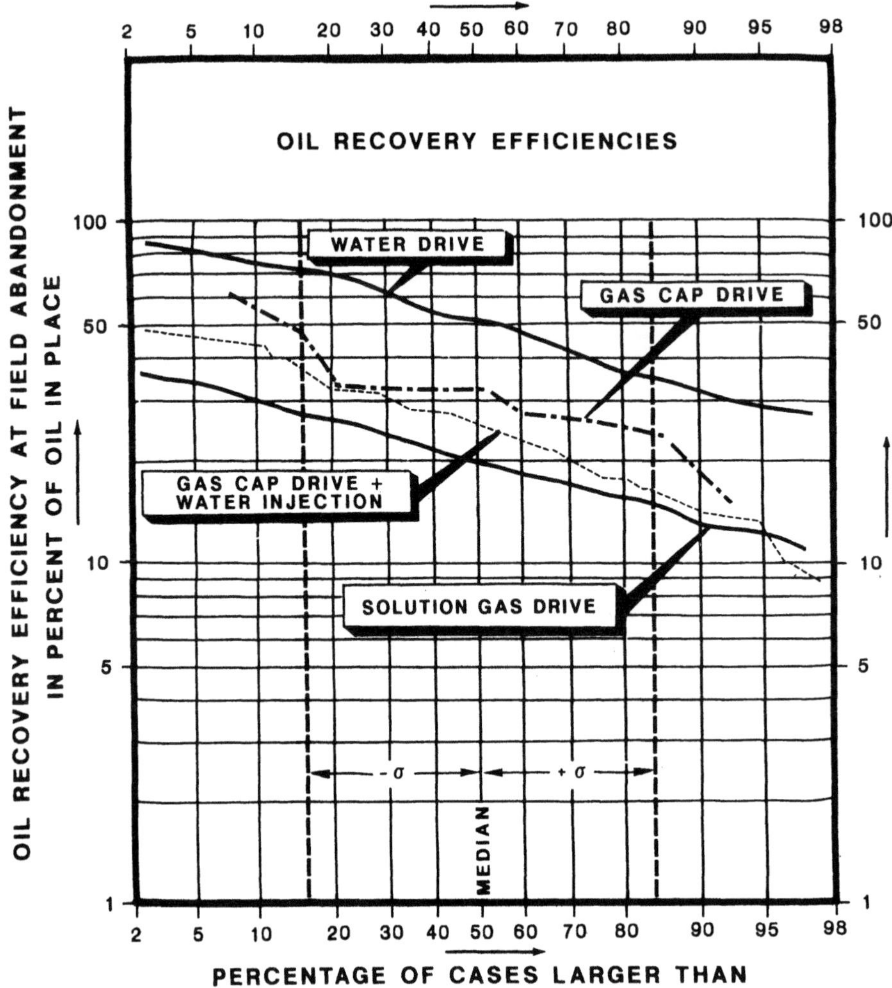

Fig. 4.7. *Log-probability distribution of percentage oil recovery factors for various drive mechanisms. (Based on a statistical analysis of: 72 reservoirs producing under water drive; 80 under solution gas drive without fluid injection; 75 under solution gas drive plus water injection; and 13 under gas cap expansion.) (From Ref. 1, reprinted with permission of the American Petroleum Institute)*

Where there is water drive, the residual gas saturation S_{gr} behind the water front[5] must be taken into account.

In the limiting case of a gas reservoir at abandonment pressure, which has been completely swept with water,

$$\bar{E}_{R,g} = 1 - \bar{S}_{gr} \frac{p_a/z_a}{p_i/z_i} . \tag{4.9b}$$

In sandstone reservoirs, S_{gr} lies between 0.1 and 0.3, with a most probable value of 0.23; in carbonates, the range is 0.1–0.23, with a most probable value of 0.18 (see Ref. 5).

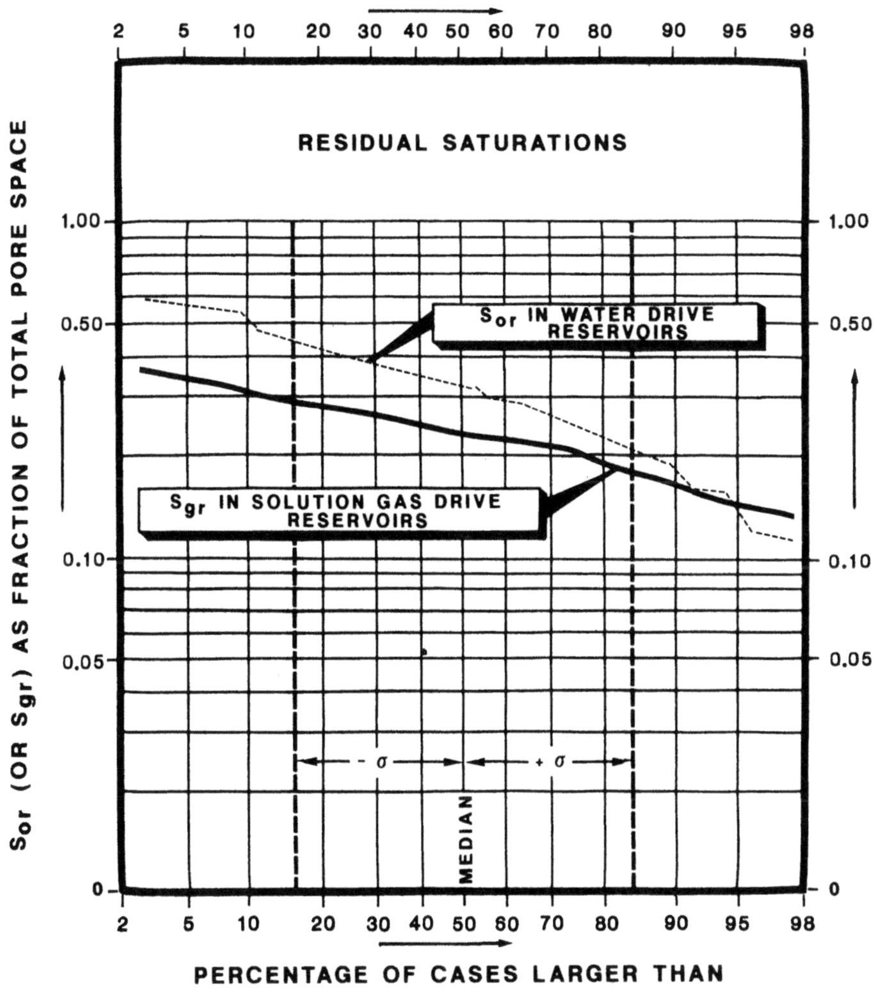

Fig. 4.8. *Log-probability distribution of the residual oil saturation for reservoirs produced by water drive (72 cases), and of the final free gas saturation in solution gas drive reservoirs (80 cases). (From Ref. 1, reprinted with permission of the American Petroleum Institute)*

4.4 Volumetric Method for the Evaluation of Reserves

There is a choice of a deterministic or a probabilistic approach to reserves evaluation by the volumetric method.

Deterministic: the reserves are classified as proven, probable or possible according to the criteria described in Sect. 4.2. For each category, a deterministic calculation is performed using Eq. (4.1) with values obtained by the methods explained in Sect. 4.3.

Probabilistic: the probabilistic distribution of each parameter in Eq. (4.1) is incorporated in a Monte Carlo analysis.

The deterministic approach can be developed further in two ways: the method of "mean values", and the "equivalent hydrocarbon column (EHC)".

4.4.1 Deterministic Method

4.4.1.1 Method of Mean Values

The mean values $\bar{h}_n$, $\bar{\phi}$, $\bar{S}_w$, $\bar{B}_o$ (or $\bar{B}_g$), and $\bar{E}_{R,o}$ (or $\bar{E}_{R,g}$) are calculated for each area A by the methods described in Sects. 4.3.2–4.3.7. The following two equations then give the volume of reserves – proven, probable or possible – for oil and gas respectively:

$$N_{pa} = \frac{A\bar{h}_n\bar{\phi}(1 - \bar{S}_w)}{\bar{B}_o}\,\bar{E}_{R,o}\,, \tag{4.10a}$$

$$G_{pa} = \frac{A\bar{h}_n\bar{\phi}(1 - \bar{S}_w)}{\bar{B}_g}\,\bar{E}_{R,g}\,. \tag{4.10b}$$

4.4.1.2 Equivalent Hydrocarbon Column

Starting from the values of h_n, ϕ_w and $S_{w,w}$ determined in each well (Sects. 4.3.3–4.3.5), we calculate the hypothetical thickness of the column of hydrocarbon that would result from removing the rock matrix and pore water (the local equivalent hydrocarbon column, or EHC):

$$\text{EHC} = h_n\phi_w(1 - S_{w,w})\,. \tag{4.11}$$

The EHC of each well is plotted on the reservoir map, and the iso-EHC contours are constructed (Fig. 4.9).

The total volume of hydrocarbons V_H at reservoir p and T is calculated from

$$V_H = \iint_A \text{EHC}\,dx\,dy\,. \tag{4.12}$$

This integration must be performed separately for the proven, probable and possible reserves areas, following the procedure outlined in Sect. 4.3.3 for V_R.

We then have:

$$N_{pa} = \frac{V_H}{\bar{B}_o}\,\bar{E}_{R,o}\,, \tag{4.13a}$$

$$G_{pa} = \frac{V_H}{\bar{B}_g}\,\bar{E}_{R,g}\,. \tag{4.13b}$$

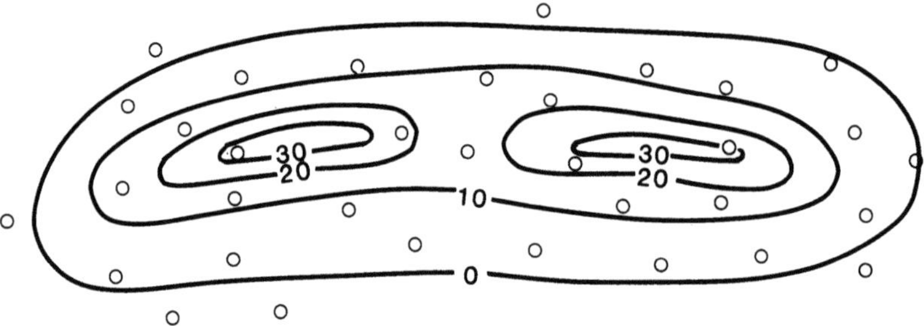

Fig. 4.9. *Iso-EHC contour map*

Being the quicker of the two methods, the EHC is a preferable approach from a practical point of view.

4.4.2 Probabilistic, or Monte Carlo, Method

The probabilistic, or Monte Carlo,[4,8,9] method replaces the need for the rigid definitions of "proven", "probable" and "possible", described in Sect. 4.2, with the concept of the *probability* of reserves having a certain value.

The distribution curve shown in Fig. 4.10 expresses the probability that the reserves will have a volume equal to or greater than any chosen value on the x-axis.

This curve itself is based on the probability distribution of each of the parameters appearing in Eq. (4.10). The likely range of values that each parameter could assume must be decided, and a probability assigned to the values within this range. The *probability density* curve, f, in Fig. 4.11 has been normalised so that there is unit area beneath it. F, the integral of f, is the actual probability distribution of the parameter in question.

Note that it is possible for certain parameters to be related: for example, within a given reservoir unit it is quite common to find an inverse correlation between S_{wi} and ϕ.

When this is the case, we need only consider the probability curve for *one* of the interrelated parameters (ϕ in this example); the dependent variable is represented by the appropriate correlating equation (e.g. $S_{wi} = a/\phi$) or diagram.

In applying the Monte Carlo technique, a table of random numbers is drawn up for each *independent* parameter in Eq. (4.10). In each table, the maximum and minimum values of the numbers, and their probability distribution, correspond

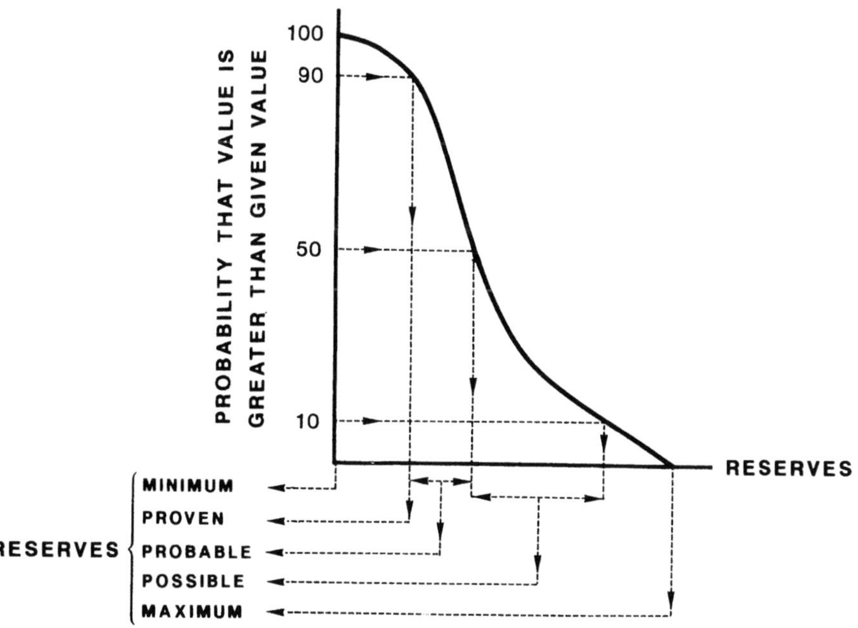

Fig. 4.10. *Probability distribution for reserves volume, calculated for a particular reservoir by the Monte Carlo method. Classifications according to probability are also shown. From Ref. 4, reprinted with permission of Kluwer Academic Publishers and of Professors Archer and Wall*

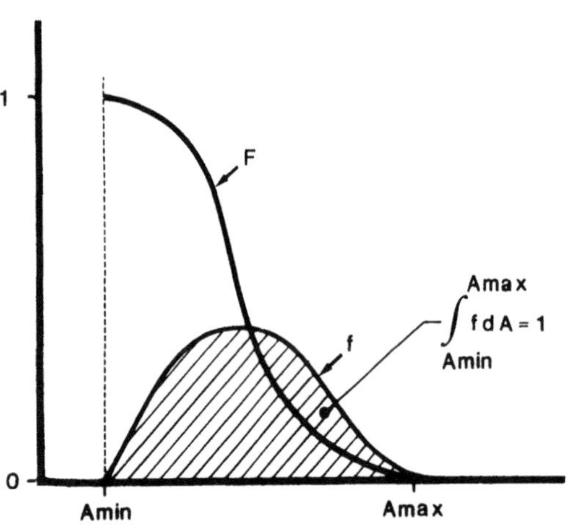

Fig. 4.11. *An example of the probability distribution for reservoir area, as input to the Monte Carlo reserves calculation*

with those assumed for the parameter itself. Random number generation can be handled rapidly by readily available computer programs.

A number is then extracted at random from each table, the corresponding values for any dependent (correlated) variables are calculated, and N_{pa} (or G_{pa}) are computed for that case.

This is repeated many times (at least 5000), each case being calculated for a randomly selected set of values from the tables.

At the conclusion of the analysis, we have a large number of reserves estimates, each corresponding to a random combination of parameter values. The results are arranged in ascending order and the probability density determined. The probability distribution for the reserves can then be computed from this, as in Fig. 4.10.

The reserves categories are conventionally defined as follows:

proven reserves: reserves volume corresponding to 90% probability on the distribution curve,

probable reserves: reserves volume corresponding to the difference between 50 and 90% probability on the distribution curve,

possible reserves: reserves volume corresponding to the difference between 10 and 50% probability on the distribution curve.

The probabilistic method has the advantage of reducing the degree of subjectivity in the criteria employed in evaluating the reservoir area. However, the assessment of the probability densities specified for each of the input parameters remains a highly subjective process.

4.5 Classification of Fields in Terms of Hydrocarbon Volume

Two sets of standards are in current use for the classification of fields according to the volume of reserves they contain. The first, proposed by the API, is applied to fields of small to medium size. The second, which has been in common use for many years although it has never been recognised officially by any regulatory body, applies to "giant" fields.

The different field types are summarised below:

API Classification[2]

Major field:	oilfield with reserves in excess of 16×10^6 sm^3 (100 Mbbl, or 14 Mt); gas field with reserves in excess of 17×10^9 sm^3 (600 Gcuft).

Major field: oilfield with reserves in excess of 16×10^6 sm^3 (100 Mbbl, or 14 Mt); gas field with reserves in excess of 17×10^9 sm^3 (600 Gcuft).

Class A field: oilfield with reserves between 8 and 16×10^6 sm^3 (50–100 Mbbl, or 7–14 Mt); gas field with reserves between 8.5 and 17×10^9 sm^3 (300–600 Gcuft).

Class B field: oilfield with reserves between 4 and 8×10^6 sm^3 (25–50 Mbbl, or 3.5–7 Mt); gas field with reserves between 4.2 and 8.5×10^9 sm^3 (150–300 Gcuft).

Class C field: oilfield with reserves between 1.6 and 4×10^6 sm^3 (10–25 Mbbl, or 1.4–3.5 Mt); gas field with reserves between 1.7 and 4.2×10^9 sm^3 (60–150 Gcuft).

Class D field: oilfield with reserves between 0.16 and 1.6×10^6 sm^3 (1–10 Mbbl, or 0.14–1.4 Mt); gas field with reserves between 0.17 and 1.7×10^9 sm^3 (6–60 Gcuft).

Class E field: oilfield with reserves less than 0.16×10^6 sm^3 (1 Mbbl, or 0.14 Mt); gas field with reserves less than 0.17×10^9 sm^3 (6 Gcuft).

Class F field: any field abandoned during the year of its discovery, even if it had been completed for production.

Fields in category D and above are referred to as "significant".

Classification[6] *of "Giant Fields"*

Super-giant field   oilfield with reserves in excess of 800×10^6 sm^3 (5 Gbbl, or 700 Mt); gas field with reserves in excess of 850×10^9 sm^3 (30 000 Gcuft).

Giant field: oilfield with reserves between 80 and 800×10^6 sm^3 (0.5–5 Gbbl, or 70–700 Mt); gas field with reserves between 85 and 850×10^9 sm^3 (3000–30 000 Gcuft).

Potentially giant field: an oilfield which could be made into a giant field by means of further development or improved recovery. Potential giants can be either "probable giants" or "possible giants", depending on the degree of uncertainty associated with the assessment of their potential.

"Combination" giant oilfield   a field containing at least 40×10^6 sm^3 (250 Mbbl, or 35 Mt) of oil reserves, plus at least 80×10^6 sm^3 (500 Mbbl, or 70 Mt) of hydrocarbon recoverable as condensate, or its calculated liquid equivalent. This would be an oil reservoir with an overlying gas cap.

Figure 4.12 provides a qualitative idea of the worldwide occurrence of oilfields of different sizes.

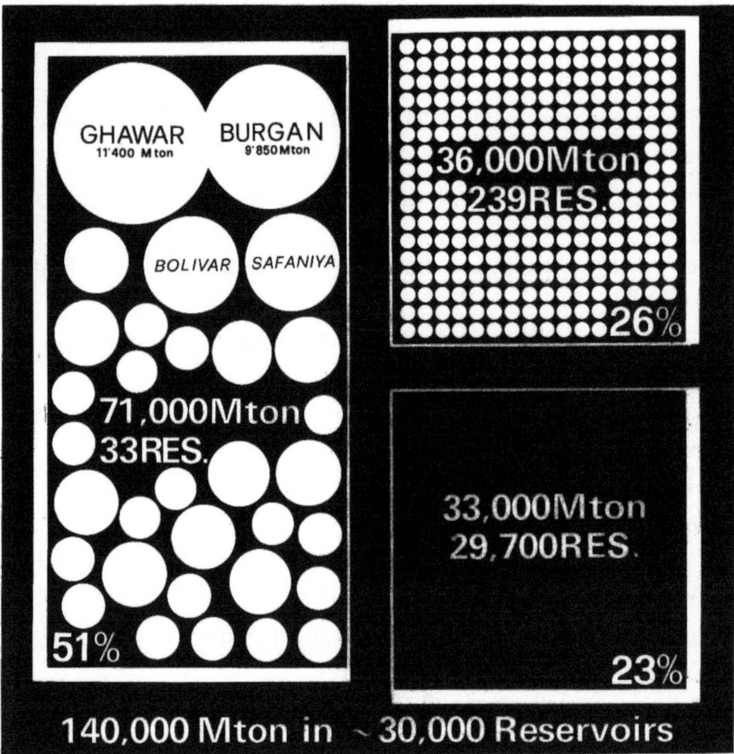

Fig. 4.12. *Distribution of the world's approximately 30 000 known oilfields in terms of the size of their reserves. More than half the world's reserves are contained in only 33 oilfields (0.1% of the 30 000 total); a further 26% of the reserves are found in roughly 0.8% of the total. Overall, 0.91% of the world's oilfields hold 76.4% of the total oil reserves*

References

1. API (1967) A statistical study of recovery efficiency. American Petroleum Institute, Bull D14, 1st edn. Dallas
2. API (1972) Instructions for completing individual well tickets for US drilling statistics. Part I. Cards 1&2 (API); Part II. Card 3 (AAPG-CSD). Americal Petroleum Institute, Washington, DC
3. API (1984) Statistical analysis of crude oil recovery efficiency. American Petroleum Institute, Bull D14, 2nd edn.
4. Archer JS, Wall CG (1986) Petroleum engineering principles and practice. Graham & Trotman, London
5. Chierici GL, Ciucci GM, Long G (1963) Experimental research on gas saturation behind the water front in gas reservoirs subjected to water drive. Proc 6th World Petr Congr Sect II, Frankfurt, Verein zur Förderung des 6. Welt Erdöl Kongress, Hamburg, pp 483–498
6. Nehring R (1978) Giant oil fields and world oil resources. Rand, Santa Monica
7. Society of Petroleum Engineers (1987) Definitions for oil and gas reserves. J Petrol Tech (May): 577–578
8. Walstrom JE, Mueller TD, McFarlane RC (1967) Evaluating uncertainty in engineering calculations. J Petrol Tech (Dec): 1595–1603; Trans AIME 240
9. World Petroleum Congresses (1987) Classification and nomenclature systems for petroleum and petroleum reserves. Study Group Rep 12th World Petrol Congr, Houston 26th April–1st May

EXERCISES

Exercise 4.1

The following data are available for an oil reservoir:

Reservoir rock

- oil-bearing volume $\qquad V_R = 0.85 \times 10^9 \ m^3$
- average porosity $\qquad \bar{\phi} = 0.24$
- average permeability $\qquad \bar{k} = 150 \ md$
- average initial water saturation $\quad \bar{S}_{wi} = 0.18$
- average pore compressibility $\qquad \bar{c}_p = 5 \times 10^{-4} \ MPa^{-1}$

Reservoir conditions

- initial pressure $\quad p_i = 45.6 \ MPa$
- temperature $\qquad T_R = 172 \ °C$

Fluid thermodynamic properties

oil (see Ex. 2.3)
- bubble point pressure $\qquad p_b = 36.4 \ MPa$
- volume factor at initial pressure $\quad B_{oi} = 1.682$
- undersaturated oil compressibility $\quad c_o = 2.6 \times 10^{-3} \ MPa^{-1}$
- viscosity at initial pressure $\qquad \mu_{oi} = 0.234 \ mPa \ s$

water (interstitial and aquifer)
- compressibility under reservoir conditions $\quad c_w = 4.3 \times 10^{-4} \ MPa^{-1}$
- viscosity under reservoir conditions $\qquad \mu_w = 0.25 \ mPa \ s$

Calculate the volume of oil in place (in m^3 and bbl) under reservoir and stock tank conditions. Using the correlations published by the API, estimate the expected recovery factor $\bar{E}_{R,o}$ for production by solution gas drive, assuming an abandonment pressure $p_a = 2 \ MPa$; and by water injection at constant pressure.

Then calculate the oil reserves corresponding to these production mechanisms.

Solution

The volume of oil in the reservoir is equal to the oil-bearing pore volume. Therefore,

$$NB_{oi} = V_R \bar{\phi} (1 - \bar{S}_{wi}) \, ,$$

where N is the volume of oil in place expressed under stock tank conditions.

Given that

$$1 \ bbl = 0.1589873 \ m^3 \, ,$$

we have

$$NB_{oi} = 0.85 \times 10^9 \times 0.24 \times (1 - 0.18) \ m^3 = 167.3 \times 10^6 \ m^3 = 1.052 \times 10^9 \ bbl$$

and

$$N = 99.5 \times 10^6 \ m^3 = 625 \times 10^6 \ bbl \, ,$$

To calculate the recovery factor for solution gas drive, we need first to evaluate the recovery factor for liquid expansion that occurs between reservoir initial pressure and bubble point pressure.

For a decrease in reservoir pressure of 1 MPa, we will observe the following changes in 1 m^3 of pore space:

- the pore volume will decrease by an amount equal to c_p as the effective geostatic pressure is increased by 1 MPa,

– the volume of oil per unit pore volume, $(1 - \bar{S}_{wi})$, will expand by $c_o(1 - \bar{S}_{wi})$,
– the volume of the interstitial water per unit pore volume, $\bar{S}_{wi}$, will expand by $c_w\bar{S}_{wi}$.

Taking these terms into account, the volume of oil expelled from this $1\ m^3$ of pore volume as a result of a 1 MPa reduction in pore pressure will be

$$\Delta V = c_p + (1 - \bar{S}_{wi})c_o + \bar{S}_{wi}c_w \ .$$

Since $(1 - \bar{S}_{wi})$ is the volume of oil contained in unit pore volume, the volume of oil expelled, per 1 MPa pressure decrease, for every m^3 of *oil* in the reservoir is

$$c_{o,e} = \frac{c_p + c_w\bar{S}_{wi} + c_o(1 - \bar{S}_{wi})}{1 - \bar{S}_{wi}} \ ,$$

where $c_{o,e}$ is the *effective compressibility* of the oil.
In this example,

$$c_{o,e} = \frac{5 \times 10^{-4} + 0.18 \times 4.3 \times 10^{-4} + (1 - 0.18) \times 2.6 \times 10^{-3}}{1 - 0.18} = 3.3 \times 10^{-3}\ MPa^{-1} \ .$$

Consequently, the recovery factor for the expansion down to bubble point will be

$$E_{R,o}(p_i \rightarrow p_b) = (45.6 - 36.4) \times 3.3 \times 10^{-3} = 0.030 = 3\% \ .$$

The recovery factor for solution gas drive from p_b down to abandonment at $p_a = 2$ MPa can be calculated using Eq. (4.8a). We have

$$E_{R,o}(p_b \rightarrow p_a) = 0.251$$

and therefore the overall recovery factor is

$$\bar{E}_{R,o} = E_{R,o}(p_i \rightarrow p_b) + E_{R,o}(p_b \rightarrow p_a) = 0.281 = 28.1\% \ .$$

We can use Eq. (4.8b) to estimate the recovery factor for water drive (water injection). This gives

$$\bar{E}_{R,o} = 0.604 = 60.4\% \ .$$

Therefore,

recovery factor for solution gas drive $(\bar{E}_{R,o})_{DD} = 0.281$,
recovery factor for water injection $(\bar{E}_{R,o})_{WD} = 0.604$,

and consequently:

Oil reserves
a. if the reservoir is produced by solution gas drive alone:

$$N_{pa} = 0.281 \times 99.5 \times 10^6\ m^3 = 28 \times 10^6\ m^3 = 176 \times 10^6\ bbl \ ,$$

b. if the reservoir is produced by water injection:

$$N_{pa} = 0.604 \times 99.5 \times 10^6\ m^3 = 60 \times 10^6\ m^3 = 378 \times 10^6\ bbl \ .$$

◇ ◇ ◇

Exercise 4.2

The following information is from a gas reservoir:

Reservoir rock
– gas-bearing volume $V_R = 0.55 \times 10^9\ m^3$
– average porosity $\bar{\phi} = 0.195$
– average initial water saturation $\bar{S}_{wi} = 0.225$

Reservoir conditions
– initial pressure $p_i = 25.0$ MPa
– temperature $T_R = 100\,°C$

The gas composition is listed in Ex. 2.1.

Calculate:

- the initial volume of gas in place, in sm^3 and scuft,
- the gas reserves corresponding to the three different drive mechanisms postulated below:
  1. Gas expansion alone, with abandonment pressure at 2 MPa.
  2. Water drive, sweeping the entire reservoir, and leaving a residual gas saturation $S_{gr} = 0.25$. Abandonment pressure in this case is $p_a = 15$ MPa.
  3. Partial water drive, with only 60% of the reservoir volume swept by water at the time of abandonment at $p_a = 2$ MPa.

Solution

From Ex. 2.1, the reservoir gas has the following properties:

- pseudo-critical pressure $p_{pc} = 4.60$ MPa
- pseudo-critical temperature $T_{pc} = 216.1$ K

At reservoir conditions ($T_R = 100\,°C$) we will have:

- pseudo-reduced temperature:

$$T_{pr} = \frac{T_R}{T_{pc}} = \frac{100 + 273.2}{216.1} = 1.727 \; .$$

We first calculate the values of the compressibility factor z, and the volume factor B_g, at initial pressure p_i, and at the different abandonment pressures specified:

p (MPa)	p_{pr} (dim'less)	z (dim'less)	B_g (dim'less)
25	5.43	0.8851	4.646×10^{-3}
15	3.26	0.8643	7.561×10^{-3}
2	0.43	0.9773	64.119×10^{-3}

where z has been calculated using Eq. (2.12), and B_g using Eq. (2.8b).

The initial volume of gas in place is

$$G = \frac{V_R \bar{\phi}(1 - \bar{S}_{wi})}{B_{gi}} = \frac{0.55 \times 10^9 \times 0.195 \times (1 - 0.225)}{4.646 \times 10^{-3}}$$

$$= 17.89 \times 10^9 \; sm^3 = 632 \times 10^9 \; scuft$$

since $1\,m^3 = 35.314662$ cuft.

Under the producing conditions described in case 1 (gas expansion without water drive, to an abandonment pressure of 2 MPa), if we ignore the change in pore and interstitial water volumes resulting from the reduction in pressure, we have:

$$\text{residual gas at abandonment} = \frac{V_R \bar{\phi}(1 - \bar{S}_{wi})}{B_g(2\,\text{MPa})}$$

$$= \frac{0.55 \times 10^9 \times 0.195 \times 0.775}{64.119 \times 10^{-3}}$$

$$= 1.3 \times 10^9 \; sm^3 \times 45.8 \times 10^9 \; scuft \; .$$

Therefore, the volume of producible gas – the gas reserves – will be

$$G_p = (17.89 - 1.30) \times 10^9 \; sm^3 = 16.6 \times 10^9 \; sm^3 = 586 \times 10^9 \; scuft \; .$$

In case 2 (total sweep of water, $S_{gr} = 0.25$ and abandonment pressure $p_a = 15$ MPa), the volume of pore volume still occupied by gas will be

$$V_{p,a} = 0.25 V_R \bar{\phi} = 0.25 \times 0.55 \times 10^9 \times 0.195 = 26.81 \times 10^6 \; m^3$$

corresponding to a gas volume at standard conditions of

$$\frac{26.81 \times 10^6}{B_g(15\ \text{MPa})} = \frac{26.81 \times 10^6}{7.561 \times 10^{-3}} = 3.55 \times 10^9\ \text{sm}^3 = 125 \times 10^9\ \text{scuft}\ .$$

Therefore, for case 2 the gas reserves are

$$G_{pa} = (17.89 - 3.55) \times 10^9\ \text{sm}^3 = 14.3 \times 10^9\ \text{sm}^3 = 506 \times 10^9\ \text{scuft}\ .$$

Finally, for case 3 (60% swept volume at abandonment pressure 2 MPa), the volume of pore volume still occupied by gas will be

$$V_{p,a} = 0.60 V_R \bar{\phi} S_{gr} + 0.40 V_R \bar{\phi}(1 - S_{wi})$$

$$= 0.60 \times 0.55 \times 10^9 \times 0.195 \times 0.25 + 0.40 \times 0.55 \times 10^9 \times 0.195 \times 0.775$$

$$= 49.34 \times 10^6\ \text{m}^3$$

corresponding to a gas volume at standard conditions of

$$\frac{49.34 \times 10^6}{B_g(2\ \text{MPa})} = \frac{49.34 \times 10^6}{64.119 \times 10^{-3}} = 0.77 \times 10^9\ \text{sm}^3 = 27 \times 10^9\ \text{scuft}\ .$$

Therefore, for case 3 the gas reserves are

$$G_{pa} = (17.89 - 0.77) \times 10^9\ \text{sm}^3 = 17.1 \times 10^9\ \text{sm}^3 = 605 \times 10^9\ \text{scuft}\ .$$

The following table summarises the results for the three cases:

Case	Reserves	
	$(\text{sm}^3 \times 10^9)$	$(\text{scuft} \times 10^9)$
1	16.6	586
2	14.3	506
3	17.1	605

Evidently in this example (and, in fact, in the majority of real cases), maximum benefit can be derived from water drive by producing the reservoir at a rate sufficiently high to induce partial flooding, with the lowest possible abandonment pressure that is consistent with well productivity.

5 Radial Flow Through Porous Media: Slightly Compressible Fluids

5.1 Introduction

The flow pattern in the strata in the vicinity of a well producing oil or gas can be considered as essentially horizontal radial (if we limit the discussion initially to near vertical wells and non-dipping formations). An understanding of horizontal radial flow is therefore important if we are to explain or predict the productive performance of a well.

The behaviour of *slightly* compressible fluids (undersaturated oil, water), whose properties are little affected by changes in pressure; and *highly* compressible fluids (dry and wet gas, condensates), which are very sensitive to the pressure, will be treated separately.

In this chapter, we will examine the bottom-hole pressure behaviour observed during the flow of the fluid through the formation towards the well, or vice versa, (production or injection tests); and during the period following the shutting in of the well – i.e. the termination of flow (buildup or fall off tests).

5.2 Equation for Single Phase Radial Flow

The basic assumptions made in the theory of single phase horizontal radial flow are:

- the porous medium is homogeneous in ϕ and k, and is horizontal and of uniform thickness; the permeability is isotropic ($k_v = k_h$),
- there are no saturation gradients: in the case of the flow of oil, its saturation is constant and equal to $(1 - S_{iw})$ everywhere, the water phase being immobile; in the case of the flow of water, its saturation is everywhere $\geq (1 - S_{or})$, and the oil phase (if present) is immobile,
- the pressure throughout the reservoir is above the bubble point of the fluid,
- the fluid properties are uniform over the thickness of the reservoir, so that gravitational effects can be ignored,
- the well has been perforated over the entire reservoir thickness, so that inflow is horizontal radial.

Consider the segment of a cylindrical reservoir shown in Fig. 5.1. An incremental volume (shaded) is defined by an arc at a distance r from the axis of the well, and is of thickness dr, with fluid flowing across it as indicated. Applying the principle of the conservation of mass to this volume:

(mass flow rate in) $-$ (mass flow rate out) = (rate of change of mass in the incremental volume).

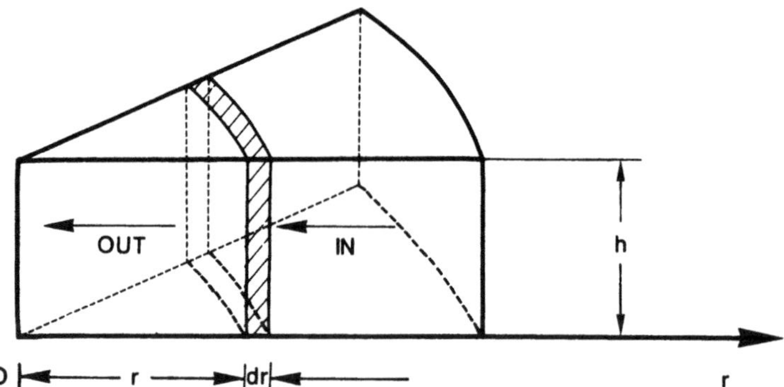

Fig. 5.1. *Diagram showing the conservation of mass in radial flow*

Therefore,

$$(q\rho)_{r+dr} - (q\rho)_r = 2\pi r h\, dr\, \frac{\partial}{\partial t}(\phi\rho) \,, \tag{5.1}$$

where $(2\pi r h\, dr)$ is the volume of the increment, ρ is the density of the mobile fluid, ϕ is the porosity. As was explained in Sect. 3.4.3, ϕ is pressure-dependent, given that the geostatic stress $\bar\sigma$ remains constant.

Recalling that:

$$(q\rho)_{r+dr} = (q\rho)_r + \frac{\partial(q\rho)}{\partial r} dr \,.$$

equation (5.1) becomes:

$$\frac{\partial(q\rho)}{\partial r} = 2\pi r h\frac{\partial(\phi\rho)}{\partial t} \,. \tag{5.2}$$

For a cross-sectional area to flow of $2\pi r h$, and a pressure gradient of $\partial p/\partial r$, Darcy's law can be written as:

$$q = \frac{2\pi r h k}{\mu} \frac{\partial p}{\partial r} \,. \tag{5.3}$$

Furthermore, we have:

$$\frac{\partial}{\partial t}(\phi\rho) = \frac{d(\phi\rho)}{dp} \frac{\partial p}{\partial t} \tag{5.4a}$$

with:

$$\frac{d(\phi\rho)}{dp} = \phi\frac{d\rho}{dp} + \rho\frac{d\phi}{dp} = \rho\phi\left[\frac{1}{\rho}\frac{d\rho}{dp} + \frac{1}{\phi}\frac{d\phi}{dp}\right] \,. \tag{5.4b}$$

If we define the compressibility of the pore fluid as:

$$c = \frac{1}{\rho}\frac{d\rho}{dp} \,, \tag{5.4c}$$

we have, by substituting from Eq. (3.8c):

$$\frac{d(\rho\phi)}{dp} = \rho\phi(c + c_f) = c_t\rho\phi \,. \tag{5.5a}$$

Here, c_t represents the *total system compressibility* (rock + pore fluid). If the pores contain only water ($S_w = 1$), $c = c_w$, and we have:

$$c_t = c_w + c_f \quad (S_w = 1) \,. \tag{5.5b}$$

If, on the other hand, there is oil in the presence of irreducible (and non-mobile) water, we can write the following reasonable approximation:

$$c = c_o S_o + c_w S_{iw} = c_o + (c_w - c_o) S_{iw} \,.$$

From these two cases we can derive a more general form for c_t:

$$c_t = c_o S_o + c_w S_w + c_f \,. \tag{5.5c}$$

From Eqs. (5.2), (5.3), (5.4a) and (5.5a) it follows that:

$$\frac{\partial}{\partial r}\left(\rho \frac{2\pi r h k}{\mu} \frac{\partial p}{\partial r}\right) = 2\pi r h \rho \phi c_t \frac{\partial p}{\partial t} \,,$$

which simplifies to:

$$\frac{1}{r} \frac{\partial}{\partial r}\left(\frac{k}{\mu}\rho r \frac{\partial p}{\partial r}\right) = \rho \phi c_t \frac{\partial p}{\partial t} \,, \tag{5.6}$$

where ρ and μ are the density and viscosity of the mobile single phase fluid.

Equation (5.6) describes the pressure at any time and at any point in the porous medium for a single phase fluid in horizontal radial flow. It is called the *general diffusivity equation*.

This is a *non-linear* partial differential equation – non-linear because ϕ, k, c_t, ρ and μ are all pressure-dependent. As a consequence, no analytical solution is possible without "linearising" the diffusivity equation so as to remove the pressure dependence of the parameters.

5.3 Linearisation of the Diffusivity Equation for Horizontal Radial Flow – Case Where the Rock-Fluid Diffusivity is Independent of the Pressure

Equation (5.6) can be expanded as:

$$\frac{1}{r}\left[\rho r \frac{d}{dp}\left(\frac{k}{\mu}\right)\left(\frac{\partial p}{\partial r}\right)^2 + \frac{k}{\mu}r\frac{d\rho}{dp}\left(\frac{\partial p}{\partial r}\right)^2 + \frac{k}{\mu}\rho \frac{\partial p}{\partial r} + \frac{k}{\mu}\rho r \frac{\partial^2 p}{\partial r^2}\right] = \rho \phi c_t \frac{\partial p}{\partial t} \,. \tag{5.7}$$

Since $\partial p/\partial r$ is small,* we can assume its square term to be negligible:

$$\left(\frac{\partial p}{\partial r}\right)^2 \ll \frac{\partial p}{\partial r}; \frac{\partial^2 p}{\partial r^2} \,. \tag{5.8}$$

Equation (5.7) then reduces to:

$$\frac{k}{\mu}\left[\frac{\partial^2 p}{\partial r^2} + \frac{1}{r}\frac{\partial p}{\partial r}\right] = \phi c_t \frac{\partial p}{\partial t} \,. \tag{5.9}$$

---

* This assumption is not necessarily valid close to the wellbore, where neglecting the term $(\partial p/\partial r)^2$ may cause errors in the local estimate of $p(r, t)$.

Rearranging:

$$\frac{1}{r}\frac{\partial}{\partial r}\left(r\frac{\partial p}{\partial r}\right) = \frac{\phi\mu c_t}{k}\frac{\partial p}{\partial t}, \tag{5.10}$$

where:

$$\frac{k}{\phi\mu c_t} = \eta = \text{hydraulic diffusivity} \tag{5.11}$$

is assumed to be constant for small changes of pressure, since k, ϕ and μ increase with pressure, while c_t decreases. The dimensions of η are $[L^2 t^{-1}]$.

Equation (5.10) has now been linearised (provided the hydraulic diffusivity can be considered constant), and is referred to in this form as the *radial diffusivity equation*.

This can now be solved analytically. One approach is to take advantage of the similarity between it and the thermal diffusivity equation:

$$\frac{1}{r}\frac{\partial}{\partial r}\left(r\frac{\partial T}{\partial r}\right) = \frac{1}{K}\frac{\partial T}{\partial t}, \tag{5.12}$$

where K = thermal diffusivity (m²/s),
$\quad\quad T$ = temperature (K).

Solutions to Eq. (5.12) for a wide variety of initial and boundary conditions have been published in the definitive work by Carslaw and Jäger, *Conduction of Heat in Solids.*[2]

Dranchuk and Quon[7] have shown that the linearisation process just described is only valid when:

$$c_t p \ll 1 . \tag{5.13}$$

5.4 Dimensionless Form of the Radial Diffusivity Equation

The usefulness of a dimensionless version of the radial diffusivity equation will become apparent in Chaps. 6 and 7.

In fact, any dimensionless analytical solution for a particular set of initial and boundary conditions can be converted immediately into practical units to suit the real case in question.

The dimensionless variables are defined as follows:

Dimensionless radius: $\quad\quad r_D = \dfrac{r}{r_w} .$ $\hfill (5.14a)$

Dimensionless time: $\quad\quad t_D = \dfrac{k}{\phi\mu c_t r_w^2}t = \dfrac{\eta}{r_w^2}t .$ $\hfill (5.14b)$

Dimensionless pressure: $\quad p_D(r_D, t_D) = \dfrac{2\pi kh}{q\mu}(p_i - p_{r,t}) .$ $\hfill (5.14c)$

where p_i is the initial static reservoir pressure; $p_{r,t}$ is the pressure at time t and radius r; q is the flow rate *at reservoir conditions*; and r_w is the wellbore radius.

Using the terms defined in Eq. (5.14), Eq. (5.10) can be rewritten:

$$\frac{1}{r_{\mathrm{D}}}\frac{\partial}{\partial r_{\mathrm{D}}}\left(r_{\mathrm{D}}\frac{\partial p_{\mathrm{D}}}{\partial r_{\mathrm{D}}}\right) = \frac{\partial p_{\mathrm{D}}}{\partial t_{\mathrm{D}}} , \tag{5.15}$$

which is the *dimensionless form of the radial diffusivity equation.*

If p_{wf} is the bottom hole flowing pressure ($r = r_{\mathrm{w}}$, $r_{\mathrm{D}} = 1$), we have:

$$p_{\mathrm{D}}(1, t_{\mathrm{D}}) = p_{\mathrm{D,w}}(t_{\mathrm{D}}) = \frac{2\pi kh}{q\mu}(p_{\mathrm{i}} - p_{\mathrm{wf}}) \tag{5.16a}$$

and, in the absence of any additional pressure drop due to skin effect (to be covered later):

$$p_{\mathrm{wf}} = p_{\mathrm{i}} - \frac{q\mu}{2\pi kh}p_{\mathrm{D,w}}(t_{\mathrm{D}}) . \tag{5.16b}$$

This equation shows clearly the importance of having available $p_{\mathrm{D}}(t_{\mathrm{D}})$ solutions for the flow of undersaturated oil towards the wellbore, with various initial and boundary conditions.

5.5 Behaviour with Time Under Flowing Conditions

Equations such as (5.10) or (5.15) are always associated with a set of initial and boundary conditions.

We will first consider qualitatively the idealised case of a cylindrical reservoir of uniform thickness h, with a sealing ("no-flow") external boundary of radius r_{e}. It is initially in a static condition, with a uniform pressure $p = p_{\mathrm{i}}$ throughout.

There is a well of radius r_{w} at its centre. From time $t = 0$, oil is flowed from the well at a constant rate q (at reservoir conditions).

The assumptions listed in Sect. 5.2 regarding the nature of the well and reservoir are applied here.

The conclusions we shall draw about this single well will also be valid for the case where there are *other* wells present in the reservoir, provided their production rates – which may be different – do not change with time. In this situation, there is an area around each well within which the fluid movement is towards that well. This is referred to as the *drainage area, A.* Its shape and size depend on a number of factors, and the radius of the *equivalent circular area* is the *drainage radius* r_{e} of the well.

If the production rates from any of the wells change with time, their drainage areas (and consequently those of neighbouring wells) will expand or contract, and the pressure behaviour will differ from that predicted for a constant rate case.

We will use the dimensionless form of the radial diffusivity equation for convenience.

Consider a circular drainage area, with production starting at time $t_{\mathrm{D}} = 0$, and continuing until:

$$t_{\mathrm{D,tr}} \cong 0.06\left(\frac{r_{\mathrm{e}}}{r_{\mathrm{w}}}\right)^{2} . \tag{5.17a}$$

From the definition of t_D, this corresponds to a real time of:

$$t_{tr} \cong 0.06\frac{\phi\mu c_t}{k}r_e^2 \ . \tag{5.17b}$$

Up to this time, the pressure disturbance invoked by flowing the well is contained within the drainage area of radius r_e (Fig. 5.2).

During this period, since $p_D(r_D, t_D)$ has not yet been affected by the non-permeable outer boundary at $r = r_e$, the reservoir is effectively infinite in size as far as the pressure is concerned. This is referred to as *infinite acting*.

For t_D between zero and $t_{D,tr}$, the flow regime is said to be *transient* or *infinite acting*.

Still in the same circular geometry, let us now look at the period:

$$0.06\left(\frac{r_e}{r_w}\right)^2 \le t_D \lesssim 0.1\left(\frac{r_e}{r_w}\right)^2 \ . \tag{5.17c}$$

This is a transition period between early transient and the next mode, and the flow regime is referred to as *late transient*.

Note that the t_D period during which late transient flow occurs is different – in both start time and duration – for different drainage area geometries; Fig. 5.7 lists the relevant values for non-circular shapes.

In a circular geometry, when $t_D > 0.1\,(r_e/r_w)^2$, the pressure decreases at a constant rate at all points in the drainage area:

$$\frac{\partial p_D}{\partial t_D} = \text{constant for any } r_D \text{ or } t_D \ . \tag{5.18a}$$

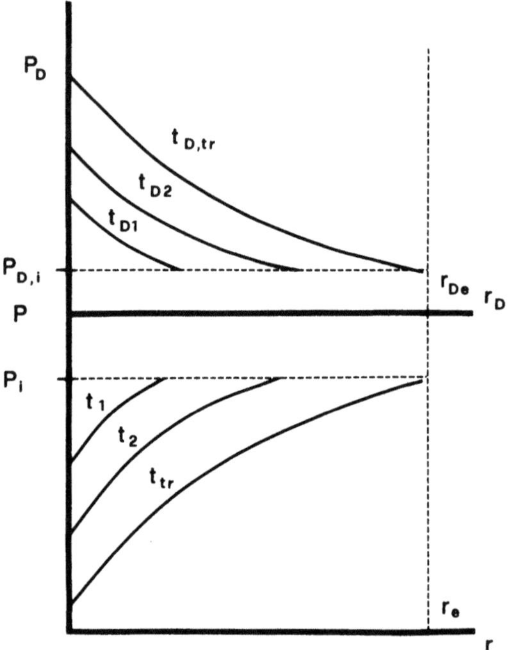

Fig. 5.2. *Outward diffusion of a pressure disturbance around a well: early transient period*

At the outer boundary of the drainage area (no-flow condition) for any r_D we have:

$$\left(\frac{\partial p_D}{\partial r_D}\right)_{r_{De}} = 0 \; . \tag{5.18b}$$

In other words, the pressure profile $p_D(r_D)$ keeps the same *shape*, and the whole profile declines at a steady rate with increasing time. Figure 5.3 shows a series of these parallel profiles corresponding to different times, in both real and dimensionless terms. The flow regime is now *pseudo-steady state* (also called *semi-steady state*).

In order for mass to be conserved at any radius r:

$$\frac{dp}{dt} = -\frac{q}{\pi r_e^2 h \phi c_t} \; , \tag{5.19a}$$

which, in dimensionless form is:

$$\frac{dp_D}{dt_D} = 2\left(\frac{r_w}{r_e}\right)^2 \; . \tag{5.19b}$$

Equation (5.19) allows us to calculate the pressure distribution $p(r)$ in the reservoir for any value of $t > t_{tr}$, provided we have one pressure profile, also measured at a $t > t_{tr}$, to start from.

The variation of the bottom hole flowing pressure p_{wf} during these three flow regime periods is shown schematically in Fig. 5.4.

Pseudo-steady state flow occurs when we have a "no-flow" (i.e. non-permeable, or closed) outer boundary condition. An alternative condition is the "constant pressure boundary" (i.e. fluid is replenished across the boundary, so that the pressure p_e remains constant). In this case the flow regime goes to *steady state* after the late transient period.

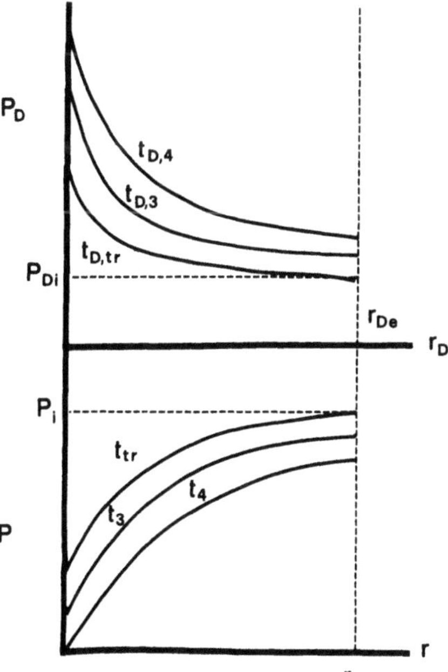

Fig. 5.3. *Pseudo-steady state pressure behaviour in the drainage area of a producing well with a no-flow boundary at $r = r_e$*

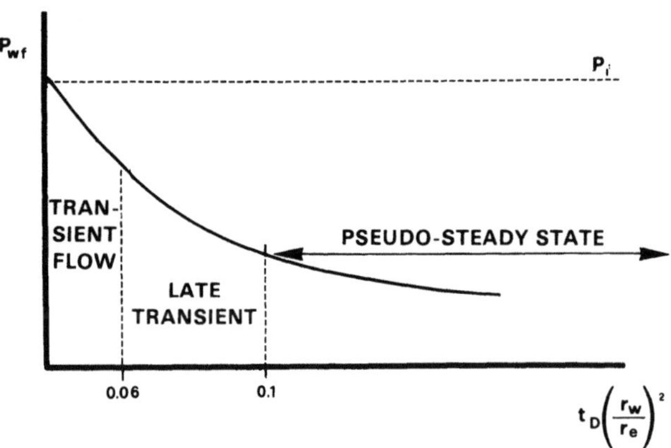

Fig. 5.4. *Bottom hole flowing pressure p_{wf} versus time*

From Darcy's equation:

$$q = \frac{2\pi rhk}{\mu}\frac{dp}{dr} ,$$ (5.20a)

where $2\pi rh$ is the cross-sectional area to flow.

Integrating Eq. (5.20a) from r_w to r:

$$p - p_w = \frac{q\mu}{2\pi hk}\ln\frac{r}{r_w} ,$$ (5.20b)

which is, in dimensionless form, for any value of t:

$$p_D(1) - p_D(r_D) = \ln r_D ,$$ (5.20c)

where $r_D = 1$ is by definition [Eq. (5.14a)] the dimensionless wellbore radius.

5.6 Solutions to the Radial Diffusivity Equation for a Fluid of Constant Compressibility

5.6.1 Transient Flow

5.6.1.1 Treatment for an Ideal Well

Earlier in this chapter it was shown that during a certain initial period when the well is put on production ($t \le t_{tr}$), the reservoir is *infinite acting* and the flow regime is *transient*.

The initial conditions are:

$$p = p_i \quad \text{at } t = 0 \text{ for all } r$$ (5.21a)

The boundary conditions are:

$$p = p_i \quad \text{at } r = \infty \text{ for all } t$$ (5.21b)

$$q = \frac{2\pi khr_w}{\mu}\left(\frac{\partial p}{\partial r}\right)_{r_w} = \text{constant for all } t .$$ (5.21c)

If we assume that r_w is negligibly small, Eq. (5.21c) simplifies to:

$$\lim_{r \to 0} \left(r \frac{\partial p}{\partial r} \right) = \frac{q\mu}{2\pi kh} = \text{constant} \,. \tag{5.21d}$$

The solution to the radial diffusivity equation under these conditions is referred to as the *line source solution for constant terminal rate*.

Using the Boltzmann transform

$$s = \frac{r^2}{4(k/\phi\mu c_t)t} = \frac{\phi\mu c_t r^2}{4kt} \tag{5.22a}$$

so that

$$\frac{\partial s}{\partial r} = \frac{\phi\mu c_t r}{2kt} \tag{5.22b}$$

and:

$$\frac{\partial s}{\partial t} = -\frac{\phi\mu c_t r^2}{4kt^2} \,, \tag{5.22c}$$

the radial diffusivity equation:

$$\frac{1}{r} \frac{\partial}{\partial r} \left(r \frac{\partial p}{\partial r} \right) = \frac{\phi\mu c_t}{k} \frac{\partial p}{\partial t} \tag{5.10}$$

becomes, through transforming the independent variables (r, t) to the independent variable s:

$$\frac{1}{r} \frac{\mathrm{d}}{\mathrm{d}s} \left(r \frac{\mathrm{d}p}{\mathrm{d}s} \frac{\partial s}{\partial r} \right) \frac{\partial s}{\partial r} = \frac{\phi\mu c_t}{k} \frac{\mathrm{d}p}{\mathrm{d}s} \frac{\partial s}{\partial t} \,, \tag{5.23a}$$

where $p = p(r, t)$ is now only a function of $s = s(r, t)$.

Substituting from Eqs. (5.22b) and (5.22c), Eq. (5.23a) becomes:

$$\frac{1}{r} \frac{\phi\mu c_t r}{2kt} \frac{\mathrm{d}}{\mathrm{d}s} \left(\frac{\phi\mu c_t r^2}{2kt} \frac{\mathrm{d}p}{\mathrm{d}s} \right) = -\frac{\phi\mu c_t}{k} \frac{\phi\mu c_t r^2}{4kt^2} \frac{\mathrm{d}p}{\mathrm{d}s} \tag{5.23b}$$

which, substituting from Equation (5.22a)

$$\frac{\mathrm{d}}{\mathrm{d}s} \left(s \frac{\mathrm{d}p}{\mathrm{d}s} \right) = -s \frac{\mathrm{d}p}{\mathrm{d}s} \tag{5.23c}$$

or:

$$\frac{\mathrm{d}p}{\mathrm{d}s} + s \frac{\mathrm{d}}{\mathrm{d}s} \left(\frac{\mathrm{d}p}{\mathrm{d}s} \right) = -s \frac{\mathrm{d}p}{\mathrm{d}s} \,. \tag{5.23d}$$

Equation (5.10), which is a partial differential equation in r and t, has been converted by means of Boltzmann's transform to an ordinary differential equation which is easy to solve. If we write:

$$\frac{\mathrm{d}p}{\mathrm{d}s} = p' \,. \tag{5.24a}$$

equation (5.23d) is now:

$$p' + s\frac{dp'}{ds} = -sp' \tag{5.24b}$$

or:

$$\frac{dp'}{p'} = -ds - \frac{ds}{s}, \tag{5.24c}$$

which, when integrated, results in the general equation:

$$\ln p' = -s - \ln s + C_1 \tag{5.25a}$$

so that:

$$p' = C_2\frac{e^{-s}}{s}, \tag{5.25b}$$

where $C_2 = e^{C_1}$.

For the calculation of C_2, we need to refer to the boundary conditions at the well. We already have that:

$$r\frac{\partial p}{\partial r} = r\frac{dp}{ds}\frac{\partial s}{\partial r} = r\frac{2\phi\mu c_t r}{4kt}\frac{dp}{ds} = 2s\frac{dp}{ds} = 2sC_2\frac{e^{-s}}{s} = 2C_2 e^{-s} \tag{5.26}$$

and, from Eq. (5.21d):

$$\lim_{s\to 0}(2C_2 e^{-s}) = 2C_2 = \frac{q\mu}{2\pi kh}, \tag{5.27}$$

since when $r \to 0$, $s \to 0$ as well.

Substituting from Eq. (5.25b) into Eq. (5.27) we now have:

$$p' = \frac{q\mu}{4\pi kh}\frac{e^{-s}}{s}. \tag{5.28}$$

Equation (5.28) is integrated, at radius r, between the pressures $p_i\,(t = 0)$ and $p(r, t)$ at time t. The corresponding values of s are [Eq. (5.22a)]:

$$s(t = 0) = \infty$$

and

$$s(r, t) = \frac{\phi\mu c_t r^2}{4kt} = x.$$

Therefore:

$$\int_{p_i}^{p(r,t)} dp = \frac{q\mu}{4\pi kh}\int_{\infty}^{x}\frac{e^{-s}}{s}ds \tag{5.29}$$

from which:

$$\frac{4\pi kh}{q\mu}[p_i - p(r, t)] = \int_{x}^{\infty}\frac{e^{-s}}{s}ds. \tag{5.30}$$

The integral in the right hand term of Eq. (5.30) is the well-known *exponential integral ei(x)*, whose behaviour is shown in Fig. 5.5.

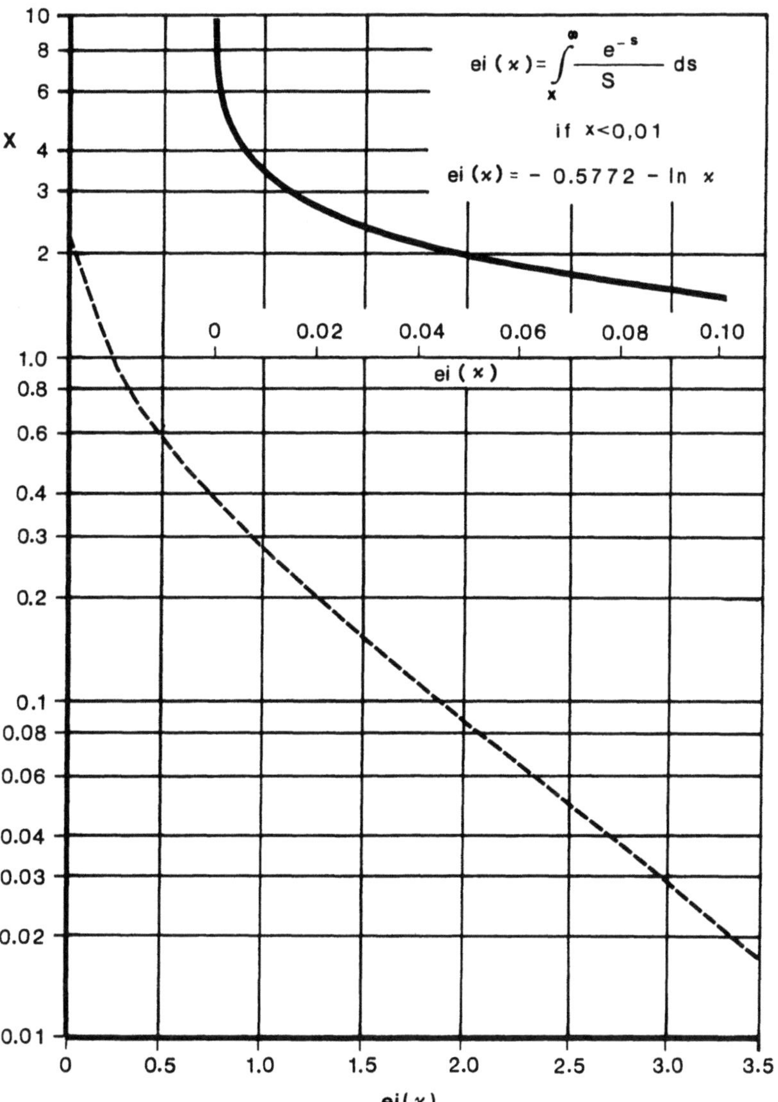

Fig. 5.5. *The exponential integral function ei(x), for the range x = 0.01 to x = 10*

$ei(x)$ can be developed as a series:

$$ei(x) = -0.57721 - \ln x - \sum_{n=1}^{\infty} \frac{(-1)^n x^n}{n\, n!} \, . \tag{5.31a}$$

For $x < 0.01$, ignoring the summation introduces only a very small error,[11] and the equation reduces to:

$$ei(x) = -0.57721 - \ln x \quad (x < 0.01) \, , \tag{5.31b}$$

where the 0.57721 is *Euler's constant*, and

$$e^{0.57721} = 1.781 = \gamma \, .$$

We can express Eq. (5.31b) alternatively as:

$$ei(x) = -\ln(\gamma x) \quad \text{for } x < 0.01 \, . \tag{5.31c}$$

Substituting this into Eq. (5.30) we obtain:

$$p_i - p(r,t) = -\frac{q\mu}{4\pi kh}\ln\frac{\gamma\phi\mu c_t r^2}{4kt} = \frac{q\mu}{4\pi kh}\ln\frac{4kt}{\gamma\phi\mu c_t r^2} .\tag{5.32a}$$

In dimensionless terms, this is:

$$p_D = \frac{1}{2}\ln\left[\frac{4}{\gamma}\left(\frac{r_w}{r}\right)^2 t_D\right] = \frac{1}{2}(\ln t_D - 2\ln r_D + 0.809) ,\tag{5.32b}$$

where:

$$\ln\frac{4}{\gamma} = \ln 2.246 = 0.809 .\tag{5.32c}$$

At the well, where $r_D = 1$ $(r = r_w)$, the relationship between pressure and time is given by:

$$p_D(1,t_D) = 0.5(\ln t_D + 0.809) ,\tag{5.33a}$$

which, in real terms, is:

$$p_{wf} = p_i - \frac{q\mu}{4\pi kh}\left(\ln\frac{kt}{\phi\mu c_t r_w^2} + 0.809\right) .\tag{5.33b}$$

It is important to realise that Eqs. (5.33a) and (5.33b) are based on an approximation [Eq. (5.31b)] which is only valid when x (or its equivalent in the terms of the equations) < 0.01, i.e.:

$$\frac{\phi\mu c_t r_w^2}{4kt} < 0.01\tag{5.34a}$$

otherwise stated as:

$$\frac{kt}{\phi\mu c_t r_w^2} > 25 .\tag{5.34b}$$

For values of t less than that required to satisfy Eq. (5.34), the solution containing the full exponential integral should be used.

In most cases, Eq. (5.34) is usually satisfied after a few minutes, or even seconds, after a well starts flowing.

5.6.1.2 Treatment for a Real Well – Skin Effect

Among the initial assumptions upon which we have based the derivation of the above equations, two are of particular importance:

– the porous medium is homogeneous and isotropic in permeability,
– the well is open to production over the entire thickness of the reservoir.

These conditions are not met in the following situations:

– damage to the near-wellbore formation by mud filtrate invasion, causing a reduction in permeability from k to k_s out to a distance r_s,
– well only drilled through a portion of the reservoir thickness ("partial penetration"),

- well completed with casing and cement, but only perforated over a portion of the reservoir interval,
- the reservoir interval consists of a number of layers of different permeabilities; this situation is further complicated if the well is only perforated in one or some of the layers,
- presence of fractures which intersect the well, resulting in a local increase in permeability.

Each of these factors will modify the flow pattern around the wellbore, so that Eq. (5.33) will not be a true representation of the situation (Fig. 5.6). Their influence is accounted for by introducing a quantity S, the *skin factor*, into the equation:

$$p_D(1, t_D) = 0.5(\ln t_D + 0.809) + S \tag{5.35}$$

from which we derive the following classical equations:

$$p_D(1, t_D) = 0.5(\ln t_D + 0.809 + 2S), \tag{5.36a}$$

$$p_{wf} = p_i - \frac{q\mu}{4\pi kh}\left(\ln\frac{kt}{\phi\mu c_t r_w^2} + 0.809 + 2S\right). \tag{5.36b}$$

S is a dimensionless parameter, and therefore *just a number*. Depending on conditions, the additional pressure drop Δp_s caused by the skin:

$$\Delta p_s = \frac{q\mu}{2\pi kh}S \tag{5.37}$$

may be *positive* $[(p_i - p_{wf})$ in the real well is *greater* than $(p_i - p_{wf})$ in the ideal well], or *negative* $[(p_i - p_{wf})$ in the real well is *less* than $(p_i - p_{wf})$ in the ideal well].

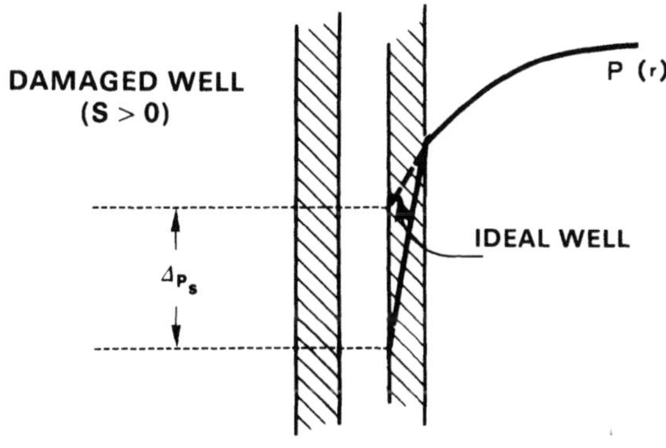

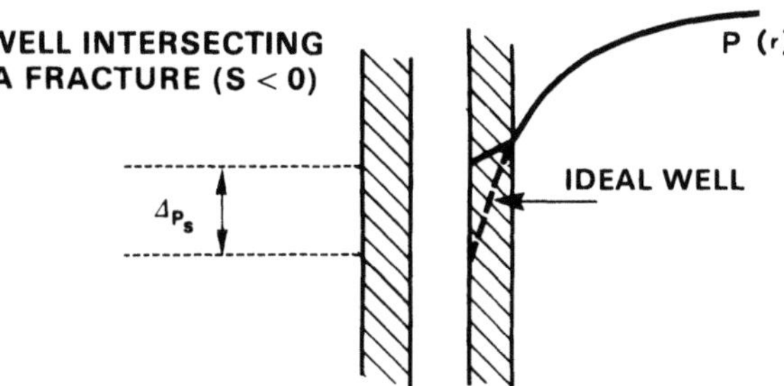

Fig. 5.6. *The pressure profile close to the wellbore in the case of formation damage (S > 0), and in the case of improved productivity due to fracturing (S < 0)*

Δp_s is positive in the case of blockage by mud filtrate,[10] partial penetration of the reservoir interval,[1,9] and partial perforation of the casing.[8]

In the case of filtrate damage, we have:

$$\Delta p_s = \frac{q\mu}{2\pi kh} \frac{k - k_s}{k_s} \ln\frac{r_s}{r_w} , \tag{5.38a}$$

which means that:

$$S = \frac{k - k_s}{k_s} \ln\frac{r_s}{r_w} . \tag{5.38b}$$

This damage can be reduced by means of surfactant treatment, which facilitates the removal of the filtrate.

In a reservoir section consisting of zones of differing permeabilities, only some of which have been perforated, Δp_s may be positive or negative[3] depending on whether the zones open to flow have an average permeability which is lower or higher than the average of the entire interval.

Δp_s is always negative where the well is intercepted by fractures.[4]

Where the well has been drilled through an oil-bearing formation and an underlying aquifer, the section will obviously be only partially perforated, so as to avoid producing water. We might therefore expect the skin to be positive ($\Delta p_s > 0$).

However, we have two fluids present: oil (which flows towards the well), and water (which, although it does not flow, still transmits the pressure disturbance).

If the oil has a higher viscosity than the water, Chierici[3] has shown that it is possible to find a negative skin factor (Δp_s).

We shall return to the skin factor, and its estimation from well tests, in Chap. 6.

5.6.2 Pseudo-Steady State Flow

Once the pressure disturbance has reached the outer limits of the drainage area of the well, and after a relatively short "late transient" period which follows this, the flow regime enters pseudo-steady state, during which for $q = $ constant, dp/dt is constant at all points in the drainage area (see Sect. 5.5).

To satisfy the conservation of mass:

(mass of fluid leaving the well per unit time) = (change per unit time of mass of fluid in the drainage area of the well)

we have the following relationship, which also allows for the rock compressibility:

$$q\rho = -\pi r_e^2 h \frac{\partial}{\partial t}(\rho\phi) = -\pi r_e^2 h \frac{d(\rho\phi)}{dp} \frac{\partial p}{\partial t} = -\pi r_e^2 h \phi \rho c_t \frac{\partial p}{\partial t} . \tag{5.39a}$$

from which it follows that:

$$\frac{\partial p}{\partial t} = -\frac{q}{\pi r_e^2 h \phi c_t} , \tag{5.39b}$$

From Eqs. (5.10) and (5.39b) we have:

$$\frac{1}{r} \frac{\partial}{\partial r}\left(r \frac{\partial p}{\partial r}\right) = -\frac{q\mu}{2\pi r_e^2 hk} \tag{5.40}$$

and, after first integrating with respect to r:

$$r\frac{\partial p}{\partial r} = -\frac{q\mu}{2\pi kh}\left(\frac{r}{r_e}\right)^2 + C_1 \, . \tag{5.41}$$

At $r = r_e$, $(\partial p/\partial r)_{r_e} = 0$, so that:

$$C_1 = \frac{q\mu}{2\pi kh} \, . \tag{5.42}$$

We therefore have:

$$\frac{\partial p}{\partial r} = \frac{q\mu}{2\pi kh}\left(\frac{1}{r} - \frac{r}{r_e^2}\right) , \tag{5.43}$$

which when integrated between r_w and r (equivalent to integration between p_{wf} and p) gives:

$$p - p_{wf} = \frac{q\mu}{2\pi kh}\left(\ln\frac{r}{r_w} - \frac{r^2 - r_w^2}{2r_e^2}\right) . \tag{5.44a}$$

The term $(r_w/r_e)^2$ is negligibly small, so that we can write, after introducing the skin factor S (see Sect. 5.6.1.2):

$$p_e - p_{wf} = \frac{q\mu}{2\pi kh}\left(\ln\frac{r_e}{r_w} - \frac{1}{2} + S\right) . \tag{5.44b}$$

This equation would enable us to calculate the bottom hole flowing pressure p_{wf} in a producing well at any time, given the outer boundary pressure p_e at the same instant. This latter term is not usually known, and it is more useful to write the equation in terms of the *average* drainage area pressure $\bar{p}$, since this can usually be estimated.

From the radial symmetry of the system we have:

$$\bar{p} = \frac{\int_{r_w}^{r_e} 2\pi rhp\,\mathrm{d}r}{\pi(r_e^2 - r_w^2)h} \tag{5.45a}$$

or, ignoring the r_w^2 term (negligible relative to r_e^2) in the denominator:

$$\bar{p} = \frac{2}{r_e^2}\int_{r_w}^{r_e} pr\,\mathrm{d}r \, . \tag{5.45b}$$

Substituting for p from Eq. (5.44a):

$$\bar{p} = \frac{2}{r_e^2}\left\{p_{wf}\int_{r_w}^{r_e} r\,\mathrm{d}r + \frac{q\mu}{2\pi kh}\left[\int_{r_w}^{r_e} r\ln\frac{r}{r_w}\,\mathrm{d}r - \frac{1}{2r_e^2}\int_{r_w}^{r_e} r^3\,\mathrm{d}r\right]\right\} , \tag{5.45c}$$

which, when we ignore r_w^2, becomes:

$$\bar{p} = p_{wf} + \frac{q\mu}{2\pi kh\,r_e^2}\frac{2}{1}\left(\frac{r_e^2}{2}\ln\frac{r_e}{r_w} - \frac{r_e^2}{4} - \frac{r_e^2}{8}\right) . \tag{5.45d}$$

Finally, after simplifying and adding the skin factor we get:

$$\bar{p} = p_{wf} + \frac{q\mu}{2\pi kh}\left(\ln\frac{r_e}{r_w} - \frac{3}{4} + S\right) . \tag{5.46}$$

Now, Eq. (5.46) can be used to calculate the average pressure in a circular drainage area in pseudo-steady state flow. Other geometries can be handled by rewriting Eq. (5.46) in a more general form:

$$\bar{p} = p_{wf} + \frac{q\mu}{2\pi kh}\left[\frac{1}{2}\ln\left(\frac{r_e}{r_w}\right)^2 - \frac{1}{2}\ln(e^{3/2}) + S\right]$$

$$= p_{wf} + \frac{q\mu}{4\pi kh}\left[\ln\frac{\pi r_e^2}{\pi e^{3/2} r_w^2} + 2S\right]$$

$$= p_{wf} + \frac{q\mu}{4\pi kh}\left(\ln\frac{A}{31.62\, r_w^2} + 0.809 + 2S\right), \tag{5.47}$$

where A is the area of a circle of radius r_e, and the quantity 31.62 is the value of the *Dietz shape factor* C_A for a circular geometry.

We now assume that A represents any drainage area, regardless of shape. The geometry is taken into account via the shape factor; some of the values of C_A calculated by Dietz[6] for a wide range of geometries are listed in Fig. 5.7. In the same table there are also reported values of $t_D\, r_w^2/A = t_{DA}$, the dimensionless time at which, for a given geometry, pseudo-steady state flow can be assumed to have developed.

With Eq. (5.47) we are able to calculate $\bar{p}$ in fields where well placement leads to non-circular drainage areas, provided of course that we know S in the well under test (see Chap. 6).

For any distribution of wells in a field, once pseudo-steady state flow conditions are established, the size of the drainage area of each well is proportional to its production offtake per unit pay thickness, q/h. This assumes that there is no variation in permeability between the wells.

Consider the case of two wells a distance d apart (Fig. 5.8):

$$\frac{r_1}{q_1/h_1} = \frac{r_2}{q_2/h_2} \tag{5.48a}$$

and

$$r_1 + r_2 = d, \tag{5.48b}$$

$$r_1 = \frac{q_1 h_2}{q_1 h_2 + q_2 h_1}d, \tag{5.49a}$$

$$r_2 = \frac{q_2 h_1}{q_1 h_2 + q_2 h_1}d. \tag{5.49b}$$

The average pressure of the reservoir $\bar{p}_R$ can be calculated as the volume-weighted mean of the average pressures $\bar{p}_i$ of the drainage areas of the wells, the volume being the volume of oil $V_{H,i}$ present in each area:

$$\bar{p}_R = \frac{\sum_i \bar{p}_i V_{H,i}}{\sum_i V_{H,i}}. \tag{5.50a}$$

To a first approximation, the production rate q_i from the ith well is proportional to the volume $V_{H,i}$ contained in its drainage area, so that we can simplify Eq. (5.50a)

Shape	C_A	$\ln C_A$	Exact for $t_{DA}>$	Less than 1% error for $t_{DA}>$	Use infinite system solution with less than 1% error for $t_{DA}<$	Shape	C_A	$\ln C_A$	Exact for $t_{DA}>$	Less than 1% error for $t_{DA}>$	Use infinite system solution with less than 1% error for $t_{DA}<$
	31.62	3.4538	0.1	0.06	0.10	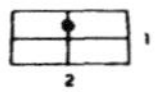	10.8374	2.3830	0.4	0.15	0.025
	31.6	3.4532	0.1	0.06	0.10	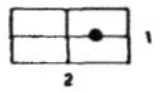	4.5141	1.5072	1.5	0.50	0.06
	27.6	3.3178	0.2	0.07	0.09	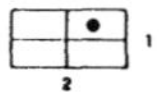	2.0769	0.7309	1.7	0.50	0.02
	27.1	3.2995	0.2	0.07	0.09		3.1573	1.1497	0.4	0.15	0.005
	21.9	3.0865	0.4	0.12	0.08	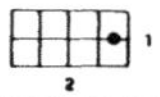	0.5813	-0.5425	2.0	0.60	0.02
	0.098	-2.3227	0.9	0.60	0.015	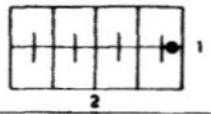	0.1109	-2.1991	3.0	0.60	0.005
	30.8828	3.4302	0.1	0.05	0.09	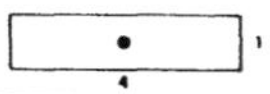	5.3790	1.6825	0.8	0.30	0.01
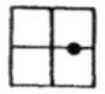	12.9851	2.5638	0.7	0.25	0.03		2.6896	0.9894	0.8	0.30	0.01
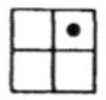	4.5132	1.5070	0.6	0.30	0.025		0.2318	-1.4619	4.0	2.00	0.03
	3.3351	1.2045	0.7	0.25	0.01	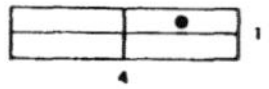	0.1155	-2.1585	4.0	2.00	0.01
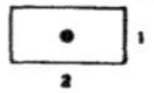	21.8369	3.0836	0.3	0.15	0.025	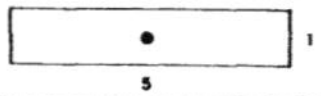	2.3606	0.8589	1.0	0.40	0.025

Fig. 5.7. *Values of the shape factor for different drainage area shapes, and the time limits to the validity of Eqs. (5.36) and (5.47). From Ref. 6, 1965, Society of Petroleum Engineers of AIME, reprinted by permission of the SPE*

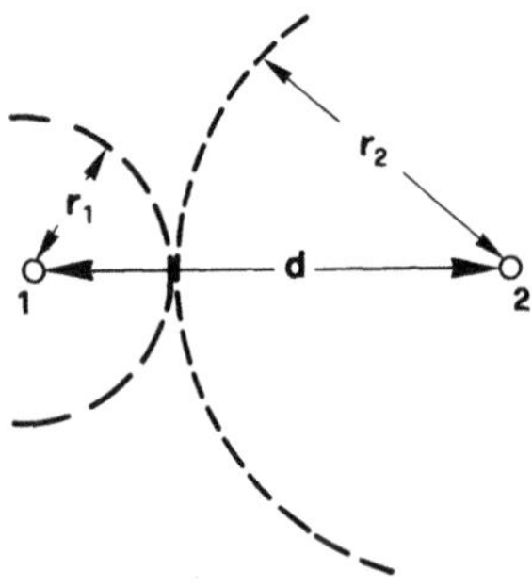

Fig. 5.8. *Drainage areas of two adjacent wells*

to:

$$\bar{p}_R = \frac{\sum_i \bar{p}_i q_i}{\sum_i q_i} \, . \tag{5.50b}$$

5.6.3 Steady State Flow

Steady state flow conditions will occur in a reservoir only when there is an influx of water from a flanking aquifer sufficient to maintain the outer boundary pressure p_e constant with time; or when the reservoir is developed with injection of water or other fluid in such a way that the total volume of fluid (oil + injected fluid) is kept constant.

The boundary conditions applicable at the end of the transient period are:

$$p = p_e = \text{constant} \quad \text{at } r = r_e \tag{5.51a}$$

$$\frac{\partial p}{\partial t} = 0 \quad \text{for all } r \text{ and } t \tag{5.51b}$$

When these conditions apply, Eq. (5.10) becomes:

$$\frac{1}{r} \frac{d}{dr}\left(r \frac{dp}{dr}\right) = 0 \, , \tag{5.52a}$$

$$r \frac{dp}{dr} = \frac{q\mu}{2\pi kh} = \text{const} \, . \tag{5.52b}$$

Integrating, and introducing the skin term:

$$p - p_{wf} = \frac{q\mu}{2\pi kh}\left(\ln\frac{r}{r_w} + S\right) . \tag{5.53}$$

Following the same procedure as for pseudo-steady state flow in Sect. 5.6.2, we obtain the following equation:

$$\bar{p} = p_{wf} + \frac{q\mu}{2\pi kh}\left(\ln\frac{r_e}{r_w} - \frac{1}{2} + S\right) . \tag{5.54}$$

Table 5.1 summarises the radial inflow equations for steady and pseudo-steady state flow introduced in the preceding pages.

The C term on the right-hand side is the units conversion coefficient.

If we adhere strictly to the SI units convention, with p in MPa, μ in mPa s, then the value of C would be $10^{-9}/2\pi$. For practical metric and oilfield units systems, the values of C are listed in Table 5.1, along with the units for each parameter.

Table 5.1. *Radial inflow equations*

	Flow regime	
	Steady state	Pseudo-steady state
General equation	$p - p_{wf} = C\dfrac{q\mu}{kh}\left(\ln\dfrac{r}{r_w} + S\right)$	$p - p_{wf} = C\dfrac{q\mu}{kh}\left(\ln\dfrac{r}{r_w} - \dfrac{r^2}{2r_e^2} + S\right)$
Equation in terms of p_e	$p_e - p_{wf} = C\dfrac{q\mu}{kh}\left(\ln\dfrac{r_e}{r_w} + S\right)$	$p_e - p_{wf} = C\dfrac{q\mu}{kh}\left(\ln\dfrac{r_e}{r_w} - \dfrac{1}{2} + S\right)$
Equation in terms of $\bar{p}$	$\bar{p} - p_{wf} = C\dfrac{q\mu}{kh}\left(\ln\dfrac{r_e}{r_w} - \dfrac{1}{2} + S\right)$	$\bar{p} - p_{wf} = C\dfrac{q\mu}{kh}\left(\ln\dfrac{r_e}{r_w} - \dfrac{3}{4} + S\right)$

Parameter	Symbol	Units			
		Practical metric		Oilfield (US)	
Pay thickness	h	metres	(m)	feet	(ft)
Permeability	k	millidarcy	(md)	millidarcy	(md)
Downhole flow rate	q	cubic metres/day	(m³/d)	barrels/day	(bbl/d)
Radii	r, r_e, r_w	metres	(m)	feet	(ft)
Viscosity	μ	centipoise	(cP)	centipoise	(cP)
Pressure	$p, p_e, p_{wf}, \bar{p}$	kilogram/square cm	(kg/cm²)	pounds/square inch	(psi)
Constant	C	19.03		141.2	

5.7 The Principle of Superposition Applied to the Solution of the Diffusivity Equation

The linearised diffusivity equation, Eq. (5.10), and indeed all differential equations with constant coefficients, are amenable to Duhamel's theorem, which states that "a linear combination of the solutions of a differential equation is itself a solution to that equation".

In other words, if we combine – in a linear fashion – solutions to the diffusivity equation corresponding to different initial and boundary conditions, we will obtain another solution to the diffusivity equation. This is the so-called *principle of superposition*, which is so widely used in mathematical physics.

Consider the case of a well which is initially shut in ($q = 0$ at $t = 0$), with a constant pressure p_i over the whole of its drainage area.

The well is now put on production at a rate q, which is varied with time (Fig. 5.9) instead of being held constant. The flow schedule is as follows:

Flow rate	Duration
q_1	$t_1 - 0$
q_2	$t_2 - t_1$
q_3	$t_3 - t_2$
...	...
q_{j-1}	$t_{j-1} - t_{j-2}$
q_j	$t_j - t_{j-1}$
...	...
q_n	$t_n - t_{n-1}$

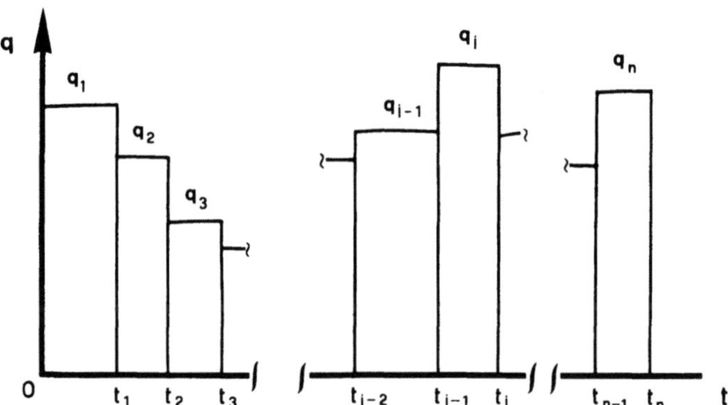

Fig. 5.9. *Schematic presentation of flow rate variations with time for use with the principle of superposition*

The same information can be presented in a different way:

Flow rate	Duration
q_1	$t_n - 0$
$q_2 - q_1$	$t_n - t_1$
$q_3 - q_2$	$t_n - t_2$
...	...
$q_j - q_{j-1}$	$t_n - t_{j-1}$
...	...
$q_n - q_{n-1}$	$t_n - t_{n-1}$

Using the general solution of diffusivity equation [Eq. (5.16)], written for convenience in dimensionless form, we have:

$$\frac{2\pi kh}{\mu}[p_i - p_{wf}(t_n)] = (q_1 - 0)[p_{Dw}(t_{Dn} - 0) + S]$$

$$+ (q_2 - q_1)[p_{Dw}(t_{Dn} - t_{D1}) + S]$$

$$+ (q_3 - q_2)[p_{Dw}(t_{Dn} - t_{D2}) + S]$$

$$\vdots$$

$$+ (q_j - q_{j-1})[p_{Dw}(t_{Dn} - t_{D,j-1}) + S]$$

$$\vdots$$

$$+ (q_n - q_{n-1})[p_{Dw}(t_{Dn} - t_{D,n-1}) + S] \qquad (5.55a)$$

or:

$$p_{wf}(t_n) = p_i - \frac{\mu}{2\pi kh}\left\{\sum_{j=1}^{n}[(q_j - q_{j-1})p_{Dw}(t_{Dn} - t_{D,j-1})] + q_n S\right\}. \qquad (5.55b)$$

Equation (5.55b) provides a generalised (and theoretically rigorous) means of interpreting the bottom hole flowing pressure response when the well is producing at any rate, or when it has been shut in after a period of production. It is also applicable (by using a *negative q*) to the equivalent injectivity and pressure fall-off tests in wells where water is being injected for pressure maintenance or water flooding.

Equation (5.55) can therefore be used to evaluate kh, S and the drainage radius (or the distance to an impermeable barrier) from the bottom-hole flowing/shut-in pressure response and a knowledge of the rate variations.

Chapter 6 will deal exclusively with this topic. By way of an introduction to this, we will round off the present chapter with a summary of the most common types of production and injection tests in current use.

5.8 Testing Production and Injection Wells

Production wells are almost invariably tested under flowing conditions $(q > 0)$. Only if the reservoir pressure is low enough for production to require artificial lift (downhole pump, gas lift), might it be preferable to conduct an injection test $(q < 0)$. This would be in order to avoid mechanical difficulties associated with having the pressure gauge in the well while the artificial lift equipment is still operational, and is generally achieved by pumping down diesel or reservoir crude.

Water injection wells are usually tested by injection, although very occasionally production (back flow) tests may be performed.

Starting from stabilised conditions $(p = p_i$ over the whole drainage area at $t = 0$; $q = 0)$, a well test will involve putting the well on production or injection either at a constant rate q (at downhole conditions) or, as operating conditions often dictate, at a sequence of rates $q_1, q_2, \ldots, q_n$ for time periods $t_1, (t_2 - t_1), \ldots, (t_n - t_{n-1})$.

The former case is referred to as a *constant rate test*, the latter as a *multi-rate test*.

With a well on production, the measurement of the bottom hole flowing pressure $p_{wf}(t)$ – a decreasing trend - constitutes a *pressure drawdown test*; while in the case of injection, the measurement – in this case an increasing trend – constitutes an *injectivity test*.

The well is usually shut in $(q = 0)$ at the end of the production or injection period, and the bottom hole pressure measurement is continued.

When a producing well is shut in, the bottom hole pressure $p_{ws}(t)$ rises, and the pressure measurement is referred to as a *buildup test* (or PBU); while the declining pressure response which follows the shutting in of an injection well is called a *pressure fall off test* (or PFO).

The methods of obtaining kh, S and r_e or the distance to a permeability boundary from pressure measurements during production or injection (p_{wf}), and during shut-in (p_{ws}), are described in detail in Chap. 6.

It is also common practice to calculate the following three additional parameters from the test data:

Well productivity index, J:

$$J = \frac{q}{\bar{p} - p_{wf}} . \qquad (5.56a)$$

Flow efficiency (FE), *or well completion factor* (CF):

$$\mathrm{CF} = \frac{J \ \mathrm{real}\,(S \neq 0)}{J \ \mathrm{ideal}\,(S = 0)} \tag{5.56b}$$

$$= \frac{\dfrac{q}{\bar{p} - p_{\mathrm{wf}}}}{\dfrac{q}{\bar{p} - p_{\mathrm{wf}} - \Delta p_{\mathrm{s}}}} = 1 - \frac{\Delta p_{\mathrm{s}}}{\bar{p} - p_{\mathrm{wf}}} \ . \tag{5.56b'}$$

Damage ratio, F_{s}:

$$F_{\mathrm{s}} = 1 - \mathrm{CF} = \frac{\Delta p_{\mathrm{s}}}{\bar{p} - p_{\mathrm{wf}}} \ . \tag{5.56c}$$

References

1. Brons F, Marting VE (1961) The effect of restricted fluid entry on well productivity. J Petrol Tech (Feb): 172–174; Trans AIME 222
2. Carslaw HS, Jaeger JC (1947) Conduction of heat in solids. Clarendon Press, Oxford
3. Chierici GL, Ciucci GM, Pizzi G (1965) Quelques cas de remontées de pression dans des couches hétérogenes avec pénétration partielle. Etude par analyseur électrique. Rev Inst Fr Pétrole (Dec): 20 (12): 1811–1846
4. Cinco Ley H, Samaniego F, Dominguez N (1978) Transient pressure behaviour for a well with a finite conductivity vertical fracture. Soc Petrol Eng J (Aug): 253–264
5. Dake LP (1978) Fundamentals of petroleum engineering. Elsevier, Amsterdam
6. Dietz DN (1965) Determination of average reservoir pressure from buildup surveys. J Petrol Tech (Aug): 955–959; Trans AIME 234
7. Dranchuk PM, Quon D (1967) Analysis of the Darcy continuity equation. Prod Monthly (Oct): 25–28
8. Harris MH (1966) The effect of perforating on well productivity. J Petrol Tech (April): 518–528; Trans AIME 237
9. Nisle RG (1958) The effect of partial penetration on pressure buildup in oil wells. Trans AIME 213: 85–90
10. van Everdingen AF (1953) The skin effect and its influence on the productive capacity of a well. Trans AIME 193: 171–176
11. van Everdingen AF, Hurst W (1949) The application of the Laplace transformation to flow problems in reservoirs. Trans AIME 179: 305–324

EXERCISES

Exercise 5.1

An oil well has been put on production at a rate $q_0 = 300\,\mathrm{m}^3/\mathrm{d}$, measured under standard conditions. The reservoir characteristics are:

– thickness:	h	$= 20\,\mathrm{m}$,
– porosity:	ϕ	$= 0.25$,
– permeability:	k	$= 350\,\mathrm{md}$,
– water saturation:	S_{iw}	$= 0.15$,
– gas saturation:	S_{g}	$= 0.00$,
– well radius:	r_{w}	$= 6''$,
– static pressure:	p_{i}	$= 250\,\mathrm{kg/cm}^2$.

The reservoir fluid properties (under reservoir conditions) were as follows:

- oil volume factor: $B_o = 1.8$,
- oil viscosity: $\mu_o = 0.5\,\text{cP}$,
- oil compressibility: $c_o = 2.5 \times 10^{-4}\,\text{cm}^2/\text{kg}$,
- water compressibility: $c_w = 3.0 \times 10^{-5}\,\text{cm}^2/\text{kg}$.

In addition:

- pore compressibility: $c_f = 1.4 \times 10^{-4}\,\text{cm}^2/\text{kg}$.

Calculate the time at which the approximation

$$ei(x) = -\ln(\gamma x) \tag{5.31c}$$

becomes valid for the calculation of the wellbore flowing pressure, and the pressures at a distance of 50, 100 and 150 m from the well.

Derive the flowing pressure response during the first day of production, assuming the absence of any skin factor or interference from neighbouring wells.

Solution

It was explained in Sect. 5.6.1.1 that the logarithmic approximation to the exponential integral [Eq. (5.31c)] is valid when $x < 0.01$, where x is the argument of the "exponential integral".

In this example:

$$x = \frac{\phi \mu_o c_t r^2}{4kt}$$

and therefore the condition for validity of Eq. (5.31c) is:

$$\frac{\phi \mu_o c_t r^2}{4kt} < 0.01$$

or, in other words:

$$t > 25 \frac{\phi \mu_o c_t r^2}{k} \ .$$

All quantities are expressed in SI units in these equations. Converting to practical metric units, we have:

$$t(\text{min}) \times 60 > 25 \, \frac{\phi[\mu_o(\text{cP}) \times 10^{-3}]\left[\dfrac{c_t(\text{cm}^2/\text{kg})}{9.80665 \times 10^4}\right] r^2(\text{m}^2)}{k(\text{md}) \times 9.86923 \times 10^{-16}} \ ,$$

which gives:

$$t(\text{min}) > 4.305 \times 10^6 \, \frac{\phi \mu_o(\text{cP}) c_t(\text{cm}^2/\text{kg}) r^2(\text{m}^2)}{k(\text{md})} \ .$$

Using Eq. (5.5c) for total compressibility:

$$c_t = c_o S_o + c_w S_{iw} + c_f$$

$$= 2.5 \times 10^{-4}(1 - 0.15) + 3.0 \times 10^{-5} \times 0.15 + 1.4 \times 10^{-4}$$

$$= 3.57 \times 10^{-4}\,\text{cm}^2/\text{kg}$$

Therefore,

$$t(\text{min}) > 4.305 \times 10^6 \, \frac{[0.25 \times 0.5 \times 3.57 \times 10^{-4} \times r^2(\text{m}^2)]}{350}$$

i.e:

$$t(\text{min}) > 0.549 \, r^2(\text{m}^2) \ .$$

For a well radius of 6″ (0.1524 m), the onset of the validity of the logarithmic approximation will be as follows:

$r\,(\mathrm{m})$	$t\,(\mathrm{min})$	
$0.1524 = r_\mathrm{w}$	1.3×10^{-2}	$= 0.76\,\mathrm{s}$
50	1373	$\cong 1$ day
100	5490	$\cong 4$ days
200	21960	$\cong 15$ days

To calculate the bottom hole flowing pressure (in the absence of skin and interference, and assuming transient radial flow) we use Eq. (5.33b):

$$p_\mathrm{wf} = p_\mathrm{i} - \frac{q\mu_\mathrm{o}}{4\pi kh}\left(\ln\frac{kt}{\phi\mu_\mathrm{o}c_\mathrm{t}r_\mathrm{w}^2} + 0.809\right).$$

Converted from SI to practical metric units, the diffusivity term is:

$$\frac{k\,(\mathrm{m}^2)\,t\,(\mathrm{s})}{\phi\mu_\mathrm{o}(\mathrm{Pa\,s})c_\mathrm{t}(\mathrm{Pa}^{-1})r_\mathrm{w}^2\,(\mathrm{m}^2)} = \frac{[k\,(\mathrm{md})\times 9.869233\times 10^{-16}]\times[t\,(\mathrm{min})\times 60]}{\phi\,[\mu_\mathrm{o}(\mathrm{cP})\times 10^{-3}]\left[\dfrac{c_\mathrm{t}(\mathrm{cm}^2/\mathrm{kg})}{9.80665\cdot 10^4}\right]r_\mathrm{w}^2\,(\mathrm{m}^2)}$$

$$= 5.807\times 10^{-6}\,\frac{k\,(\mathrm{md})t\,(\mathrm{min})}{\phi\mu_\mathrm{o}(\mathrm{cP})c_\mathrm{t}(\mathrm{cm}^2/\mathrm{kg})r_\mathrm{w}^2\,(\mathrm{m}^2)}\,,$$

while:

$$\frac{q\,(\mathrm{m}^3/\mathrm{s})\mu_\mathrm{o}(\mathrm{Pa\,s})}{4\pi k\,(\mathrm{m}^2)h\,(\mathrm{m})} = \frac{\left[\dfrac{q\,(\mathrm{m}^3/\mathrm{d})}{86\,400}\right][\mu_\mathrm{o}(\mathrm{cP})\times 10^{-3}]}{4\pi\,[k\,(\mathrm{md})\times 9.869233\times 10^{-16}]\,[h\,(\mathrm{m})]}$$

$$= 933\,240\,\frac{q\,(\mathrm{m}^3/\mathrm{d})\mu_\mathrm{o}(\mathrm{cP})}{k\,(\mathrm{md})h\,(\mathrm{m})}\,.$$

Using the relationship:

$$1\,\mathrm{kg/cm^2} = 9.80665\times 10^4\,\mathrm{Pa}\,,$$

Equation (5.33) becomes:

$$p_\mathrm{wf}\,(\mathrm{kg/cm^2}) = p_\mathrm{i}\,(\mathrm{kg/cm^2}) - 9.516\,\frac{q_\mathrm{o}\,(\mathrm{m}^3/\mathrm{d})\,B_\mathrm{o}\mu_\mathrm{o}\,(\mathrm{cP})}{k\,(\mathrm{md})h\,(\mathrm{m})}$$

$$\times\left(\ln 5.807\times 10^{-6}\,\frac{k\,(\mathrm{md})t\,(\mathrm{min})}{\phi\mu_\mathrm{o}(\mathrm{cP})c_\mathrm{t}(\mathrm{cm}^2/\mathrm{kg})r_\mathrm{w}^2\,(\mathrm{m}^2)} + 0.809\right).$$

Here, we have introduced the volume factor B_0 to convert the surface oil production rate q_0 to a downhole rate. Implementing the units coefficients, and changing from the natural logarithm to $\log_{10}$, the equation can now be written:

$$p_\mathrm{wf} = p_\mathrm{i} - 21.912\,\frac{q_\mathrm{o}B_\mathrm{o}\mu_\mathrm{o}}{kh}\left[\log\frac{kt}{\phi\mu_\mathrm{o}c_\mathrm{t}r_\mathrm{w}^2} - 4.885\right].$$

For the well in this exercise, we have:

$$p_\mathrm{wf} = 250 - 21.912\,\frac{300\times 1.8\times 0.5}{350\times 20}\left[\log\frac{350\cdot t}{0.25\times 0.5\times 3.57\times 10^{-4}\times 0.1524^2} - 4.885\right].$$

so that:

$$p_\mathrm{wf}(\mathrm{kg/cm^2}) = 250 - 0.8451\,[\log(3.38\times 10^8\,t) - 4.885]\,.$$

At various times, this works out as follows:

t	$p_{wf}\,(\mathrm{kg/cm^2})$
1′	246.92
10′	246.08
30′	245.67
1 h	245.42
3 h	245.01
6 h	244.76
12 h	244.51
18 h	244.36
24 h	244.25

$$\diamond\,\diamond\,\diamond$$

Exercise 5.2

A second well is drilled 200 m from the well of Ex. 5.1.

Geological evidence has lead to some doubt as to whether the reservoir is continuous between the two wells; there is the possibility that a sealing fault is preventing hydraulic communication.

An extended production test is performed in the first (Ex. 5.1) well, at a surface rate of 300 m³/d. The second well is shut in, and its bottom hole pressure is recorded to ascertain whether any pressure signal can be detected from the producing well, indicating at least some degree of communication between the two.

Assuming the pressure gauge used has a resolution of 0.2 kg/cm², and that the properties of the reservoir rock are constant, estimate how long the test must be run in order to determine that there is (perfect) hydraulic continuity between the wells.

Solution

From Eq. (5.30):

$$\frac{4\pi kh}{q\mu_o}[p_i - p(r,t)] = \int_x^\infty \frac{e^{-s}}{s}\,ds = ei(x)\,,$$

we have, after converting from SI to practical metric units:

$$\frac{4\pi\,[k\,(\mathrm{md})\times 9.869233\times 10^{-16}]\,h\,(\mathrm{m})}{\dfrac{q_o\,(\mathrm{m^3/d})\,B_o}{86\,400}}\left[p_i - p(r,t)\right](\mathrm{kg/cm^2})\times 9.80665\times 10^4 = ei(x)\,,$$

which gives:

$$0.1051\,\frac{k\,(\mathrm{md})\,h\,(\mathrm{m})}{q_o\,(\mathrm{m^3/d})\,B_o\mu_o\,(\mathrm{cP})}[p_i - p(r,t)]\,(\mathrm{kg/cm^2}) = ei(x)\,.$$

We want to see a pressure decrease of 0.2 kg/cm² at a distance of $r = 200$ m from the producing well.

In other words:

$$p_i - p(200,t) = 0.2\,\mathrm{kg/cm^2}$$

Using the data from Ex. 5.1, we therefore have:

$$ei(x) = 0.1051\,\frac{350\times 20\times 0.2}{300\times 1.8\times 0.5} = 0.545\,.$$

From Fig. 5.5 we have, for $ei(x) = 0.545$:

$$x = 0.51$$

and, from Eq. (5.22a):

$$s = x = \frac{\phi \mu_o c_t r^2}{4kt},$$

which, in practical metric units is:

$$x = \frac{\phi [\mu_o (\text{cP}) \times 10^{-3}] \left[\dfrac{c_t (\text{cm}^2/\text{kg})}{9.80665 \cdot 10^4} \right] r^2 (\text{m}^2)}{4 [k (\text{md}) \times 9.869233 \times 10^{-16}] [t (\text{min}) \times 60]}$$

$$= 43\,051 \frac{\phi \mu_o (\text{cP}) c_t (\text{cm}^2/\text{kg}) r^2 (\text{m}^2)}{k (\text{md}) t (\text{min})}.$$

so that:

$$x = 0.51 = 43\,051 \left[\frac{0.25 \times 0.5 \times 3.57 \times 10^{-4} \times 200^2}{350\, t (\text{min})} \right].$$

Therefore:

$$t = 430 \, \text{min} \cong 7 \, \text{h}$$

If there is perfect hydraulic continuity between the two wells, we should measure a decrease of $0.2 \, \text{kg/cm}^2$ in the shut in well after 7 h of production from the other well. In practical terms, we should have a measurable signal exceeding gauge error in under 12 h.

Referring to Ex. 5.1, note that this time is considerably less than that required for the logarithmic approximation to be valid at $r = 200 \, \text{m}$.

If instead of Eq. (5.30) we had written:

$$\frac{4\pi kh}{q\mu_o} [p_i - p(r, t)] = \ln \frac{kt}{\phi \mu_o c_t r^2} + 0.809 ,$$

we would therefore have had a serious error in our calculated time.

Exercise 5.3

A field has been developed by drilling wells in a regular pattern, such that each well lies at the centre of a rectangle measuring 100 m by 400 m.

The drainage area of each well is therefore rectangular in shape, $(100 \times 400)\,\text{m}^2$, with the well at the centre.

The oil from this field has the following properties:

- bubble point pressure (at reservoir T): $p_b = 100 \, \text{kg/cm}^2$,
- volume factor at current average pressure: $B_o = 1.15$,
- viscosity at current average pressure: $\mu_o = 35 \, \text{cP}$.

A production test is run in one of the wells, completed with a $9\text{-}\frac{5}{8}''$ casing perforated across the entire reservoir interval. In this well, the formation characteristics are:

- thickness: $h = 120 \, \text{m}$,
- permeability: $k = 850 \, \text{md}$.

There is a certain amount of mud filtrate damage to the near-wellbore formation, which has reduced the local permeability to $k_s = 200 \, \text{md}$ out to a radius of 1 m.

The well test, an extended drawdown in pseudo-steady state flow, gave the following results:

- oil flow rate (surface): $\qquad$ q_o = 150 m³/d (standard conditions) ,
- bottom hole flowing pressure: $\qquad$ p_{wf} = 185 kg/cm² .

Calculate the productivity index J, the completion factor CF, and the damage ratio F_s.

Solution

The first thing to note is that the bottom hole flowing pressure is higher than the bubble point, so we have single phase flow of undersaturated oil.

The skin damage effect is quantified as (Eq. (5.38b)):

$$S = \frac{k - k_s}{k_s} \ln \frac{r_s}{r_w} \ .$$

Since

$$r_w = 0.5(9\tfrac{5}{8}'') = 0.122 \, \text{m} \ ,$$

we have,

$$S = \frac{850 - 200}{200} \ln \frac{1}{0.122} = 6.84 \ .$$

With the rate expressed at reservoir conditions, Eq. (5.47) is:

$$\bar{p} - p_{wf} = \frac{q B_0 \mu_0}{4\pi k h}\left[\ln \frac{A}{C_A r_w^2} + 0.809 + 2S \right] \ .$$

Converting from SI units to practical metric:

$$[\bar{p} - p_{wf}](\text{kg/cm}^2) \times 9.80665 \times 10^4 = \frac{\dfrac{q\,(\text{m}^3/\text{d})}{86\,400} B_0 [\mu_0\,(\text{cP}) \times 10^{-3}]}{4\pi[k\,(\text{md}) \times 9.869233 \times 10^{-16}]h\,(\text{m})}$$

$$\times \left[\ln \frac{A}{C_A r_w^2} + 0.809 + 2S \right] \ ,$$

which, replacing the natural log by $\log_{10}$, is:

$$[\bar{p} - p_{wf}](\text{kg/cm}^2) = 21.912 \frac{q\,(\text{m}^3/\text{d}) B_0 \mu_0\,(\text{cP})}{k\,(\text{md})h\,(\text{m})}\left[\log \frac{A}{C_A r_w^2} + 0.35134 + 0.869\,S \right] \ .$$

Referring to Fig. 5.7, the Dietz shape factor for the drainage area (4:1 rectangle) with the well in the centre is:

$$C_A = 5.379 \ .$$

Therefore:

$$\bar{p} - p_{wf} = 21.912 \frac{150 \times 1.15 \times 35}{850 \times 120}\left[\log \frac{400 \times 100}{5.379 \times 0.122^2} + 0.35134 + 0.869 \times 6.84 \right] = 15.56 \, \text{kg/cm}^2 \ ,$$

so that the average pressure of the drainage area is:

$$\bar{p} = p_{wf} + 15.56 = 185 + 15.56$$

$$= 200.56 \, \text{kg/cm}^2 \ .$$

The pressure drop Δp_s caused by the skin damage zone is defined in Eq. (5.37). In practical metric units, this is:

$$\Delta p_s\,(\text{kg/cm}^2) = 19.033 \frac{q_o\,(\text{m}^3/\text{d}) B_0 \mu_0\,(\text{cP})}{k\,(\text{md})h\,(\text{m})} S$$

so that, in this example:

$$\Delta p_{\rm s} = 19.033 \frac{(150 \times 1.15 \times 35)}{850 \times 120} 6.84 = 7.71 \,\text{kg/cm}^2 \ .$$

Finally, we can calculate:

Productivity index:

$$J = \frac{q}{\bar{p} - p_{\rm wf}} = \frac{150}{15.56} = 9.64 \frac{\text{m}^3}{\text{d} \times \text{kg/cm}^2} \ .$$

Completion factor:

$$\text{CF} = 1 - \frac{\Delta p_{\rm s}}{\bar{p} - p_{\rm wf}} = 1 - \frac{7.71}{15.56} = 50.4\% \ .$$

Damage ratio:

$$F_{\rm s} = 1 - \text{CF} = 49.6\% \ .$$

Note that the productivity index could be almost *doubled* by eliminating the skin damage (e.g. surfactant treatment of the damaged zone). As a consequence, *twice the production rate could be obtained for the same bottom-hole flowing pressure.*

6 The Interpretation of Production Tests in Oil Wells

6.1 Introduction

Data obtained from a production test, if correctly interpreted, represent the most important source of information about the flow geometry and the *dynamic* properties (effective permeability) of the reservoir rock in the drainage area of the well, in addition to its average pressure.

Possible damage to the permeability of the near-wellbore rock (the skin effect caused by mud filtrate or completion fluid invasion) can also be evaluated, with a view to designing a suitable remedial treatment, where necessary, to improve the productivity of the well.

The results of a conventional well test refer to the *whole* volume of the reservoir rock within the area investigated, under the prevailing local geostatic conditions at the time of the test, with the flow of in situ reservoir fluid or fluids. Since it is the flowing response of the entire thickness of the pay zone that is measured, it is not possible to evaluate the permeability profile across the interval, nor the permeability of individual layers; this would require highly specialised "multi-layer test" techniques.

In terms of systems analysis, a well test constitutes the measurement of the response [bottom hole flowing pressure] of a system [well + reservoir rock + reservoir fluid] to an input signal [flow rate $q(t)$].

From the behaviour of $p_{wf}(t)$ in response to $q(t)$, we attempt to determine the transform $p_D(t_D)$ for the system, as well as (if possible) *unique* values of the characteristic constants (k, S) and initial conditions (p_i).

6.2 Measurement of Flow Rate and Pressure

Flow rates are usually measured by conventional field techniques. Produced fluids are fed into a test separator kept at constant temperature and pressure. The flow rates of oil, gas (and water if present) are measured at the separator outlets, either directly using a calibrated flow meter, or volumetrically.

The oil, now at atmospheric pressure, is then fed into a vertical tank of uniform cross section, and the oil rate is estimated from periodic measurements of the level of oil in the tank as a function of time.

Several different types of gauge (or "bomb") are used to measure the bottom hole flowing pressure.

Mechanical gauges (Amerada, Kuster) are of a very simple design, and are highly robust. They have been in use for many years, and still find applications in low-cost operations, or where conditions (notably high temperature) preclude the use of more sophisticated gauges.

The gauge is run in-hole on a fine steel cable ("slick line"), spooled from a winch. Pressure is recorded inside the gauge by means of a stylus attached to one end of a Bourdon tube manometer, which inscribes a trace on a length of film on the surface of a cylinder rotating at a constant speed (Fig. 6.1).

The test record is read using a microscope – the deflection of the trace in the vertical direction is proportional to the pressure, and in the horizontal travel is proportional to the time.

If well calibrated, a mechanical gauge can have an absolute accuracy of 0.2% (e.g. 20 psi at 10 000 psi), with a reading resolution of perhaps 5–10 psi.

By the mid-1970s, equipment technology had improved considerably, and with the advent of "surface read-out", gauge data could be transmitted to surface in real time via a conductor wire in the cable. This requires a thicker cable (because of the conductor wire), and bulky surface equipment, which necessitate a special transport unit. Owing to the complexity of such systems, surface read out tends to be an expensive service.

The advantage of surface read-out is, of course, that the bottom hole pressure behaviour can be observed *during the test*, allowing alterations to be made to the test conditions if the pressure response so indicates (change flow rate, extend or shorten test times), and tool malfunction to be identified immediately.

For slick line applications, where the relatively expensive advantages offered by surface read-out were not required, the "down-hole memory" system

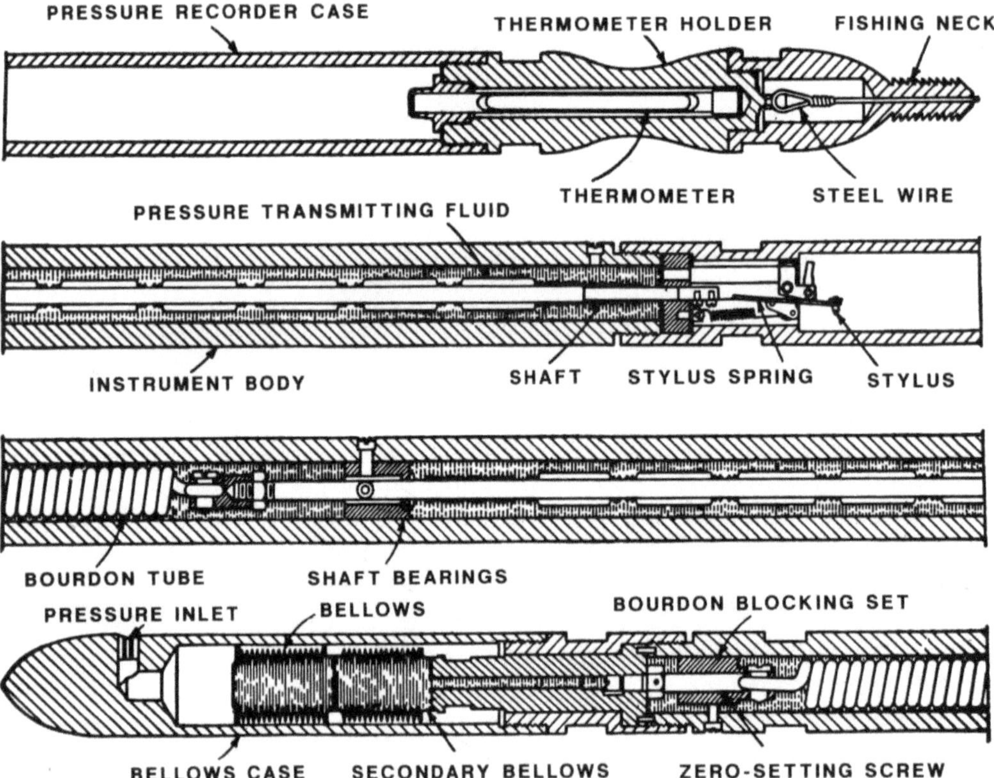

Fig. 6.1. *Schematic of the sections of an Amerada gauge used for the measurement of bottom hole flowing pressure*

was introduced. Here, the data is stored in a memory chip (EPROM) which is interrogated after the test when the equipment has been brought to surface.

Advances in sensor design also saw the introduction of the resistive strain gauge (Sperry Sun, Schlumberger, Atlas, GRC), the capacitance gauge (Panex, GRC), the field effect transistor and, most importantly, the quartz crystal gauge (Schlumberger/Flopetrol, Geoservices, Hewlett Packard, TerraQuartz, etc). These new gauges could be run with down hole memory acquisition or surface read-out.

The quartz gauge has a measurement resolution of 0.01 psi, and is capable of an absolute accuracy of 0.025% (e.g. ± 1.25 psi on a 5000 psi reading). The equivalent figures for a typical strain gauge are 0.1 psi resolution and 0.2% accuracy (e.g. ± 10.0 psi at 5000 psi); and for the capacitance gauge 0.01 psi resolution and 0.04% accuracy (e.g. ± 2.0 psi at 5000 psi). These absolute accuracy ratings assume of course that the gauge has been properly calibrated in the workshop before the test, and that an accurate measurement of the temperature (which strongly affects the sensor response) is available during the test.

Because of their technical complexity, these gauges lack the ruggedness of the purely mechanical gauge, and are usually restricted to maximum operating temperatures of 150–175 °C (compared with over 300 °C for some variants of the mechanical variety).

6.3 The Problem of Non-Uniqueness

We will start with a simple case of a well test in a well producing at a constant surface rate q_{sc} measured at standard conditions (i.e. $q_{sc}B_o$ at reservoir conditions). We assume that the reservoir was initially at a uniform pressure p_i, and that there are no other wells on production in the field.

Taking Eq. (5.16), adding the skin factor and replacing $p_{D,w}$ by p_D we have

$$\frac{2\pi k_o h}{q_{sc} B_o \mu_o} [p_i - p_{wf}(t)] = p_D(t_D) + S \,, \tag{6.1}$$

where

$$t_D = \frac{k_o}{\phi \mu_o c_t r_w^2} t \,, \tag{5.14b}$$

Equation (6.1) contains *three* unknowns: k_o, S and the $p_D(t_D)$ function [provided we know B_o, μ_o (Chap. 2), and c_t, ϕ (Chap. 3)].

As long as the flow regime is transient (that is, until the pressure disturbance has reached the outer boundaries of the reservoir), we can write, for radial flow:

$$p_D(t_D) = \tfrac{1}{2}(\ln t_D + 0.809) \,, \tag{5.33a}$$

so that

$$\frac{2\pi k_o h}{q_{sc} B_o \mu_o} [p_i - p_{wf}(t)] = \tfrac{1}{2}(\ln t_D + 0.809 + 2S)$$

$$= \tfrac{1}{2}\left(\ln t + \ln \frac{k_o}{\phi \mu_o c_t r_w^2} + 0.809 + 2S\right) \,, \tag{6.2}$$

from which

$$-\frac{\mathrm{d}p_{\mathrm{wf}}}{\mathrm{d}\ln t} = \frac{q_{\mathrm{sc}}B_{\mathrm{o}}\mu_{\mathrm{o}}}{4\pi k_{\mathrm{o}}h} = m \; . \tag{6.3}$$

Eqution (6.3) reveals two very important facts:

1. While the reservoir flow is transient, a "semi-log" plot of p_{wf} versus $\ln t$ is a *straight line* (Fig. 6.2).
2. The slope of the straight line can be used to calculate $k_{\mathrm{o}}h/\mu$, and therefore k_{o}.

Once we have computed k_{o}, we can get S directly from Eq. (6.2).

Suppose now that at a time t the flow rate is altered: more precisely, suppose the well is shut in. The produced rate is now zero – this is equivalent to annulling the production rate $+ q_{\mathrm{sc}}\, B_{\mathrm{o}}$ by superposing an equal and opposite rate $- q_{\mathrm{sc}} B_{\mathrm{o}}$ (Fig. 6.3) at time t.

We now apply the superposition theorem introduced in Sect. 5.7, representing the bottom hole pressure at a time Δt after shut in by $p_{\mathrm{ws}}(\Delta t)$, so that

$$\frac{2\pi k_{\mathrm{o}} h}{q_{\mathrm{sc}} B_{\mathrm{o}} \mu_{\mathrm{o}}}[p_{\mathrm{i}} - p_{\mathrm{ws}}(\Delta t)] = p_{\mathrm{D}}(t_{\mathrm{D}} + \Delta t_{\mathrm{D}}) + S - [p_{\mathrm{D}}(\Delta t_{\mathrm{D}}) + S]$$

$$= p_{\mathrm{D}}(t_{\mathrm{D}} + \Delta t_{\mathrm{D}}) - p_{\mathrm{D}}(\Delta t_{\mathrm{D}}) \; . \tag{6.4}$$

If the flow is radial and is still transient throughout the drainage area at time $(t + \Delta t)$, we can introduce the logarithmic expression for p_{D} (Eq. (5.33a)) and

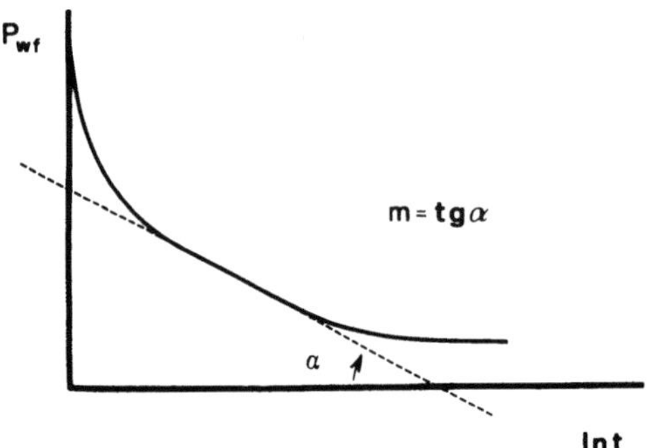

Fig. 6.2. *Behaviour of the bottom hole flowing pressure (semi-log plot) in a producing well during the transient and late transient periods*

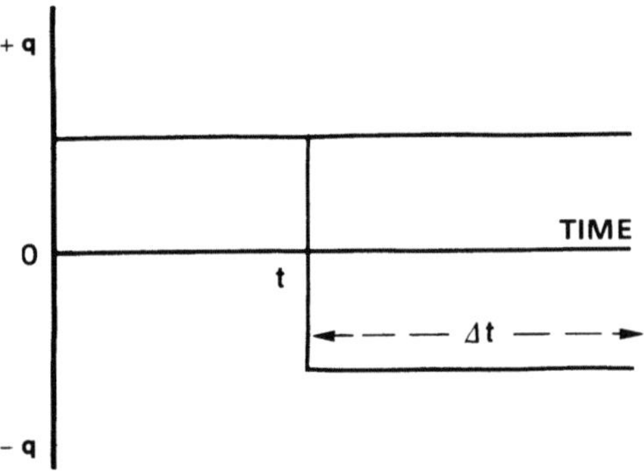

Fig. 6.3. *Representation of the shutting in of a well by superposing a negative flow rate*

write

$$\frac{2\pi k_o h}{q_{sc} B_o \mu_o} [p_i - p_{ws}(\Delta t)] = \tfrac{1}{2} \ln \frac{t + \Delta t}{\Delta t} \,, \tag{6.5}$$

from which we get

$$\frac{dp_{ws}}{d\left[\ln \dfrac{\Delta t}{t + \Delta t} \right]} = \frac{q_{sc} B_o \mu_o}{4\pi k_o h} = m \,. \tag{6.6}$$

This is similar in form to the drawdown equation [Eq. (6.3)], and it follows that k_o can be calculated from the slope of the straight line on a semi-log plot of p_{ws} versus $\ln[\Delta t/(t + \Delta t)]$.

But there is a problem here: *we have no proof or even an indication that the flow in the reservoir is still transient after shutting in the well.*

In particular, if the drawdown plot of p_{wf} versus $\ln t$ shows that the transient period has in fact ended (Fig. 6.2) *before* the well was shut in, then Eq. (6.5) is certainly not valid.

If t_{lim} is the duration of transient flow in drawdown [to the end of the straight-line trend $p_{wf} = f(\ln t)$], we can state, for $\Delta t_D < t_{D,\,lim}$:

$$\frac{2\pi k_o h}{q_{sc} B_o \mu_o} [p_i - p_{ws}(\Delta t)] = p_D(t_D + \Delta t_D) - \tfrac{1}{2}(\ln \Delta t_D + 0.809) \,. \tag{6.7}$$

For each $p_D(t_D)$ function that might be used to describe the flow after the transient period (corresponding to different assumptions about the shape and size of the drainage area of the well), *there exists a different solution to Eq. (6.7) in terms of reservoir parameters (k, S).*

This is precisely the problem of non-uniqueness in well test interpretation. This is a problem which, although present in numerous aspects of reservoir engineering, is frequently overlooked, under the erroneous assumption that the reservoir is "perfect".

This dilemma will arise in *any* well test where the flow rate is varied in whatever manner during the test – it is not restricted to the case of a pressure buildup following a shut in discussed already. It is especially significant in variable rate or multi-rate drawdown testing, where the flow rate is varied with time.

6.4 Calculation of the $p_D(t_D)$ Function for Non-Circular Geometries

Equation (6.4) described the behaviour of the bottom hole shut-in pressure p_{ws}, in terms of the time Δt elapsed since closing the well. Since a generalised $p_D(t_D)$ function is used in this equation, no assumption has been made about geometry or the nature of the flow regime in the drainage area.

$$\frac{2\pi k_o h}{q_{sc} B_o \mu_o} (p_i - p_{ws}) = p_D(t_D + \Delta t_D) - p_D(\Delta t_D) \,. \tag{6.4}$$

For small values of Δt_D, we have for the transient due to the hypothetical rate $-q_{sc}B_o$ which we superposed on the producing rate $+q_{sc}B_o$ to simulate well closure:

$$p_D(\Delta t_D) = \tfrac{1}{2}[\ln(\Delta t_D) + 0.809] \, , \tag{6.8}$$

Adding and subtracting the term $\tfrac{1}{2}\ln(t_D + \Delta t_D)$ from the right-hand side of Eq. (6.4), we get

$$\frac{2\pi k_o h}{q_{sc}B_o \mu_o}(p_i - p_{ws}) = \tfrac{1}{2}\ln\frac{t + \Delta t}{\Delta t} + p_D(t_D + \Delta t_D)$$

$$- \tfrac{1}{2}[\ln(t_D + \Delta t_D) + 0.809] \, . \tag{6.9}$$

Since $\Delta t_D \ll t_D$ we can write, to a good approximation

$$p_D(t_D + \Delta t_D) \cong p_D(t_D) \tag{6.10a}$$

and

$$\ln(t_D + \Delta t_D) \cong \ln t_D \, , \tag{6.10b}$$

so that Eq. (6.9) becomes

$$\frac{2\pi k_o h}{q_{sc}B_o \mu_o}(p_i - p_{ws}) = \tfrac{1}{2}\ln\frac{t + \Delta t}{\Delta t} + p_D(t_D) - \tfrac{1}{2}[\ln t_D + 0.809] \, . \tag{6.11}$$

Equation (6.11) is only valid for small values of $\Delta t \ll t$, and therefore only describes the straight-line portion AB of the pressure buildup in Fig. 6.4. Equation (6.11) can be extrapolated to infinite time (Fig. 6.4). This is well outside its range of validity, and the pressure at $\Delta t = \infty$, referred to as p^*, has no direct physical significance.

Since

$$\lim_{\Delta t \to \infty} \ln\frac{t + \Delta t}{\Delta t} = 0 \, , \tag{6.12}$$

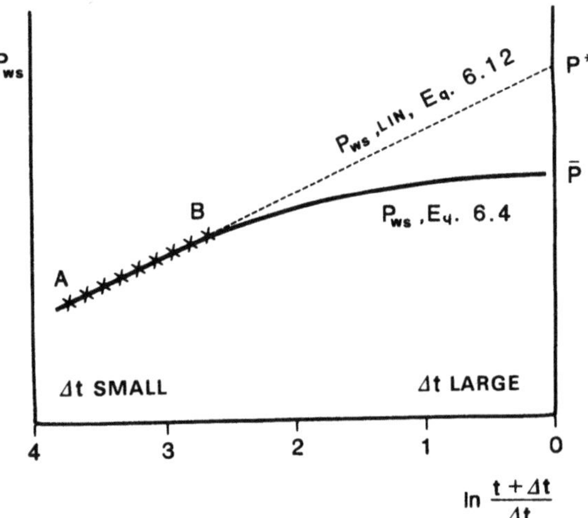

Fig. 6.4. *Pressure build up after a pseudo-steady state drawdown period, showing the significance of p**

we have

$$\frac{2\pi k_o h}{q_{sc}B_o\mu_o}(p_i - p^*) = p_D(t_D) - \tfrac{1}{2}(\ln t_D + 0.809) \, . \tag{6.13}$$

If the true equilibrium shut-in pressure for the drainage area (after an infinite closure period) is $\bar{p}$, then, allowing for the volume $q_{sc}B_o t$ of oil that was removed during the flowing period, we can write the material balance:

$$Ah\phi c_t(p_i - \bar{p}) = q_{sc}B_o t \, , \tag{6.14}$$

where A is the drainage area of the well.

Multiplying the terms in Eq. (6.14) by $2\pi k_o/\mu_o$ and rearranging, we get

$$\frac{2\pi k_o h}{q_{sc}B_o\mu_o}(p_i - \bar{p}) = 2\pi\frac{k_o}{\phi\mu_o c_t A}t = 2\pi t_{DA} \, , \tag{6.15}$$

where t_{DA} is the *dimensionless time in terms of the drainage area A, regardless of shape*:

$$t_{DA} = \frac{k}{\phi\mu c_t A}t \, . \tag{6.16}$$

This compares with the other form of the dimensionless time:

$$t_D = \frac{k}{\phi\mu c_t r_w^2}t \, . \tag{5.14b}$$

Subtracting Eq. (6.13) from Eq. (6.15) gives

$$\frac{4\pi k_o h}{q_{sc}B_o\mu_o}(p^* - \bar{p}) = 4\pi t_{DA} + \ln t_D - 2p_D(t_D) + 0.809 \, . \tag{6.17}$$

The term $(p^* - \bar{p})$ represents *the difference between the average pressure that the reservoir would attain after the period t of production, if the drainage area were infinitely large* (entirely transient pressure buildup), *and the average pressure attained in a real reservoir of finite size where pressure equilibrium was reached* ($p = \bar{p} = $ const. over the whole drainage area).

Note that all the terms on the right-hand side of Eq. (6.17) are dimensionless. We define, for convenience,

$$p_{D(MBH)}(t_{DA}) = \frac{4\pi k_o h}{q_{sc}B_o\mu_o}(p^* - \bar{p}) \, . \tag{6.18}$$

Equations (6.16) and (5.14b) give the relationship

$$t_D = t_{DA}\frac{A}{r_w^2} \, , \tag{6.19a}$$

so that

$$\ln t_D = \ln t_{DA} + \ln\frac{A}{r_w^2} \tag{6.19b}$$

and from Eq. (6.17):

$$p_D(t_{DA}) = 2\pi t_{DA} + \frac{1}{2}\ln t_{DA} - \frac{1}{2}p_{D(MBH)}(t_{DA}) + \frac{1}{2}\ln\frac{A}{r_w^2} + 0.405 \, . \tag{6.20}$$

Equation (6.20), proposed by Cobb and Dowdle,[4] is an extemely important one, in that it can be used to calculate $p_D(t_{DA})$ for any size and shape of drainage area, once the value of $p_{D(MBH)}(t_{DA})$ is known for the same geometry.

Matthews et al. [17,18] have published the values of $p_{D(MBH)}(t_{DA})$ for a wide range of geometries, which they calculated using the *method of images*. This method, much used in the study of potential fields, is based on spatial superposition theory and can be studied in the original publication,[18] listed at the end of this chapter.

Figures 6.5–6.11 are MBH diagrams [$p_{D(MBH)}(t_{DA})$ versus t_{DA}] for various geometries: circular, hexagonal, square, rectangular, rhomboidal, equilateral and right-angled triangles – all with the well at the centre – and square and rectangular with the well off centre.

In practice, to calculate the value of $p_D(t_{DA})$ for a given geometry and drainage area A, the following steps should be followed:

1. From the slope of the straight-line portion of the transient pressure data on a plot of p_{wf} versus $\ln t$ [(drawdown), or p_{ws} versus $\ln[\Delta t/(t + \Delta t)]$ (buildup), calculate k_o/μ_o using Eq. (6.3) or (6.6)].
2. Calculate the constant $(k_o/\phi\mu_o c_t A))$ to convert t to dimensionless time t_{DA}.
3. For different values of t (and therefore of t_{DA}), read the corresponding value of $p_{D(MBH)}(t_{DA})$ from the MBH diagram for the appropriate geometry.
4. Calculate $p_D(t_{DA})$ for the different t_{DA} values using Eq. (6.20), and construct the curve.

As soon as the flow goes to pseudo-steady state in the drainage area, we have[17]

$$p_{D(MBH)}(t_{DA}) = \ln(C_A t_{DA}) \tag{6.21}$$

The values at which Eq. (6.21) becomes valid, and the related shape factors C_A are listed for a variety of geometries studied by Matthews et al. in Fig. 5.7.

In pseudo-steady state flow, we can substitute Eq. (6.21) into Eq. (6.20) to get:

$$p_D(t_{DA}) = 2\pi t_{DA} + \tfrac{1}{2}\ln\frac{A}{C_A r_w^2} + 0.405 , \tag{6.22}$$

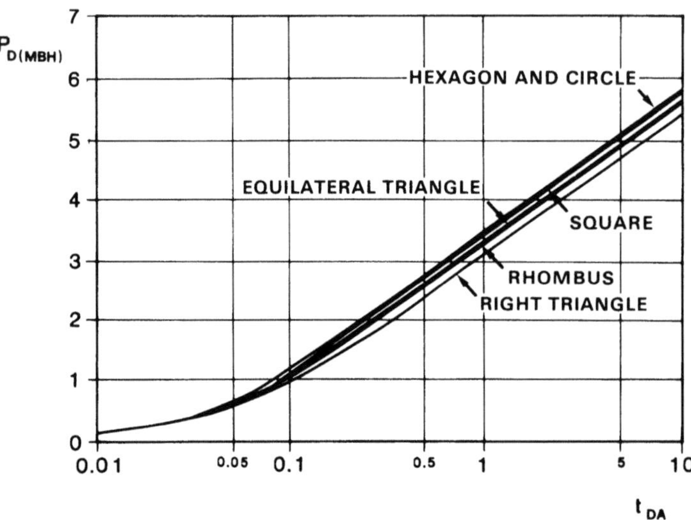

Fig. 6.5. $p_{D(MBH)} = f(t_{DA})$ *for various drainage area geometries. From Ref. 18, 1954, Society of Petroleum Engineers of AIME. reprinted with permission of the SPE*

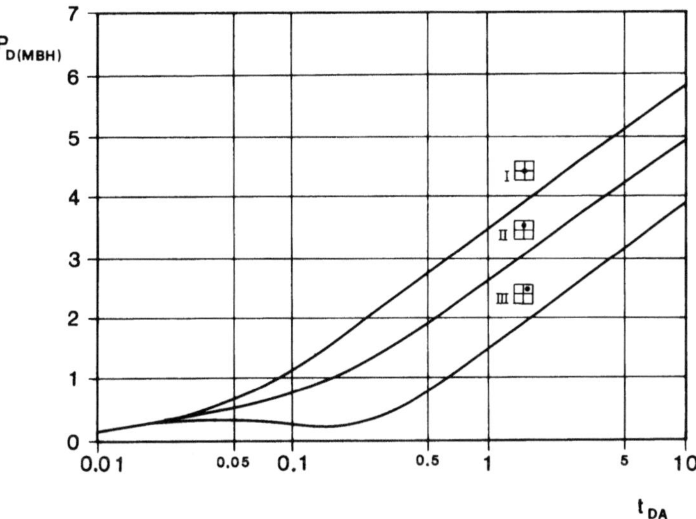

Fig. 6.6. $p_{D(MBH)} = f(t_{DA})$ *for a well in various positions within a square drainage area. From Ref. 18, 1954, Society of Petroleum Engineers of AIME, reprinted with permission of the SPE*

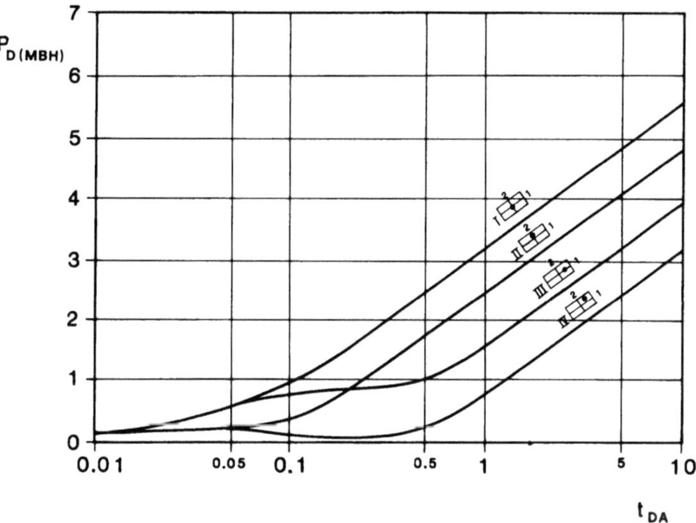

Fig. 6.7. $p_{D(MBH)} = f(t_{DA})$ *for a well in various positions within a 2:1 rectangular drainage area. From Ref. 18, 1954, Society of Petroleum Engineers of AIME, reprinted with permission of the SPE*

which, in real units, with the skin factor included, becomes

$$p_i - p_{wf} = \frac{q_{sc}B_o}{Ah\phi c_t}t + \frac{q_{sc}B_o\mu_o}{4\pi k_o h}\left[\ln\frac{A}{C_A r_w^2} + 0.809 + 2S\right].\tag{6.23}$$

From this it is easy to deduce the material balance relationship

$$-\frac{dp_{wf}}{dt} = \frac{q_{sc}B_o}{Ah\phi c_t}.\tag{6.24}$$

An important application of Eq. (6.22) [or (6.23)] is the extrapolation of the pressure response p_{wf} beyond the transient period, particularly in the analysis of pressure buildup tests.

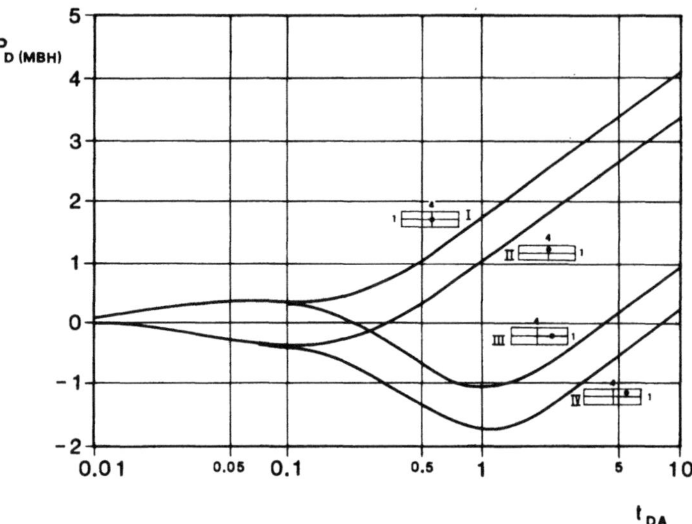

Fig. 6.8. $p_{D(MBH)} = f(t_{DA})$ *for a well in various positions within a 4 : 1 rectangular drainage area. From Ref. 18, 1954, Society of Petroleum Engineers of AIME, reprinted with permission of the SPE*

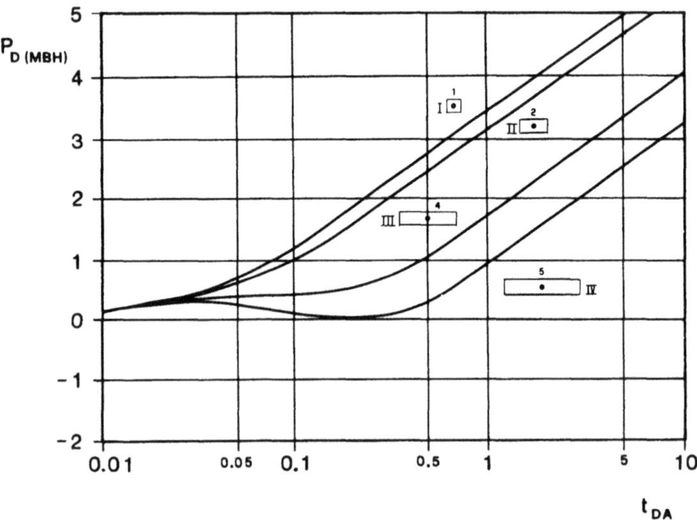

Fig. 6.9. $p_{D(MBH)} = f(t_{DA})$ *for a well in the centre of a rectangular drainage area with sides in various ratios. From Ref. 18, 1954, Society of Petroleum Engineers of AIME, reprinted with permission of the SPE*

6.5 The Interpretation of Pressure Drawdown Tests

6.5.1 Overview

Ideally, when a pressure drawdown test is to be performed, the drainage area should be at a uniform (static) pressure before the well is put on production at $t = 0$.

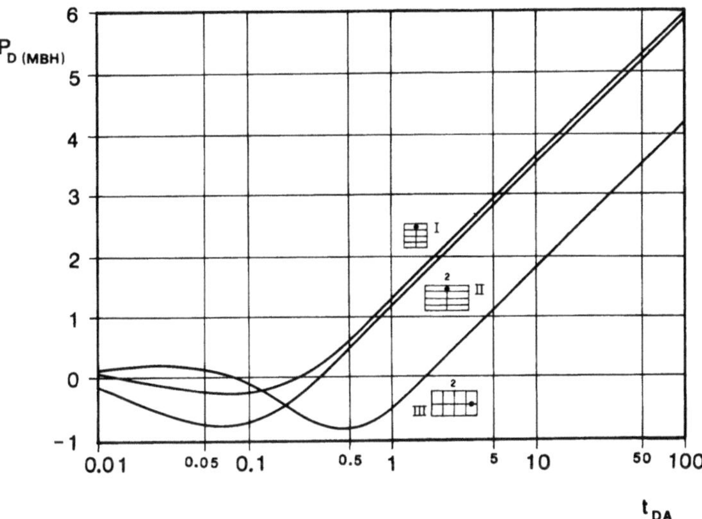

Fig. 6.10. $p_{D(MBH)} = f(t_{DA})$ *for a well located 1/8 of the way from the middle of one side of a square drainage area (I), 1/8 from the middle of the longer side of a 2:1 rectangle (II), and 1/8 from the middle of the shorter side of a 2:1 rectangle (III). From Ref. 18, 1954, Society of Petroleum Engineers of AIME, reprinted with permission of the SPE*

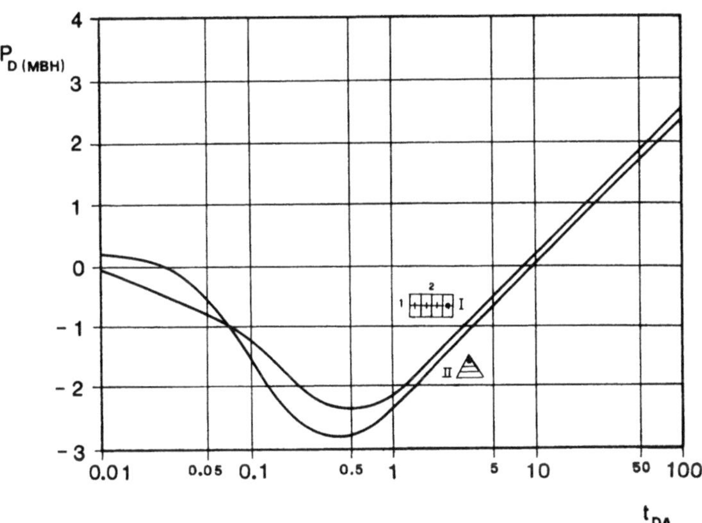

Fig. 6.11. $p_{D(MBH)} = f(t_{DA})$ *for a well located 1/8 of the way from the middle of the shorter side of a 2:1 rectangle (I), and 1/6 of the vertical height from the apex of an equilateral triangle (II). From Ref. 18, 1954, Society of Petroleum Engineers of AIME, reprinted with permission of the SPE*

The objectives of a pressure drawdown test are:

– determination of the average effective permeability of the drainage area;
– determination of the skin factor, S;
– estimation of the volume of hydrocarbons in the drainage area.

The best candidates for this sort of test are:

– newly completed wells in a part of the reservoir not yet in production;
– wells which have been shut in long enough for the pressure to have stabilised throughout the drainage area;

– wells which cannot be shut-in (and therefore are not amenable to buildup testing), either for mechanical reasons or because the loss of production could not be tolerated.

The equation describing the bottom hole flowing pressure behaviour with time during a drawdown test, starting from a uniform drainage area pressure p_i, with a constant surface production rate $q_{sc}(q_{sc}B_o$ at bottom-hole conditions), is

$$\frac{4\pi k_o h}{q_{sc}B_o\mu_o}(p_i - p_{wf}) = \ln t_D + 0.809 + 2S \;. \tag{5.36b}$$

This is valid *as long as the reservoir is infinite-acting*, in other words, until the pressure disturbance caused by the production has reached the outer boundaries of the drainage area.

In practice, the response is complicated by an early transient phenomenon associated with the production of reservoir fluids into the wellbore: either filling up of the casing and tubing, if initially empty; or displacing any completion fluid, drilling mud, oil from a preceding test, etc. that may be present.

In addition, if the damaged zone is deep (notably from the effects of perforation fluid), there will be a transient period associated with this skin zone which must be taken into account.

The analysis of this situation will be developed using classical theory, *with all terms expressed in SI units*. The equations to be used to evaluate a drawdown test will then be presented in Table 6.1, in both oilfield and practical metric units.

In a subsequent section, we will examine some more recent developments in well test analysis – the "type curve" and the pressure derivative.

6.5.2 Constant Rate Drawdown Tests

If q_{sc} = constant for the duration of the test, the following equation describes the pressure behaviour:

$$p_i - p_{wf} = \frac{q_{sc}B_o\mu_o}{4\pi k_o h}\left(\ln t + \ln \frac{k_o}{\phi\mu_o c_t r_w^2} + 0.809 + 2S\right) \tag{6.25}$$

and

$$-\frac{dp_{wf}}{d\ln t} = \frac{q_{sc}B_o\mu_o}{4\pi k_o h} = m \;. \tag{6.26}$$

Therefore, by measuring the slope m of the straight line portion of the semi-log plot (Fig. 6.12) of p_{wf} versus $\ln t$ (transient flow in an infinite-acting reservoir), we can calculate

$$\frac{k_o h}{\mu_o} = \frac{q_{sc}B_o}{4\pi m} \;, \tag{6.27}$$

k_o can then be estimated if h and μ_o are known.

If we introduce this value of k_o into Eq. (6.16) we have

$$t_{DA} = \frac{k_o}{\phi\mu_o c_t A}t \;, \tag{6.16}$$

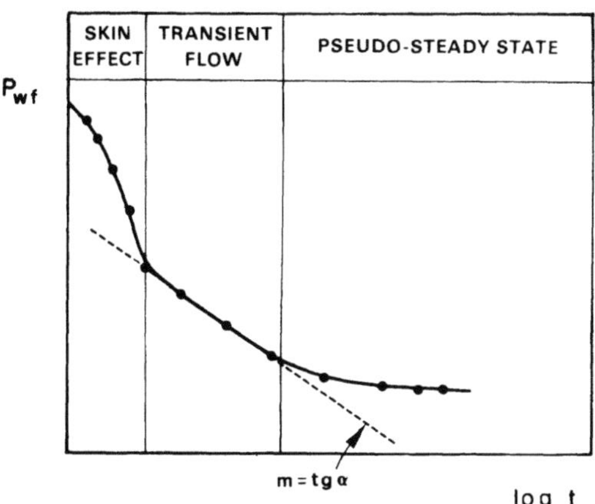

Fig. 6.12. *Semi-log plot of a constant rate pressure drawdown test. The* straight line *portion corresponds to transient flow*

where A is the assumed drainage area of the well. The dimensionless time $t_{DA,\max}$ corresponding to $t_{\max}$ at the end of the transient (straight line) period can now be calculated. Referring to Fig. 5.7, $t_{DA,\max}$ should now be compared with the entry for the appropriate geometry in the column headed "Use infinite system solution with less than 1% error if $t_{DA} <$ ", to see if the data selected for the straight line are consistent with the supposed geometry (or vice versa).

Once k_o is known, S can be calculated from Eq. (6.25):

$$S = \frac{2\pi k_o h}{q_{sc} B_o \mu_o}(p_i - p_{1h}) - \frac{1}{2}\left(\ln 3600 + \ln\frac{k_o}{\phi\mu_o c_t r_w^2} + 0.809\right), \tag{6.28}$$

where p_{1h} is the pressure *on the straight line portion* of the semi-log plot (extrapolated if necessary) at a time of 1 h after the well is put on production. Note that p_{1h} will only correspond to an actual *measured* pressure if the wellbore and skin zone transients have completely dissipated; otherwise it is a hypothetical pressure obtained by extrapolation of the transient straight line. Substituting m [(Eq. 6.26)] into Eq. (6.28), we get

$$S = \frac{p_i - p_{1h}}{2m} - \frac{1}{2}\left(\ln\frac{k_o}{\phi\mu_o c_t r_w^2} + 8.998\right), \tag{6.29}$$

which can be used to solve for S.

If the flowing test lasts long enough for the pressure disturbance to reach the outer boundaries of the drainage area, two features will be observed in the pressure behaviour:

a. From a certain time t_{tr} marking the end of the transient period, the pressure will diverge from the linear trend on the plot of $p_{wf} = f(\ln t)$,
b. From time t_{tr}, p_{wf} tends towards, and eventually becomes, a linear function of t (Fig. 6.13), and is described by Eq. (5.19a).

For $t > t_{tr}$, then, we have.

$$-\frac{dp_{wf}}{dt} = \frac{q_{sc} B_o}{Ah\phi c_t}. \tag{6.24}$$

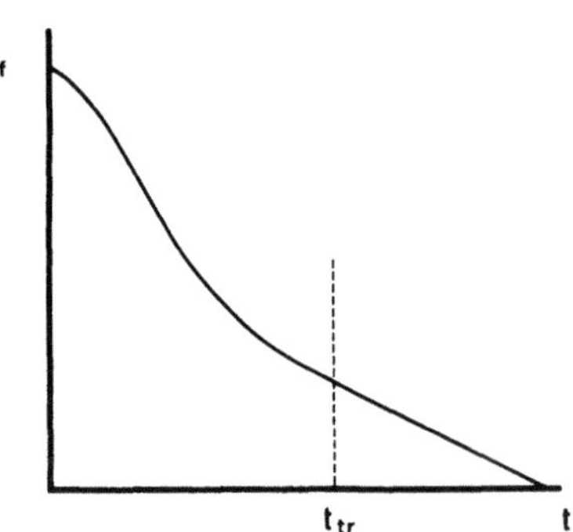

Fig. 6.13. *Pressure drawdown plotted as p_{wf} versus t, showing the end of the transient period*

Replacing $Ah\phi$ by V_p, the connected pore volume in the drainage area, this becomes

$$V_p = \frac{q_{sc}B_o}{c_t}\left(-\frac{dp_{wf}}{dt}\right)^{-1}_{t>tr} \qquad (6.30)$$

from which we can estimate V_p.

If $t_{DA,lim}$ is the value of t_{DA} for the appropriate geometry in the column headed "Use infinite system solution with less than 1% error if $t_{DA} <$ " in Fig. 5.7, then

$$\frac{k_o}{\phi\mu_o c_t A}t_{tr} = t_{DA,lim} , \qquad (6.31a)$$

so that

$$A = \frac{k_o}{\phi\mu_o c_t}\,\frac{t_{tr}}{t_{DA,lim}} . \qquad (6.31b)$$

Since $V_p = \overline{A}h\phi$, Eqs. (6.30) and (6.31) can be combined to give

$$\overline{h} = \frac{q_{sc}B_o\mu_o}{k_o}\,\frac{t_{DA,lim}}{t_{tr}}\left(-\frac{dp_{wf}}{dt}\right)^{-1}_{t>tr} . \qquad (6.32)$$

In an *extended* drawdown test, whose duration is long enough for pseudo-steady state flow to develop in the drainage area, Eqs. (6.26), (6.29), (6.30), (6.31) and (6.32) can be used to calculate, respectively

k_o: average effective permeability to oil in the drainage area
S: skin factor
V_p: connected pore volume in the drainage area
$\underline{A}$:   drainage area
$\overline{h}$:   average net pay thickness

At this point, it is worth recalling the more important of the simplifying assumptions that were made in Sect. 5.2 when solving the diffusivity equation. These assumptions limit the validity of the working equations that have been cited:

1. *The reservoir fluid is single phase oil* (plus the non-mobile connate water at saturation S_{iw}). The equations are therefore *not valid for a reservoir with a primary or secondary gas cap, or in any other way below the bubble point of the oil.*
2. *The permeability of the reservoir rock is constant* (or at least *statistically* constant, with a scatter of heterogeneity throughout the drainage area).

The extended drawdown test, which is designed to investigate the outer bounds of the drainage area, is referred to as a "reservoir limit test".

6.5.3 Drawdown Tests with Slowly Varying Flow Rates

It is often extremely difficult to maintain a perfectly constant production rate during a drawdown test, especially in newly completed wells which have just been put on production. The flow rate typically decreases slowly with time. This phenomenon is also observed in small reservoirs, where there is an appreciable reduction in bottom hole flowing pressure with time as the test proceeds. This is also the case, but for a very different reason, in a low permeability reservoir.

Winestock and Colpitts[25] have presented the following procedure for handling this situation. Firstly, we rewrite Eq. (6.25) as

$$\frac{p_i - p_{wf}}{q_{sc}} = \frac{B_o \mu_o}{4\pi k_o h} \left(\ln t + \ln \frac{k_o}{\phi \mu_o c_t r_w^2} + 0.809 + 2S \right)$$

$$+ \text{ second order terms (negligible)} . \tag{6.25a}$$

From this, we get

$$\frac{d}{d \ln t} \left(\frac{p_i - p_{wf}}{q_{sc}} \right) = \frac{B_o \mu_o}{4\pi k_o h} = m' \tag{6.26a}$$

from which

$$\frac{k_o h}{\mu_o} = \frac{B_o}{4\pi m'} . \tag{6.27a}$$

This means that all we have to do is plot the values of $(p_i - p_{wf})/q_{sc}$ against $\ln t$ (Fig. 6.14), and then follow the procedure outlined in Sect. 6.5.2, applying the check that the line has been fitted through the transient data.

The skin factor is calculated by arranging Eq. (6.25a) in the same way as Eq. (6.25)

$$S = \frac{1}{2m'} \left(\frac{p_i - p_{wf}}{q_{sc}} \right)_{1h} - \frac{1}{2} \left(\ln \frac{k_o}{\phi \mu_o c_t r_w^2} + 8.998 \right) . \tag{6.29a}$$

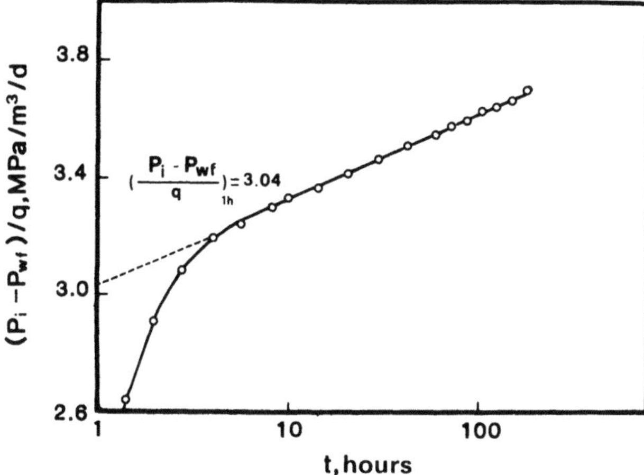

Fig. 6.14. *Semi-log plot for the interpretation of a drawdown test with slowly varying flow rate. From Ref. 15, 1982, Society of Petroleum Engineers of AIME, reprinted with permission of the SPE*

Table 6.1. *Equations for the analysis of pressure drawdown tests*

Equations	Values of the constants "C"		
	SI units	Practical metric units	Oilfield units (US)
$p_i - p_{wf} = m\left(\log\dfrac{k_o t}{\phi \mu_o c_t r_w^2} + C_1 + 0.869S\right)$	0.351	-4.885	-1.069
$m = C_2 \dfrac{q_{sc} B_o \mu_o}{k_o h}$	0.183	21.907	162.59
$t_D = C_3 \dfrac{k_o}{\phi \mu_o c_t r_w^2} t$	1	5.807×10^{-6}	3.797×10^{-2}
$t_{DA} = C_3 \dfrac{k_o}{\phi \mu_o c_t A} t$	1	5.807×10^{-6}	6.053×10^{-9}
$S = 1.151\left(\dfrac{p_i - p_{1h}}{m} - \log\dfrac{k_o}{\phi \mu_o c_t r_w^2} + C_4\right)$	-3.908	3.107	1.069
$V_p = C_5 \dfrac{q_{sc} B_o}{c_t}\left(-\dfrac{dp_{wf}}{dt}\right)^{-1}_{t > tr}$	1	6.944×10^{-4}	0.0417
$\dfrac{p_i - p_{wf}}{q_{sc}} = m'\left(\log\dfrac{k_o t}{\phi \mu_o c_t r_w^2} + C_6 + 0.869S\right)$	0.351	-4.885	-1.069
$m' = C_7 \dfrac{B_o \mu_o}{k_o h}$	0.183	21.907	162.59
$S = 1.151\left(\dfrac{p_i - p_{1h}}{m' q_{sc}} - \log\dfrac{k_o}{\phi \mu_o c_t r_w^2} + C_8\right)$	-3.908	3.107	1.069

Parameter	Symbol	Unit		
		SI	Practical metric	Oilfield (US)
Drainage area	A	m^2	m^2	acres
Oil volume factor (flash)	B_o	dimensionless	dimensionless	dimensionless
Total system compressibility	c_t	Pa^{-1}	cm^2/kg	psi^{-1}
Net pay thickness	h	m	m	ft
Effective permeability to oil	k_o	m^2	md	md
Slope of line $(-dp_{wf}/d\log t)$	m	Pa/cycle	kg/cm^2 cycle	psi/cycle
Slope of line $\dfrac{d}{d\log t}\left(\dfrac{p_i - p_{wf}}{q_{sc}}\right)$	m'	$\dfrac{Pas}{m^3}/\text{cycle}$	$\dfrac{kg}{cm^2}\dfrac{d}{m^3}/\text{cycle}$	$\dfrac{psi}{bbl/d}/\text{cycle}$
Initial pressure of drainage area	p_i	Pa	kg/cm^2	psi
Bottom hole flowing pressure	p_{wf}	Pa	kg/cm^2	psi
Bottom hole pressure at $t = 1\,h$	p_{1h}	Pa	kg/cm^2	psi
Oil production rate at surface conditions	q_{sc}	m^3/s	m^3/d	bbl/d
Wellbore radius	r_w	m	m	in
Skin factor	S	dimensionless	dimensionless	dimensionless
Time	t	s	min	hr
Pore volume in drainage area	V_p	m^3	m^3	bbl
Oil viscosity (reservoir conditions)	μ_o	Pas	$cP = mPas$	$cP = mPas$
Porosity	ϕ	dimensionless	dimensionless	dimensionless

A summary of the equations used for interpreting pressure drawdown in wells producing at constant, or slowly varying, flow rate is presented in Table 6.1.

6.5.4 Multi-Rate Pressure Drawdown Tests

In order to determine the maximum delivery that a well would be capable of, it is common practice to perform a series of production tests at successively higher rates. This provides the following information:

1. The production rate at which the flowing pressure will fall below the bubble point of the oil (the liberation of gas in the near-wellbore formation will cause an increase in the skin factor),
2. The pressure drop $(p_{wf} - p_{wh})$ from bottom hole to wellhead at different production rates. The sum of the pressure drop $(p_i - p_{wf})$ in the reservoir, plus this wellbore pressure drop, must be less than p_i if the well is to produce oil to surface $(p_{wh} > 0)$ without artificial lift.

The correct procedure to adopt when conducting a multi-rate test is to open the well at the first rate q_1 for the desired time period, then to shut the well in long enough for restabilisation to occur, so that there is uniform pressure throughout the drainage area (as a rule-of-thumb, this means shut-in time equal to flowing time). The well is then opened at the second rate q_2, then shut in, and so on. This is the essence of the "flow-after-flow test".

The intermediate shut ins (if allowed to stabilise) ensure that successive flowing tests are not perturbed by superposition of transients from preceding flow rates. Thus, each flowing period can be analysed as a constant rate drawdown, as described in Sect. 6.5.2.

In practice, there are usually time constraints (the cost of retaining the drilling rig if it is still in place, production lost by closing the well in) which dictate that the flow rates should follow one after the other, without intermediate shut ins $(q_1, q_2, q_3, \ldots, q_n)$. This more conventional "multi-rate pressure drawdown test" is shown in Fig. 6.15.

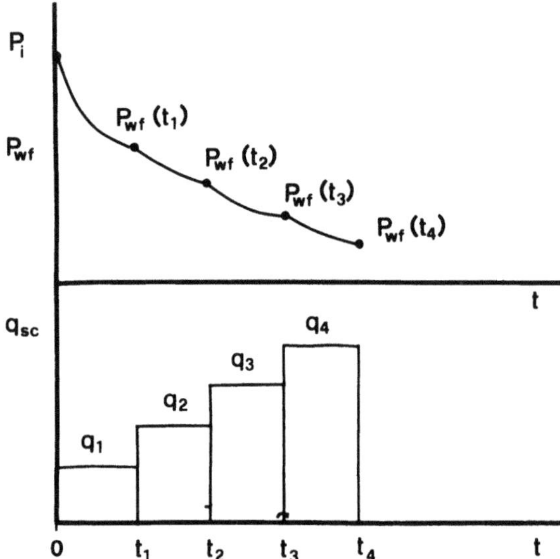

Fig. 6.15. *Production rates and flowing pressures during a multi-rate test*

We define $p_{wf,n}$ as the bottom hole flowing pressure at the end of the nth flowing period (at time $t_{D,n}$), and we assume S to be independent of the flow rate. Applying the superposition theory developed in Sect. 5.7, we can write Eq. (5.55b), after dividing both sides by q_n, as

$$\frac{2\pi k_o h}{\mu_o B_o} \frac{p_i - p_{wf,n}}{q_n} = \sum_{j=1}^{n} \left[\frac{q_j - q_{j-1}}{q_n} p_D(t_{D,n} - t_{D,j-1}) \right] + S . \tag{6.33}$$

Equation (6.33) provides the means of analysing any multi-rate drawdown test, employing the following procedure:

1. For each flow rate q_r, note the flowing pressure $p_{wf,r}$ at time t_r at the *end* of that flowing period, and calculate

$$Y_r = \frac{p_i - p_{wf,r}}{q_r} . \tag{6.34a}$$

2. For each flow rate q_r, calculate the superposition time function

$$X_r = \sum_{j=1}^{r} \frac{q_j - q_{j-1}}{q_r} p_D(t_{D,r} - t_{D,j-1}) . \tag{6.34b}$$

corresponding to time t_r.
3. Plot the data (X_r, Y_r) on a cartesian grid, as shown in Fig. 6.16. There should be a straight line trend with a slope

$$m'' = \frac{\mu_o B_o}{2\pi k_o h} \tag{6.34c}$$

from which we can calculate

$$\frac{k_o h}{\mu_o} = \frac{B_o}{2\pi m''} . \tag{6.34d}$$

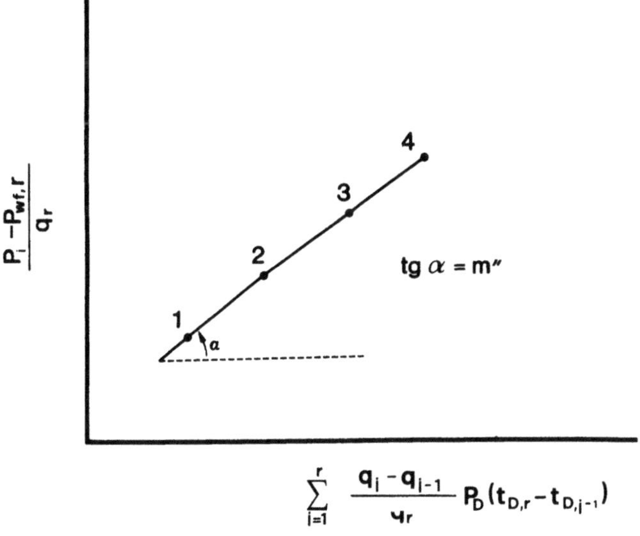

$$\sum_{i=1}^{r} \frac{q_i - q_{i-1}}{q_r} p_D(t_{D,r} - t_{D,i-1})$$

Fig. 6.16. *Superposition plot used for the analysis of multi-rate drawdown tests*

The skin factor S can be calculated from Eq. (6.33) using the value of Y_r on the straight line at $X_r = 0$.

This procedure has one obvious disadvantage: we need to know the $p_D(t_D)$ function describing the flow model. This in turn requires a knowledge of k_o and the geometry and size of the drainage area (Sect. 6.4).

If the first flowing period (q_1) lasts long enough for the early wellbore and damaged zone transients to dissipate, and the reservoir transient period to be clearly observed, k_o can be estimated from analysis of this data, as described in Sect. 6.5.2.

The evaluation of $p_D(t_D)$ and its effects on the subsequent data is best tackled by means of "type-curves", which will be covered in Sect. 6.8.

Odeh and Jones[22] have proposed an interpretation procedure based on Eq. (6.33) and the proviso (often forgotten by users of this method) that *all n successive flow rates remain transient for the duration t_n of the entire test sequence*. When this is the case, we can write (for radial flow)

$$p_D(t_{D,n} - t_{D,j-1}) = \frac{1}{2}\left\{\ln\left[\frac{k_o}{\phi\mu_o c_t r_w^2}(t_n - t_{j-1})\right] + 0.809\right\},\qquad(6.35)$$

so that

$$\frac{4\pi k_o h}{\mu_o B_o}\frac{p_i - p_{wf,n}}{q_n} = \sum_{j=1}^{n}\left[\frac{q_j - q_{j-1}}{q_n}\ln(t_n - t_{j-1})\right]$$

$$+ \ln\frac{k_o}{\phi\mu_o c_t r_w^2} + 0.809 + 2S .\qquad(6.36)$$

k_o is then calculated by the method just described, starting from Eq. (6.33) with

$$X_r = \sum_{j=1}^{r}\frac{q_j - q_{j-1}}{q_r}\ln(t_r - t_{j-1}) .\qquad(6.34b')$$

The skin factor S can be calculated from Eq. (6.36) in the same way as for Eq. (6.33), using the value of Y_r on the straight line at $X_r = 0$.

In fact, it is practically impossible to know ahead of time if pseudo-steady state flow will be reached in the drainage area during a test, invalidating Eq. (6.36). *The Odeh–Jones method must, therefore, be used with extreme caution.*

If the well has been flowing for a long time, and if for operational reasons the production rate has not been kept constant, any well test should be preceded by a long period of steady production (or, in the limiting case, zero production) so that the subsequent test data can be interpreted.

Suppose (Fig. 6.17) we have a well which has been producing at a varying rate up to time t_{N-1} (we may not even know the rate history), and is kept at a constant rate q_N for a time $(t_N - t_{N-1})$.

We take Eq. (5.55b), which expresses the theory of superposition, and divide it into two parts on either side of t_N (the start of the multi-rate test in Fig. 6.17). If we define

$$\delta t_n = t_n - t_N\qquad(6.37a)$$

and

$$\delta t_{j-1} = t_{j-1} - t_N \quad\text{for } j > N + 1 ,\qquad(6.37b)$$

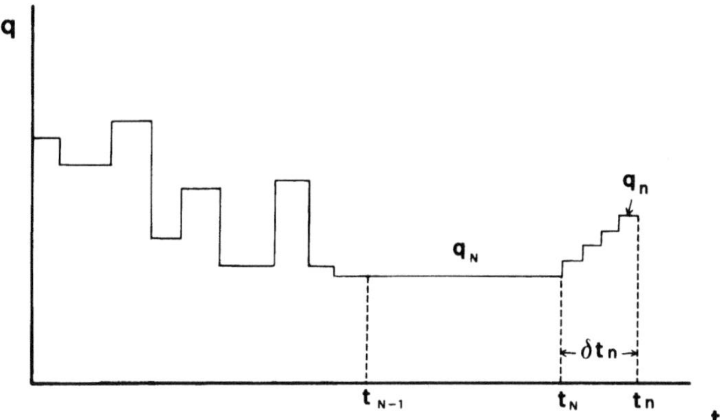

Fig. 6.17. *In a well with a production history like the one shown, a prolonged period of constant production is necessary prior to the multi-rate test*

we have

$$\frac{2\pi k_o h}{\mu_o B_o}(p_i - p_{wf,n}) = \sum_{j=1}^{N} [(q_j - q_{j-1})p_D(t_{DN} + \delta t_{D,n} - t_{D,j-1})] + q_N S$$

$$+ \sum_{j=N+1}^{n} [(q_j - q_{j-1})p_D(\delta t_{D,n} - \delta t_{D,j-1})] + (q_n - q_N)S \quad (6.38)$$

Provided $(t_N - t_{N-1}) \gg \delta t_n$, so that $t_N \gg \delta t_n$, we can write

$$\sum_{j=1}^{N} (q_j - q_{j-1})p_D(t_{D,N} + \delta t_{D,n} - t_{D,j-1})$$

$$\cong \sum_{j=1}^{N} (q_j - q_{j-1})p_D(t_{D,N} - t_{D,j-1}) . \quad (6.39)$$

Since

$$\frac{2\pi k_o h}{\mu_o B_o}(p_i - p_{wf,N}) = \sum_{j=1}^{N} [(q_j - q_{j-1})p_D(t_{D,N} - t_{D,j-1})] + q_N S , \quad (6.40)$$

we can reduce Eq. (6.38), by substituting Eq. (6.39) and subtracting Eq. (6.40), to

$$\frac{2\pi k_o h}{\mu_o B_o}(p_{wf,N} - p_{wf,n}) = \sum_{j=N+1}^{n} [(q_j - q_{j-1})p_D(\delta t_{D,n} - \delta t_{D,j-1})] + (q_n - q_N)S \quad (6.41)$$

or, dividing through by $(q_n - q_N)$:

$$\frac{2\pi k_o h}{\mu_o B_o}\frac{p_{wf,N} - p_{wf,n}}{q_n - q_N} = \sum_{j=N+1}^{n} \left[\frac{q_j - q_{j-1}}{q_n - q_N}p_D(\delta t_{D,n} - \delta t_{D,j-1}) \right] + S . \quad (6.42)$$

Equation (6.42) is identical to Eq. (6.33) if we make the following equivalences:

- time at start of test: $t_o = t_N$,
- initial pressure: $p_i = p_{wf,N}$,
- flow rate during test: $q = q_n - q_N$.

In other words, as long as the stabilisation period before the test is sufficiently long, *the multi-rate production test can be analysed as if it started from the shut in condition, provided the interpretation uses the changes in flow rate and pressure relative to their values at the end of the constant rate stabilisation period, rather than the absolute flow rate and flowing pressure.*

6.6 Interpretation of Pressure Buildup Tests

6.6.1 Overview

The theory of superposition, or of Duhamel, can be regarded as modelling the closure of a well by superposing a fictitious negative flow rate $-q_{sc}$ on to the producing rate $+q_{sc}$ so as to achieve a total flow rate of zero from the instant of shut in.

The well closure leads to an increase of bottom hole pressure $p_{ws}(t)$, or *pressure buildup*, with increasing time.

From the point of view of interpretation, a pressure buildup offers the advantage that *during the early part of the test, the p_{ws} data will correspond to the transient response to the constant rate $-q_{sc}$.* This is especially the case after long flowing periods, and removes some of the uncertainty that would otherwise be present in a variable rate test.

The following information can be derived from pressure buildup analysis:

- the effective permeability to oil in the drainage area,
- the total skin factor S (consisting of contributions from the damaged zone around the wellbore, partial penetration, well inclination (drilling) and partial completion (perforation) in the producing interval, heterogeneity of the reservoir rock – especially when it consists of layers with highly contrasting permeabilities),
- the average pressure $\bar{p}$ in the drainage area at the instant of shut in.

To start with, we will consider the simplistic case of a well that has been producing at a constant rate q_{sc} for its entire life, and is shut in at time t. The theory for this situation is straightforward to develop – we can then pass on to the more complex, and more realistic, case of a well that has a history of varying rates, which may include earlier periods of closure.

6.6.2 Well with a Constant Rate History

The following equation was presented in Sect. 6.4:

$$\frac{2\pi k_o h}{q_{sc} B_o \mu_o}(p_i - p_{ws}) = \frac{1}{2}\ln\frac{t + \Delta t}{\Delta t} + p_D(t_D) - \frac{1}{2}[\ln t_D + 0.809]\,, \tag{6.11}$$

where q_{sc} is the constant surface production rate prior to shut in, t (and its dimensionless form t_D) is the total flowing time, Δt is the time elapsed since shut in, and $p_D(t_D)$ is the solution to the diffusivity equation at the well ($r = r_w$) for the particular drainage area geometry.

Equation (6.11) is only valid for small $\Delta t(\ll t)$, i.e. *during the transient period associated with the fictitious flow rate $-q_{sc}$.*

From Eq: (6.11) we can derive

$$-\frac{dp_{ws}}{d\ln\dfrac{t+\Delta t}{\Delta t}} = \frac{q_{sc}B_o\mu_o}{4\pi k_o h} = m\,.\tag{6.43}$$

In other words, if we plot p_{ws} against $\ln[(t+\Delta t)/\Delta t]$, we can calculate a value for k_o from the slope of the straight line portion, as shown in Fig. 6.18, using Eq. (6.43).

Note the early behaviour of the shut-in pressure, which is not at all a straight line. This is due to wellbore fluid compression as production from the sand face dies away, (afterflow), and the associated decrease in the skin effect. For the purposes of semi-log analysis (Horner plot), this portion of the data should be recognised and avoided.

Similarly, the late portion of the data may have to be excluded. A steady reduction in slope (if present) indicates the onset of pseudo-steady flow for the rate $-q_{sc}$.

The approach just described is the *Horner method*[11] of evaluating k_o. For completeness, it is worth noting that there is an alternative approach to this problem, which differs only in that p_{ws} is plotted against $\ln \Delta t$ instead of the Horner time function. k_o is calculated from the slope m of the straight line portion using an identical equation to Eq. (6.43).

This is the *MDH method*, proposed by Miller et al.[20] It has enjoyed considerable popularity because it was widely believed (erroneously, according to Ramey[24]) that the straight line portion is more clearly developed by plotting the data in this way (and, of course, it is easier to plot).

In fact, when to use the Horner or MDH methods depends on the conditions: where superposition effects from rate history are significant, a superposition approach should be used – in this context, the Horner plot correctly accounts for a buildup following a single flowing period, both of which are in radial transient flow.

The MDH plot does not include superposition, and therefore is only correct when superposition effects are negligible ($\Delta t \ll t$, or drawdown in pseudo-steady state flow in a very large reservoir, or a reservoir with a constant pressure boundary already seen by the drawdown before shut in, for instance).

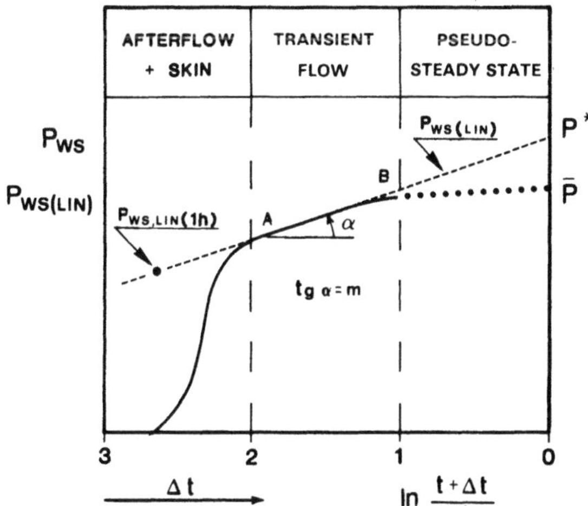

Fig. 6.18. *Analysis of a pressure buildup in a well with a constant rate history: the Horner plot*

As has already been explained, where the time function used is not correct, the pressure behaviour will deviate from the straight line, and great caution must be exercised in interpreting the post-transient data.

In Fig. 6.18, the term $p_{ws(LIN)}$ represents all values of p_{ws} lying on the straight line of slope m drawn through the transient part of the data. Obviously, $p_{ws(LIN)} = p_{ws}$ where the transient data actually falls on the line (portion AB), i.e. where $p_{ws} = f\{\ln[(t + \Delta t)/\Delta t]\}$.

The equation of this line (representing a hypothetical transient flow in an infinite acting medium, with no after flow or skin effects) must be [recalling Eq. (6.11)]:

$$\frac{2\pi k_o h}{q_{sc} B_o \mu_o} [p_i - p_{ws(LIN)}] = \frac{1}{2}\ln\frac{t + \Delta t}{\Delta t} + p_D(t_D) - \frac{1}{2}\ln t_D - 0.405 . \tag{6.44}$$

Furthermore, we know that the bottom hole flowing pressure p_{wf} at the moment of shut in is [see also Eq. (5.16)]:

$$\frac{2\pi k_o h}{q_{sc} B_o \mu_o} (p_i - p_{wf}) = p_D(t_D) + S . \tag{6.45}$$

If we subtract Eq. (6.44) from Eq. (6.45) we are left with

$$S = \frac{2\pi k_o h}{q_{sc} B_o \mu_o} [p_{ws(LIN)} - p_{wf}] - \frac{1}{2}\left(\ln t_D + 0.809 - \ln\frac{t + \Delta t}{\Delta t}\right) \tag{6.46}$$

In the same way as for drawdown analysis, we next read or calculate the value of $p_{ws(LIN)}(1\ h)$ at $\Delta t = 1\ h$ on the straight line.

In SI units, with t and Δt in seconds, we can write

$$\ln\frac{t + \Delta t}{\Delta t} = \ln\left(\frac{t}{\Delta t} + 1\right)$$

$$\cong \ln\frac{t}{3600} = \ln t - 8.1887$$

$$= \ln t_D - \ln\frac{k_o}{\phi\mu_o c_t r_w^2} - 8.1887 , \tag{6.47}$$

which, when substituted into Eq. (6.46), gives

$$S = \frac{2\pi k_o h}{q_{sc} B_o \mu_o} [p_{ws(LIN)}(1\ h) - p_{wf}] - \frac{1}{2}\left(\ln\frac{k_o}{\phi\mu_o c_t r_w^2} + 8.998\right) \tag{6.48a}$$

or, in terms of the slope m:

$$S = \frac{p_{ws(LIN)}(1\ h) - p_{wf}}{2m} - \frac{1}{2}\left(\ln\frac{k_o}{\phi\mu_o c_t r_w^2} + 8.998\right) . \tag{6.48b}$$

This is identical in form to Eq. (6.29).

The calculation of $\bar{p}$ then follows immediately. This has been described already in Sect. 6.4, but we will cover the procedure once more for completeness.

Firstly, we need an estimate of the drainage area A, and its shape and disposition relative to the well.

Starting from the assumption that the volume V_i of reservoir rock drained by a well is proportional to the production rate $q_{sc,i}$ if $q_{sc,TOT}$, is the *total* production rate from the reservoir in the period immediately preceding the closure of the ith well, and V_R is the total reservoir volume, we have

$$V_i = \frac{q_{sc,i}}{q_{sc,TOT}} V_R \, . \tag{6.49}$$

Dividing V_i by the net reservoir thickness, we get A_i. By repeating this calculation for neighbouring wells, we can obtain a reasonable idea of the position of the well within its drainage area. Based on this, we next select the $p_{D(MBH)}(t_{DA})$ curve from Figs. 6.5–6.11 in Sect. 6.4, for the appropriate geometry.

Now calculate:

$$t_{DA} = \frac{k_o}{\phi \mu_o c_t A} t \, , \tag{6.16}$$

where t is the total production time prior to shut in.

Read the value of $p_{D(MBH)}(t_{DA})$ corresponding to this t_{DA} from the selected curve.

$$p_{D(MBH)}(t_{DA}) = \frac{4\pi k_o h}{q_{sc} B_o \mu_o} (p^* - \bar{p}) \, . \tag{6.18}$$

As explained in Sect. 6.4, p^* is the value of p_{ws} obtained by extrapolating the straight line on the Horner plot (Fig. 6.18) to infinite time [where $\ln[(t + \Delta t)/\Delta t] = 0$]. Rearranging Eq. (6.18), we have

$$\bar{p} = p^* - \frac{q_{sc} B_o \mu_o}{4\pi k_o h} p_{D(MBH)}(t_{DA}) \, . \tag{6.50}$$

If the flowing pressure is observed to be in pseudo-steady state before the well is shut in, the value of $p_{D(MBH)}(t_{DA})$ can be calculated directly via Eq. (6.21). Equation (6.50) then becomes

$$\bar{p} = p^* - \frac{q_{sc} B_o \mu_o}{4\pi k_o h} \ln(C_A t_{DA}) = p^* - \frac{q_{sc} B_o \mu_o}{4\pi k_o h} \left(\ln t + \ln \frac{k_o}{\phi \mu_o c_t A} + \ln C_A \right), \tag{6.51}$$

where the value of C_A is read from Fig. 5.7 for the appropriate geometry. A summary of the equations used for interpreting pressure buildups in wells which produced at constant flow rate is presented in Table 6.2.

6.6.3 Well with a Variable Rate History

We now consider the case of a well which has been producing at various rates before being shut in. Referring to Fig. 6.19, t_n is the total production time, and q_n is the final rate before shut in. As before, the well closure is equivalent to superposing a rate $-q_n$ so that the total becomes zero.

Applying superposition theory, and denoting the bottom hole shut-in pressure as p_{ws}, with Δt as the elapsed time since closure, we have

$$\frac{2\pi k_o h}{\mu_o B_o} (p_i - p_{ws}) = \sum_{j=1}^{n} [(q_j - q_{j-1}) p_D(t_{D,n} + \Delta t_D - t_{D,j-1})]$$

$$- q_n p_D(\Delta t_D) \, . \tag{6.52}$$

Table 6.2. *Equations for the analysis of pressure buildup tests*

Equations	Values of the constants "C"		
	SI units	Practical metric units	Oilfield units (US)
$m = -\dfrac{dp_{ws}}{d\left(\log\dfrac{t+\Delta t}{\Delta t}\right)} = C_2\dfrac{q_{sc}B_o\mu_o}{k_oh}$	0.183	21.907	162.59
$t_D = C_3\dfrac{k_o}{\phi\mu_o c_t r^2}t$	1	5.807×10^{-6}	3.797×10^{-2}
$t_{DA} = C_3\dfrac{k_o}{\phi\mu_o c_t A}t$	1	5.807×10^{-6}	6.053×10^{-9}
$S = 1.151\left(\dfrac{p_{ws(LIN)}(1\,h) - p_{wf}}{m} - \log\dfrac{k_o}{\phi\mu_o c_t r_w^2} + C_4\right)$	-3.908	3.107	1.069
$\bar{p} = p^* - C_9\dfrac{q_{sc}B_o\mu_o}{k_oh}\,p_{D(MBH)}(t_{DA})$	7.958×10^{-2}	9.516	70.63
$\bar{p} = p^* - m\left(\log t + \log\dfrac{k_o}{\phi\mu_o c_t A} + \log C_A + C_{10}\right)$	zero	-5.236	-8.218

Parameter	Symbol	Unit		
		SI	Practical metric	Oilfield (US)
Drainage area	A	m^2	m^2	acres
Oil volume factor (flash)	B_o	dimensionless	dimensionless	dimensionless
Total system compressibility	c_t	Pa^{-1}	cm^2/kg	psi^{-1}
Dietz shape factor	C_A	dimensionless	dimensionless	dimensionless
Net pay thickness	h	m	m	ft
Effective permeability to oil	k_o	m^2	md	md
Slope of line $-dp_{ws}/d\log[(t + \Delta t)/\Delta t]$	m	Pa/cycle	kg/cm^2 cycle	psi/cycle
Matthews et al. dimensionless pressure	$p_{D(MHB)}$	dimensionless	dimensionless	dimensionless
Bottom hole flowing pressure	p_{wf}	Pa	kg/cm^2	psi
Bottom hole shut-in pressure	p_{ws}	Pa	kg/cm^2	psi
Pressure on the Horner straight line (extrapolated beyond transient data)	$p_{w(LIN)}$	Pa	kg/cm^2	psi
Average pressure of drainage area	$\bar{p}$	Pa	kg/cm^2	psi
Pressure on the Horner straight line extrapolated to $\log[(t + \Delta t)/\Delta t] = 0$	p^*	Pa	kg/cm^2	psi
Oil production rate at surface conditions	q_{sc}	m^3/s	m^3/d	bbl/d
Wellbore radius	r_w	m	m	in
Skin factor	S	dimensionless	dimensionless	dimensionless
Production time	$t, t_n, \bar{t}$	s	min	hr
Shut-in time	Δt	s	min	hr
Oil viscosity (reservoir conditions)	μ_o	Pa s	cP = mPa s	cP = mPa s
Porosity	ϕ	dimensionless	dimensionless	dimensionless

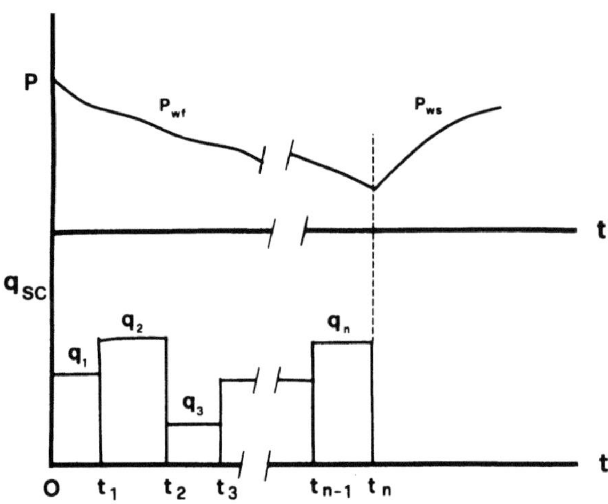

Fig. 6.19. *Pressure and flow rates for a well with a varying rate history prior to shut in*

During the shut-in period, the transient response $p_D(\Delta t_D)$ due to the fictitious flow rate $- q_n$ is expressed by Eq. (5.33a). If we divide Eq. (6.52) through by q_n, add and subtract the term $\frac{1}{2}\ln(t_{D,n} + \Delta t_D)$ from the right-hand side, and ignore $\Delta t_D \ll t_{D,n}$ where it occurs in the term $(t_{D,n} + \Delta t_D)$ we arrive at

$$\frac{2\pi k_o h}{q_n B_o \mu_o}(p_i - p_{ws}) = \frac{1}{2}\ln\frac{t_n + \Delta t}{\Delta t} + \sum_{j=1}^{n}\left[\frac{q_j - q_{j-1}}{q_n}p_D(t_{D,n} - t_{D,j-1})\right]$$

$$-\frac{1}{2}(\ln t_{D,n} + 0.809) . \tag{6.53a}$$

The second and third terms on the right-hand side of Eq. (6.53a) *are constants, since (provided $\Delta t \ll t_n$) they are functions only of t_n.*

Therefore,

$$-\frac{\mathrm{d}p_{ws}}{\mathrm{d}\ln\dfrac{t_n + \Delta t}{\Delta t}} = \frac{q_n B_o \mu_o}{4\pi k_o h} = m , \tag{6.54}$$

where, as usual, the m is the slope of the straight line portion of the buildup data.

In other words, the analysis is performed as if the well had been producing for the whole period t_n at the last flow rate q_n (on condition that $\Delta t \ll t_n$).

The calculations of S and $\bar{p}$ are performed in the same way as for the constant rate case (Sect. 6.6.2).

As an alternative approach, an *effective producing time $\bar{t}$* is used instead of the true time t_n. It is defined as the cumulative volume of fluid produced from the well, divided by the last flow rate:

$$\bar{t} = \frac{N_p}{q_n} \tag{6.55}$$

The analysis is then performed as if for a constant rate buildup, with flow rate q_n constant for the time $\bar{t}$.

In the absence of superposition effects, Eq. (6.53a) simplifies in this case to

$$\frac{2\pi k_o h}{q_n B_o \mu_o}(p_i - p_{ws}) = \frac{1}{2}\ln\frac{\bar{t} + \Delta t}{\Delta t} + p_D(\bar{t}_D) - \frac{1}{2}(\ln \bar{t}_D + 0.809) . \tag{6.53b}$$

still, of course, subject to the condition $\Delta t \ll \bar{t}$.

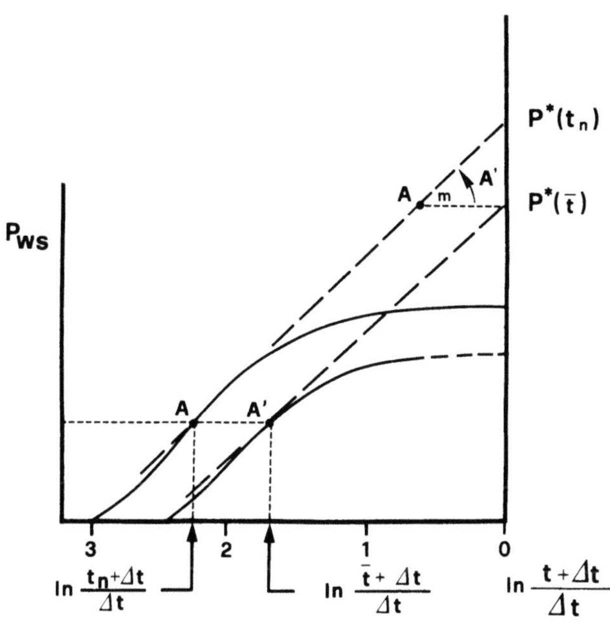

Fig. 6.20. *Interpretation of a pressure buildup test in a well with a varying rate history [Eqs. (6.54) and (6.57)]*

p_{ws} is plotted against $\ln\left[(\bar{t} + \Delta t)/\Delta t\right]$ in the same way as for $\ln\left[(t_n + \Delta t)/\Delta t\right]$ in the case of Eq. (6.53a).

The shift introduced in the x-direction by this modification is

$$\ln\frac{t_n + \Delta t}{\Delta t} - \ln\frac{\bar{t} + \Delta t}{\Delta t} \cong \ln\frac{t_n}{\bar{t}} \,. \tag{6.56}$$

The straight lines from the two methods are parallel,[5] and the lateral shift translates to a vertical displacement (Fig. 6.20):

$$p^*(t_n) - p^*(\bar{t}) = m\ln\frac{t_n}{\bar{t}} \,. \tag{6.57}$$

This should be allowed for when calculating $\bar{p}$ by the procedure outlined in Sect. 6.6.2. It is particularly significant when $\bar{t}$ is very different from t_n (for instance, very small or very large last flow rate q_n).

6.7 The Derivative of Pressure with Respect to Time

Differentiating the terms in Eq. (6.2) with respect to time, we find that *for transient flow* $t(dp_{wf}/dt) = -m = $ const. Similarly, for the derivative of Eq. (6.5) with respect to time we have to a good approximation – again *in transient flow* $\Delta t(dp_{ws}/d\Delta t) = $ constant.

This observation is the basis of a diagnostic method for identifying the flow model appropriate for the drainage area during a pressure drawdown or buildup test, from the behaviour of p_{wf} or p_{ws}.

The method entered into common use in the early 1980s, following the introduction of high resolution bottom hole pressure gauges and surface read out. It is a particularly useful aid in recognising that part of the pressure data which corresponds to transient flow. This is, of course, the data which is used to estimate $k_o h/\mu_o$.

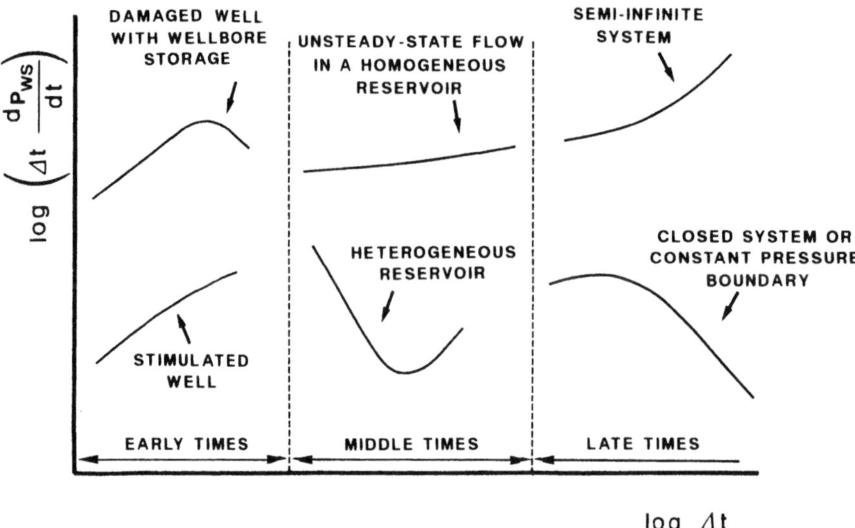

Fig. 6.21. *Log–log plot of the pressure derivative, showing characteristic shapes in some common situations. From Ref. 7, 1987, Society of Petroleum Engineers of AIME, reprinted with permission of the SPE*

The pressure derivative is usually plotted on a log–log grid (Fig. 6.21) – for instance, $\log[\Delta t(\mathrm{d}p_{ws}/\mathrm{d}t)]$ versus $\log \Delta t$. With this presentation, the various flow regimes – after flow, skin effect, transient and pseudo-steady state, etc. – can be clearly distinguished.

With surface read out, the data can be plotted in this way in real time during a well test. By observing the shape of the derivative, it is possible to ensure that test is not run any longer than is necessary (for instance, by terminating a buildup test as soon as sufficient transient data has been acquired to evaluate permeability and skin as described in Sect. 6.6).

6.8 Using Type-Curves

Since the early 1970s, *type curves* have become widely used in the analysis of pressure drawdown and buildup data,[7,8] particularly for the more complex situations such as naturally fractured, layered or areally heterogeneous reservoirs, or prolonged afterflow (continued flow of fluid from the sand face into the well for some time after shut-in).

The following illustration of the use of type curves is based on the simple example of a well flowing at a constant rate, with bottom hole flowing pressure p_{wf}. Although this is a drawdown test, this procedure can easily be extended to buildup tests.

A type curve is a plot of the dimensionless pressure response p_D with dimensionless time t_D for a well-defined set of initial and boundary conditions and flow model. In general, a whole family of curves is presented on the one diagram, each curve corresponding to a particular value of a parameter (e.g. ratio of fracture conductivity to reservoir permeability × thickness for one particular type of "fractured well" model).

Type curves are calculated analytically, or generated by numerical methods, depending on the flow model concerned. Curves have been published for the following principal cases (many other models have been studied):

- wells with prolonged afterflow, where a declining fluid production continues *downhole* for some time after the well has been shut in, or picks up correspondingly slowly when the well has been opened on surface. This masks part (or, in extreme cases, all) of the formation transient response. The curves are defined in terms of the skin factor S, and the *wellbore storage coefficient C* (a function of the velocity of the rising fluid column if it is a *wellbore fill-up*, or *slug*, test[27]; or of the fluid volume and compressibility if the wellbore was already full at the start of the test) (e.g. Ramey,[23] McKinley,[19] and, more recently, Gringarten[28] for homogeneous, isotropic reservoirs).
- layered reservoir with partial well penetration. These curves are defined in terms of the number of layers, of the layer which is open to production, and the ratios of kh among the layers (Yeh[26]).
- a well that has been hydraulically fractured, where a single vertical fracture radiates (symmetrically) into the reservoir (this model might also be valid for a naturally fractured reservoir where a vertical fracture plane intercepts the well). Curves are published for fractures of infinite[29] and finite[30] conductivity, and uniform flux[31] of fluid through the fracture face. Figure 6.22 (from Lee[15]) contains a set of type curves for a fractured well plotted for different ratios of drainage radius x_e/ fracture half-length L_f.

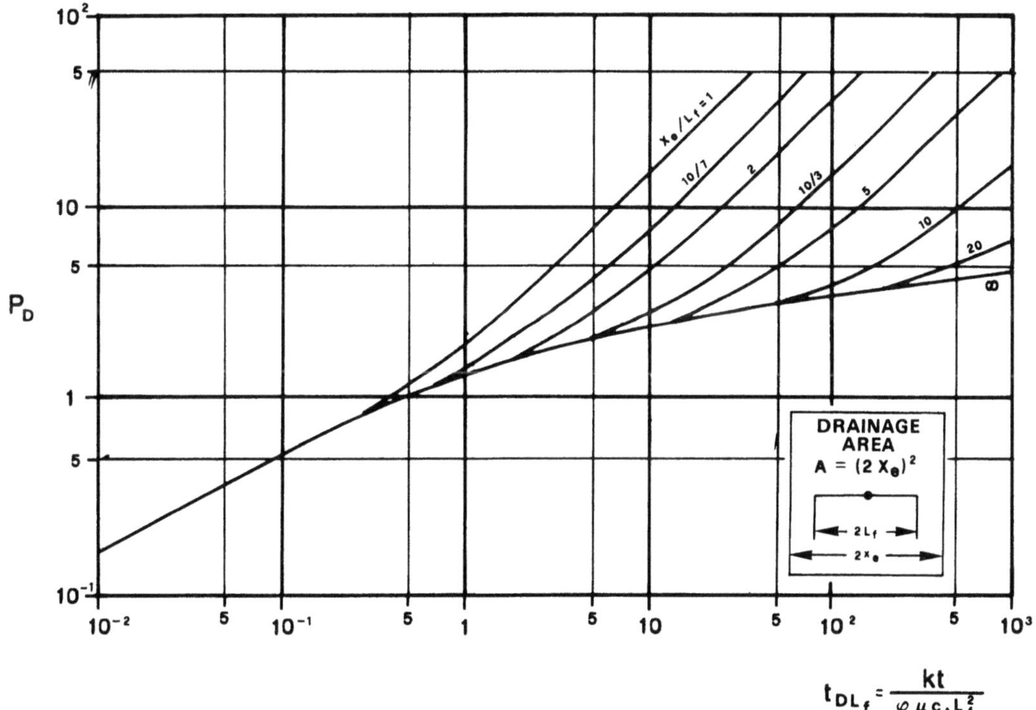

$$t_{DL_f} = \frac{kt}{\varphi \mu c_t L_f^2}$$

Fig. 6.22. *Type curves for a well intercepted by a single vertical fracture, for different values of the ratio x_e/L_f. From Ref. 15, 1982, Society of Petroleum Engineers of AIME, reprinted with permission of the SPE*

– naturally fractured reservoir, where production is through a network of highly conductive fissures which is fed by relatively tight matrix blocks containing most of the hydrocarbon.[36] The curves of Bourdet et al.[32, 33] include the effects of afterflow and skin, and are defined in terms of the bulk porosities of the fissures and matrix (storativity ratio), and the matrix block size, shape and permeability (interporosity flow coefficient).

Suppose we have flowing pressure data from a drawdown test, and wish to do an analysis with an appropriate set of type curves. For the example we will assume we have a hydraulically fractured well, and will use the type curves of Fig. 6.22. The curves are plotted as p_D versus t_{DLf} (defined in Fig. 6.22), and each corresponds to a different value of x_e/L_f To do the analysis by hand, we should have this set of curves on transparent paper since it is based on an overlay technique. (These days, this procedure is performed with ease on a computer screen – however, the principle is the same.)

Recalling the definitions of t_D and p_D in Eqs. (5.14b) and (5.14c), we have

$$\log p_D = \log \frac{2\pi k_o h}{q_{sc} B_o \mu_o} + \log(p_i - p_{wf}) \,, \tag{6.58a}$$

$$\log t_D = \log \frac{k_o}{\phi \mu_o c_t L_f^2} + \log t \,. \tag{6.58b}$$

On a log–log grid with the same scaling as Fig. 6.22, plot the values of $(p_i - p_{wf}) = f(t)$ from the test data. The transparent type-curve sheet is laid over this.

If the assumption that the data conform to the single vertical fracture model is correct, then one of the type-curves should conform to the shape of the plotted data (being careful not to skew one set of axes relative to the other of course, while moving the type-curve sheet over the data sheet). From the resulting match, we obtain immediately the value of x_e/L_f, from which we can estimate the fracture half-length L_f if we know the drainage area $(2x_e)$.[2]

In order to match one of the curves on the data, the type-curve sheet will have been displaced an amount Δy in the y-direction and Δx in the x-direction, relative to the data plot. Δy is expressed by

$$\Delta y = \log p_D - \log(p_i - p_{wf}) = \log \frac{2\pi k_o h}{q_{sc} B_o \mu_o} \tag{6.59a}$$

from which

$$k_o = \frac{q_{sc} B_o \mu_o}{2\pi h} \text{antilog}(\Delta y) \,. \tag{6.59b}$$

For Δx we have

$$\Delta x = \log t_D - \log t = \log \frac{k_o}{\phi \mu_o c_t L_f^2} \tag{6.60a}$$

from which

$$\phi = \frac{k_o}{\mu_o c_t L_f^2 \,\text{antilog}(\Delta x)} \,. \tag{6.60b}$$

Obviously, if none of the type curves, or an interpolation between a pair of curves, overlays the data, then the reservoir flow model (in this case, a single vertical fracture) is incorrect. A set of curves corresponding to a different model would then be required.

With the profusion of desktop computers or workstations, with high resolution graphics monitors, to be found in many oil company offices these days, most petroleum engineers will have easy access to libraries of published and experimental type curves through well test analysis software which will enable them to have selected type curves displayed on screen, and perform the matching procedure described, either using the keyboard or mouse, or automatically through a regression routine. The appropriate reservoir parameters will be calculated from the match (k_o, S, x_f, etc.), plus the standard deviation of the final match (i.e. the suitability of the assumed model) in the case of regression.

The dimensionless pressure derivative $p'_D = dp_D/dt_D$ is usually displayed along with p_D in the type curves. By matching to the derivative of the test data, this makes the identification of the appropriate type curve much simpler and more certain.[34] Use of the pressure derivative is obviously better suited to computerised analysis than plots on paper by hand.

In this type of "inverse problem" solving, where we attempt to describe the internal structure of a system (well with or without afterflow, coupled to a reservoir with or without skin effect, fractures, layered permeability, constant pressure or no-flow outer boundaries, etc.) from the pressure drawdown or buildup response of that system, *we cannot be certain that solution obtained by type-curve analysis is unique.*[7,12] In other words, *there may exist more than one well/reservoir configuration which would exhibit the same pressure behaviour as that measured during the well test, in response to the same change in flow rate.*

In order to reduce the indeterminacy associated with analysis of pressure gauge data in isolation, we need to bring in information from other sources to provide a first idea of the appropriate model, or to corroborate or eliminate a possible solution. Such information includes well log interpretations, core analyses, stratigraphic maps from seismic surveys, production data, and other well tests in the same field.

Once a suitable model has been deduced, it remains to make careful use of the relevant set of type curves to obtain an accurate determination of the various parameters from the pressure drawdown or buildup data.

6.9 The Influence of Reservoir Heterogeneity on Bottom-Hole Pressure Behaviour

Within the well + reservoir system, there may exist any number of different kinds of heterogeneity which will have an effect on the bottom hole flowing pressure when the well is put on production and subsequently shut in. Often, the objective of the reservoir engineer to evaluate these heterogeneities from drawdown and buildup data is thwarted by the problem of non-uniqueness, as explained in Sect. 6.8.

The following subsections will describe some of the more common types of heterogeneity and their influence on pressure behaviour.

6.9.1 Skin Effect Due to Well Geometry

As was mentioned in Sect. 5.6.1.2, one of the components of the "skin effect" is the additional pressure drop caused by partial penetration of the pay zone, either because the well was not drilled through the entire interval, or because only a part of the well has been perforated and is open to flow.

The skin factor contribution from partial penetration has been quantified for the case of a homogeneous reservoir by Nisle,[21] and by Brons and Marting.[1] Figure 6.23 is from the work of these last two authors. It presents the value of S as a function of the "penetration ratio" b (the fraction of the net pay thickness h which is open to flow), and the ratio h/r_w.

When only part of the interval is open to flow, the situation is far more complicated if the reservoir is stratified. In the first study of this problem, Chierici et al.[3] showed that it was possible to have positive or negative skins, and that the permeability k deduced from the transient analysis of drawdown or buildup data was the thickness-averaged permeability of all the layers making up the reservoir interval.

The geometrical skin factor encountered in a cased well where only part of the casing through the reservoir section has been perforated (or, at least, only part of the perforations are actually flowing) may be quite large.[9] This is particularly true in a naturally fissured reservoir, where not all the fractures will be able to communicate with the wellbore through the perforations.

The geometrical component of the skin factor must be taken into account when planning well stimulation (acidising, hydraulic fracturing). These operations are intended precisely to reduce the skin, but cannot affect the geometrical contribution.

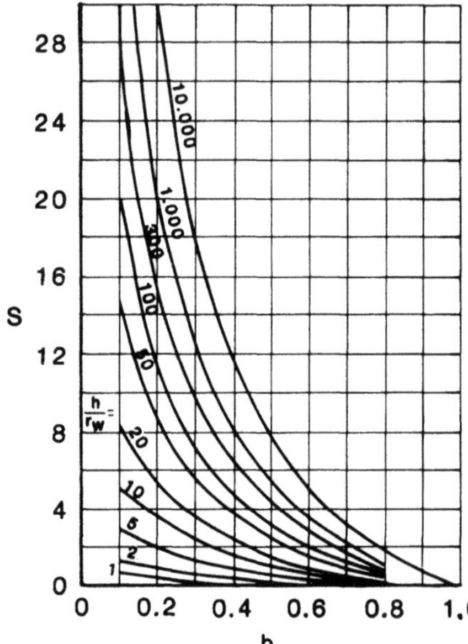

Fig. 6.23. *Skin factor, S, as a function of the fraction b of the hydrocarbon-bearing interval which is open to flow, and h/r_w. From Ref. 1, 1961, Society of Petroleum Engineers of AIME, reprinted with permission of the SPE*

6.9.2   Layered Reservoir or Reservoir with an Underlying Aquifer

When the reservoir rock consists of strata of different permeabilities, k_i, and of different thicknesses h_i, the literature customarily distinguishes two extreme cases:

1. There is no vertical communication (cross-flow) within the reservoir between adjacent layers; communication only occurs in the wellbore. In this case,[17] the pressure disturbance propagates in each layer independently, and the flow rate q_i from each layer is proportional to the product $k_i h_i$;
2. There is unimpeded vertical communication within the reservoir between adjacent layers, so that the pressure disturbance is, notwithstanding some localised disturbances, essentially a function only of r and t, and does not depend on the vertical position z.

In both cases, the interpretation of drawdown and buildup tests can be performed in the same way as for a single homogeneous reservoir (see Sects. 6.5 and 6.6), using the *total* production rate for q_{sc}, provided all the layers are in transient flow and the well is perforated over the whole thickness of the pay.

The permeability $\bar{k}$ obtained from the straight line portion of the data is related to the k_i and h_i of the strata as follows:

$$\bar{k} = \frac{\sum\limits_{i=1}^{n} k_i h_i}{\sum\limits_{i=1}^{n} h_i} . \tag{6.61}$$

The special case where there is only a partial degree of vertical communication between the layers, flow being impeded by, for instance, thin non-continuous shale strata, is worth investigating.

In this situation, the semi-log plot of the pressure drawdown or buildup response will exhibit two or more straight line sections with different slopes, with almost flat transitions between them.[17] Figure 6.24 illustrates the case of a pressure buildup test for a reservoir consisting of two layers of contrasting permeabilities in partial vertical communication. This example, from Matthews and Russel[17], can be

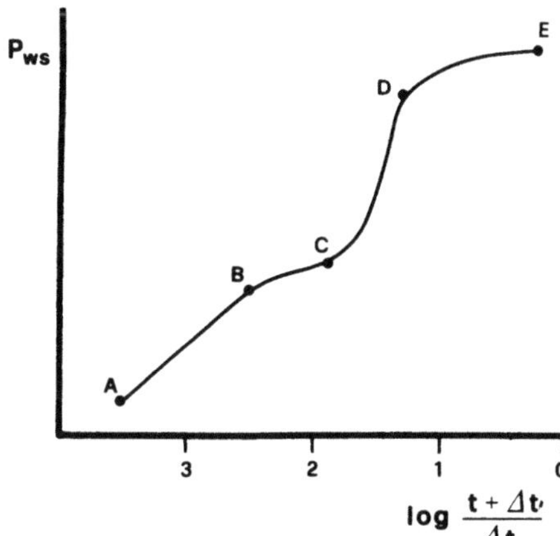

Fig. 6.24. *Typical pressure buildup behaviour in a reservoir consisting of two layers of very different permeabilities, in partial vertical communication. From Ref. 17, 1967, Society of Petroleum Engineers of AIME, reprinted with permission of the SPE*

interpreted as follows:

- the first straight line section AB corresponds to the pressure buildup in the high permeability layer (this was in fact preceded by a curved section attributed to afterflow and skin effects, not shown in the figure),
- during the almost flat section BC, the pressure in the high permeability layer begins to stabilise,
- the steeper portion CD is caused by flow of fluid from the low permeability layer towards the other[17] (Fig. 6.25), impeded by the poor vertical communication between the two. This cross flow results from the fact that, some distance out from the wellbore, the pressure in the tight layer is still relatively higher, partly because this region has not yet produced, and partly because the pressure has not yet had time to balance out completely in this layer.
- the final portion DE corresponds to stabilisation of the pressure in the system as the two layers reach equilibrium. The cross flow will of course have died away when this occurs.

In this relatively simple case of just two layers in partial vertical communication, which produce a drawdown or buildup plot with two straight line portions, an interpretation of the data is quite feasible.

In a reservoir consisting of many layers of different permeabilities in poor vertical communication, the well test behaviour – a whole series of straight line sections with different slopes occurring at different times – may be too complex to attempt a quantitative interpretation.

An interesting case, and one which is often overlooked, is where the hydrocarbon-bearing interval overlies, and is in perfect communication with, an aquifer (bottom water drive).

Consider for simplicity the single homogeneous, isotropic layer, of permeability k, containing oil of viscosity μ_o over an upper thickness h_o shown in Fig. 6.26. The lower part of this layer is aquifer, thickness h_w, water viscosity μ_w.

In order to avoid producing water, the well would normally either only penetrate the oil-bearing section, or would only be perforated in that section. In fact, if water coning was expected (see Chap. 12), only the *upper* part of the pay

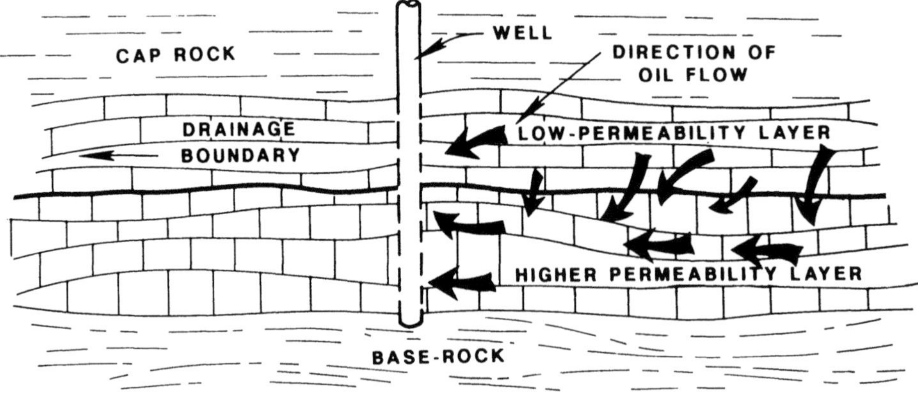

Fig. 6.25. *The flow of oil from a tight layer into a high permeability layer during a pressure buildup test. The high productivity from the more permeable layer has resulted in its becoming depleted relative to the tight zone. From Ref. 17, 1967, Society of Petroleum Engineers of AIME, reprinted with permission of the SPE*

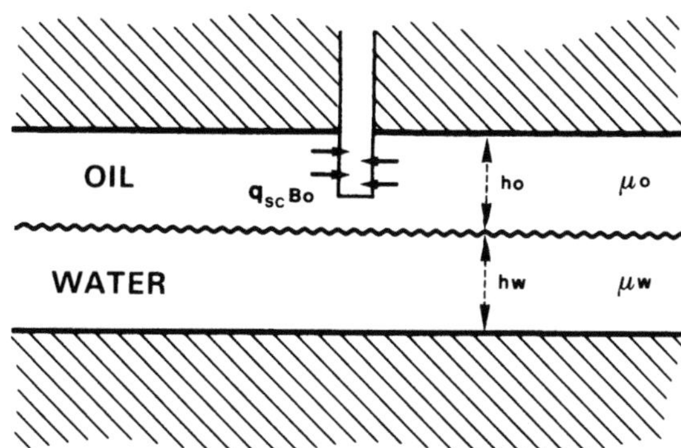

Fig. 6.26. *Production from an oil zone in contact with an underlying aquifer*

would be open, as shown in the figure, so that communication to the wellbore is through only a fraction of h_o.

Although the well may be producing water-free oil, the pressure disturbance initiated by production is propagating through the *total* system (oil + water) in the reservoir.

In other words, it is equivalent to having two layers with the same permeability k, but containing fluids of different viscosities μ_o and μ_w.

Applying the theory for multi-layered systems, as demonstrated for the first time by Chierici et al.,[3] the slope m of the pressure drawdown data [Eq. (6.26)], or of the buildup data [Eq. (6.43)], during the transient phase is related to the layer characteristics by

$$k = \frac{q_{sc}B_o}{4\pi m} \frac{1}{(h_o/\mu_o) + (h_w/\mu_w)}. \tag{6.62}$$

If the aquifer term h_w/μ_w were to be overlooked, the value of k calculated from m would be too large (the more so if $\mu_o \gg \mu_w$), leading to an overestimation of the productivity J_{ideal} (Sect. 5.8).

It has sometimes happened that well stimulation has failed to produce any improvement in performance, because the well was not in fact "damaged", but the well test data had been wrongly interpreted.

6.9.3 The Presence of a Permeability Barrier in the Drainage Area

Suppose we have a reservoir that is infinitely large (so that the flow regime will always be transient), of constant thickness h, homogeneous and isotropic, with permeability k_o. The full thickness of the reservoir is penetrated by a well.

Now suppose that at a distance L from this well there is a permeability barrier (Fig. 6.27) – perhaps a sealing fault, a pinch-out, or a change from permeable to impermeable facies. We will assume this barrier is linear, at least where it traverses the drainage area of the well.

Since fluid movement across the barrier is not possible, there is a "no-flow" boundary condition along its length.

This situation can be modelled mathematically by locating a fictitious "image well" on the other side of the barrier, symmetrical with the real well. If the production from the image well is made identical to that from the real well, then, as the following explanation will show, the effect will be to create a condition of *zero*

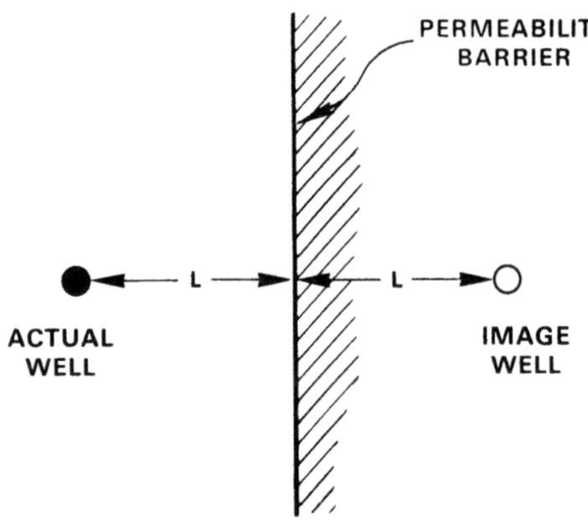

Fig. 6.27. *Modelling a permeability barrier by the method of images*

flow along a line midway between the two wells. We have therefore reproduced the effect of the physical barrier and can dispense with it, treating the reservoir as if it were infinite acting once more.

This technique is the well-known "method of images", which can be used whenever we wish to model conditions of zero flow across any surface (for instance, the drainage boundaries of the well).

Starting from a stabilised initial pressure p_i at $t = 0$, we impose a flow rate q_{sc} at the real well, and the same at the image well in order to maintain the no flow condition at the barrier.

By the principle of superposition (Sect. 5.7), the pressure drawdown $(p_i - p_{wf})$ at the real well will be the sum of the pressure decrease caused by production from the well itself, *plus* the pressure decrease transmitted from the image well at a distance $2L$. Referring to Eq. (5.32), this is

$$p_i - p_{wf} = \frac{q_{sc}B_o\mu_o}{4\pi k_o h}\left[\ln\frac{4k_o t}{\gamma\phi\mu_o c_t r_w^2} + \ln\frac{4k_o t}{\gamma\phi\mu_o c_t (2L)^2}\right]. \tag{6.63}$$

As long as the image well disturbance is negligibly small – as it is initially – the pressure at the real well behaves as if there were no barrier.

If the time taken for the image well pressure disturbance to travel the distance $2L$ to the real well is t_{2L}, then for $t < t_{2L}$ the image well effectively does not exist as far as the real well pressure response is concerned.

Therefore, for $t < t_{2L}$, the slope of the pressure drawdown will be described by [Eq. (6.26)]

$$m = -\frac{dp_{wf}}{d\ln t} = \frac{q_{sc}B_o\mu_o}{4\pi k_o h}. \tag{6.26}$$

For $t > t_{2L}$, the image well pressure disturbance will become significant when added to that of the real well.

Eq. (6.63) can be written as

$$p_i - p_{wf} = \frac{q_{sc}B_o\mu_o}{4\pi k_o h}\left[\ln\frac{4k_o t}{\gamma\phi\mu_o c_t r_w^2} + \ln\left(\frac{4k_o t}{\gamma\phi\mu_o c_t r_w^2}\times\frac{r_w^2}{4L^2}\right)\right]$$

$$= \frac{q_{sc}B_o\mu_o}{2\pi k_o h}\left[\ln\frac{4k_o t}{\gamma\phi\mu_o c_t r_w^2} + \ln\frac{r_w}{2L}\right]$$

$$= \frac{q_{sc}B_o\mu_o}{2\pi k_o h}\left[\ln\frac{k_o t}{\phi\mu_o c_t r_w^2} + 0.809 + \ln\frac{r_w}{2L}\right]. \tag{6.64}$$

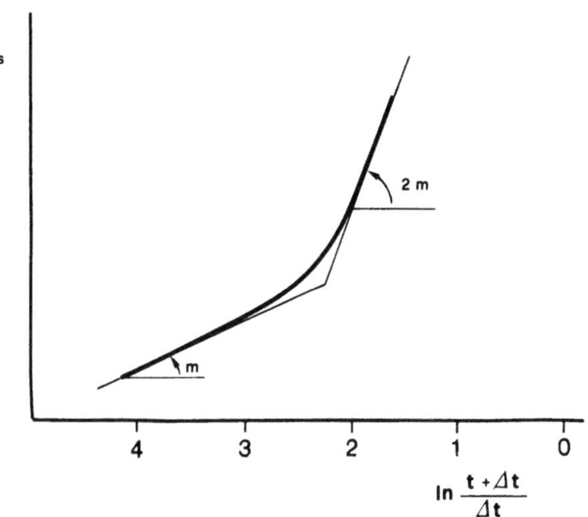

Fig. 6.28. *Doubling of slope of the pressure buildup response caused by the presence of a permeability barrier*

Now, for $t > t_{2L}$:

$$m' = -\frac{\mathrm{d}p_{wf}}{\mathrm{d}\ln t} = \frac{q_{sc}B_o\mu_o}{2\pi k_o h} = 2m \ . \tag{6.65}$$

In other words, the slope $\mathrm{d}p_{wf}/\mathrm{d}(\ln t)$ doubles in value once the well "sees" the presence of the permeability barrier (i.e. the image well disturbance reaches the well).

It is straightforward to show that this same doubling of slope will be observed in pressure buildup data in the same situation (Fig. 6.28).

The conclusion to be drawn from this above discussion is that *the presence of a permeability barrier within the drainage area of the well can be deduced from the fact that the slope m of the plot*:

$p_{wf} = f(\ln t)$ in the case of a constant rate drawdown

$p_{ws} = f[\ln(t + \Delta t)/\Delta t]$ for pressure buildups

doubles a certain time after the start of production or buildup.

The time t_{2L}, or Δt_{2L}, corresponding to the time at which the slope doubles, gives an approximate indication of the distance L of the permeability barrier from the well.

In SI units:

$$L \cong 0.75 \sqrt{\frac{k_o t_{2L}}{\phi\mu_o c_t}} \ . \tag{6.66}$$

In practical metric units (Table 6.1), the numerical coefficient in Eq. (6.66) becomes 1.806×10^{-3}, while in oilfield units (Table 6.1) it is[6] 1.217×10^{-2}.

6.10 Comments on the Interpretation of Injection Tests

In oil reservoirs which are produced under water injection, it is common practice to perform well tests in the injector wells in order to estimate layer permeability, average pressure, and possible skin factor, by changing the water injection rate and measuring the bottom-hole pressure response.

Injection tests can of course be run in producing wells, but this is rarely done.
During an injection test, the pressure p_{wf} increases with time, while during a shut-in period, p_{ws} will decrease ("pressure fall-off"). This behaviour is, not surprisingly, the opposite of what happens during a flowing/shut-in test in a producer.

There are three basic types of water injection test:

1. Injection of water into a flanking or underlying aquifer: in this case the flow in the reservoir is strictly monophasic. The data from an injection or pressure fall-off test can be interpreted by the same methods (and the same equations) as for drawdowns and buildups in wells producing oil.

2. Injection of water directly into the oil-bearing formation (no free gas present): In this case we have single phase flow through the region immediately surrounding the well, with immobile oil at residual saturation S_{or}; and two-phase flow (oil/water) beyond it (Fig. 6.29). Injection test analysis is still quite straightforward if the mobility ($k\,k_{rw}/\mu_o$) of the oil is not too different from that of the water, ($k\,k_{rw}/\mu_w$). It becomes more risky when the oil mobility is lower, as this frequently leads to fingering of the injected water into the oil ahead of the main front.

3. Injection of water directly into the oil-bearing formation (with free gas present): this gives rise to the phenomenon of "fill-up", or reservoir recompression, during which the free gas redissolves in the oil. If the free gas is dispersed in the oil-bearing section, we have a three-phase situation, and well test data is practically uninterpretable. If there is good vertical permeability and the gas has segregated to form a gas cap, the situation is actually biphasic with a constant pressure boundary at the top. In this case, test data are difficult to interpret but useful results can be obtained.

The following notes will describe in some detail the simplest – and most common – case of the injection of water into the aquifer. This will also cover the injection of water into the oil zone if the mobility ratio λ_w/λ_o is close to 1.0 .

The pressure behaviour observed during a constant rate injection/fall-off test is shown in Fig. 6.30.

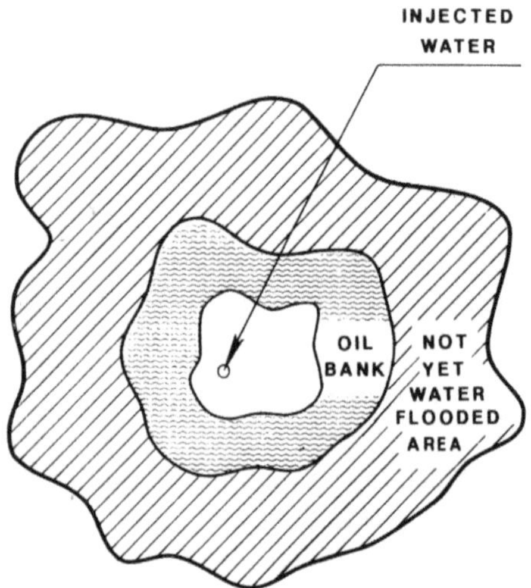

Fig. 6.29. *Distribution of fluids around a water injection well*

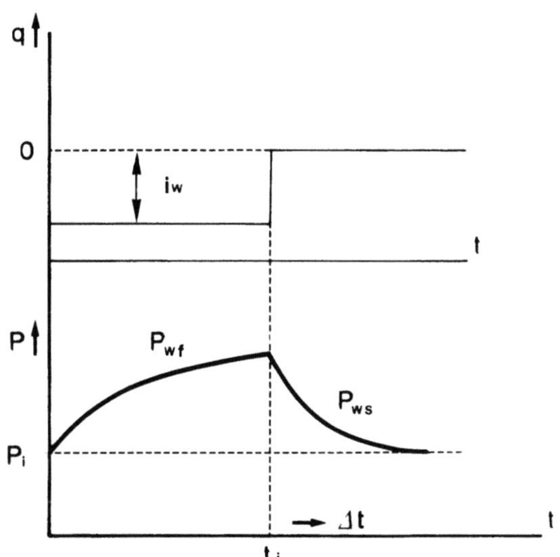

Fig. 6.30. *Rate and pressure variations during a constant rate injection test*

Starting from initial static conditions ($p = p_i$ over the entire aquifer), the bottom-hole flowing pressure p_{wf} for an injection rate i_w is [from Eq. (6.25) for transient flow]:

$$p_{wf} - p_i = \frac{i_w B_w \mu_w}{4\pi k_w h}\left(\ln t + \ln\frac{k_w}{\phi\mu_w c_t r_w^2} + 0.809 + 2S\right),\qquad(6.67)$$

where

$$c_t = c_w + c_f$$

so that

$$m = \frac{dp_{wf}}{d\ln t} = \frac{i_w B_w \mu_w}{4\pi k_w h},\qquad(6.68)$$

Therefore

$$\frac{k_w h}{\mu_w} = \frac{i_w B_w}{4\pi m}.\qquad(6.69)$$

The skin factor calculation is analogous to that for the constant rate drawdown developed in Sect. 6.5.2:

$$S = \frac{p_{1h} - p_i}{2m} - \frac{1}{2}\left(\ln\frac{k_w}{\phi\mu_w c_t r_w^2} + 8.998\right).\qquad(6.70)$$

We now look at the special case where the water has been injected directly into the oil zone for a short period of time, resulting in a bank of cold water of *limited extent* around the well.

In this situation, the transient behaviour is determined by the properties of the *oil*, so that μ_o, and c_o should be used in Eqs. (6.67)–(6.70) instead of μ_w and c_w.

Analysis of the injection test will yield an estimate of k_o (instead of k_w), and the presence of the localised bank of water around the well will appear as a skin effect which will be negative if $\mu_o > \mu_w$ (or, rather, $\lambda_w > \lambda_o$).

The equations which describe the pressure fall-off behaviour observed when the injection well is shut in are identical to those used in Sect. 6.6 for the pressure buildup of a producer. Of course, shutting in a producing well is followed by an *increase* in the bottom-hole pressure, whereas shutting in an injector results in a *decrease*.

Using the terminology of Fig. 6.30 in Eq. (6.5) we have

$$p_{\text{ws}} - p_{\text{i}} = \frac{i_{\text{w}} B_{\text{w}} \mu_{\text{w}}}{4\pi k_{\text{w}} h} \ln \frac{t_{\text{i}} + \Delta t}{\Delta t} , \tag{6.71}$$

$$\frac{\mathrm{d}p_{\text{ws}}}{\mathrm{d} \ln \left[(t_{\text{i}} + \Delta t)/\Delta t \right]} = \frac{i_{\text{w}} B_{\text{w}} \mu_{\text{w}}}{4\pi k_{\text{w}} h} = m , \tag{6.72}$$

$$\frac{k_{\text{w}} h}{\mu_{\text{w}}} = \frac{i_{\text{w}} B_{\text{w}}}{4\pi m} . \tag{6.73}$$

To calculate S, we can refer to Eq. (6.48b):

$$S = \frac{p_{\text{wf}}(\Delta t = 0) - p_{\text{ws(LIN)}}(1\,\text{h})}{2m} - \frac{1}{2}\left(\ln \frac{k_{\text{w}}}{\phi \mu_{\text{w}} c_{\text{t}} r_{\text{w}}^2} + 8.998 \right), \tag{6.74}$$

where $p_{\text{ws(LIN)}}$ (1 h) is the value of p_{ws} read from the straight line section of the plot of p_{ws} versus $\ln \left[(t_{\text{i}} + \Delta t)/\Delta t \right]$ (extrapolated if necessary) at $\Delta t = 1$ h from the instant of shut in (Fig. 6.31).

The average reservoir pressure $\bar{p}$ in the "drainage" area of the well (better referred to as the "area of influence" in the case of injection) is calculated in the manner outlined in Sect. 6.4, and in Sect. 6.6.2 for the case of a pressure buildup.

However, it is important to realise that, where water injection is concerned, there can never be a no-flow condition at the outer limit of the area of influence of the well, an hypothesis that was fundamental to the calculation of the $p_{\text{D(MBH)}} = f(t_{\text{D}A})$ curves of Figs. 6.5–6.11 used in estimating $\bar{p}$ in Sect. 6.6.2.

It is quite common fields to be drilled in a "five spot" pattern, locating injection wells actually within the oil-bearing region of the reservoir. This consists of a network of squares with a producer located at each corner, and one injector at the

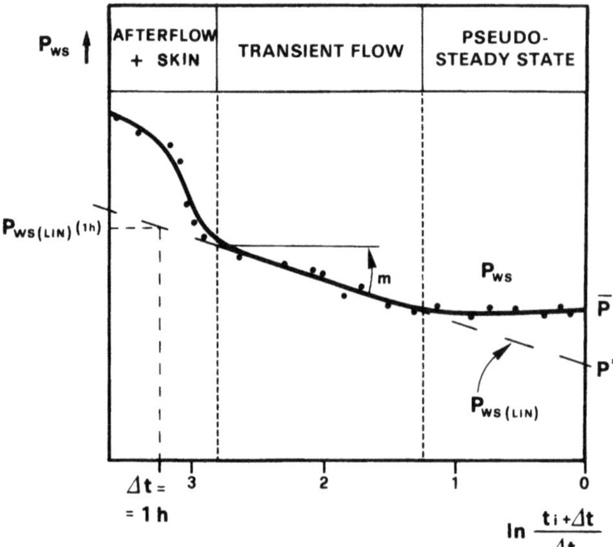

Fig. 6.31. *Horner plot of a pressure fall-off in a water injection well*

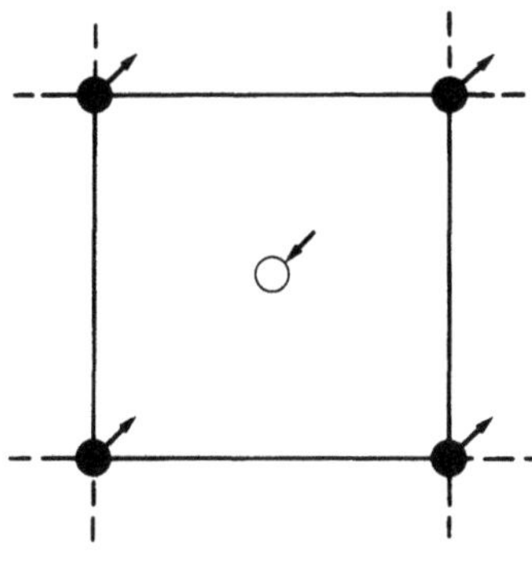

Fig. 6.32. *The five spot pattern of production and injection wells*

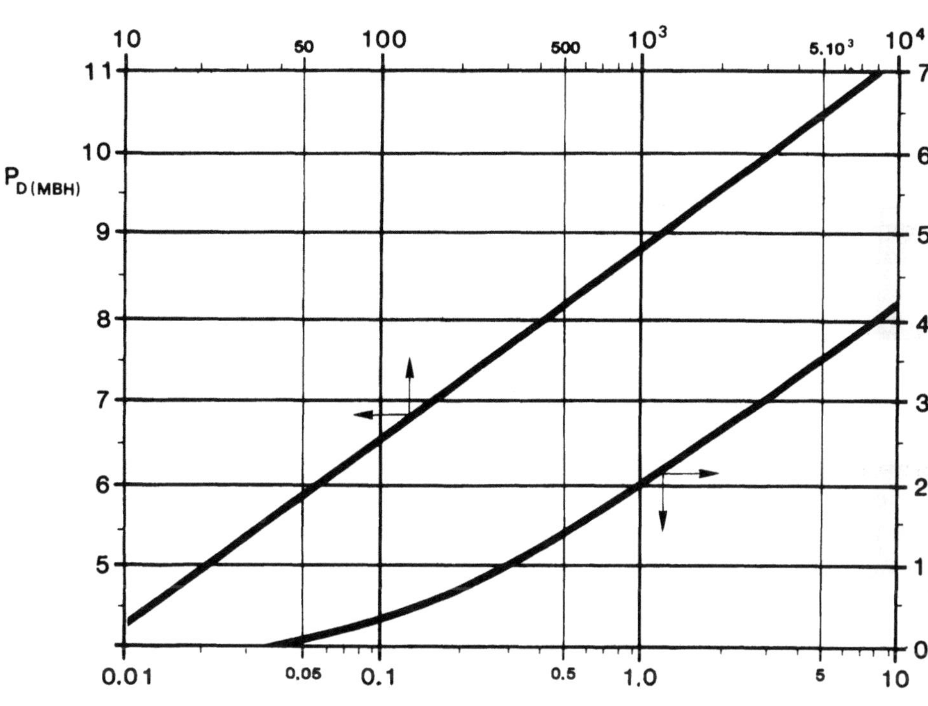

Fig. 6.33. $p_{D(MBH)} = f(t_{DA})$ *for a five spot configuration. From Ref. 17, 1967, Society of Petroleum Engineers of AIME, reprinted with permission of the SPE*

centre of each square (Fig. 6.32). The $p_{D(MBH)} = f(t_{DA})$ curve for this configuration is presented in Fig. 6.33.

The calculation procedure for $\bar{p}$ is similar to that described in Sect. 6.6.2:

1. Extrapolate the straight line $p_{ws(LIN)} = f\{\ln[(t_i + \Delta t)/\Delta t]\}$ to $\Delta t = \infty$ (i.e. $\ln[(t_i + \Delta t)/\Delta t] = 0$), to obtain the fictitious mean pressure p^* (Fig. 6.31).
2. Calculate a value for:

$$t_{DA} = \frac{k_w}{\phi \mu_w c_t A} t_i \tag{6.75}$$

where

$$t_{\mathrm{i}} = \frac{W_{\mathrm{i}}}{i_{\mathrm{w}}} = \frac{\text{cumulative volume of water injected}}{\text{last injection rate}} \,,$$

A = area of a single five spot array

and $k_{\mathrm{w}}/\mu_{\mathrm{w}}$ is estimated from the pressure fall-off analysis.

3. Enter the curve in Fig. 6.33 with $t_{\mathrm{D}A}$, and read off the corresponding value of

$$p_{\mathrm{D(MBH)}}(t_{\mathrm{D}A}) = \frac{4\pi k_{\mathrm{w}} h}{i_{\mathrm{w}} B_{\mathrm{w}} \mu_{\mathrm{w}}} (\bar{p} - p^*) \,. \tag{6.76a}$$

Then calculate

$$\bar{p} = p^* + \frac{i_{\mathrm{w}} B_{\mathrm{w}} \mu_{\mathrm{w}}}{4\pi k_{\mathrm{w}} h} p_{\mathrm{D(MBH)}}(t_{\mathrm{D}A}) \,. \tag{6.76b}$$

6.11   Interference Testing Between Wells

Conventional drawdown and buildup testing of producing wells provides important information about the near-wellbore skin effect and the nature of the reservoir (flow model, permeability, barriers, etc.) within the drainage area of the well. The distance out to which any well test will actually "see" depends on the diffusivity $k/\phi\mu c$, the duration of the test, and the resolution of the pressure gauge itself (i.e. its ability to distinguish *small* changes in pressure caused, for instance, by a *distant* permeability barrier).

The region investigated with these tests, however, is confined to the drainage area of the well concerned, and cannot give any information about the degree of hydraulic continuity with neighbouring wells, nor about the properties of the rock between them.

The measurement of inter-well reservoir characteristics can be achieved, in principle, by means of *interference testing*.[6]

Basically, interference testing involves changing the production rate at "active" or "input" well, and measuring the change in pressure behaviour (if any) observed at a neighbouring "observation" well or wells.

If a number of such tests is performed successively among appropriate groups of wells, it is possible to map variations in the areal distribution of permeability over as much of the reservoir as can be covered. Knowledge of permeability anisotropy and regions of high and low permeability are fundamental to the correct planning of recovery by water injection, where maximising the sweep efficiency requires avoiding irregular frontal advance.

The ideal condition for this type of test would be to have all the wells in the field shut in, and to put only one well on production. Obviously, it is unlikely that oilfield economics would support this way of testing.

A more acceptable alternative is to have all the wells on production at constant rates, and to shut only one well in. In this case, we should see an increase in the pressure at neighbouring wells which are in communication with the shut in well.

We will first consider the simplest case of two wells, both shut in, one of which is put on production at a constant rate q_{sc} at time $t = 0$.

Initially, there is a uniform pressure p_i in the region around the two wells. We will make the usual assumptions about the reservoir rock: homogeneous and isotropic, constant thickness h, single phase oil in the presence of (immobile) water at its irreducible saturation S_{iw}.

Referring to Eqs. (5.14b), (5.14c), (5.22a) and (5.30), the dimensionless pressure p_D at a distance $r_D = r/r_w$ from the axis of the producing well, at time t_D, is

$$p_D(r_D, t_D) = \frac{1}{2}\text{ei}\left(\frac{\phi\mu_o c_t r^2}{4kt}\right)$$

$$= \frac{1}{2}\text{ei}\left(\frac{\phi\mu_o c_t r_w^2}{4kt}\frac{r^2}{r_w^2}\right)$$

$$= \frac{1}{2}\text{ei}\left(\frac{r_D^2}{4t_D}\right). \tag{6.77}$$

Equation (6.77) is the one used for the analysis of interference tests by type curves, It is valid provided the reservoir boundaries (or any permeability contrast, such as a fault) are far enough away that the flow in the inter-well region remains transient throughout the test.

When this is the case, the appropriate type curve is the *ei-function*, shown in Fig. 6.34 as a function of t_D/r_D^2.

To interpret interference test data, plot the values of the bottom-hole pressure $p_{ws}(t)$ in the observation well as $[p_i - p_{ws}(t)]$ against t on a log–log grid with the same scales as Fig. 6.34.

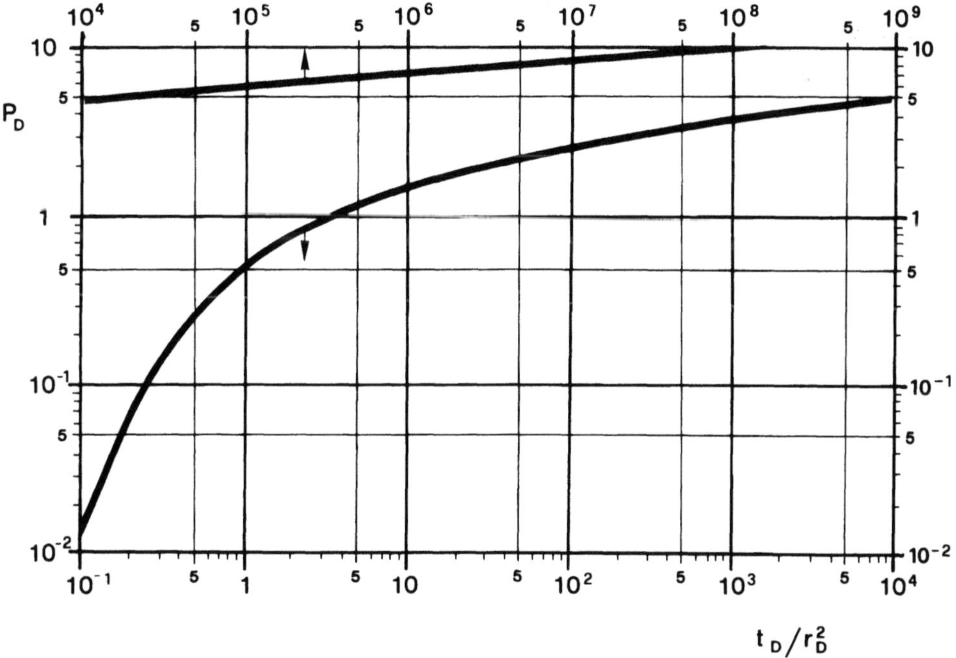

Fig. 6.34. *Interference test type-curve $p_D(r_D, t_D)$ (exponential integral function). From Ref. 6, 1977, Society of Petroleum Engineers of AIME, reprinted with permission of the SPE*

Overlay the data plot on the type curve in Fig. 6.34 and, keeping the axes correctly aligned, move it until a match is obtained somewhere on the curve.

Once the two curves have been overlain, mark an arbitrary point (anywhere will do) and read off the coordinates of this *same point* on both the type curve and the data plot. Then, from the pressure axis values of the point:

$$p_D(r_D, t_D) = \frac{2\pi k_o h}{q_{sc} B_o \mu_o} [p_i - p_{ws}(r, t)] , \tag{6.78}$$

from which

$$k_o = \frac{q_{sc} B_o \mu_o}{2\pi h} \frac{p_D(r_D, t_D)}{p_i - p_{ws}(r, t)} . \tag{6.79}$$

From the time axis values of the point:

$$\frac{t_D}{r_D^2} = \frac{k_o t}{\phi \mu_o c_t r_w^2} \frac{r_w^2}{r^2} = \frac{k_o t}{\phi \mu_o c_t r^2} , \tag{6.80}$$

so that

$$\phi c_t = \frac{k_o}{\mu_o r^2} \frac{t}{t_D/r_D^2} , \tag{6.81}$$

where r is the distance between the active and observation wells, and k_o is derived from Eq. (6.79).

The pressure $p_{ws}(t)$ measured in an observation well at a distance r from a producer can be analysed using Eqs. (6.79) and (6.81), and we can determine the permeability to oil and the porosity-compressibility product ϕc_t (which is in fact the volume of oil produced from a unit volume of reservoir for a unit decrease in pressure).

In Sect. 5.6.1.1 we found that for $x < 0.01$, the approximation:

$$ei(x) = -0.57721 - \ln x . \tag{5.31b}$$

was valid. Therefore, referring to Eq. (6.77), if

$$\frac{r_D^2}{4 t_D} < 0.01 \tag{6.82a}$$

or, inverting

$$\frac{t_D}{r_D^2} > 25 \tag{6.82b}$$

then Eq. (6.77) can be expressed in the following form

$$\begin{aligned}
p_D(r_D, t_D) &= \frac{1}{2}\left[-0.57721 - \ln \frac{r_D^2}{4 t_D} \right] \\
&= \frac{1}{2} \ln \left[\frac{4}{\exp 0.57721} \frac{t_D}{r_D^2} \right] \\
&= \frac{1}{2}\left[\ln \frac{t_D}{r_D^2} + 0.809 \right]
\end{aligned} \tag{6.83a}$$

from which we can conclude that

$$p_i - p_{ws}(r, t) = \frac{q_{sc} B_o \mu_o}{4\pi k_o h} \left[\ln \frac{k_o t}{\phi \mu_o c_t r_w^2} - 2 \ln \frac{r}{r_w} + 0.809 \right]. \tag{6.83b}$$

We have deduced that if we plot the observation well pressure p_{ws} against $\ln t$, then the slope of the straight line portion (if t_D is large enough for this to be present) is

$$-\frac{\mathrm{d} p_{ws}}{\mathrm{d} \ln t} = \frac{q_{sc} B_o \mu_o}{4\pi k_o h} = m\,, \tag{6.84a}$$

so that

$$k_o = \frac{q_{sc} B_o \mu_o}{4\pi m h}\,. \tag{6.84b}$$

These are the same as Eqs. (6.26) and (6.27) for the pressure drawdown test. There are however two fundamental differences between the two types of test:

1. The skin factor cannot be determined from interference test data, because the flow rate close to the observation well bore is almost zero.
2. The value of k_o determined with Eq. (6.84b) is the *average permeability of the region between the wells*.

This above explanation of interference testing is necessarily simplistic, in view of the extremely qualitative nature of this kind of test.

In fact, there are more complex approaches to interference testing – for example, a succession of flow rates at the producing well, followed by a shut in; or keeping all the wells on production at a constant rate and closing in one of them.

We have assumed so far that k_o is constant in the area of investigation of the test, an area corresponding roughly to a circle with a radius at least equal to the distance between the two wells.

If the active well is surrounded by several observation wells, an interpretation of the data from each well can provide an idea of reservoir heterogeneity, since a different permeability may be found from each. None of these will correspond exactly to a real permeability, because the theory assumes a constant and isotropic k_o.

A proper interpretation of interference testing in heterogeneous reservoirs requires the use of a numerical simulator coupled with a linear optimiser, as described for instance by Jacquard and Jain.[12]

However, this is another instance of an "inverse problem", where we are trying to determine the internal characteristics of a system, or "black box", by analysing its response to an external stimulus, and once again we cannot be certain that any one solution is unique, and therefore that a particular permeability distribution determined from a number of interference tests represents the real one.

In any case, interference testing is always useful, because it can provide a broad picture of the anisotropy of the reservoir rock, and the directions of maximum and minimum permeability.

6.12   Pulse Testing

The point has already been made several times that from a theoretical viewpoint, a reservoir can be considered as a "black box" whose internal structure we attempt to describe from an analysis of its response to an external stimulus.

Looked at in this way, a well test consists of an input signal (production from a well) applied to a system (the reservoir), and the measurement of its response at the well (bottom-hole pressure drawdown or buildup) and/or at neighbouring wells (interference test).

In system theory, it is recognised that the quality of the results obtained depends on the nature of the signal applied. One of the implications of this statement is that a constant signal (a constant production or injection rate), is the least suitable kind of stimulus with which to investigate the internal nature of a "black box" system. The important factor is that a constant signal consists of a single frequency (zero).

On the other hand, a sequence of square waves (successive periods of production and shut in, as in Fig. 6.35) is the most effective type of signal to input to the system. This is because it contains a very wide spectrum of frequencies (as can be seen from its representation as a Fourier series).

This fact was first presented by Johnson et al.[13] in the context of well testing, and by Chierici et al.[2] for the case of gas reservoirs producing under water drive.

In a pulse test, the input to the active well is a sequence of cycles of equal duration Δt_c. Each cycle consists of a shut-in period ($q = 0$) lasting a time Δt_p, followed by a flowing period ($\Delta t_c - \Delta t_p$). These times are the same for every cycle.

This wave train propagates out into the reservoir at a characteristic velocity, and is attenuated as it progresses.

The bottom-hole pressure is measured in the observation well or wells using high resolution pressure gauges capable of detecting changes of as little as 10 Pa.

Referring to Fig. 6.35, two features of the pressure pulse sequence are used in the interpretation: the *amplitude* Δp_i of the pulse following the ith flow period (this will

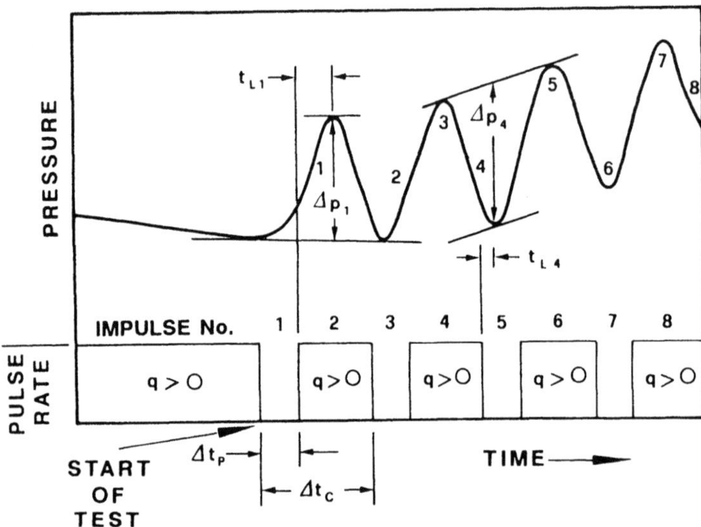

Fig. 6.35. *A pulse test, showing the input flow rate at the active well, and the pressure response at the observation well. From Ref. 6, 1977, Society of Petroleum Engineers of AIME, reprinted with permission of the SPE*

be a peak after a shut-in period, a trough after a flowing period); and the *time lag* t_{Li} between the end of each period and the arrival of its peak or trough.

The following notes on the quantitative analysis of pulse test data will be limited to main points of a method proposed by Kamal and Brigham,[14] which is well suited to manual interpretation.

To demonstrate the principle, we will consider the very first pressure pulse ($i = 1$) arriving at the observation well, which is a distance r from the pulsing well.

The amplitude of this pulse (in this case a peak) is Δp_1, and it arrives at a time $t_{L,1}$ after the end of the first shut-in period (Fig. 6.35).

The following two terms are first calculated:

$$F' = \frac{\Delta t_p}{\Delta t_c}$$

(6.85a)

and

$$F''_1 = \frac{t_{L1}}{\Delta t_c} ,$$

(6.85b)

where Δt_c is the duration of each cycle, and Δt_p is the duration of the shut-in.

Kamal and Brigham[14] have published diagrams for the analysis of each of the pulses, in a series labelled "first odd pulse", "first even pulse", "second odd pulse", and so on. On the diagram shown in Fig. 6.36, which is for the very first pulse ("first odd"), we select the curve corresponding to F' [Eq. (6.85a)] and read off the y-axis value of $[\Delta p_D (t_L/\Delta t_c)^2]_{\text{Fig}}$ corresponding to F'' (Eq. (6.85b)) on the x-axis.

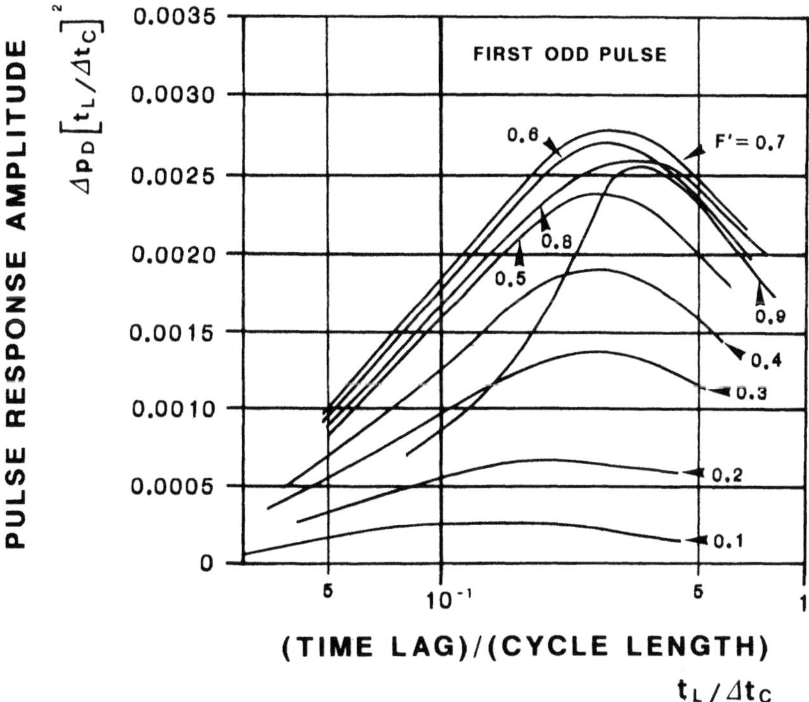

Fig. 6.36. *Pulse testing: diagram for the calculation of k (Eq. (6.86a)) for the very first (first odd) pulse arriving at the observation well. From Ref. 6, 1977, Society of Petroleum Engineers of AIME, reprinted with permission of the SPE*

We then have

$$k_\mathrm{o} = \frac{q_\mathrm{sc} B_\mathrm{o} \mu_\mathrm{o} [\Delta p_\mathrm{D}(t_L/\Delta t_\mathrm{c})^2]_\mathrm{Fig}}{h\,\Delta p_1 (t_{L1}/\Delta t_\mathrm{c})^2} \,. \qquad (6.86a)$$

The first odd pulse ($i = 1$) can also be analysed using the appropriate curve in Fig. 6.37. In this case, we read off a value for $[(t_L)_\mathrm{D}/r_\mathrm{D}^2]_\mathrm{Fig}$ from the y-axis and calculate:

$$\phi c_\mathrm{t} = \frac{k_\mathrm{o} t_{L1}}{\mu_\mathrm{o} r^2 [(t_L)_\mathrm{D}/r_\mathrm{D}^2]_\mathrm{Fig}} \,. \qquad (6.86b)$$

A full interpretation would involve as many of the cycles as possible, finally taking an average of the computed values.

The pulse test technique requires a much shorter overall shut-in time than a conventional interference test. By using highly sensitive pressure gauges, it is possible to keep the observation wells on production throughout the pulse test, provided their rates are held precisely constant so as not to superpose additional transient signals on the measured response. In this way, the pulsing well is the only one requiring shutting in – and, even then, only intermittently – so that lost production is minimised.

For completeness, a second type of pulse test should be included here: sometimes referred to as the *vertical interference/pulse test*.[35] This is run in a *single well between two zones*, one zone being pulsed (flow/shut in), the other being under observation.

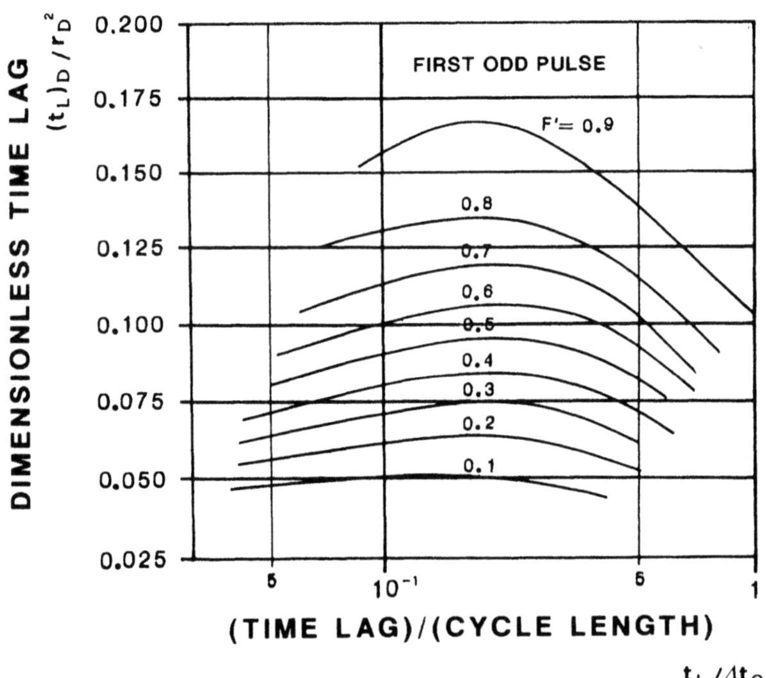

Fig. 6.37. *Pulse testing: diagram for the calculation of ϕc_t (Eq. (6.86b)) for the very first (first odd) pulse arriving at the observation well. From Ref. 6, 1977, Society of Petroleum Engineers of AIME, reprinted with permission of the SPE*

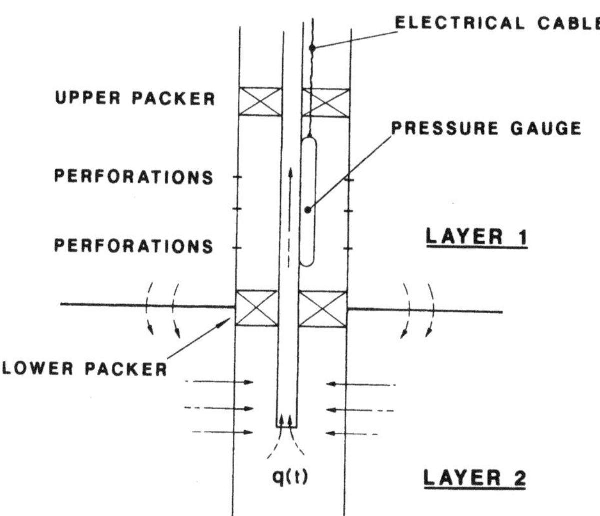

Fig. 6.38. *Completion design for the execution of a vertical interference or pulse test between two zones in the same well*

The two zones are of course isolated in the well bore by a packer or, better still, by a straddle packer (Fig. 6.38) which isolates the observation zone from the wellbore above.

The theoretical background to this type of test has been developed by Hirasaki.[10] The objective is to determine the vertical permeability (if any) between two zones in the same reservoir. When there are impermeable intercalations, such as thin shales, of unknown lateral extent, the degree of communication between sedimentary units is very much an unknown quantity. This is a particularly important factor in reservoirs where the principle production mechanism is gravity segregation.

For the results from vertical testing to have any meaning, it is essential that the cement in the casing/formation annulus should provide a perfect hydraulic seal, at least above, below and between the two zones. There is almost always some doubt about this aspect of the test when a high vertical permeability is indicated from the results.

References

1. Brons F, Marting VE (1961) The effect of restricted fluid entry on well productivity. J Petrol Tech (Feb): 172–174; Trans AIME 222
2. Chierici GL, Ciucci GM (1969) Water-drive gas reservoirs: influence of pulse-testing on the indetermination range of reserve estimates. J Petrol Tech (Dec): 1521–1527
3. Chierici GL, Ciucci GM, Pizzi G (1965) Quelques cas de remontées de pression dans des couches hétérogènes avec pénétration partielle. Etude par analyseur électrique. Rev Inst Fr Pétrole (Dec): 1811–1846
4. Cobb WM, Dowdle WL (1973) A simple method or determining well pressures in closed rectangular reservoirs. J Petrol Tech (Nov): 1305–1306
5. Dake LP (1978) Fundamentals of reservoir engineering. Elsevier Amsterdam
6. Earlougher RC Jr (1977) Advances in well testing analysis. Society of Petroleum Engineers, Monograph 5, Dallas

7. Gringarten AC (1987) Type-curve analysis: what it can and cannot do. J Petrol Tech (Jan): 11–13
8. Gringarten AC, Ramey HJ Jr, Raghavan R (1974) Unsteady-state pressure distribution created by a well with a single infinite-conductivity vertical fracture, Soc Petrol Eng J (Aug): 347–360; Trans AIME 257
9. Harris HM (1966) The effect of perforating on well productivity. J Petrol Tech (April): 518–528; Trans AIME 237
10. Hirasaki GJ (1974) Pulse tests and other early transient pressure analyses for in-situ estimation of vertical premeability. Soc Petrol Eng J (Feb): 75–90; Trans AIME 257
11. Horner DR (1951) Pressure build up in wells. Proc 3rd World Petrol Congr, vol II, Brill EJ, Leiden pp 503–523
12. Jacquard P, Jain C (1965) Permeability distribution from field pressure data. Soc Petrol Eng J (Dec): 281–294; Trans AIME 234
13. Johnson CR, Greenkorn RA, Woods EG (1966) Pulse-testing: a new method for describing reservoir flow properties between wells. J Petrol Tech (Dec): 1599–1604; Trans AIME 237
14. Kamal M, Brigham WE (1975) Pulse testing response for unequal pulse and shut-in periods. Soc Petrol Eng J (Oct): 399–410; Trans AIME 259
15. Lee J (1982) Well testing., Society of Petroleum Engineers. Textbook vol 1, Dallas
16. Lefkovits HC, Hazebroek P, Allen EE, Matthews CS (1961) A study of the behaviour of bounded reservoirs composed of stratified layers. Soc Petrol Eng J (March): 43–58; Trans AIME 222
17. Matthews CS, Russell DG (1967) Pressure buildup and flow tests in wells. Society of Petroleum Engineers, Monogr, vol 1, Dallas
18. Matthews CS, Brons F, Hazebroek P (1954) A method for the determination of average pressure in a bounded reservoir. Trans AIME 201: 182–191
19. McKinley RM (1971) Wellbore transmissibility from afterflow-dominated pressure buildup data. J Petrol Tech (July): 863–872; Trans AIME 251
20. Miller CC, Dyes AB, Hutchinson CA Jr (1950) Estimation of permeability and reservoir pressure from bottom-hole pressure build-up characterstics. Trans AIME 189: 91–104
21. Nisle RG (1958) The effect of partial penetration on pressure buildup in oil wells. Trans AIME 213: 85–90
22. Odeh AS, Jones LG (1965) Pressure drawdown analysis, variable-rate case. J Petrol Tech (Aug): 960–964; Trans AIME 234
23. Ramey HJ Jr (1970) Short-time well test data interpretation in the presence of skin effect and wellbore storage. J Petrol Tech (Jan): 97–104; Trans AIME 249
24. Ramey HJ Jr, Cobb WM (1971) A general pressure buildup theory for a well in a closed drainage area. Trans AIME 251: 1493–1505
25. Winestock AG, Colpitts GP (1965) Advances in estimating gas well deliverability. J Can Petrol Tech (July-Sept): 111–119
26. Yeh NS (1986) Analysis of well test pressure data from a restricted entry well in a layered reservoir. PhD Thesis, University of Tulsa, Oklahoma
27. Ramey HJ Jr, Agarwal RG, Martin I (1975) Analysis of slug test or DST flow period data. J Can Petrol Tech (July-Sept): 37–42*
28. Gringarten AC, Bourdet DP, Landel PA, Kniazeff VJ (1979) A comparison between different skin and wellbore storage type-curves for early-time transient analysis. Pap SPE 8205, presented at the 54th Annu Fall Meet of the SPE, Las Vegas, Sept 23–26, 1979
29. Gringarten AC, Ramey HJ Jr, Raghavan RJ (1972) Unsteady state pressure distributions created by a well with a single infinite-conductivity vertical fracture. J Petrol Tech (Aug): 347–360; Trans AIME 222
30. Wong DW, Harrington AG, Cinco-Ley H (1986) Application of the pressure derivative function in the pressure transient testing of fractured wells. SPE Formation Evaluation (Oct), pp ,470–480
31. Gringarten AC, Ramey HJ Jr, Raghavan RJ (1975) Applied pressure analysis for fractured wells. J Petrol Tech (July): 887–892
32. Bourdet DP, Alagoa A, Ayoub JA, Pirard YM (1984) New type-curves aid analysis of fissured zone well tests. World Oil (April) 197(5): 77–87
33. Bourdet DP, Ayoub JA, Whittle TM, Douglas AA, Pirard YM, Kniazeff V (1983) Interpreting well tests in fractured reservoirs. World Oil (Oct)

---

* Refs. 27 to 36 were provided by Peter J. Westaway.

34. Bourdet DP, Ayoub JA, Pirard YM (1984) Use of the pressure derivative in welltest interpretation. Pap SPE 12777, presented at the SPE California Regional Meet, Long Beach, April 1984
35. Ehlig-Economides C, Ayoub JA (1986) Vertical interference testing across a low permeability zone. SPE Formation Evaluation (Oct), 497–510
36. Stewart G, Ascharsobbi F (1988) Well test interpretation for naturally fractured reservoirs. Pap SPE 18173, presented at the 63rd Annu Technical Conference, Houston, October 1988

EXERCISES

Exercise 6.1

A well completed with a $9\frac{5}{8}''$ casing is producing from a drainage area of $160\,000\,\mathrm{m}^2$. Calculate the $p_{\mathrm{D}}(t_{\mathrm{D}A})$ function for the following geometries:

1. Circular area, with the well at the centre.
2. 2:1 rectangular area, with the well at the centre.
3. 2:1 rectangular area, with the well closer to one side as shown in the Fig. E6/1.0.
4. 4:1 rectangular area, with the well at the centre.

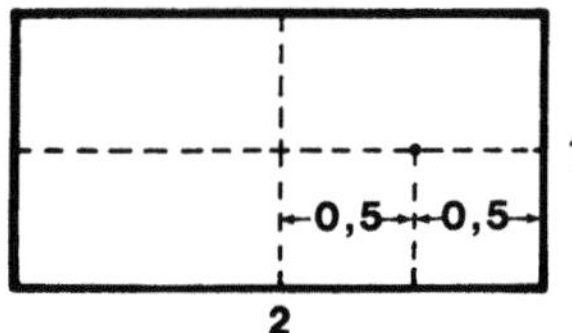

Fig. E6/1.0

Solution

From Eq. (6.20), we have

$$p_{\mathrm{D}}(t_{\mathrm{D}A}) = 2\pi t_{\mathrm{D}A} + \frac{1}{2}\ln t_{\mathrm{D}A} - \frac{1}{2}p_{\mathrm{D\,(MBH)}}(t_{\mathrm{D}A}) + \frac{1}{2}\ln\frac{A}{r_{\mathrm{w}}^2} + 0.405 \,, \qquad (6/1.1)$$

where, if there is pseudo-steady state flow in the drainage area, we know that

$$p_{\mathrm{D(MBH)}}(t_{\mathrm{D}A}) = \ln(C_A t_{\mathrm{D}A}) \,. \qquad (6/1.2)$$

For the geometries specified, the following values should be used:

Geometry	$p_{\mathrm{D(MBH)}} = (t_{\mathrm{D}A})$ Fig. No.	Pseudo-steady state flow if $t_{\mathrm{D}A} >$ (Fig. 5.7)	C_A (Fig. 5.7)
1	6.5	0.1	31.62
2	6.7 – I	0.3	21.84
3	6.7 – III	1.5	4.51
4	6.8 – I	0.8	5.38

Since

$$r_{\mathrm{w}} = \tfrac{1}{2}(9\tfrac{5}{8}'') = 0.12224\,\mathrm{m} \,,$$

we can calculate

$$\frac{1}{2}\ln\frac{A}{r_{\mathrm{w}}^2} + 0.405 = \frac{1}{2}\ln\frac{160\,000}{(0.12224)^2} + 0.405 = 8.498$$

Putting this into Eq. (6/1.1), we get

$$p_{\mathrm{D}}(t_{\mathrm{D}A}) = 2\pi t_{\mathrm{D}A} + \tfrac{1}{2}\ln t_{\mathrm{D}A} + 8.498 - \tfrac{1}{2}p_{\mathrm{D(MBH)}}(t_{\mathrm{D}A})\ . \qquad (6/1.3)$$

If we define

$$\alpha(t_{\mathrm{D}A}) = 2\pi t_{\mathrm{D}A} + \tfrac{1}{2}\ln t_{\mathrm{D}A} + 8.498\ , \qquad (6/1.4)$$

then

$$p_{\mathrm{D}}(t_{\mathrm{D}A}) = \alpha(t_{\mathrm{D}A}) - \tfrac{1}{2}p_{\mathrm{D(MBH)}}(t_{\mathrm{D}A})\ . \qquad (6/1.5)$$

In pseudo-steady state flow conditions, we can substitute Eq. (6/1.2) into Eq. 6/1.3), giving

$$p_{\mathrm{D}}(t_{\mathrm{D}A}) = 2\pi t_{\mathrm{D}A} + \tfrac{1}{2}\ln t_{\mathrm{D}A} + 8.498 - \tfrac{1}{2}\ln C_A - \tfrac{1}{2}\ln t_{\mathrm{D}A}\ ,$$

so that

$$p_{\mathrm{D}}(t_{\mathrm{D}A}) = 8.498 - \tfrac{1}{2}\ln C_A + 2\pi t_{\mathrm{D}A}\ . \qquad (6/1.6)$$

If we define

$$\beta(C_A) = 8.498 - \tfrac{1}{2}\ln C_A\ ,$$

then

$$p_{\mathrm{D}}(t_{\mathrm{D}A}) = \beta(C_A) + 2\pi t_{\mathrm{D}A}\ . \qquad (6/1.7)$$

The calculated values of $p_{\mathrm{D}}(t_{\mathrm{D}A})$ for the four geometries are presented in the following two tables. The first table lists the range of $t_{\mathrm{D}A}$ during which the flow is transient; the second covers pseudo-steady state flow.

Calculated values of $p_{\mathrm{D}}(t_{\mathrm{D}A})$ in transient flow

$t_{\mathrm{D}A}$	$\alpha(t_{\mathrm{D}A})$	$p_{\mathrm{D(MBH)}}(t_{\mathrm{D}A})$ for the four shapes				$p_{\mathrm{D}}(t_{\mathrm{D}A})$ for the four shapes			
		1	2	3	4	1	2	3	4
0.01	6.258	0.12	0.12	0.12	0.12	6.198	6.198	6.198	6.198
0.02	6.668	0.25	0.30	0.30	0.25	6.538	6.518	6.518	6.538
0.03	6.933	0.37	0.40	0.40	0.28	6.743	6.733	6.733	6.793
0.04	7.140	0.50	0.50	0.50	0.31	6.890	6.890	6.890	6.985
0.05	7.314	0.62	0.62	0.62	0.35	7.004	7.004	7.004	7.139
0.06	7.468	0.75	0.68	0.67	0.35	7.093	7.128	7.133	7.293
0.07	7.608	0.87	0.78	0.72	0.34	7.173	7.218	7.248	7.438
0.08	7.738	0.98	0.85	0.75	0.30	7.248	7.313	7.363	7.588
0.09	7.859	1.10	0.90	0.78	0.29	7.309	7.409	7.469	7.714
0.1	7.975	1.24	1.00	0.80	0.28	7.355	7.475	7.575	7.835
0.2	8.950	—	1.55	0.92	0.45	—	8.172	8.490	8.725
0.3	9.781	—	1.95	0.95	0.63	—	8.806	9.306	9.466
0.4	10.553	—	—	0.99	0.80	—	—	10.058	10.153
0.5	11.293	—	—	1.05	1.00	—	—	10.790	10.793
0.6	12.012	—	—	1.15	1.20	—	—	11.437	11.412
0.7	12.718	—	—	1.25	1.30	—	—	12.093	12.068
0.8	13.413	—	—	1.35	1.48	—	—	12.738	12.673
0.9	14.100	—	—	1.48	—	—	—	13.360	—
1.0	14.781	—	—	1.60	—	—	—	13.981	—

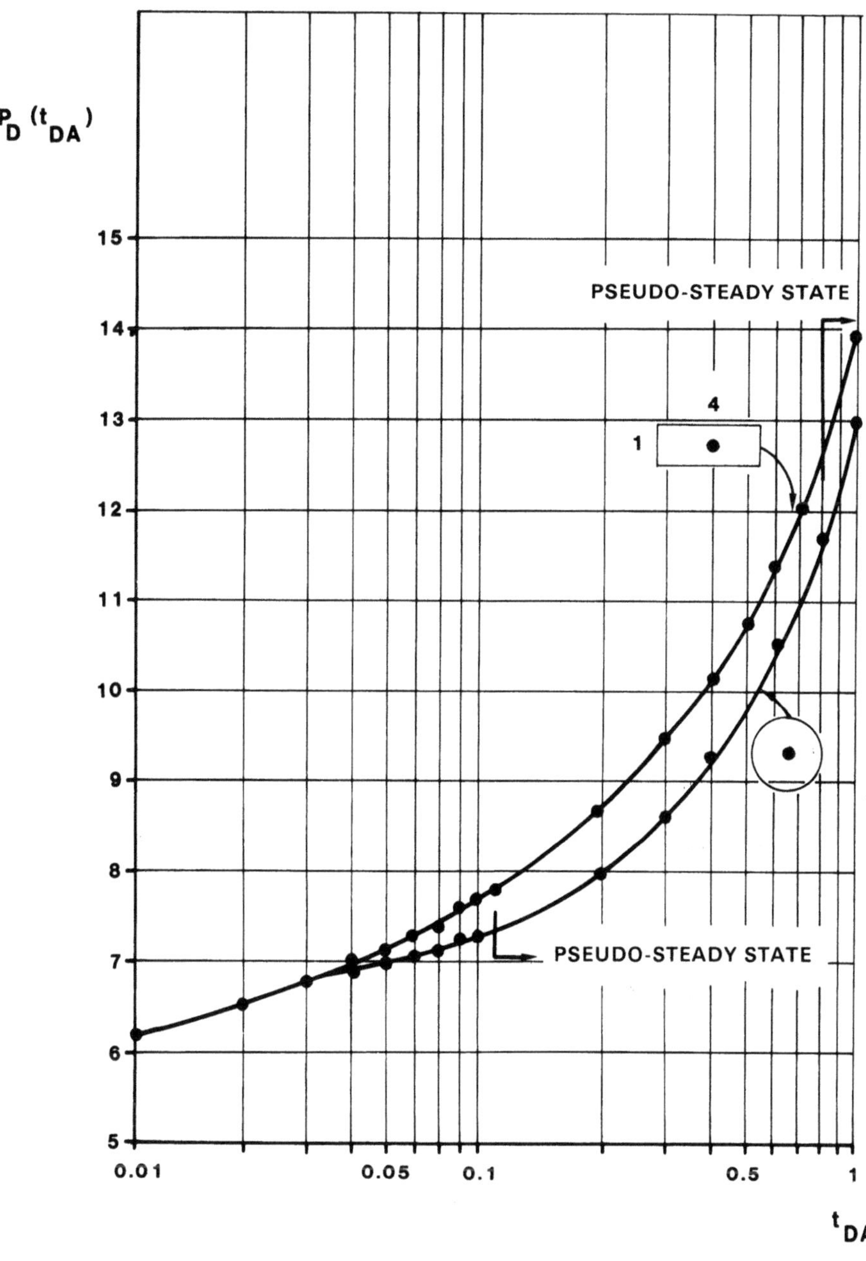

Fig. E6/1.1

Calculated values of $p_D(t_{DA})$ in pseudo-steady state flow

$$\beta(C_A) = 8.498 - \tfrac{1}{2}\ln C_A = 6.771 \text{ for geometry } 1$$
$$= 6.956 \text{ for geometry } 2$$
$$= 7.745 \text{ for geometry } 3$$
$$= 7.657 \text{ for geometry } 4$$

t_{DA}	$p_D(t_{DA})$ for shape No.			
	1	2	3	4
0.2	8.028	—	—	—
0.3	8.656	—	—	—
0.4	9.284	9.469	—	—
0.6	10.541	10.726	—	—
0.8	11.798	11.983	—	—
1.0	13.054	13.239	—	13.940
2.0	19.337	19.522	20.311	20.223
4.0	31.904	32.089	32.878	32.790
6.0	44.470	44.655	45.444	45.356
8.0	57.036	57.221	58.010	57.922
10	69.603	69.788	70.567	70.489
20	132.43	132.62	133.41	133.32
40	258.10	258.28	259.07	258.98

The values of $p_D(t_{DA})$ for $0.01 \leq t_{DA} \leq 1$ are plotted in Fig. E6/1.1 for geometries 1 and 4. The other two geometries lie between these two extremes.

◇ ◇ ◇

Exercise 6.2

An exploration well has found oil in a formation which, from the available geophysical evidence, is believed to be a modest sized lens. Log data for the oil-bearing interval, from 4450 – 4550 m, is as follows:

net pay thickness: $h = 58$ m ,
porosity: $\phi = 25\%$,
water saturation: $S_{iw} = 15\%$.

The reservoir pressure at discovery was

initial pressure: $p_i = 465$ kg/cm^2 .

From PVT analysis of a reservoir oil sample the following data is available:

bubble-point pressure: $p_b = 371$ kg/cm^2 ,
volume factor at p_i: $B_{oi} = 1.68$,
compressibility at p_i: $c_o = 2.6 \times 10^{-4}$ cm^2/kg ,
viscosity at p_i: $\mu_{oi} = 0.54$ cP

and based on correlations:

compressibility of reservoir water: $c_w = 2.9 \times 10^{-5}$ cm^2/kg
pore compressibility: $c_f = 1.5 \times 10^{-4}$ cm^2/kg

A $9\frac{5}{8}''$ casing was run, and perforated across the entire oil-bearing interval.

An extended drawdown test – 1 month of production at a constant $q_{sc} = 550$ m^3/d (stock tank) was performed in order to determine the size, and, if possible, the shape of the reservoir.

During the test, the bottom-hole flowing pressure p_{wf} was recorded with time from the start of production. The data are listed below:

Time	p_{wf} (kg/cm^2)	Time	p_{wf} (kg/cm^2)
1 min	464.80	2 d	450.18
5 min	461.92	4 d	448.10
10 min	458.30	6 d	445.92
15 min	455.00	8 d	443.82
30 min	454.16	10 d	441.72
45 min	453.73	15 d	436.48
1 h	453.61	20 d	431.25
2 h	453.30	25 d	426.00
4 h	453.00	30 d	420.76
8 h	452.46		
16 h	451.86		
1 d	451.33		

Interpret the test data to estimate

- the production characteristics of the well and reservoir (permeability, skin factor, PI and completion factor)
- volume of oil in place
- the maximum likely extent of the reservoir and its shape.

Solution

A plot of p_{wf} against $\log t$ (Fig. E6/2.1) has a straight line (transient flow) portion between $t = 45$ min and $t = 240$ min.

The data prior to $t = 45$ min corresponds to fluid segregation in the wellbore (oil replacing water and mud, etc.), while the data after $t = 240$ min exhibits late transient behaviour.

In the straight line portion, we have

$$m = \frac{-\,\mathrm{d}p_{wf}}{\mathrm{d}\log t} = 1.02\,\text{kg/cm}^2 \times \text{cycle}$$

Using practical metric units (Table 6.1)

$$m = 21.907\,\frac{q_{sc}B_o\mu_o}{k_o h}$$

so that, in this example

$$k_o = 21.907\,\frac{550 \times 1.68 \times 0.54}{1.02 \times 58} = 185\,\text{md}\ .$$

The equation for the calculation of the skin factor (Table 6.1) is

$$S = 1.151\left(\frac{p_i - p_{1\,h}}{m} - \log \frac{k_o}{\phi\mu_o c_t r_w^2} + 3.107\right).$$

Since

$$c_t = c_o S_o + c_w S_w + c_f\ ,$$

we have

$$c_t = (2.6 \times 0.85 + 0.29 \times 0.15 + 1.5) \times 10^{-4} = 3.75 \times 10^{-4}\,\text{cm/kg}\ .$$

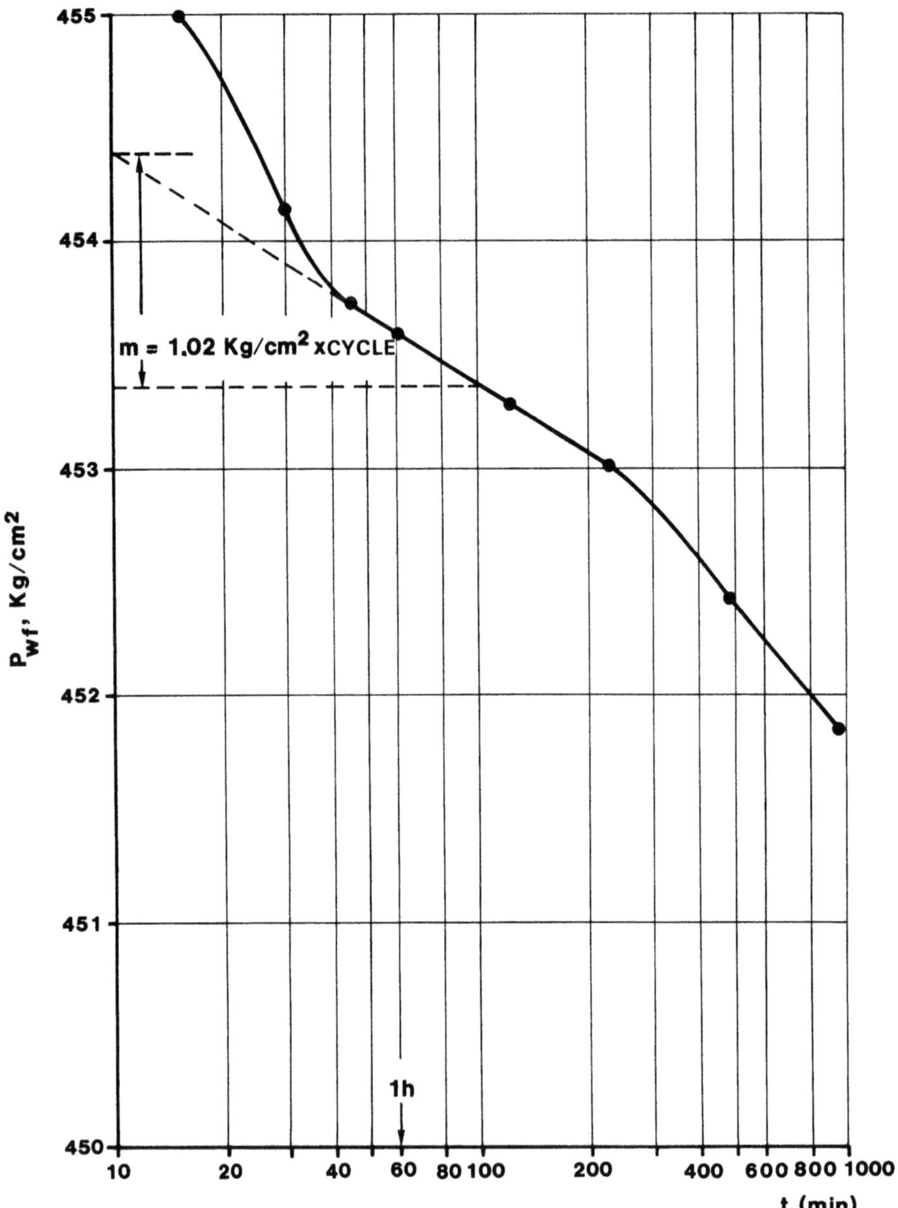

Fig. E6/2.1

With the values

$$p_i = 465.00 \text{ kg/cm}^2 \ ,$$

$$p_{1\,h} = 453.61 \text{ kg/cm}^2 \ ,$$

$$r_w = \tfrac{1}{2}(9\tfrac{5}{8}'') = 0.12224 \text{ m} \ ,$$

we compute

$$S = 1.151 \left[\frac{465 - 453.61}{1.02} - \log \frac{185}{0.25 \times 0.54 \times 3.75 \times 10^{-4} \times 0.12224^2} + 3.107 \right] = 6.77 \ .$$

The skin pressure drop, in practical metric units, is

$$\Delta p_s = 0.869 \, m \, S \ .$$

Therefore

$$\Delta p_s = 0.869 \times 1.02 \times 6.77 = 6.00 \text{ kg/cm}^2 \ .$$

So after 1 d of production we will have

Productivity index:

$$\text{PI} = J = \frac{550}{465 - 451.33} = 40.2 \, \frac{\text{m}^3}{\text{d} \times \text{kg/cm}^2}$$

Completion factor

$$\text{CF} = \frac{550/(465 - 451.33)}{550/(465 - 451.33 - 6)} = 56.1\% \ .$$

To summarise the information deduced so far from the production test (which satisfies the first of the three tasks specified), we have

- permeability to oil in the vicinity of the well: 185 md ,
- skin factor: 6.77 ,
- productivity index: 40.2 m^3/(d × kg/cm^2) ,
- well completion factor: 56.1%
- well damage factor = (1 − CF):   43.9% .

Referring to the behaviour of p_{wf} in Fig. E6/2.2 after the transient period, we can see that it varies linearly with t after about 4 d. The slope is

$$-\frac{\mathrm{d}p_{wf}}{\mathrm{d}t} = \frac{441.72 - 420.76}{30 - 10} = 1.048 \, \frac{\text{kg}}{\text{cm}^2 \times \text{d}} \ .$$

This linear trend is indicative of pseudo-steady state flow in the reservoir.
From Eq. (6.24) we have

$$Ah\phi = \text{pore volume of reservoir} = V_p = \frac{q_{sc} B_o}{(-\,\mathrm{d}p_{wf}/\mathrm{d}t)\,c_t} \ ,$$

therefore

$$V_p = \frac{550 \times 1.68}{1.048 \times 3.75 \times 10^{-4}} = 2.35 \times 10^6 \text{ m}^3$$

from which, since

$$N = V_p \frac{(1 - S_w)}{B_o} \ ,$$

we get

$$N = \frac{2.35 \times 0.85}{1.68} = 1.19 \times 10^6 \text{ m}^3 \text{ (stock tank) } .$$

This is apparently a rather small accumulation, containing about a million standard cubic metres of *oil in place.*
If we can assume h and ϕ are uniform throughout the reservoir, we can write

$$A = \frac{V_p}{h\phi} = \frac{2.35 \times 10^6}{58 \times 0.25} = 162\,000 \text{ m}^2 \ .$$

We can make an approximate evaluation of reservoir geometry as follows. We use the equation (in SI units)

$$p_i - p_{wf}(t) = \frac{q_{sc} B_o \mu_o}{2\pi k h} \left[p_D(t_D) + S \right] \ .$$

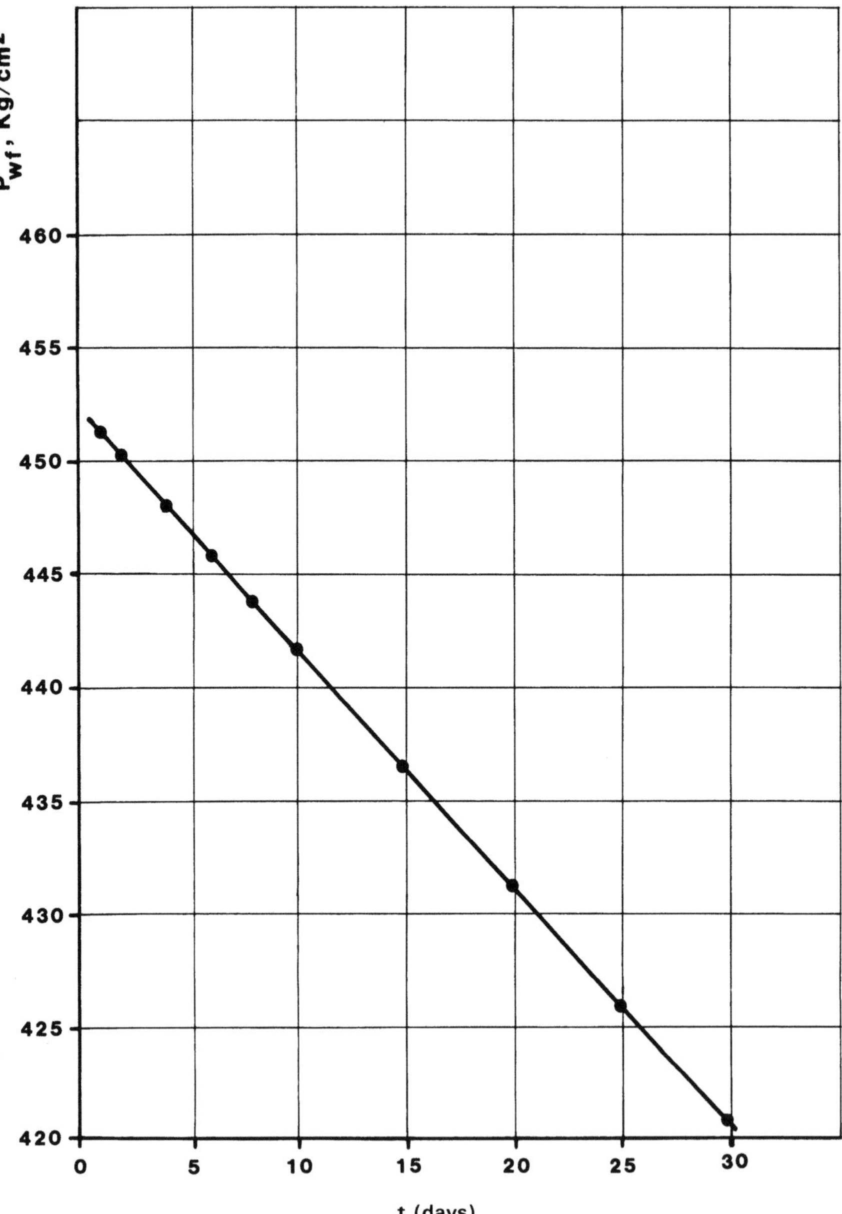

Fig. E6/2.2

In practical metric units, this becomes

$$p_D(t_D) = 5.2541 \times 10^{-2} \frac{kh}{q_{sc} B_o \mu_o} [p_i - p_{wf}(t)] - S \; .$$

In Fig. E6/2.2, we choose a point that is well into the pseudo-steady state flow period on the straight line – for example,

$$\begin{cases} t = 30 \text{ d} \; , \\ p_{wf} = 420.76 \text{ kg/cm}^2 \; , \end{cases}$$

Then

$$p_D(30 \text{ d}) = 5.2541 \times 10^{-2} \frac{185 \times 58}{550 \times 1.68 \times 0.54} [465 - 420.76] - 6.77$$

which gives

$$p_D(30\ d) = 43.216 \ .$$

From the equation in Table 6.1, in metric units:

$$t_{DA} = 5.807 \times 10^{-6} \frac{k_o t}{\phi \mu_o c_t A} \ ,$$

we have

$$t_{DA}(30\,d) = 5.807 \times 10^{-6} \frac{185 \times 30 \times 1440}{0.25 \times 0.54 \times 3.75 \times 10^{-4} \times 162\,000} = 5.66 \ .$$

Equation (6.22), which is valid in pseudo-steady state flow:

$$p_D(t_{DA}) = 2\pi t_{DA} + \frac{1}{2}\ln \frac{A}{C_A r_w^2} + 0.405 \tag{6.22}$$

can easily be rewritten as follows:

$$\ln C_A = 4\pi t_{DA} + \ln \frac{A}{r_w^2} - 2p_D(t_{DA}) + 0.809 \ ,$$

so that, in this example

$$\ln C_A = 4\pi \times 5.66 + \ln \frac{162\,000}{0.12224^2} - 2 \times 43.216 + 0.809 \ .$$

This gives us

$$C_A = 5.48 \ .$$

Referring to Fig. 5.7, we might reasonably deduce that the reservoir is rectangular in shape (viewed from above), with the sides in an approximately 4:1 ratio and the well in the centre ($C_A = 5.38$); however, other possibilities might be a square, or a 2:1 rectangle, with the well off centre – both of which have $C_A = 4.51$.

Exercise 6.3

The well which was the subject of Ex. 6.2 was kept on production for a further 43 d. A cumulative total of 33 720 m^3 (stock tank) of oil was produced for the whole test. The well was then shut in, and the pressure buildup was measured with a high resolution down hole gauge.

Just before shut in, the well was producing at a stable rate of 562 m^3/d (stock tank), at a bottom-hole flowing pressure $p_{wf} = 391.74$ kg/cm^2. The following data were recorded during the buildup test:

Δt	p_{ws}, kg/cm^2	Δt	p_{ws}, kg/cm^2
1 min	392.30	3 h	398.76
5 min	394.26	4 h	398.87
10 min	395.38	5 h	398.96
20 min	396.66	6 h	399.03
30 min	397.62	8 h	399.14
40 min	398.06	10 h	399.20
1 h	398.32	12 h	399.24
1 h 30 min	398.48	18 h	399.28
2 h	398.60	24 h	399.34

With the help of the data and results from Ex. 6.2, as well as the PVT properties of the oil below the initial reservoir pressure provided in Ex. 2.3, analyse the pressure buildup data and calculate the well and drainage area characteristics.

Solution

Since the well has produced 33 720 m^3 (stock tank) of oil, and the stable flow rate prior to shut in was 562 m^3/d (stock tank), the equivalent production time t to be used in the buildup analysis is [Eq. (6.55)]

$$\bar{t} = \frac{33\,720}{562} = 60\,\text{d} \ .$$

As can be seen from the behaviour of p_{ws}, the pressure in the drainage area has built up to about 400 kg/cm^2, and we should reevaluate B_o and μ_o at this new pressure.

From the data in Ex. 2.3 we have

$$at\ 400\ kg/cm^2 : B_o = 1.71 \ ,$$
$$\mu = 0.50\ \text{cP}$$

Using

$$\bar{t} = 60\ \text{d} = 86\,400\ \text{min}$$

the Horner time functions for each data point are

Δt	$(\bar{t} + \Delta t)/\Delta t$	p_{ws}, kg/cm^2
1 min	86 401	392.30
5 min	17 281	394.26
10 min	8 641	395.38
20 min	4 321	396.66
30 min	2 881	397.62
40 min	2 161	398.06
1 h	1 441	398.32
1 h 30 min	961	398.48
2 h	721	398.60
3 h	481	398.76
4 h	361	398.87
5 h	289	398.96
6 h	241	399.03
8 h	181	399.14
10 h	145	399.20
12 h	121	399.24
18 h	81	399.28
24 h	61	399.34

Fig. E.6/3.1 is the Horner plot of p_{ws} against $\log[(t + \Delta t)/\Delta t]$. There is a straight line portion of the data between $\Delta t = 1$ h and $\Delta t = 8$ h.

For $\Delta t < 1$ h, the pressure behaviour is dominated by afterflow and the effect of the skin, which mask the reservoir transient.

For $\Delta t > 8$ h, the buildup data deviate from the linear trend, as a consequence of the limited size of the area of drainage.

For the straight line portion of the buildup,

$$m = -\frac{dp_{ws}}{d\log[(\bar{t} + \Delta t)/\Delta t]} = 0.9067\frac{\text{kg/cm}^2}{\text{cycle}} \ .$$

P$_{WS}$ (Kg/cm2)
401
400
399
398
397
P*= 401.18 Kg/cm2
$\bar{P}$
$\frac{t+\triangle t}{\triangle t}$
5
10
5
10^2
5
10^3
1h
5
10^4
401
400
399
398
397
P$_{WS}$ (Kg/cm2)

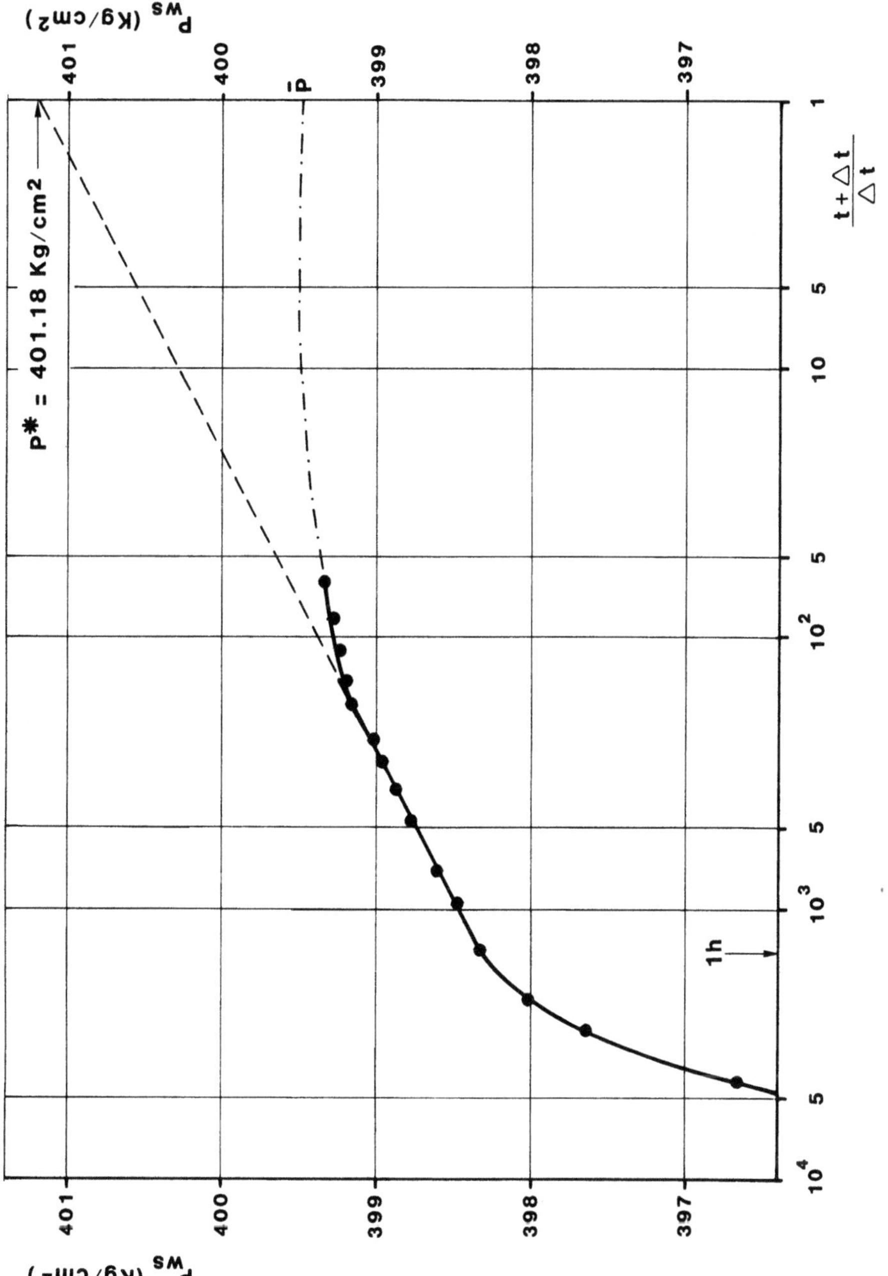

Fig. E6/3.1

In practical metric units:

$$m = 21.907 \frac{q_{sc} B_o \mu_o}{k_o h}$$

and we calculate

$$k_o = 21.907 \frac{562 \times 1.71 \times 0.50}{0.9067 \times 58} = 200 \text{ md} .$$

This is in good agreement with the value of 185 md calculated from the pressure drawdown data in Ex. 6.2.

The equation for the calculation of the skin factor is (in practical metric units)

$$S = 1.151 \left(\frac{p_{ws(LIN)}(1h) - p_{wf}}{m} - \log \frac{k_o}{\phi \mu_o c_t r_w^2} + 3.107 \right) .$$

In this example, we have

$$p_{ws(LIN)}(1 \text{ h}) \equiv p_{ws}(1 \text{ h}) = 398.32 \text{ kg/cm}^2 ,$$

$$p_{wf} = 391.74 \text{ kg/cm}^2 ,$$

$$m = 0.9067 \text{ kg/cm}^2/\text{cycle} ,$$

$$r_w = 0.12224 \text{ m} ,$$

$$\log \frac{k_o}{\phi \mu_o c_t r_w^2} = \log \frac{200}{0.25 \times 0.50 \times 3.75 \times 10^{-4} \times 0.12224^2} = 8.45566 ,$$

so that

$$S = 1.151 \left(\frac{398.32 - 391.74}{0.9067} - 5.349 \right) = 2.20 .$$

The drawdown analysis (Ex. 6.2) gave a markedly higher estimate ($S = 6.67$) than this.

One explanation is that the prolonged flow of oil has cleaned up the near-wellbore formation by flushing out residual mud filtrate.

Extrapolating the straight line in Fig. E6/3.1 to $[(\bar{t} + \Delta t)/\Delta t] = 1$, we find that

$$p^*(\bar{t}) = 401.18 \text{ kg/cm}^2 .$$

From Eq. (6.57)

$$p^*(t_n) = p^*(\bar{t}) + m \log \frac{t_n}{\bar{t}} = 401.18 + 0.9067 \log \frac{73}{60} ,$$

so that

$$p^*(t_n) = 401.26 \text{ kg/cm}^2 .$$

In Ex. 6.2, the drainage area of the well was estimated to be $162\,000 \text{ m}^2$, with a roughly 4:1 rectangular shape with the well at the centre.

The dimensionless time t_{DA} [Eq. (6.16)] corresponding to the equivalent production time of 60 d will be (working in practical metric units)

$$t_{DA} = 5.807 \times 10^{-6} \frac{200 \times 60 \times 1440}{0.25 \times 0.5 \times 3.75 \times 10^{-4} \times 162\,000} = 13.21 .$$

From Fig. 6.9 you will notice that the curve $p_{D(MBH)} = f(t_{DA})$ has already become linear by the time $t_{DA} = 13.21$. Therefore, $p_{D(MBH)}$ can be calculated from Eq. (6.21):

$$p_{D(MBH)}(t_{DA}) = \ln(C_A t_{DA})$$

using a shape factor $C_A = 5.379$ for this geometry (Fig. 5.7).

Therefore

$$p_{D(MBH)}(13.21) = \ln(13.21 \times 5.379) = 4.263$$

The average pressure $\bar{p}$ in the area of drainage can be calculated from Eq. (6.50) (again working in metric units):

$$\bar{p} = p^*(t_n) - 9.516 \frac{q_{sc} B_o \mu_o}{k_o h} p_{D(MBH)}(t_{DA}) \, ,$$

which, in this case, becomes

$$\bar{p} = 401.26 - 9.516 \frac{(562 \times 1.71 \times 0.50)}{200 \times 58} 4.263 = 399.58 \text{ kg/cm}^2 \, .$$

This leads to the following results:

Productivity index:
$$\text{PI} = J = \frac{562}{399.58 - 391.74} = 71.7 \frac{m^3}{\text{d} \times \text{kg/cm}^2} \, ,$$

$$\Delta p_s = 0.869 \, m S = 0.869 \times 0.9067 \times 2.2 = 1.73 \text{ kg/cm}^2$$

and therefore

Completion factor:
$$\text{CF} = \frac{562/(399.58 - 391.74)}{562/(399.58 - 391.74 - 1.73)} = 77.7\% \, .$$

To summarise the results:

	From drawdown test	From buildup test
Pressure (kg/cm^2)	465 (initial)	399.58
Permeability (md)	185	200
Skin factor (dimensionless)	6.77	2.20
PI(m^3)/(d × kg/cm^2)	40.2	71.7
CF (dimensionless)	56.1%	77.7%
Drainage area (m^2)	162 000	—
Drainage area shape	4:1 rectangle; well in middle	

7 The Interpretation of Production Tests in Gas Wells

7.1 Introduction

In Section 5.3, the diffusivity equation for radial flow:

$$\frac{1}{r}\frac{\partial}{\partial r}\left(\frac{k}{\mu}\rho r\,\frac{\partial p}{\partial r}\right) = \phi\,c_t\rho\,\frac{\partial p}{\partial t} \tag{5.6}$$

was linearised for the case of slightly compressible fluids to the following form:

$$\frac{1}{r}\frac{\partial}{\partial r}\left(r\,\frac{\partial p}{\partial r}\right) = \frac{\phi\mu c_t}{k}\,\frac{\partial p}{\partial t}\,. \tag{5.10}$$

The conditions necessary in order to be able to linearise the equation in this way are:

1. μ and c_t are not pressure-dependent.
2. k is also not pressure-dependent.
3. $\partial p/\partial r$ is small, so that $(\partial p/\partial r)^2$ is negligible.

While these conditions are reasonably satisfied by liquids (water, oil above its bubble point), the first certainly does not hold true for a gas. In fact, we have

$$c_g = \frac{1}{p} - \frac{1}{z}\left(\frac{\partial z}{\partial p}\right)_T \tag{2.8a}$$

and, expanding Eq. (2.13a) as a series as far as the first term

$$\mu_g = \mu_{g,sc}\left[1 + A\left(\frac{p}{z}\right)^B\right]\,. \tag{7.1}$$

In other words, c_t, μ_g and their product $(c_t\mu_g)$ are all pressure-dependent.

Condition 2 is equally valid for liquid and gas.

Condition 3 is not always met in the case of gas, as the sometimes very high flow velocities near the wellbore mean that $(\partial p/\partial r)^2$ may not be negligibly small.

A different approach is therefore needed to linearise the radial diffusivity equation for a gas. Before describing this, we need to introduce a new parameter which is the key to achieving the linearisation.

7.2 The Real Gas Pseudo-Pressure – *m(p)*

The concept of a "real gas pseudo-pressure", denoted by "*m(p)*", was introduced by Al-Hussainy, Ramey and Crawford[1] of the Texas A&M University. They applied

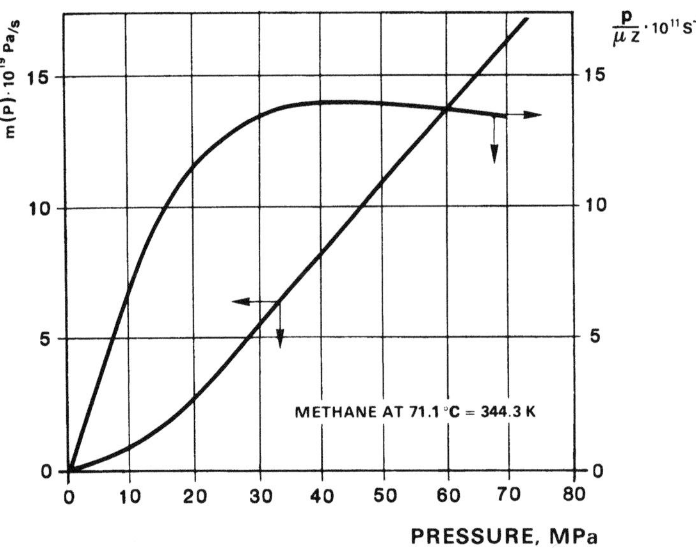

Fig. 7.1. *The variation of p/μz and m(p) with pressure for pure methane at 71.1 °C*

a version of the Leibenzon transform

$$m(p) = 2 \int_{p_0}^{p} \frac{p}{\mu z}\, dp \; , \tag{7.2}$$

where p_0 is an arbitrary reference pressure and $\mu = \mu(p, T)$; $z = z(p, T)$.

The choice of p_0 is purely arbitrary. In fact, under isothermal conditions, the change in pseudo-pressure corresponding to a change of pressure from p_1 to p_2 is

$$m(p_1) - m(p_2) = 2 \int_{p_0}^{p_1} \frac{p}{\mu z}\, dp - 2 \int_{p_0}^{p_2} \frac{p}{\mu z}\, dp$$

$$= 2 \int_{p_2}^{p_1} \frac{p}{\mu z}\, dp \; . \tag{7.3}$$

The dimensions of $m(p)$ are $(mL^{-1}t^{-3})$: the unit of measure is Pa/s. In practical metric units this is $(kg/cm^2)^2/cp$, and in oilfield units $(psi)^2/cp$.

As can be seen in Fig. 7.1, $m(p)$ is a non-linear function of p, which tends towards linearity at higher pressures (generally $p > 20$ MPa).

7.3 Linearisation of the Diffusivity Equation for a Real Gas Using the *m(p)* Function

Starting from the diffusivity equation in radial coordinates,

$$\frac{1}{r} \frac{\partial}{\partial r} \left(\frac{k}{\mu} \rho r \frac{\partial p}{\partial r} \right) = \phi c_t \rho \frac{\partial p}{\partial t} \; , \tag{5.6}$$

we substitute the following relationships:[4]

$$\rho = \frac{Mp}{zRT},$$ (7.4a)

$$\frac{\partial m}{\partial r} = \frac{dm}{dp}\frac{\partial p}{\partial r} = \frac{2p}{\mu z}\frac{\partial p}{\partial r},$$ (7.4b)

from which

$$\frac{\partial p}{\partial r} = \frac{\mu z}{2p}\frac{\partial m}{\partial r},$$ (7.4b′)

$$\frac{\partial m}{\partial t} = \frac{dm}{dp}\frac{\partial p}{\partial t} = \frac{2p}{\mu z}\frac{\partial p}{\partial t},$$ (7.4c)

from which

$$\frac{\partial p}{\partial t} = \frac{\mu z}{2p}\frac{\partial m}{\partial t}.$$ (7.4c′)

This diffusivity equation then takes the form

$$\frac{1}{r}\frac{\partial}{\partial r}\left(\frac{k}{\mu}r\frac{Mp}{zRT}\frac{\mu z}{2p}\frac{\partial m}{\partial r}\right) = \phi c_t \frac{Mp}{zRT}\frac{\mu z}{2p}\frac{\partial m}{\partial t}.$$ (7.5a)

If we assume $c_f \ll c_g$ so that $c_t = c_g$, this simplifies to

$$\frac{1}{r}\frac{\partial}{\partial r}\left(r\frac{\partial m}{\partial r}\right) = \frac{\phi \mu_g c_g}{k}\frac{\partial m}{\partial t}.$$ (7.5b)

Equation (7.5b) is the radial diffusivity equation for a real gas. It is linear on condition that:

$$\frac{\phi \mu_g c_g}{k} = \text{const}.$$ (7.6)

It is claimed, in classical textbooks, that this condition is satisfied in as much as ϕ and k are little affected by pressure, and the product $(\mu_g c_g)$ is more or less constant, with μ_g increasing with increasing pressure while c_g decreases.

In fact, $(\mu_g c_g)$ is not constant, as Fig. 7.2 (for pure methane at $T = 71.1\,^\circ\text{C} = 344.3\,\text{K}$) shows.

$(\mu_g c_g)$ decreases by almost one order of magnitude as the pressure increases from 5 to 15 MPa, and by a further half between 15 and 30 MPa.

This behaviour is easily understood when the product $(\mu_g c_g)$ is expressed in terms of Eqs. (2.8a) and (7.1). Ignoring the derivative of z, this is,

$$\mu_g c_g = \mu_{g,\text{sc}}\left[\frac{1}{p} + \frac{A}{z}\left(\frac{p}{z}\right)^{B-1}\right].$$ (7.7)

It is therefore not correct to say that Eq. (7.5b) is truly linearised, especially at low pressures.

In fact, we *assume* that, over a limited range of pressure, the term $\phi \mu_g c_g / k$ only varies slightly: it is usual practice to take the values $\mu_i c_i$ calculated at the initial pressure p_i.

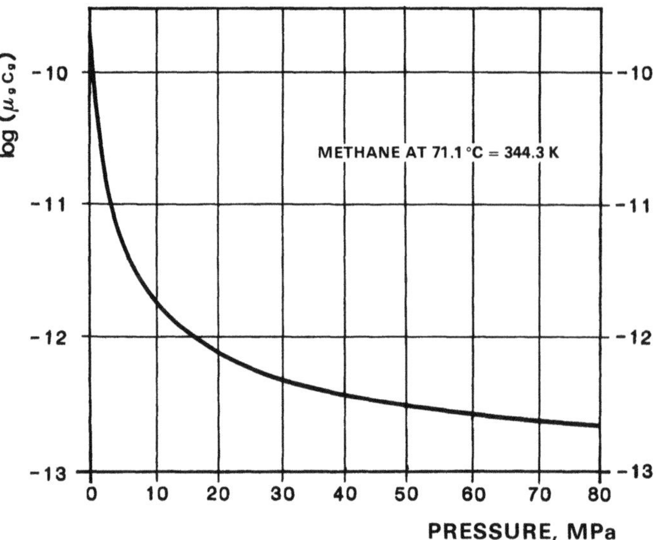

Fig. 7.2. *The variation of $\mu_g c_g$ with pressure for pure methane at 71.1 °C*

There are two consequences of the fact that Eq. (7.5b) is not rigorously linear:[2]

1. Any solution obtained through Eq. (7.5b) will be an approximation. The smaller the pressure range Δp involved, the better the approximation.
2. The principle of superposition, or Duhamel's theorem (Sect. 5.7), is not strictly applicable to the flow of gas, since it is valid only for truly linear differential equations, with *constant* coefficients.

On the other hand, a consideration that distinguishes this from the linearisation of the diffusivity equation for a fluid of constant compressibility is that, in the case of the flow of gas, the term $(\partial p/\partial r)^2$ is *not* dropped out.

Therefore, because Eq. (7.5b) takes second order terms in $(\partial p/\partial r)$ into account, *it is a better approximation to the original diffusivity Eq. (5.6) than is Eq. (5.10).*

7.4 Dimensionless Form of the Diffusivity Equation for a Gas

If the gas flow rate is q at reservoir pressure p and temperature T, and q_{sc} at standard conditions ($p_{sc} = 0.1013$ MPa, $T_{sc} = 288.2$ K), we define the following parameter:

$$\frac{2p}{\mu z}\,\frac{q\mu}{2\pi kh} = \frac{q_{sc}T}{\pi kh}\,\frac{p_{sc}}{T_{sc}}, \tag{7.8}$$

which has units of Pa/s – the same as $m(p)$. Consequently, the ratio

$$m_D(t_D, r_D) = [m(p_i) - m(p_{t,r})]\,\frac{kh}{q_{sc}T}\,\frac{\pi T_{sc}}{p_{sc}} \tag{7.9a}$$

is a dimensionless quantity called the *dimensionless real gas pseudo-pressure*.

From Eq. (7.9a), then

$$m(p_i) - m(p_{t,r}) = \frac{q_{sc}T}{kh}\,\frac{p_{sc}}{\pi T_{sc}}\,m_D(t_D, r_D)\,. \tag{7.9b}$$

We also have the following dimensionless parameters:

$$r_D = \frac{r}{r_w} \, ,$$ (5.14a)

$$t_D = \frac{k}{\phi \mu_g c_g r_w^2} t \, ,$$ (5.14b)

$$t_{DA} = \frac{k}{\phi \mu_g c_g A} t \, .$$ (6.16)

Bringing in Eqs. (7.8), (5.14a), (5.14b) and (6.16), Eq. (7.5b) can be rewritten as:

$$\frac{1}{r_D} \frac{\partial}{\partial r_D} \left(r_D \frac{\partial m_D}{\partial r_D} \right) = \frac{\partial m_D}{\partial t_D} \, .$$ (7.10)

Equation (7.10) is the dimensionless radial diffusivity equation for gas.

It is identical in form to Eq. (5.15) (valid for fluids of constant compressibility and viscosity), with the dimensionless pseudo-pressure m_D replacing the dimensionless pressure p_D.

It follows that every solution to Eq. (5.15) is also a solution to Eq. (7.10), with m_D in the place of p_D.

This is an extremely important conclusion. It means that all the solutions for transient, pseudo-steady state and steady state flow previously derived in Chapters 5 and 6 can be used for the interpretation of drawdown and buildup tests in gas wells.

Two important points already cited in Sect. 7.3 are repeated here:

– solutions to Eq. (7.10) obtained by simple substitution of the term m_D into the p_D solution to Eq. (5.15) are approximate solutions insofar as the product $(\mu_g c_g)$ is not constant. The smaller the range Δm_D covered by the test, the better the approximation.

   The μc_t term in Eqs. (5.14) and (6.16) is conventionally taken as the value of $(\mu_g c_g)$ calculated at the highest value of m_D in the test.

– the principle of superposition is not strictly applicable to the flow of gas, since Eq. (7.10) does not have constant coefficients, owing to the variation of $\mu_g c_g$ with pressure (and therefore with m_D).

7.5 Non-Darcy Flow

As has already been pointed out, the Darcy flow equation

$$\boldsymbol{u} = -\frac{[k]\rho}{\mu} \operatorname{grad} \Phi$$ (3.31a)

expresses a relationship between a *mean velocity* $\boldsymbol{u}$,

$$\boldsymbol{u} = \frac{q}{A}$$

and a potential gradient. A is the total cross-sectional area to flow, and includes the pores and grains of the rock. Statistically, the effective flow area is ϕA although, of course, this varies locally from pore to pore and, more importantly, from the throat to the centre of each pore.

We will now look briefly at the phenomenon of gas flow through a pore channel on a microscopic scale, with reference to Fig. 7.3. For simplicity, we will assume horizontal flow.

Obviously, the velocity v_1 of the gas in the narrow section 1 will be higher than v_2 in the central section 2 of the pore.

Consequently, the fluid undergoes a series of accelerations and decelerations as it passes from pore to pore; the fast moving fluid (v_1) coming through the narrow sections collides with the fluid (v_2) in the pore centres, with a loss of kinetic energy $\frac{1}{2}(v_1^2 - v_2^2)$ per unit mass.

This can be demonstrated quite simply in mathematical terms using Bernouilli's theorem and the definition of potential per unit mass cited in Eq. (3.27).

Referring to Fig. 7.3, where, for horizontal flow, $z_1 = z_2$,

$$\Delta\Phi^* = \Phi_1^* - \Phi_2^* = \int_0^{p_1} \frac{dp}{\rho} + \frac{v_1^2}{2} - \left(\int_0^{p_2} \frac{dp}{\rho} + \frac{v_2^2}{2} \right)$$

$$= \int_{p_2}^{p_1} \frac{dp}{\rho} + \frac{1}{2}(v_1^2 - v_2^2) \,. \tag{7.11}$$

This phenomenon, which is incorrectly attributed to "turbulence", is observed with any fluid (oil, water, gas) moving through a porous medium. However, it is more significant when the mean velocity $(v_1 + v_2)/2$ is large, since

$$\tfrac{1}{2}(v_1^2 - v_2^2) = (v_1 - v_2)\frac{v_1 + v_2}{2} \,. \tag{7.12}$$

For a given mass flow rate (ρv), the velocity of a gas is much higher than that of oil, because $\rho_g \ll \rho_o$. For this reason, the phenomenon described here will be more prominent in a gas reservoir, especially close to the wellbore, where the flow velocity is higher still.

In order to account for this in the equations for gas flow, Darcy's equation is commonly modified as follows:[8]

$$\frac{dp}{dr} = \frac{\mu_g}{k}u_g + \beta\rho_g u_g^2 \,, \tag{7.13a}$$

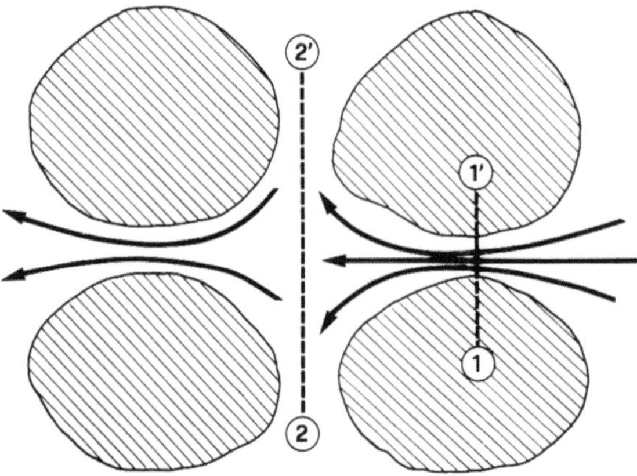

Fig. 7.3. *Inertial effects on the movement of gas from the throat to the centre of a pore*

where, as usual,

$$u_{\rm g} = \frac{q_{\rm sc}B_{\rm g}}{2\pi rh} \, . \tag{7.13b}$$

β is called the "*coefficient of inertia*", and has the dimensions of reciprocal length L^{-1}, as can be seen from Eq. (7.13a).

Equation (7.13a) is identical in form to Forcheimer's equation, and is sometimes incorrectly given this name, even though it describes a completely different phenomenon. In particular, the term $\beta\rho u^2$ accounts for an additional loss of energy in the so-called "non-Darcy flow regime", so Eq. (7.13a) could most appropriately be called the *non-Darcy flow equation*.

We will now consider the horizontal radial flow of gas towards the well, and calculate the additional pressure drop $(\Delta p)_{\rm nD}$ caused by inertial effects.

Assuming steady state flow, we can combine Eq. (7.13b) with the relationship

$$\rho_{\rm g} = \frac{\rho_{\rm sc}}{B_{\rm g}} = \frac{M}{23.645}\frac{1}{B_{\rm g}} \, , \tag{7.13c}$$

where M is the molecular weight of the gas, and rewrite the inertial term in Eq. (7.13a) as

$$(\Delta p)_{\rm nD} = -\int_{r_{\rm e}}^{r_{\rm w}} \beta\rho_{\rm g}u_{\rm g}^2 \, {\rm d}r = 1.071 \times 10^{-3} \frac{\beta M q_{\rm sc}^2}{h^2}\int_{r_{\rm w}}^{r_{\rm e}} \frac{B_{\rm g}}{r^2}\,{\rm d}r \, . \tag{7.14a}$$

We now introduce the real gas pseudo-pressure $m(p)$ defined in Sect. 7.2, to get

$$\begin{aligned}
[\Delta m(p)]_{\rm nD} &= 1.071 \times 10^{-3} \frac{\beta M q_{\rm sc}^2}{h^2}\int_{r_{\rm w}}^{r_{\rm e}} \frac{2p}{\mu_{\rm g}z}\left(\frac{101{,}325}{288.2}\frac{z}{p}T\right)\frac{{\rm d}r}{r^2} \\
&= 0.753 \frac{\beta T M q_{\rm sc}^2}{h^2}\int_{r_{\rm w}}^{r_{\rm e}} \frac{{\rm d}r}{\mu_{\rm g}r^2} \, . \tag{7.14b}
\end{aligned}$$

The viscosity $\mu_{\rm g}$ varies with pressure, and therefore with the distance from the well. Inertial effects are strongest close to the well, and this is where the inertial pressure losses are greatest. The integration in Eq. (7.14b) is simplified by making the approximation that $\mu_{\rm g}$ is constant and equal to $(\mu_{\rm g})_{\rm w}$, the gas viscosity at the bottom-hole flowing pressure $p_{\rm wf}$.

We then have

$$[\Delta m(p)]_{\rm nD} = 0.753 \frac{\beta T M q_{\rm sc}^2}{h^2(\mu_{\rm g})_{\rm w}}\left(\frac{1}{r_{\rm w}} - \frac{1}{r_{\rm e}}\right) \tag{7.15a}$$

and, assuming $1/r_{\rm e}$ to be negligibly small compared to $1/r_{\rm w}$,

$$[\Delta m(p)]_{\rm nD} = 0.753 \frac{\beta T M}{r_{\rm w}(\mu_{\rm g})_{\rm w}h^2}q_{\rm sc}^2 = Fq_{\rm sc}^2, \tag{7.15b}$$

where F has been defined as

$$F = 0.753 \frac{\beta T M}{r_{\rm w}(\mu_{\rm g})_{\rm w}h^2} \, . \tag{7.16}$$

F is the "*non-Darcy flow coefficient*", and has the dimensions $[{\rm mL}^{-7}{\rm t}^{-1}]$; in the SI system this is $({\rm Pa\ s})/{\rm m}^6$.

The value of F is determined by means of flowing tests at at least two different rates. This will be described later in this chapter. Note that the Fq_{sc}^2 term represents an additional pressure drop close to the wellbore, which is added to the drawdown corresponding to Darcy flow.

We can therefore consider it to be *an additional skin effect* which is added to the skin terms already described in Sect. 5.6.1.2. However, there is a very important difference – the inertial skin pressure drop is proportional to the flow rate squared, whereas the others are directly proportional to the flow rate.

To complete this introduction to the non-Darcy skin, note that the inertial coefficient β which appears in Eq. (7.16) can be determined from core measurements[7] made at successively higher gas flow rates.

In general, β increases with decreasing k (Fig. 7.4) according to a relationship of the form

$$\log \beta = C - n \log k \,. \tag{7.17}$$

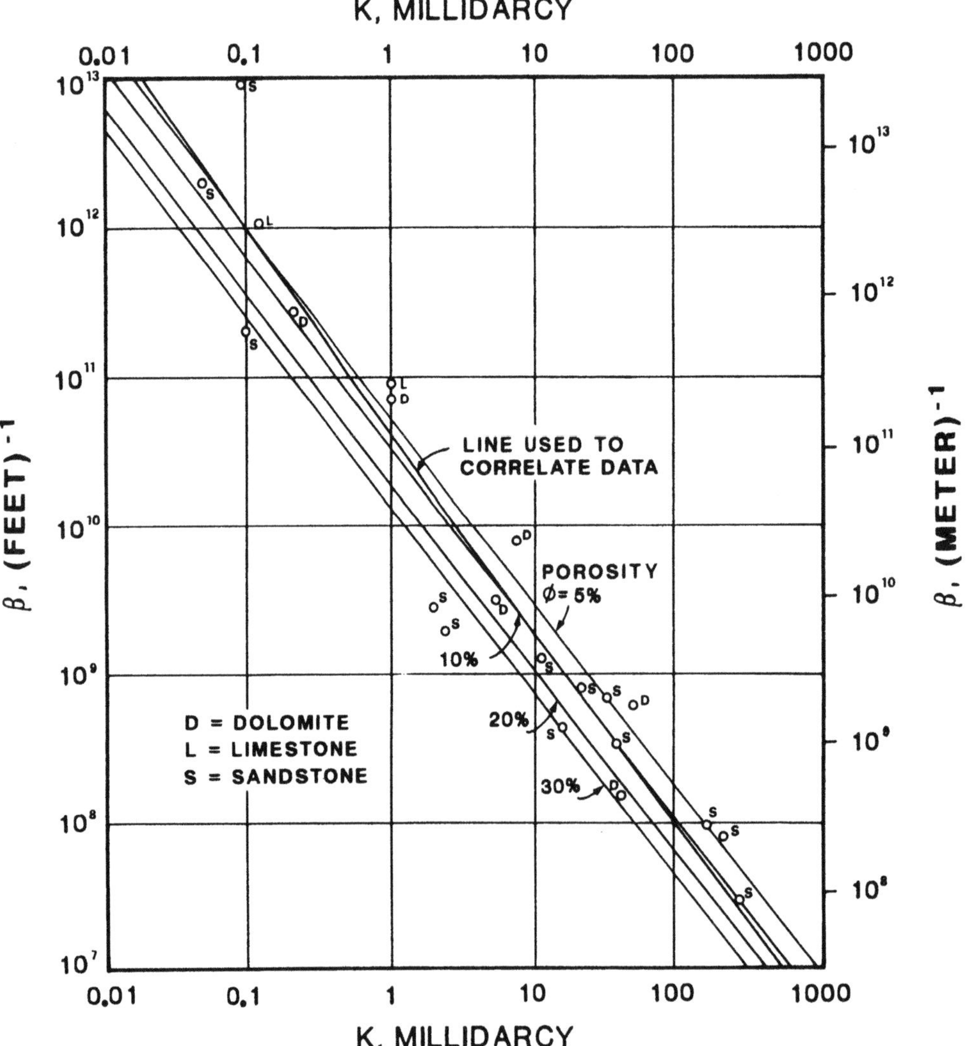

Fig. 7.4. *The dependence of the coefficient β on k and ϕ. From Ref. 8, 1959, McGraw-Hill Inc, reprinted with permission of McGraw-Hill Inc.*

We can, therefore, expect inertial skin effects to be more pronounced at lower permeabilities.

7.6 The Radial Horizontal Flow of Gas Towards the Well

7.6.1 Transient Flow

As long as we can assume $\mu_g c_g$ to be constant, there is a formal similarity between Eq. (7.10) for the radial horizontal flow of gas expressed in dimensionless terms, and the equivalent Eq. (5.15) for the radial horizontal flow of a constant compressibility fluid. This means that we can apply all the solutions derived for the transient flow of liquids (Sect. 5.6.1) to the corresponding problems involving gas.

Using Eq. (5.35), we can deduce by analogy the dimensionless equation for the transient flow of gas towards the well

$$m_D(1, t_D) = \tfrac{1}{2}(\ln t_D + 0.809) + S' , \tag{7.18}$$

where the S' term encompasses the true Darcy skin effects (S) and the additional dimensionless losses attributed to non-Darcy flow.

From Eqs. (7.18), (7.9), (5.14b) and (5.15b), in real dimensions we have:

$$m(p_i) - m(p_{wf}) = \frac{q_{sc}}{2\pi kh} \frac{p_{sc}T}{T_{sc}} \left(\ln \frac{k}{\phi \mu_g c_g r_w^2} t + 0.809 + 2S \right) + F q_{sc}^2 . \tag{7.19a}$$

In SI units this is

$$m(p_i) - m(p_{wf}) = 55.956 \frac{q_{sc}T}{kh} \left(\ln \frac{kt}{\phi \mu_g c_g r_w^2} + 0.809 + 2S + 2D q_{sc} \right) . \tag{7.19b}$$

where we have defined the rate dependent skin coefficient, D, as

$$D = \frac{\pi kh T_{sc}}{p_{sc}T} F = \beta \frac{M}{23.645} \frac{k}{2\pi r_w h} \frac{1}{(\mu_g)_w} . \tag{7.20}$$

D has dimensions $[L^{-3} t]$; in SI units this is s/m^3.

Looking at Eqs. (7.18) and (7.19b) we have

$$S' = S + D q_{sc} . \tag{7.21}$$

Note that the flow rate q_{sc} includes fluids which are produced as liquids on surface (e.g. the condensate fraction of natural gas). Suppose $(q_g)_{sc}$ is the flow rate of gas measured on surface, and $(q_L)_{sc}$ is the flow rate at stock tank conditions of liquid hydrocarbons, of density ρ_L and molecular weight M_L. Then we have

$$q_{sc} = (q_g)_{sc} + \frac{q_L \rho_L}{M_L} 23.645 . \tag{7.19c}$$

Equation (7.19) describes the behaviour of $m(p_{wf})$ throughout the transient period (Sect. 5.6), as a function of q_{sc}, the properties of the rock (ϕ, k, h, T) and of the gas, and including the Darcy skin (S) and non-Darcy skin (expressed in terms of the coefficient D).

7.6.2 Pseudo-Steady State Flow

Equation (5.44b) for the pseudo-steady state flow of a constant compressibility fluid can be written in dimensionless form as

$$p_{D,wf} - p_{D,e} = \ln r_{D,e} - \tfrac{1}{2} + S .\tag{5.44c}$$

The equivalent dimensionless equation for a gas will therefore be

$$m_D(p_{wf}) - m_D(p_e) = \ln r_{D,e} - \tfrac{1}{2} + S' .\tag{7.22}$$

Rewriting this in its fully dimensioned form, we get

$$m(p_e) - m(p_{wf}) = \frac{q_{sc}}{\pi k h}\frac{p_{sc}T}{T_{sc}}\left(\ln\frac{r_e}{r_w} - \frac{1}{2} + S + Dq_{sc} \right),\tag{7.23a}$$

or, in SI units,

$$m(p_e) - m(p_{wf}) = 111.91\frac{q_{sc}T}{kh}\left(\ln\frac{r_e}{r_w} - \frac{1}{2} + S + Dq_{sc} \right).\tag{7.23b}$$

In order to express $m(p_{wf})$ in terms of the average pseudo-pressure $m(\bar{p})$ of the drainage area, we simply draw on the analogy with Eq. (5.46) for the radial horizontal flow of a fluid of constant compressibility.

The dimensionless form of Eq. (5.46) is

$$p_D(1, t_D) - \bar{p}_D(t_D) = \ln r_{D,e} - \tfrac{3}{4} + S ,\tag{5.46b}$$

from which we can write, for a gas,

$$m_D(p_{wf}) - m_D(\bar{p}) = \ln r_{D,e} - \tfrac{3}{4} + S' .\tag{7.24}$$

Expanding this into fully dimensioned terms, we have

$$m(\bar{p}) - m(p_{wf}) = \frac{q_{sc}}{\pi k h}\frac{p_{sc}T}{T_{sc}}\left(\ln\frac{r_e}{r_w} - \frac{3}{4} + S + Dq_{sc} \right)$$

$$= 111.91\frac{q_{sc}T}{kh}\left(\ln\frac{r_e}{r_w} - \frac{3}{4} + S + Dq_{sc} \right).\tag{7.25}$$

For differently shaped drainage areas, the more generalised equation can be easily derived,

$$m(\bar{p}) - m(p_{wf}) = 55.955\frac{q_{sc}T}{kh}\left(\ln\frac{A}{C_A r_w^2} + 0.809 + 2S + 2Dq_{sc} \right),\tag{7.26}$$

where A is the drainage area of the well, and C_A is the shape factor, values of which are listed for a variety of geometries in Fig. 5.7.

7.6.3 Steady State Flow

Following the same procedure as for transient and pseudo-steady state flow, starting with Eqs. (5.53) and (5.54) for the radial horizontal flow of a fluid of constant compressibility, we can derive the equations for steady state flow, expressed here in SI units:

$$m(p_e) - m(p_{wf}) = 111.91\frac{q_{sc}T}{kh}\left(\ln\frac{r_e}{r_w} + S + Dq_{sc} \right)\tag{7.27a}$$

and:

$$m(\bar{p}) - m(p_{wf}) = 111.91 \frac{q_{sc}T}{kh}\left(\ln\frac{r_e}{r_w} - \frac{1}{2} + S + Dq_{sc}\right). \tag{7.27b}$$

These describe the radial flow of gas towards a well at the centre of a circular reservoir. A summary of the equations describing transient, pseudo-steady and steady state radial flow of gas, with quantities expressed in SI, practical metric, and oilfield units, is presented in Table 7.1.

7.6.4 General Solution to the Diffusivity Equation for a Gas

The diffusivity equation for a gas, expressed in dimensionless form, is

$$\frac{1}{r_D}\frac{\partial}{\partial r_D}\left(r_D\frac{\partial m_D}{\partial r_D}\right) = \frac{\partial m_D}{\partial t_D}. \tag{7.10}$$

The general solution, valid for any t_D, can be derived directly from Eq. (6.20) because of the direct analogy between the dimensionless diffusivity equations for a fluid of constant compressibility [Eq. (5.15)] and for gas [Eq. (7.10)].

From Eq. (6.20), we have

$$m_D(t_{DA}) = 2\pi t_{DA} + \frac{1}{2}\ln t_{DA} - \frac{1}{2}m_{D(MBH)}(t_{DA}) + \frac{1}{2}\ln\frac{A}{r_w^2} + 0.405, \tag{7.28}$$

where the term $m_{D(MBH)}(t_{DA}) \equiv p_{D(MBH)}(t_{DA})$ is read from the appropriate Matthews, Brons and Hazebroek diagram (Figs. 6.5–6.11) for the particular value of t_{DA}, in the same way as for a constant compressibility fluid in Sect. 6.4.

Equation (7.28) covers transient, late transient, pseudo-steady and steady state flow regimes, and is valid for any drainage area geometry for which there exists an MBH curve.

Once the value of $m_D(t_{DA})$ has been obtained, $m(p_{wf})$ can be calculated from the definition of m_D itself:

$$m(p_i) - m(p_{wf}) = \frac{q_{sc}}{\pi kh}\frac{p_{sc}T}{T_{sc}}m_D(t_{DA}), \tag{7.29a}$$

or, in SI units:

$$m(p_i) - m(p_{wf}) = 111.91\frac{q_{sc}T}{kh}m_D(t_{DA}). \tag{7.29b}$$

7.7 A Comparison Between the *m(p)* and *p²* Solutions to the Gas Flow Equation

By using the real gas pseudo-pressure function $m(p)$, we are able to handle, in a reasonably rigorous manner, the equations for the flow of gas in the reservoir, in the various flow regimes from transient through to steady state.

Before the concept of pseudo-pressure was introduced, it was (and still is today in some places) common practice to model the flow of gas in the reservoir *using the equations for constant compressibility fluids*, with the gas *flow rate calculated at the arithmetic mean of the maximum and minimum pressures in the test.*

Table 7.1. *Equations for the flow of gas in the reservoir*

Equations	Constants	Values of the constants "C"		
		SI Units	Practical metric units	Oilfield units (US)
Transient flow regime				
$m(p_\mathrm{i}) - m(p_\mathrm{wf}) = C_1 \dfrac{q_\mathrm{sc}T}{kh}\left(\log\dfrac{kt}{\phi\mu_\mathrm{g}c_\mathrm{g}r_\mathrm{w}^2} + C_2\right.$	C_1	128.81	0.157	1640
$\left. + 0.869S + 0.869Dq_\mathrm{sc}\right)$	C_2	0.351	-4.885	-3.228
Pseudo-steady state flow regime				
$m(p_\mathrm{e}) - m(p_\mathrm{wf}) = C_3 \dfrac{q_\mathrm{sc}T}{kh}\left(\log\dfrac{r_\mathrm{e}}{r_\mathrm{w}} - 0.217 + \right.$ $\left. + 0.434S + 0.434Dq_\mathrm{sc}\right)$	C_3	257.62	0.314	3280
$m(\bar{p}) - m(p_\mathrm{wf}) = C_4 \dfrac{q_\mathrm{sc}T}{kh}\left(\log\dfrac{r_\mathrm{e}}{r_\mathrm{w}} - 0.326 + \right.$ $\left. + 0.434S + 0.434Dq_\mathrm{sc}\right)$	C_4	257.62	0.314	3280
Steady state flow regime				
$m(p_\mathrm{e}) - m(p_\mathrm{wf}) = C_5 \dfrac{q_\mathrm{sc}T}{kh}\left(\log\dfrac{r_\mathrm{e}}{r_\mathrm{w}} + \right.$ $\left. + 0.434S + 0.434Dq_\mathrm{sc}\right)$	C_5	257.62	0.314	3280
$m(\bar{p}) - m(p_\mathrm{wf}) = C_6 \dfrac{q_\mathrm{sc}T}{kh}\left(\log\dfrac{r_\mathrm{e}}{r_\mathrm{w}} - 0.217 + \right.$ $\left. + 0.434S + 0.434Dq_\mathrm{sc}\right)$	C_6	257.62	0.314	3280

Parameter	Symbol	Unit		
		SI	Practical metric	Oilfield (US)
Isothermal gas compressibility	c_g	Pa^{-1}	$\mathrm{cm}^2/\mathrm{kg}$	psi^{-1}
Rate-dependent skin coefficient	D	$\mathrm{s/m}^3$	$\mathrm{days/m}^3$	days/Mscf
Net pay thickness	h	m	m	ft
Effective permeability to gas	k	m^2	md	md
Real gas pseudo-pressure	$m(p)$	Pa/s	$(\mathrm{kg/cm}^2)^2/\mathrm{cP}$	$(\mathrm{psi})^2/\mathrm{cP}$
Pressure at external boundary	p_e	Pa	$\mathrm{kg/cm}^2$	psi
Initial pressure	p_i	Pa	$\mathrm{kg/cm}^2$	psi
Standard pressure	p_{sc}	1.01325×10^5 Pa	$1.033\ \mathrm{kg/cm}^2$	14.7 psia
Bottom-hole flowing pressure	p_{wf}	Pa	$\mathrm{kg/cm}^2$	psi
Average pressure of drainage area	$\bar{p}$	Pa	$\mathrm{kg/cm}^2$	psi
Gas production rate at std. conditions	q_{sc}	sm^3/s	$\mathrm{Nm}^3/\mathrm{day} \equiv \mathrm{sm}^3/\mathrm{day}$	Mscf/day
External radius of drainage area	r_e	m	m	ft
Wellbore radius	r_w	m	m	ft
Skin factor (Darcy)	S	dimensionless	dimensionless	dimensionless
Time	t	s	min	hr
Temperature	T	K	K	R
Standard temperature	T_{sc}	288.2 K	288.2 K	518.8 R
Gas viscosity (reservoir conditions)	μ_g	Pa s	$\mathrm{cP} = \mathrm{mPa\ s}$	cP
Porosity	ϕ	dimensionless	dimensionless	dimensionless

The following example will demonstrate this method.

In transient flow, the equation for oil is

$$\frac{2\pi k_o h}{q_{sc} B_o \mu_o}(p_i - p_{wf}) = \frac{1}{2}\left(\ln\frac{k_o t}{\phi\mu_o c_t r_w^2} + 0.809 + 2S\right), \tag{6.2}$$

where $q_{sc} B_o = q$ is the oil flow rate under reservoir conditions.

At the *arithmetic* mean pressure:

$$\tilde{p} = \frac{p_i + p_{wf}}{2}, \tag{7.30a}$$

where the gas deviation factor $\tilde{z} = \tilde{z}(\tilde{p})$ and viscosity $\tilde{\mu}_g = \tilde{\mu}_g(\tilde{p})$, we have, in SI units

$$\tilde{B}_g = \frac{2\tilde{z}T}{p_i + p_w}\frac{101{,}325}{288.2}, \tag{7.30b}$$

so that

$$q = q_{sc}\tilde{B}_g = 703.16\frac{\tilde{z}T}{p_i + p_w}q_{sc}. \tag{7.31}$$

Substituting $q_{sc}\tilde{B}_g$ for $q_{sc}B_o$ in Eq. (6.2), and S' for S to include non-Darcy effects, we get

$$p_i^2 - p_{wf}^2 = 55.956\frac{q_{sc}\tilde{z}\tilde{\mu}_g T}{kh}\left(\ln\frac{kt}{\phi\mu_g c_g r_w^2} + 0.809 + 2S'\right). \tag{7.32}$$

This is identical to Eq. (7.19b) with the equivalence

$$m(p_i) - m(p_{wf}) = \frac{p_i^2 - p_{wf}^2}{\tilde{z}\tilde{\mu}_g}, \tag{7.33a}$$

or, in other words,

$$\int_{p_{wf}}^{p_i}\frac{2p}{\mu_g z}\,dp = \frac{p_i^2 - p_{wf}^2}{\tilde{z}\tilde{\mu}_g}, \tag{7.33b}$$

or, further:

$$\int_{p_{wf}}^{p_i}\frac{p}{\mu_g z}\,dp = (p_i - p_{wf})\frac{p_i + p_{wf}}{2\tilde{z}\tilde{\mu}_g}. \tag{7.33c}$$

The situation is presented schematically in Fig. 7.5, where $p/\mu_g z$ is plotted on the y-axis, versus p on the x-axis. A and B are the end-points of the $p/\mu_g z$ curve, corresponding to pressures p_{wf} and p_i respectively.

The integral on the left-hand side of Eq. (7.33c) is the area beneath the *curve* AB (shaded in Fig. 7.5a). The term on the right-hand side of the equation, however, corresponds to the area beneath the *straight line segment* AB (shaded in Fig. 7.5b). So the only time that Eq. (7.33c) will be exact is when the curve and the segment coincide.

In other words, in order for the p^2 solution [Eq. (7.32)] to be exactly equivalent to the $m(p)$ solution, $p/\mu_g z$ must be a linear function of p over the range p_{wf} to p_i.

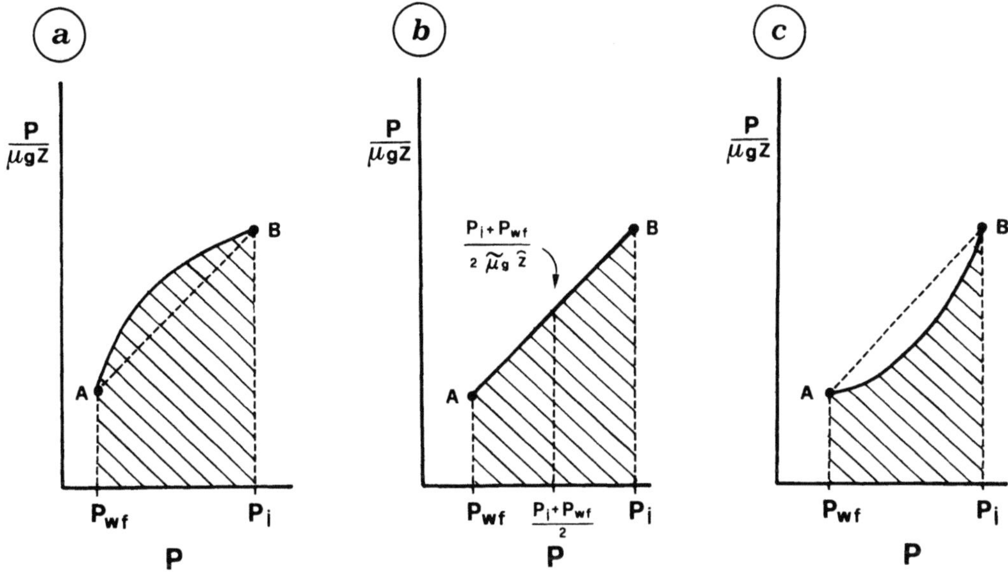

Fig. 7.5a–c. *Comparison between the solutions to the radial diffusivity equation for gas using $m(p)$ and p^2 [see Eq. (7.33)].*

This is indeed the case at low pressures (in Fig. 7.1, $1 < p \le 15$ MPa), and in this range it is equally valid to use the p^2 approximation on [Eq. (7.32)] or $m(p)$ [Eq. (7.19b)].

At sufficiently high pressures ($p > 30$ MPa in Fig. 7.1), $p/\mu_g z$ becomes practically constant. This means we can write

$$m(p_1) - m(p_2) = 2 \int_{p_2}^{p_1} \frac{p}{\mu_g z}\,\mathrm{d}p = C(p_1 - p_2)\,, \tag{7.34}$$

where $C/2 = (p/\mu_g z)_{\mathrm{av}}$ = average value of $p/\mu_g z$ in the range p_{wf} to p_{i}.

If both p_{wf} and p_{i} lie within this range, Eq. (7.19a) can be written as

$$p_{\mathrm{i}} - p_{\mathrm{wf}} = \frac{q_{\mathrm{sc}}}{4\pi kh} \frac{1}{\left(\dfrac{p}{\mu z}\right)_{\mathrm{av}}} \frac{p_{\mathrm{sc}} T}{T_{\mathrm{sc}}} \left(\ln \frac{kt}{\phi \mu_g c_g r_{\mathrm{w}}^2} + 0.809 + 2S + 2Dq_{\mathrm{sc}} \right)\,, \tag{7.19d}$$

which is very similar to Eq. (5.36b) for the flow of a fluid of constant compressibility.

In SI units, Eq. (7.19d) is

$$p_{\mathrm{i}} - p_{\mathrm{wf}} = 27.98 \frac{q_{\mathrm{sc}}}{(p/\mu_g z)_{\mathrm{av}}} \frac{T}{kh} \left(\ln \frac{kt}{\phi \mu_g c_g r_{\mathrm{w}}^2} + 0.809 + 2S + 2Dq_{\mathrm{sc}} \right)\,. \tag{7.19e}$$

At sufficiently low pressures, then, where $p/\mu_g z$ is a linear function of p, the p^2 approximation can be used [Eq. (7.32)] to describe the radial flow of gas; while at sufficiently high pressures, where $p/\mu_g z$ is roughly constant, p can be used directly [Eqs. (7.19d, e)]. Only in the intermediate pressure range is it really essential to use the rigorous $m(p)$ solution.

7.8 Gas Well Deliverability Testing

7.8.1 Introduction

One of the fundamental objectives to be met by the production engineer, where gas wells are concerned, is the optimum choice of production rate, subject to a number of operational constraints such as: a minimum acceptable wellhead or bottom-hole flowing pressure, avoidance of sand production, etc.

Substantial use will be made of data from production tests, and it is here that gas well engineering differs considerably from that for oil wells.

Remember that the time required for the onset of pseudo-steady state flow (assuming a circular or square drainage area) is given by

$$t_{\rm D} > 0.1 \left[\frac{r_{\rm e}}{r_{\rm w}} \right]^2 . \tag{5.17c}$$

In other words,

$$t > 0.1 \frac{\phi r_{\rm e}^2}{k} \mu c . \tag{7.35}$$

Since $\mu_{\rm g} \ll \mu_{\rm o}$ ($\mu_{\rm g}$ is typically of the order of 1.5×10^{-5}–3×10^{-5} Pa s = 0.015–0.03 cP – one or two orders of magnitude less than oil), we will always have

$$(\mu_{\rm g} c_{\rm g}) \ll (\mu_{\rm o} c_{\rm o}) \tag{7.36}$$

even though $c_{\rm g}$ is larger than $c_{\rm o}$.

Consequently, *for the same drainage radius $r_{\rm e}$, the time to the onset of pseudo-steady state flow is always less in a gas reservoir than in an oil reservoir.*

This time does not depend on the magnitude of the production rate that has been set for the well [Eq. (7.35)].

For these reasons, in gas reservoirs of sufficiently high permeability it is quite feasible (and, in fact, preferable) to run production tests under pseudo-steady state flow conditions. Where this is not practicable because low permeability dictates an unacceptably long time to reach pseudo-steady state, transient testing must be used.

This contrasts with oil well testing, where the tendency is to limit the test to the transient flow period, and to use the transient inflow equations to analyse the data.

This does not however mean that there is some fundamental inconsistency between the treatment of data from a gas well test and that from an oil well.

In fact, analysis of test data is far more dependable if the flow regime does not change. A test that passes through the transient and late transient flow regimes and into pseudo-steady state, although interpretable, will usually produce results with a higher degree of uncertainty than a test that remains in the same flow regime throughout.

7.8.2 Tests Conducted in Pseudo-Steady State Flow

Starting from stable conditions, with $p = \bar{p}$ throughout the drainage area, the test is begun by producing the well to a stable bottom-hole flowing pressure $p_{\rm wf,1}$, at an initial rate $q_{1,\rm sc}$ (Fig. 7.6). The rate is then increased to $q_{2,\rm sc}$, the well allowed to

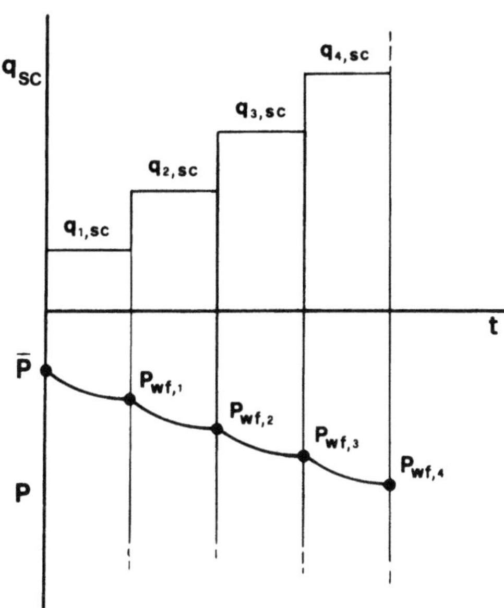

Fig. 7.6. *Flow rate schedule and pressure behaviour during a "flow-after-flow" test in a gas well*

stabilise at a flowing pressure $p_{wf,2}$, and so on for a total of at least four *increasing* rates.

Practice has shown (and this is supported theoretically) that increasing rates give better results than decreasing rates, and that the worst kind of test would be one with alternately increasing and decreasing rates.

The methods used to evaluate the production potential, or deliverability, of the well from the pairs of stable flowing data $(q_{i,sc}, p_{wf,i})$ are described in the following subsections.

This technique – a sequence of increasing flow rates *with no intermediate shut ins* – is called the *flow-after-flow*[9] test.

7.8.2.1 The Empirical "Back Pressure Test" (US Bureau of Mines)

An empirical approach to the evaluation of gas well deliverability, based on the flow-after-flow methodology, was laid down by the US Bureau of Mines[11] more than half a century ago. This 'back pressure' test procedure is still in use by some regulatory bodies.[5]

It was observed that the test measurements $(q_{i,sc}, p_{wf,i})$ almost always satisfied the relationship

$$q_{sc} = C(\bar{p}^2 - p_{wf}^2)^n , \tag{7.37a}$$

where C and n are coefficients which vary from well to well (and, possibly, from test to test). Carter et al.[3] showed that n should vary between 0.5 and 1.0: in particular, when the non-Darcy pressure losses are negligible, $n = 1$; as they become more appreciable, n approaches 0.5.

From Eq. (7.37a) we have

$$\log q_{sc} = \log C + n \log(\bar{p}^2 - p_{wf}^2) . \tag{7.37b}$$

This means that if the test data are plotted on log–log scales as $(\bar{p}^2 - p_{wf}^2)$ versus q_{sc}, they should lie on a straight line of slope n (Fig. 7.7). This is not always the case

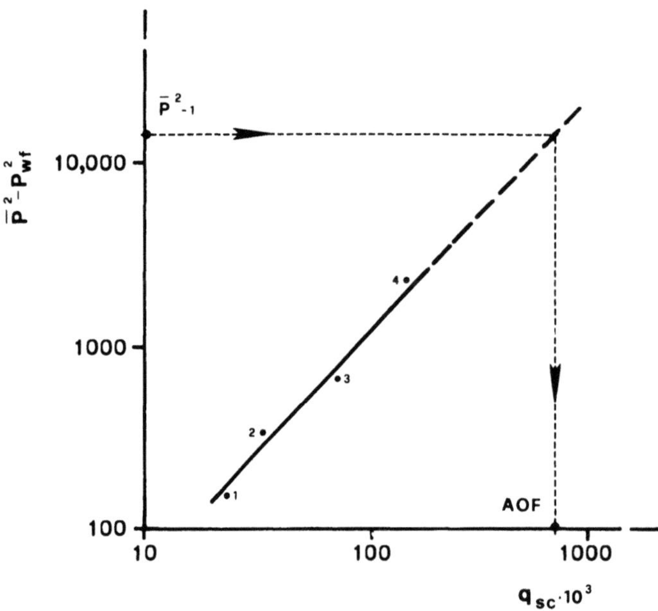

Fig. 7.7. *The empirical C-and-n plot for a back pressure test, first introduced by the USBM*

in practice, since in reality C and n both have some dependence on q_{sc} which is not allowed for in Eq. (7.37a).

The straight line is extrapolated to $(\bar{p}^2 - p^2_{sc})$, equivalent to a bottom-hole flowing pressure equal to atmospheric. The corresponding flow rate q_{sc} at this point is the *absolute open flow potential* (AOF), and represents the hypothetical maximum potential productivity of the well (which could, of course, never be realised in practice).

Once the line has been established from the test data on a C-and-n plot as in Fig. 7.7, the production rate corresponding to any attainable flowing pressure (or "back pressure") can easily be read off.

In an even more empirical variant of this technique, the tubing head flowing pressure p_{tf} is measured instead of the bottom-hole pressure. We now have:

$$q_{sc} = C'(\bar{p}^2 - p^2_{tf})^{n'} \tag{7.38}$$

which, in an approximate manner, also includes pressure losses along the production string.

Some theoretical support for Eq. (7.37a) can be derived from Eq. (7.25) for pseudo-steady state flow. Assuming Eq. (7.33a) to be valid so that we can substitute p^2 for the real gas pseudo-pressure $m(p)$, we will have

$$q_{sc} = \pi k h \frac{T_{sc}}{p_{sc}T} \frac{\bar{p}^2 - p^2_{wf}}{\tilde{z}\tilde{\mu}_g \left(\ln\dfrac{r_e}{r_w} - \dfrac{3}{4} + S + Dq_{sc} \right)} \cdot \tag{7.39}$$

For $D = 0$, Eq. (7.39) is the same as Eq. (7.37a) with

$$n = 1 \,, \tag{7.40a}$$

$$C = \frac{T_{sc}}{p_{sc}T} \frac{\pi k h}{\tilde{z}\tilde{\mu}_g \left(\ln\dfrac{r_e}{r_w} - \dfrac{3}{4} + S \right)} \cdot \tag{7.40b}$$

If $D > 0$, n must be < 1 in order for Eq. (7.39) to give the same results as Eq. (7.37a.) It is obvious from a comparison of these two equations that *a single constant value of n can only maintain this equivalence over a narrow range of pressures.*

7.8.2.2 Rigorous Interpretation of Flow-After-Flow Tests

The pseudo-steady state flow Eq. (7.25) can be expressed in the following form:

$$\frac{m(\bar{p}) - m(p_{wf})}{q_{sc}} = A + Bq_{sc} ,\qquad(7.41a)$$

where

$$A = \frac{p_{sc}T}{T_{sc}} \frac{\ln\dfrac{r_e}{r_w} - \dfrac{3}{4} + S}{\pi kh} ,\qquad(7.41b)$$

and

$$B = \frac{p_{sc}T}{T_{sc}} \frac{D}{\pi kh} .\qquad(7.41c)$$

Equation (7.41) provides the basis of a theoretically rigorous approach to the interpretation of flow-after-flow tests. A plot on Cartesian axes of $[m(\bar{p}) - m(p_{wf})]/q_{sc}$ against q_{sc} should produce a straight line trend (Fig. 7.8), with a slope equal to B and an intercept A at $q_{sc} = 0$, where

$$\lim_{q_{sc} \to 0} \frac{m(\bar{p}) - m(p_{wf})}{q_{sc}} = A .\qquad(7.41d)$$

This line can then be used to determine the value of q_{sc} corresponding to any p_{wf} [after transforming to $m(p_{wf})$]. Conversely, the value of $m(p_{wf})$ (and therefore p_{wf}) corresponding to a specified q_{sc} can be calculated.

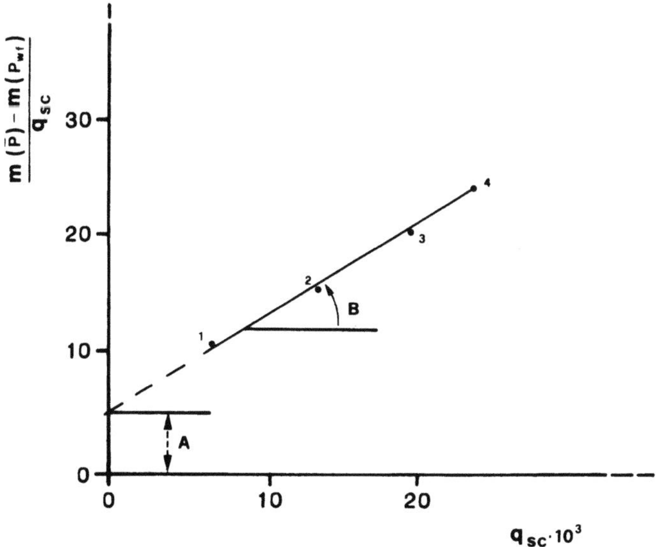

Fig. 7.8. *Interpretation of a flow-after-flow test*

The AOF is obtained by solving Eq. (7.41a), which is in fact a quadratic equation in q_{sc} [see Eq. (7.43)], with $p_{wf} = p_{sc}$

$$q_{sc}(\text{AOF}) = \frac{-A + \sqrt{A^2 + 4B[m(\bar{p}) - m(p_{wf} = p_{sc})]}}{2B}. \tag{7.42}$$

Note that the coefficients A and B estimated from the test data are not sufficient to isolate the reservoir parameters kh and r_e, nor the well parameters S and D. B gives only a qualitative indication of the importance of non-Darcy effects [the line would be horizontal ($B = 0$) if there were none at all].

It is common practice to record a buildup test at the end of the flow-after-flow sequence, and to evaluate these from the transient data.

It is important to appreciate that Eq. (7.41a), which can also be written as

$$m(\bar{p}) - m(p_{wf}) = Aq_{sc} + Bq_{sc}^2, \tag{7.43}$$

is the only equation which describes the pseudo-steady state (flow rate/pressure) relationship in a rigorously correct manner.

7.8.3 Transient Gas Well Tests

It may be that the permeability is so low (in a very deep, well-cemented reservoir rock, for instance), or for the drainage area to be so large (for example, a reservoir with a single well), that the time required to attain pseudo-steady state flow [see Eq. (7.35)] would be unacceptably long.

In this case, *and only in this case*, well tests must be conducted in transient flow.

Two techniques are used: *isochronal testing* (which incorporates flowing and shut in periods), and *multi-rate testing* (without shut ins).

Of the two, isochronal testing is the more widely used.

7.8.3.1 The Isochronal Test

Referring to Fig. 7.9, the test consists of first flowing the well at a rate $q_{1,sc}$ for a time t_p *which is less than that required to reach pseudo-steady state*; then closing it in for a time $t_{c,1}$ *sufficient to allow the pressure to build up to its initial value* p_i; then flowing it at a rate $q_{2,sc} > q_{1,sc}$ *for the same duration* t_p and shutting in for a time $t_{c,2}$ sufficient to restabilise at p_i; and so on for *increasing* rates $q_{3,sc}$, $q_{4,sc}$, etc.

Surface rates and bottom-hole flowing pressures are recorded throughout the test.

If there were n flowing periods, the test will provide data from n constant rate drawdowns ($q_{1,sc}$, $q_{2,sc}$, $\ldots$, $q_{n,sc}$) and n buildups.

The main features of the isochronal test, then, are:

- a sequence of n tests is performed at successively increasing rates $q_{i,sc}$,
- all n flowing periods have the same duration, and do not extend beyond the transient flow regime,
- the well is shut in after each flowing period and the well is allowed to build up to initial pressure p_i.

Obviously, the shut in times $t_{c,i}$ will be different and usually increase over the test.

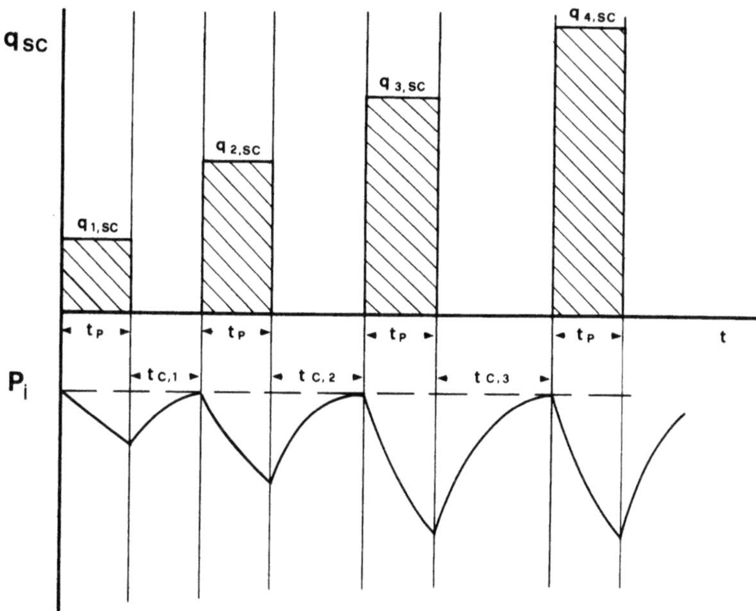

Fig. 7.9. *Flowing periods and bottom-hole flowing pressure behaviour during an isochronal test in a gas well*

Since the speed at which the pressure disturbance propagates into the reservoir *depends only on the diffusivity* $k/\phi\mu_g c_g$ *and not on the flow rate* (see Sect. 7.8.1), the radius of investigation r_i for each of n flowing periods of identical duration t_p, all in transient flow, will be the same.

The transient flow Eq. (7.19) can be written as:

$$\log[m(p_i) - m(p_{wf})] = \log q_{sc} + \log\left\{\frac{128.81\,T}{kh}\left[\log t + \log\frac{k}{\phi\mu_g c_g r_w^2} + 0.351\right.\right.$$

$$\left.\left. + 0.869(S + Dq_{sc})\right]\right\} \tag{7.44}$$

or, alternatively, as

$$\frac{m(p_i) - m(p_{wf})}{q_{sc}} = \frac{128.81\,T}{kh}\left[\log t + \log\frac{k}{\phi\mu_g c_g r_w^2}\right.$$

$$\left. + 0.351 + 0.869S\right] + Fq_{sc}\,. \tag{7.45}$$

In Eq. (7.44), note that in a particular well, *for a given t*, the term in square brackets would be the same for any flow rate if it were not for the presence of the Dq_{sc}.

In the alternative Eq. (7.45), *again for a given t*, the term in square brackets is truly the same for any flow rate, since the non-Darcy term has been moved outside the brackets as Fq_{sc}.

Interpretation of isochronal tests is usually based on Eq. (7.44): the values of $[m(p_i) - m(p_{wf})]$ *measured after the same elapsed time* in each flowing period are plotted on log–log scales against the flow rate q_{sc}. The pressures corresponding to one value of elapsed time (e.g. at $t = t_{\mathrm{p}}$, the end of each flowing period), or to several times, may be used. Figure 7.10 is an example of such a plot: pressures have been read from each flowing period at four elapsed times t_1–t_4. The data in each time group lie on roughly parallel straight lines.

Interpretation based on the alternative formulation, Eq. (7.45), involves making a Cartesian plot of values of $[m(p_i) - m(p_{\mathrm{wf}})]/q_{\mathrm{sc}}$ (again *read at the same time or times during each flowing period*) against q_{sc}. This should produce a series of exactly parallel lines (Fig. 7.11) with a slope equal to F.

In both cases, if one extended flowing period is programmed (at the end of the isochronal test, for instance), long enough to attain pseudo-steady state flow, the final flowing pressure can be plotted (point P in Figs. 7.10 and 7.11). *A parallel line drawn through P will represent the pseudo-steady state behaviour of the well.*

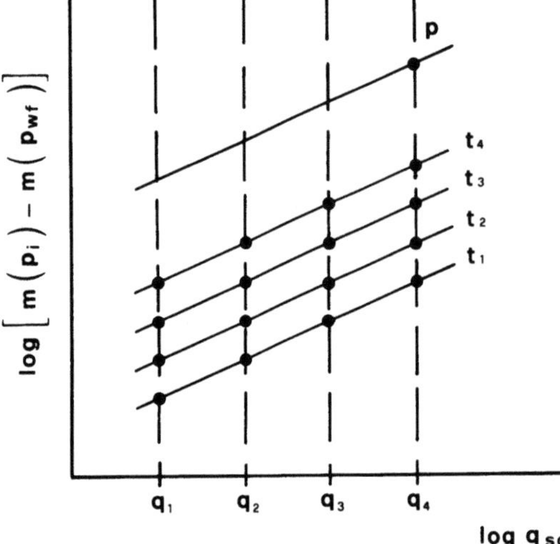

Fig. 7.10. *Interpretation of an isochronal test by plotting* $\log[m(p_i) - m(p_{\mathrm{wf}})] = f[\log q_{\mathrm{sc}}]$

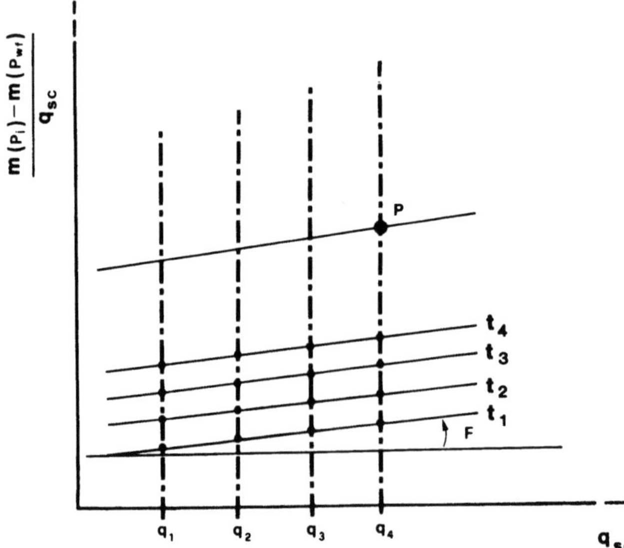

Fig. 7.11. *Interpretation of an isochronal test by plotting* $[m(p_i) - m(p_{\mathrm{wf}})]/q_{\mathrm{sc}} = f[q_{\mathrm{sc}}]$

This line can then be used to determine the flow rate q_{sc} for any specified bottom-hole flowing pressure p_{wf}, or vice versa.

The extended flowing period is always the last in the series, and often corresponds to putting the field on production proper.

Of course, each of the constant rate flowing periods and shut ins making up the isochronal sequence can be analysed as a transient drawdown or buildup test, as described in Sect. 6.5 for oil wells.

A point to note about tests in gas wells is that the presence of the non-Darcy flow term Dq_{sc} together with the true Darcy skin factor S means that data for at least two flow rates are necessary in order to evaluate the two parameters. From data for a single flow rate, only a total skin $S_t = (S + Dq_{sc})$ can be derived.

Since, in SI units

$$m(p_i) - m(p_{wf}) = 55.956 \frac{q_{sc}T}{kh} \left(\ln \frac{k}{\phi \mu_g c_g r_w^2} + \ln t \right.$$

$$\left. + 0.809 + 2S + 2Dq_{sc} \right), \tag{7.19b}$$

we have

$$b = -\frac{dm(p_{wf})}{d \ln t} = 55.956 \frac{q_{sc}T}{kh}, \tag{7.46a}$$

from which

$$kh = 55.956 \frac{q_{sc}T}{b}. \tag{7.46b}$$

If h is known, k can be calculated.

As a check, we next calculate for a drainage area A of any shape:

$$t_{DA, \max} = \frac{k}{\phi \mu_g c_g A} t_{\max}, \tag{7.16}$$

where $t_{\max}$ corresponds to the end of the straight line portion of the data used to estimate the slope b on the plot of $m(p_{wf})$ versus $\ln t$ (Fig. 7.12).

This $t_{DA, \max}$ value should now be compared with those listed in the column headed "use infinite system solution with less than 1% error for $t_{DA} < \ldots$", in Fig. 5.7, selecting the appropriate geometry. In this way we can ensure that our choice of straight line is consistent with what we know of the drainage area, and therefore that the estimate of k is reasonable.

Reading $m(p_{1h})$ from the straight line (extrapolated if necessary), corresponding to $t = 1$ h, we have

$$S + Dq_{sc} = \frac{m(p_i) - m(p_{1h})}{2b} - \frac{1}{2} \left(\ln \frac{k}{\phi \mu_g c_g r_w^2} + \ln 3600 + 0.809 \right), \tag{7.47a}$$

which becomes

$$S + Dq_{sc} = \frac{m(p_i) - m(p_{1h})}{2b} - \frac{1}{2} \left(\ln \frac{k}{\phi \mu_g c_g r_w^2} + 8.998 \right). \tag{7.47b}$$

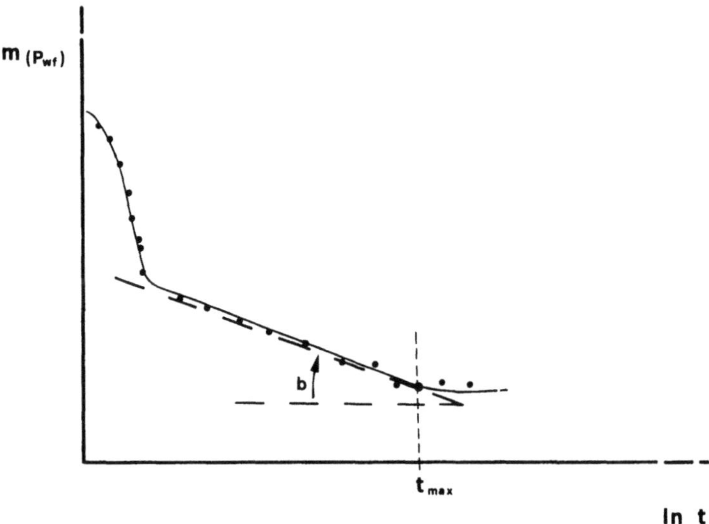

Fig. 7.12. *Checking the validity of the staight line portion of the $m(p_{wf})$ versus ln t plot*

In order to evaluate S and D, we need to evaluate the total skins $(S + Dq_{sc})$ from two different flowing periods. If their rates were $q_{1,sc}$ and $q_{2,sc}$, then

$$S + Dq_{1,sc} = S_1 ,$$
                                                                              (7.48a)

$$S + Dq_{2,sc} = S_2 .$$

Then

$$S = \frac{S_1 q_{2,sc} - S_2 q_{1,sc}}{q_{2,sc} - q_{1,sc}}$$
                                                                              (7.48b)

and

$$D = \frac{S_2 - S_1}{q_{2,sc} - q_{1,sc}} .$$

Of course, if more than two flowing tests are available at different rates, S and D can be evaluated in this manner by taking the tests in pairs. Alternatively, a plot, on Cartesian axes, of the total skins S_j [Eq. (7.47b)] against the flow rates $q_{j,sc}$ for all the tests should produce a straight line trend with a slope equal to D and an intercept S at $q_{sc} = 0$.

7.8.3.2 Multi-Rate Pressure Drawdown Tests

Unlike the flow-after-flow test, a multi-rate test is confined to transient flow conditions, either because of very low reservoir permeability, or because the drainage area is very large [Eq. (7.35)].

The Odeh–Jones method,[10] described in Sect. 6.5.4 for oil wells, is equally applicable to gas wells in spite of the complications arising from non-Darcy effects which introduce a dependence on q_{sc}.[2]

The basic theory of the method has already been explained in Sect. 6.5.4. Starting from the definitive Eq. (6.36), with a few minor modifications, we shall be able to handle the generalised gas well case:

- replace p by $m(p)$
- replace p_D by m_D, which, for radial transient flow is expressed as

$$m_D(t_D) = \tfrac{1}{2}(\ln t_D + 0.809) + S' \tag{7.18}$$

- replace S by

$$S' = S + Dq_{sc}$$

to include the effects of non-Darcy flow, especially close to the wellbore.

By making these changes, and neglecting second order terms, Eq. (6.36) becomes

$$\frac{m(p_i) - m(p_{wf,n})}{q_{n,sc}} = \frac{55.956\,T}{kh} \left\{ \sum_{j=1}^{n} \left[\frac{q_{j,sc} - q_{(j-1)sc}}{q_{n,sc}} \ln(t_n - t_{j-1}) \right] \right.$$
$$\left. + \ln\frac{k}{\phi\mu_g c_g r_w^2} + 0.809 + 2(S + Dq_{n,sc}) \right\} \tag{7.49}$$

or, substituting for D from Eq. (7.20),

$$\frac{m(p_i) - m(p_{wf,n}) - Fq_{n,sc}^2}{q_{n,sc}} = \frac{55.956\,T}{kh} \sum_{j=1}^{n} \left[\frac{q_{j,sc} - q_{(j-1)sc}}{q_{n,sc}} \ln(t_n - t_{j-1}) \right]$$
$$+ \frac{55.956\,T}{kh} \left(\ln\frac{k}{\phi\mu_g c_g r_w^2} + 0.809 + 2S \right). \tag{7.50}$$

So, by making a Cartesian plot (Fig. 7.13) of

$$\frac{m(p_i) - m(p_{wf,n}) - Fq_{n,sc}^2}{q_{n,sc}} = f\left\{ \sum_{j=1}^{n} \frac{q_{j,sc} - q_{(j-1)sc}}{q_{n,sc}} \ln(t_n - t_{j-1}) \right\}, \tag{7.51}$$

the data should fall on a straight line, which will have a slope

$$b' = \frac{1}{2\pi kh} \frac{p_{sc}T}{T_{sc}} = 55.956 \frac{T}{kh} \tag{7.52a}$$

and a y-axis intercept

$$y_o = \frac{1}{2\pi kh} \frac{p_{sc}T}{T_{sc}} \left(\ln\frac{k}{\phi\mu_g c_g r_w^2} + 0.809 + 2S \right)$$
$$= 55.956 \frac{T}{kh} \left(\ln\frac{k}{\phi\mu_g c_g r_w^2} + 0.809 + 2S \right) \tag{7.52b}$$

at $x = 0$.

kh and k can then be calculated from b', and then S from y_o.

In preparing the y-axis values for the plot [left-hand side of Eq. (7.51)], *we need to know F* – which, of course, is not the case.

This problem can be overcome by successive approximation, starting with an assumed value of F and plotting the data. If a straight line is not observed, a new value of F is tried, and so on. The value of F producing a straight line will be the correct one. In Fig. 7.13, the best F was 0.04.

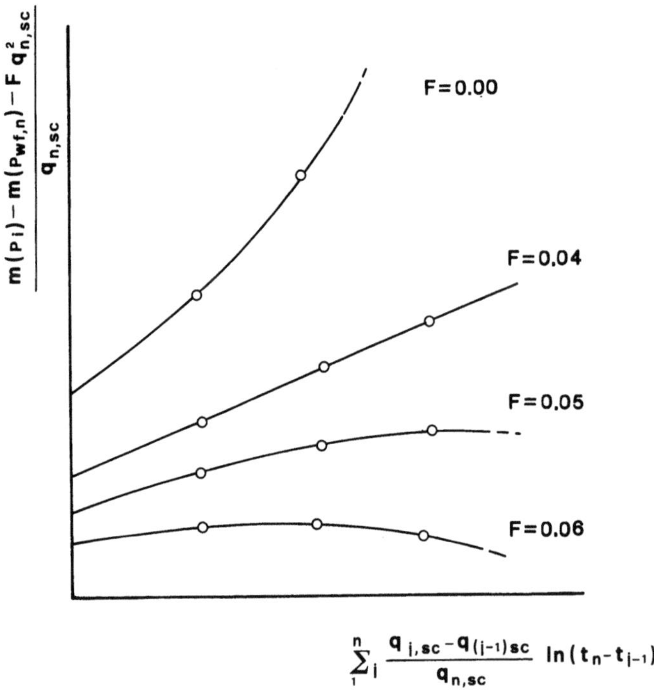

$$\sum_{1}^{n} \frac{q_{i,sc} - q_{(i-1)sc}}{q_{n,sc}} \ln(t_n - t_{i-1})$$

Fig. 7.13. *Interpretation of a multi-rate gas well test by the Essis–Thomas method. From Ref. 4, reprinted with permission of Elsevier Science Publishers and of Prof. Laurie P. Dake*

Clearly, the chances of a successful result depend very much on the quality of the data; if the measurements are in any way suspect, this method should not be attempted.

This technique, which is a variation of the well-known Odeh–Jones method, was first presented by Essis and Thomas.[6]

7.9 The Interpretation of Pressure Buildup Tests

Any pressure buildup data acquired when a gas well is shut in after a period of production *under transient flowing conditions* can be analysed by techniques similar to those outlined in Sect. 6.6 for oil wells.

The only difference is that the skin factor derived from the analysis is now the total skin $S' = S + Dq_{sc}$ instead of S. In order to evaluate S and D, it would be necessary to interpret at least two buildup tests, each following a different production rate q_{sc}. This can be done as part of the analysis of an isochronal test, where each flowing period is followed by a shut in [the final buildup (if any) following the extended flowing period should not be included, since pseudo-steady state conditions would normally have been attained prior to shut in].

A pressure buildup at the end of a flow-after-flow test requires a different approach. This was described in Sect. 6.6.3 for oil wells, and will not be repeated here.

With the usual nomenclature for time – t for the duration of the flowing period, Δt for the elapsed time from the start of the shut in period – we can apply

superposition theory to Eq. (7.19) and model the buildup response by superposing
a second drawdown at a flow rate $-q_{sc}$ at time t:

$$m(p_i) - m(p_{ws}) = \frac{q_{sc}}{2\pi kh} \frac{p_{sc}T}{T_{sc}} \ln \frac{t + \Delta t}{\Delta t} . \qquad (7.53)$$

We then define

$$b = -\frac{dm(p_{ws})}{d \ln \dfrac{t + \Delta t}{\Delta t}} = \frac{q_{sc}}{2\pi kh} \frac{p_{sc}T}{T_{sc}} = 55.956 \frac{q_{sc}T}{kh} , \qquad (7.54a)$$

so that

$$kh = 55.956 \frac{q_{sc}T}{b} \qquad (7.54b)$$

with everything expressed in SI units.

In the same way as in Sect. 6.6.2, we define $m_{LIN}(p_{ws})$ (1 h) as the value of $m(p_{ws})$
read from the straight line on the Horner plot (extrapolated if necessary) at a shut in
time $\Delta t = 1$ h (Fig. 7.14). Following the same procedure as Section 6.6.2, we can
derive from Eq. (7.19)

$$S + Dq_{sc} = \frac{m_{LIN}(p_{ws})(1\ \text{h}) - m(p_{wf})}{2b} - \frac{1}{2}\left(\ln \frac{k}{\phi \mu_g c_g r_w^2} + 8.998 \right). \qquad (7.55)$$

Once again, two buildup tests following different flow rates would be required
to separate S and D.

A summary of the equations used for the interpretation of pressure drawdown
and buildup tests in gas wells, with quantities expressed in SI, practical metric and
field units, is presented in Table 7.2.

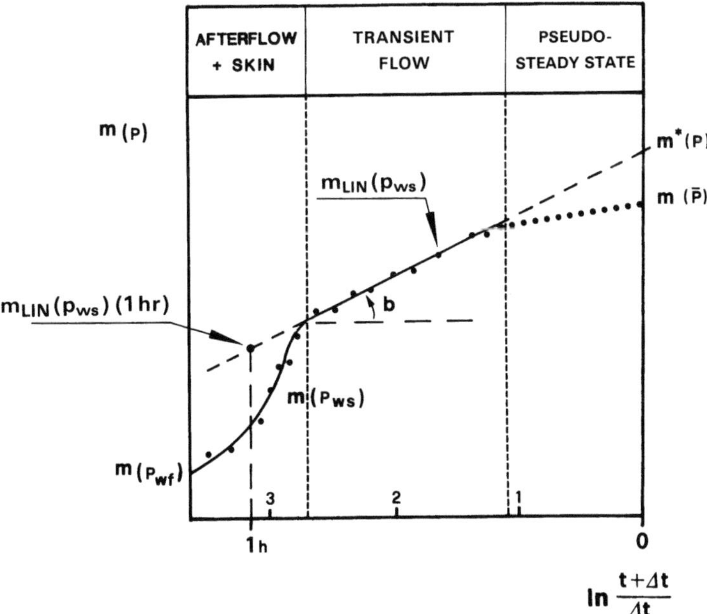

Fig. 7.14. *Horner method for the interpretiation of buildup data from a gas well*

Table 7.2. *Equations for the interpretation of drawdowns and buildup tests in gas wells.*

Equations	Constants	Values of the constants "C"		
		SI Units	Practical metric units	Oilfield units (US)
Pressure drawdown				
$m(p_i) - m(p_{wf}) = b\left[\log\dfrac{kt}{\phi\mu_g c_g r_w^2} + C_1 + 0.869\,(S + Dq_{sc})\right]$	C_1	0.351	-4.885	-3.228
$b = -\dfrac{dm(p_{wf})}{d\log t} = C_2\dfrac{q_{sc}T}{kh}$	C_2	128.81	0.157	1640
$S + Dq_{sc} = 1.151\left(\dfrac{m(p_i) - m(p_{1\,h})}{b} - \log\dfrac{k}{\phi\mu_g c_g r_w^2} - C_3\right)$	C_3	3.909	-3.107	-3.228
Pressure buildup				
$m(p_i) - m(p_{ws}) = b\log\dfrac{t + \Delta t}{\Delta t}$				
$b = -\dfrac{dm(p_{ws})}{d\log\dfrac{t + \Delta t}{\Delta t}} = C_2\dfrac{q_{sc}T}{kh}$	C_2	128.81	0.157	1640
$S + Dq_{sc} = 1.151\left(\dfrac{m_{LIN}(p_{ws})(1\,h) - m(p_{wf})}{b} - \log\dfrac{k}{\phi\mu_g c_g r_w^2} - C_3\right)$	C_3	3.909	-3.107	-3.228

N.B. The units used are specified in Table 7.1.

7.10 Final Considerations

In view of the equivalence between Eq. (5.15) [describing the radial horizontal flow of slightly compressible fluids in dimensionless terms $p_D(r_D, t_D)$] and Eq. (7.10) [describing the same flow of gas in terms of the dimensionless real gas pseudo-pressure $m_D(r_D, t_D)$], the following topics covered in Chap. 6 are also relevant to gas reservoirs, after the necessary modifications to symbols:

- use of pressure and pressure derivative type curves,
- skin factor from completion geometry,
- layered reservoirs and underlying aquifer,
- identification of boundaries within the area of drainage,
- interference testing between wells,
- pulse testing.

Please refer to Sects. 6.7, 6.8, 6.9, 6.11 and 6.12.

References

1. Al-Hussaini R, Ramey HJ Jr, Crawford PB (1966) The flow of real gases through porous media. J Petrol Tech (May): 624–636; Trans AIME 237
2. Al-Hussaini R, Ramey HJ Jr (1966) Application of real gas flow theory to well testing and deliverability forecasting. J Petrol Tech (May): 637–642; Trans AIME 237
3. Carter RD, Millers SC Jr, Riley HG (1963) Determination of stabilized well performance from short flow tests. J Petrol Tech (June): 651–658; Trans AIME 228
4. Dake LP (1978) Fundamentals of reservoir engineering. Elsevier, Amsterdam
5. Energy Resources Conservation Board (1975) Theory and practice of the testing of gas wells, 3rd edn. Rep ERCB 75-34, Calgary, Alberta, Canada
6. Essis AEB, Thomas GW (1971) The use of open flow potential test data in determining formation capacity and skin factor. J Petrol Tech (July): 879–887; Trans AIME 251
7. Geertsma J (1974) Estimating the coefficient of inertial resistance in fluid flow through porous media. Soc Petrol Eng J (Oct): 445–450
8. Katz DL, Cornell D, Kobayashi R, Poettmann F, Vary JA, Elenbaas JR, Weinaug CF (1959) Handbook of natural gas engineering. McGraw-Hill, New York
9. Lee J (1982) Well testing. Society of Petroleum Engineers, Textbook Series vol 1, Dallas
10. Odeh AS, Jones LG (1965) Pressure drawdown analysis, variable rate case. J Petrol Tech (Aug): 960–964; Trans AIME 234
11. Rawlins EL, Schellardt MA (1936) Back pressure data on natural gas wells and their application to production practice. United States Bureau of Mines, Monogr 7, Bartlesville, OK

EXERCISES

Exercise 7.1

Using the following data for pure methane at 160 °F, taken from the article "The viscosity of methane" by M.A. Gonzales, R.F. Bukaceck and A.L. Lee, published in the Soc Petrol Eng J (March 1967), pages 75–79:

Pressure (psia)	Density (g/cm^3)	Viscosity (μP)	Pressure (psia)	Density (g/cm^3)	Viscosity (μP)
14.7	0.00057	125	2000	0.08481	160
100	0.00389	130	2500	0.10618	172
200	0.00783	130	3000	0.12646	186
300	0.01183	131	3500	0.14518	202
400	0.01587	132	4000	0.16212	217
500	0.01997	133	4500	0.17734	232
600	0.02410	135	5000	0.19091	246
800	0.03253	137	6000	0.21378	274
1000	0.04105	140	7000	0.23276	299
1250	0.05191	145	8000	0.24900	323
1500	0.06283	149	9000	0.26245	345
1750	0.07384	154	10000	0.27442	366

1. Determine the behaviour of the product $\mu_g c_g$ as a function of p. Remember that $\mu_g c_g$ represents the influence of the thermodynamic properties of the gas in the diffusivity Eq. (7.5b), and that it is assumed to be a constant for the purpose of integrating this equation.
2. Calculate the value of the term $p/\mu_g z$ as a function of p, at the temperature of the tabulated data.
3. Calculate, using Eq. (7.3), the pseudo-pressure $m(p)$ of methane at the temperature of the tabulated data.

Do all the calculations in SI units.

Solution

Firstly, for the SI units conversion, we have

$$160\,°\mathrm{F} = 71.11\,°\mathrm{C}$$

$$1\,\mathrm{psi} = 6.894757 \times 10^{-3}\,\mathrm{MPa}$$

We need Eq. (2.8a) for c_g:

$$c_g = 1/p - (1/z)\left(\frac{\partial z}{\partial p}\right)_T \tag{2.8a}$$

Therefore, to calculate c_g we need to know the value of $z(p)$ and its derivative, at constant T. $z(p)$ is also needed in the calculation of $p/\mu z = f(p)$.

z is derived as follows from the gas density ρ_g data listed in the table above:
We start from the relationship

$$\rho_g(p, T) = \frac{\rho_{sc}}{B_g(p, T)} = \rho_{sc}\frac{p}{z}\frac{T_{sc}}{p_{sc}}\frac{1}{T}, \tag{7/1.1}$$

where ρ_{sc} is the density of the gas at p_{sc} and T_{sc}:

$$p_{sc} = 1.033\,\mathrm{kg/cm^2} = 14.7\,\mathrm{psia} = 0.1013\,\mathrm{MPa}$$

$$T_{sc} = 15\,°\mathrm{C} = 59\,°\mathrm{F} = 288.2\,\mathrm{K}.$$

At p_{sc} then

$$\rho_g(p_{sc}, T) = \rho_{sc}\frac{T_{sc}}{T}, \tag{7/1.2}$$

which, when substituted into Eq. (7/1.1), gives

$$z = \frac{\rho_g(p_{sc}, T)}{\rho_g(p, T)}\frac{p}{p_{sc}}.$$

(1)	(2)	(3)	(4)	(5)	(6)	(7)	(8)	(9)
p	z	$\partial z/\partial p$	c_g	μ_g	$c_g\mu_g$	$p/\mu_g z$	$2\int_{p_{j-1}}^{p_j}\dfrac{p}{\mu_g z}\,dp$	$m(p)$
(MPa)		(Pa^{-1})	(Pa^{-1})	(Pa s)	(s)	(s^{-1})	(MPa/s)	(MPa/s)
0.1013	1.0000	-5.445×10^{-9}	9.877×10^{-6}	1.25×10^{-5}	1.235×10^{-10}	8.108×10^{9}	8.212×10^{8}	8.212×10^{8}
0.689	0.9968	-7.514×10^{-9}	1.459×10^{-6}	1.30×10^{-5}	1.897×10^{-11}	5.321×10^{10}	3.604×10^{10}	3.686×10^{10}
1.379	0.9909	-9.790×10^{-9}	7.350×10^{-7}	1.30×10^{-5}	9.555×10^{-12}	1.071×10^{11}	1.106×10^{11}	1.474×10^{11}
2.068	0.9833	-9.717×10^{-9}	4.934×10^{-7}	1.31×10^{-5}	6.464×10^{-12}	1.606×10^{11}	1.844×10^{11}	3.319×10^{11}
2.758	0.9770	-9.065×10^{-9}	3.719×10^{-7}	1.32×10^{-5}	4.909×10^{-12}	2.139×10^{11}	2.584×10^{11}	5.904×10^{11}
3.447	0.9708	-8.412×10^{-9}	2.988×10^{-7}	1.33×10^{-5}	3.974×10^{-12}	2.670×10^{11}	3.314×10^{11}	9.216×10^{11}
4.137	0.9654	-8.301×10^{-9}	2.503×10^{-7}	1.35×10^{-5}	3.379×10^{-12}	3.174×10^{11}	4.032×10^{11}	1.325×10^{12}
5.516	0.9536	-7.544×10^{-9}	1.891×10^{-7}	1.37×10^{-5}	2.591×10^{-12}	4.222×10^{11}	1.022×10^{12}	2.347×10^{12}
6.895	0.9446	-5.981×10^{-9}	1.514×10^{-7}	1.40×10^{-5}	2.120×10^{-12}	5.214×10^{11}	1.298×10^{12}	3.644×10^{12}
8.618	0.9357	-5.450×10^{-9}	1.219×10^{-7}	1.45×10^{-5}	1.768×10^{-12}	6.352×10^{11}	1.994×10^{12}	5.640×10^{12}
10.34	0.9257	-3.403×10^{-9}	1.004×10^{-7}	1.49×10^{-5}	1.496×10^{-12}	7.498×10^{11}	2.385×10^{12}	8.024×10^{12}
12.07	0.9190	-3.249×10^{-9}	8.639×10^{-8}	1.54×10^{-5}	1.330×10^{-12}	8.526×10^{11}	2.772×10^{12}	1.080×10^{13}
13.79	0.9144	-2.000×10^{-9}	7.470×10^{-8}	1.60×10^{-5}	1.195×10^{-12}	9.425×10^{11}	3.088×10^{12}	1.388×10^{13}
17.24	0.9130	$+0.605\times10^{-9}$	5.734×10^{-8}	1.72×10^{-5}	9.862×10^{-13}	1.098×10^{12}	7.040×10^{12}	2.092×10^{13}
20.68	0.9199	2.843×10^{-9}	4.527×10^{-8}	1.86×10^{-5}	8.420×10^{-13}	1.209×10^{12}	7.936×10^{12}	2.886×10^{13}
24.13	0.9348	5.318×10^{-9}	3.575×10^{-8}	2.02×10^{-5}	7.222×10^{-13}	1.278×10^{12}	8.580×10^{12}	3.744×10^{13}
27.58	0.9567	6.768×10^{-9}	2.918×10^{-8}	2.17×10^{-5}	6.332×10^{-13}	1.328×10^{12}	8.992×10^{12}	4.644×10^{13}
31.03	0.9839	8.300×10^{-9}	2.379×10^{-8}	2.32×10^{-5}	5.519×10^{-13}	1.359×10^{12}	9.272×10^{12}	5.572×10^{13}
34.47	1.0155	1.048×10^{-8}	1.869×10^{-8}	2.46×10^{-5}	4.598×10^{-13}	1.380×10^{12}	9.424×10^{12}	6.512×10^{13}
41.37	1.0883	1.092×10^{-8}	1.414×10^{-8}	2.74×10^{-5}	3.874×10^{-13}	1.387×10^{12}	1.909×10^{13}	8.420×10^{13}
48.26	1.1661	1.142×10^{-8}	1.093×10^{-8}	2.99×10^{-5}	3.268×10^{-13}	1.384×10^{12}	1.909×10^{13}	1.033×10^{14}
55.16	1.2458	1.186×10^{-8}	8.609×10^{-9}	3.23×10^{-5}	2.781×10^{-13}	1.371×10^{12}	1.901×10^{13}	1.223×10^{14}
62.05	1.3297	1.212×10^{-8}	7.001×10^{-9}	3.45×10^{-5}	2.415×10^{-13}	1.353×10^{12}	1.877×10^{13}	1.411×10^{14}
68.95	1.4130	1.231×10^{-8}	5.791×10^{-9}	3.66×10^{-5}	2.120×10^{-13}	1.333×10^{12}	1.853×10^{13}	1.596×10^{14}

This equation allows us to calculate z from $\rho_g(p, T)$. Note that $\rho_g(p_{sc}, T)$ is reported in the first row of the data table, for a pressure of 14.7 psia $[\rho_g(14.7 \text{ psia}, 160\,^\circ\text{F}) = 0.00057 \text{ g/cm}^3]$.

The determination of $(\partial z/\partial p)$ needed in Eq. (2.8) to evaluate c_g can now be performed graphically from a plot of $z(p)$ versus p at $T = 160\,^\circ\text{F} = 71.1\,^\circ\text{C}$.

The various stages in the subsequent calculations are listed in the table on page 263.

The values of $z(p)$ appearing in column 2 have been plotted in Fig. E7/1.1, while $c_g\mu_g$ from column 6 is plotted in Fig. E7/1.2.

It can be clearly seen from the second figure that $c_g\mu_g$ is not a constant: in this case it decreases by an order of magnitude when the pressure increases from 3 to 32 MPa!

Therefore, strictly speaking, the gas diffusivity equation (7.5b) is not linear, because the expression $\phi\mu_g c_g/k$ is a function of the pressure p [*and of the pseudo-pressure $m(p)$*]; *consequently, the principle of superposition (Duhamel's theorem) is not applicable.*

However, in a practical situation, such rigorous considerations are overlooked in favour of the simplifying assumption that $c_g\mu_g = $ constant.

Several interesting observations can be made about the behaviour of the curve $p/\mu_g z = f(p)$ illustrated in Fig. 7.1 of the main text, with reference to the values listed in column 7 of the table.

The curve can be divided into three sections:

Section I: Between $p = 0$ and $p \approx 15$ MPa: the curve has an almost constant slope, indicating that $\mu_g z$ is constant in this pressure range.

Using the values in the table, it is easy to verify that $\mu_g z$ increases from a minimum of 1.25×10^{-5} Pa s to a maximum of only 1.50×10^{-5} Pa s between 0 and 15 MPa.

Section II: Between $p \approx 15$ MPa and $p \approx 30$ MPa: the slope of the curve decreases steadily with increasing pressure, and eventually levels off at about 30 MPa pressure.

Section III: $p > 30$ MPa: $p/\mu_g z$ is roughly constant, and essentially independent of p.

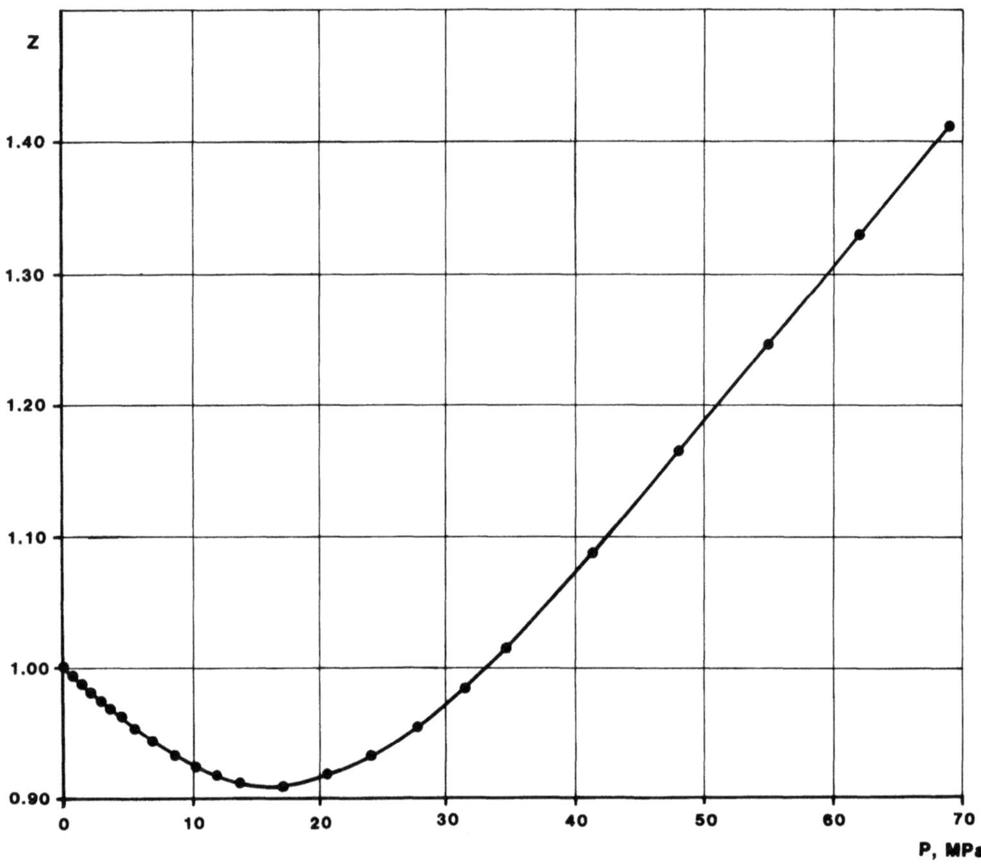

Fig. E7/1.1

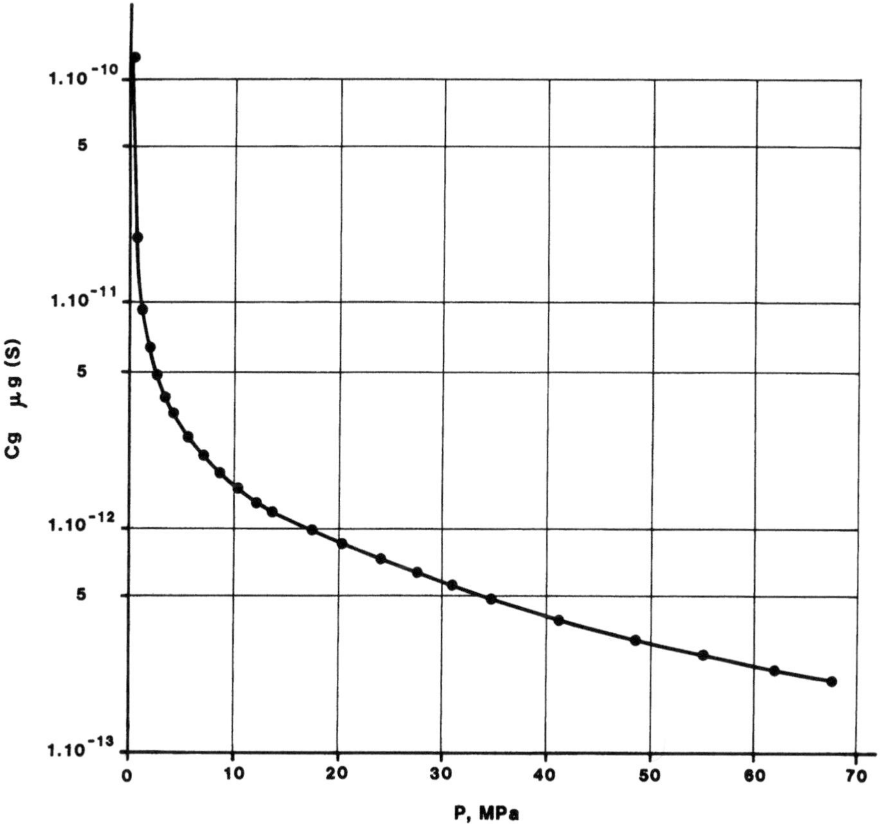

Fig. E7/1.2

From the definition of the real gas pseudo-pressure,

$$m(p) = 2 \int_{p_0}^{p} \frac{p}{\mu_g z} \, dp \, , \tag{7.2}$$

we have

$$\frac{dm}{dp} = 2 \frac{p}{\mu_g z} \, .$$

In section I of the curve ($0 \leq p \leq 15$ MPa), where $\mu_g z \approx$ constant, we can write

$$m(p) = \frac{2}{\mu_g z} \int_0^p p \, dp = \frac{p^2}{\mu_g z} \, ,$$

having set $p_0 = 0$.

In section III of the curve ($p > 30$ MPa), if we define

$$\frac{p}{\mu_g z} = \frac{C}{2} = \text{constant}$$

we can write

$$m(p) = 2 \frac{C}{2} \int_0^p dp = Cp \, ,$$

and therefore:

$$\frac{\partial m}{\partial r} = \frac{dm}{dp} \frac{\partial p}{\partial r} = C \frac{\partial p}{\partial r} \, ,$$

$$\frac{\partial m}{\partial t} = \frac{dm}{dp}\frac{\partial p}{\partial t} = C\frac{\partial p}{\partial t}.$$

We can now draw the following conclusions:

1. *At sufficiently low pressures* (in this case, less than ~ 15 MPa), the diffusivity Eq. (7.5b) can be expressed as

$$\frac{1}{r}\frac{\partial}{\partial r}\left(r\frac{\partial p^2}{\partial r}\right) = \frac{\phi\mu_g c_g}{k}\frac{\partial p^2}{\partial t} \tag{7.5c}$$

by replacing $m(p)$ with p^2.

2. *At sufficiently high pressures* (above ~ 30 MPa in this example), Eq. (7.5b) can be stated as

$$\frac{1}{r}\frac{\partial}{\partial r}\left(r\frac{\partial p}{\partial r}\right) = \frac{\phi\mu_g c_g}{k}\frac{\partial p}{\partial t} \tag{7.5d}$$

and takes the same form as Eq. (5.10) for liquids. *In other words, at high pressures, gases can be treated like liquids as far as the diffusivity equation is concerned*, with $m(p)$ being replaced by p.

3. *Only in the intermediate pressure range* (15 MPa $< p <$ 30 MPa in the present case) is it essential to use the $m(p)$ function.

◇ ◇ ◇

Exercise 7.2

A flow-after-flow test (described in Sect. 7.8.2) has just been performed in a discovery well in a gas reservoir.

The initial pressure of the virgin reservoir is 125 kg/cm^2 (abs).

The well was tested at five flow rates, increasing from the first to the last. The stabilised (pseudo-steady state) rates and bottom-hole flowing pressures were as follows:

Test no.	Flow rate (Nm^3/day)	Bottom hole pressure p_{wf} (kg/cm^2 abs)
1	120 000	124.41
2	240 000	123.80
3	360 000	123.16
4	480 000	122.50
5	600 000	121.82

Interpret this production data, either using the "back pressure test" Eq. (7.37a), or the quadratic Eq. (7.43), and calculate the AOF with both equations.

Given that the reservoir pressure is less than 150 atm, the conclusions derived in Ex. 7.1 would suggest that the p^2 approximation can be used instead of $m(p)$.

Solution

The relevant data for the back pressure test method is listed in this table:

Test no.	q_{sc} (m^3/day)	$\bar{p}^2 - p_{wf}^2$ (kg/cm^2)^2	$\log q$	$\log(\bar{p}^2 - p_{wf}^2)$
1	1.2×10^5	147.152	5.07918	2.16777
2	2.4×10^5	298.560	5.38021	2.47503
3	3.6×10^5	456.615	5.55630	2.65955
4	4.8×10^5	618.750	5.68124	2.79152
5	6.0×10^5	784.888	5.77815	2.89481

A least-squares fit of Eq. (7.37b) to this data gives the following results:

$$\log C = 2.92282$$

$$n = 0.99285$$

so that

$$C = 837.18$$

From this we can calculate

$$\text{AOF (m}^3\text{/day)} = 837.18\,(125^2 - 1.033^2)^{0.99285}$$

Therefore

$$\text{AOF} = 12.2 \times 10^6 \text{ Nm}^3\text{/day} \; .$$

The curve $\log q_{\text{sc}} = f\,[\log(\bar{p}^2 - p_{\text{wf}}^2)]$ is presented in Fig. E7/2.1. The five test points do indeed lie on an approximately straight line.

To interpret the data using the quadratic equation, we can replace the $m(p)$ terms in Eq. (7.41a) by p^2.

The following table lists the relevant quantities, plotted in Fig. E7/2.2:

Test no.	q_{sc} (Nm^3/day)	$(\bar{p}^2 - p_{\text{wf}}^2)/q_{\text{sc}}$ (kg/cm^2)^2/m^3/day
1	1.2×10^5	1.2263×10^{-3}
2	2.4×10^5	1.2440×10^{-3}
3	3.6×10^5	1.2684×10^{-3}
4	4.8×10^5	1.2891×10^{-3}
5	6.0×10^5	1.3081×10^{-3}

A least-squares fit to these data gives

$$A = 1.205 \times 10^{-3}$$

$$B = 2 \times 10^{-10}$$

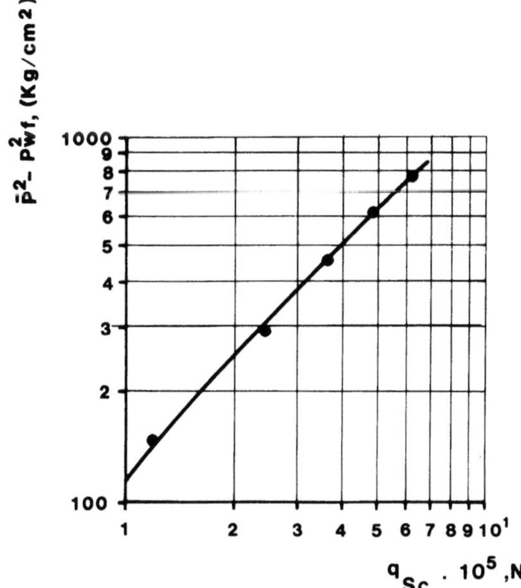

Fig. E7/2.1

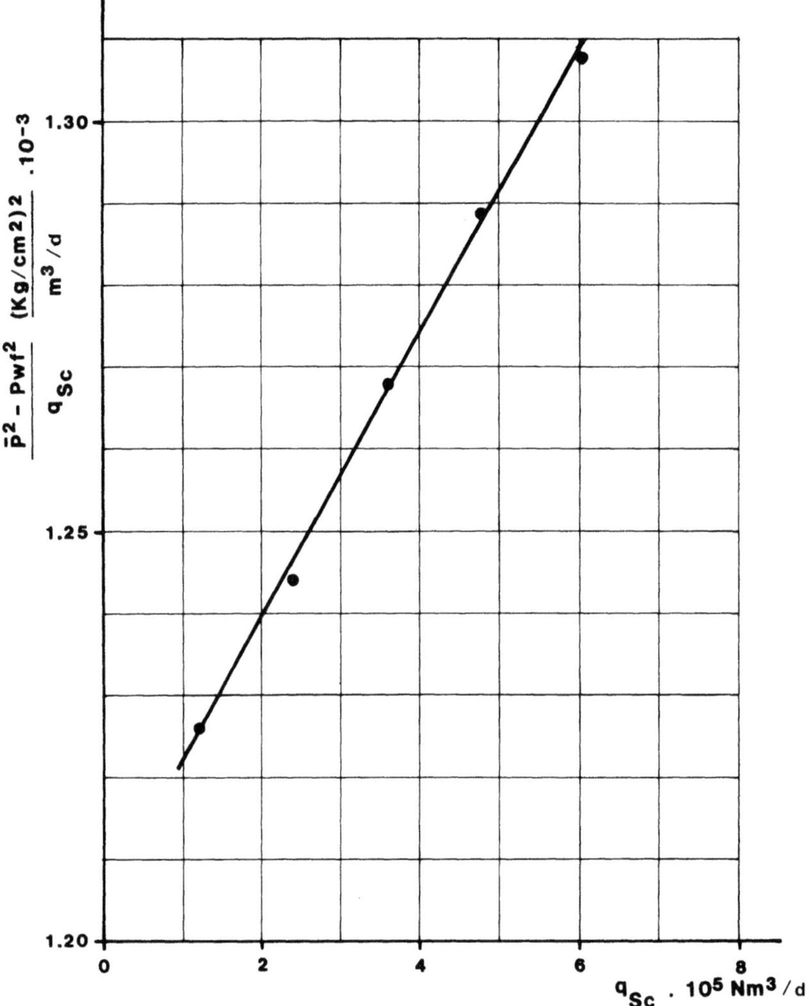

Fig. E7/2.2

The AOF is calculated from

$$125^2 - 1.033^2 = 1.205 \times 10^{-3}(AOF) + 2 \times 10^{-10}(AOF)^2$$

Therefore,

$$(AOF)^2 + 6.025 \times 10^6(AOF) - 7.812 \times 10^{13} = 0$$

so that,

$$AOF = 6.4 \times 10^6\ Nm^3/day\ .$$

Note that this value, calculated by the more theoretically rigorous quadratic method, is little more than half the AOF derived by the empirical back-pressure test method.

◇ ◇ ◇

Exercise 7.3

An isochronal test (Sect. 7.8.3.1) is performed on a gas well. The test consisted of three flowing periods, each lasting 3 h, at rates of 100 000; 200 000 and 300 000 Nm³/day, with 9-h shut in periods between.

At the end of each shut in period the pressure had built up to the value it had prior to the isochronal test.

The following bottom-hole flowing pressures were measured:

Flow rate (Nm^3/day)	100 000	200 000	300 000
Time (h)	Pressure (kg/cm^2 abs)		
0.5	189.66	173.39	156.19
1.0	189.05	172.17	154.35
1.5	188.69	171.45	153.28
2.0	188.44	170.95	152.52
2.5	188.24	170.55	151.93
3.0	188.08	170.23	151.44

The last flowing period ($q_{sc} = 300\,000$ Nm^3/day) was extended until pseudo-steady state flow was attained. The stabilised flowing pressure was $p_{wf} = 145.7$ kg/cm^2 (abs).

The produced gas was almost pure methane.

The reservoir conditions were:

– initial pressure: 205 kg/cm^2 (abs)
– temperature: 71.1 °C.

The producing formation was 12 m thick, with a porosity of 15%.
The well had been completed with a $9\frac{5}{8}''$ casing.
From the test data, calculate:

1. The flow equation and open flow potential,
2. The permeability to gas,
3. The skin factor S, and the non-Darcy coefficient D.

Solution

Referring to the conclusions of Ex. 7.1, we see that the current test pressures are in the range where the rigorous use of $m(p)$ is required. Accordingly, the calculations will all be performed in terms of the real gas pseudo-pressure.

Since the produced gas is almost pure methane, we can use the $m(p)$ function already computed in Ex. 7.1, which was computed at 71.1 °C – the reservoir temperature in the current test.

From Ex. 7.1 we have:

p (MPa)	(kg/cm^2 abs)	$m(p)$ (MPa/s)
13.79	140.62	1.388×10^{13}
17.24	175.80	2.092×10^{13}
20.68	210.88	2.886×10^{13}

Over the pressure range observed in this test, $m(p)$ is almost a linear function of p. A least-squares fit gives

$$m(p) \text{ [MPa/s]} = 2.132 \times 10^{11} p \text{ [}kg/cm^2 \text{ abs]} - 1.6254 \times 10^{13} \tag{7/3.1}$$

To interpret the test, we plot the values of $[m(p_i) - m(p_{wf})]/q_{sc}$ against q_{sc} for each t, as in Fig. 7.11.

We have established via Eq. (7/3.1) that:

$$[m(p_i) - m(p_{wf})] \text{ [MPa/s]} = 2.132 \times 10^{11}(p_i - p_{wf}) \text{ [}kg/cm^2\text{]}$$

and using this we can derive the following tables:

$q_{sc} = 100\,000$ Nm^3/day

t (h)	$[m(p_i) - m(p_{wf})]$ (MPa/s)	$[m(p_i) - m(p_{wf})]/q_{sc}$
0.5	3.27049×10^{12}	3.27049×10^7
1.0	3.40054×10^{12}	3.40054×10^7
1.5	3.47729×10^{12}	3.47729×10^7
2.0	3.53059×10^{12}	3.53059×10^7
2.5	3.57323×10^{12}	3.57323×10^7
3.0	3.60734×10^{12}	3.60734×10^7

$q_{sc} = 200\,000$ Nm^3/day

t (h)	$[m(p_i) - m(p_{wf})]$ (MPa/s)	$[m(p_i) - m(p_{wf})]/q_{sc}$
0.5	6.739252×10^{12}	3.369626×10^7
1.0	6.999356×10^{12}	3.499678×10^7
1.5	7.152860×10^{12}	3.576430×10^7
2.0	7.259460×10^{12}	3.629730×10^7
2.5	7.344740×10^{12}	3.672370×10^7
3.0	7.412964×10^{12}	3.706482×10^7

$q_{sc} = 300\,000$ Nm^3/day

t (h)	$[m(p_i) - m(p_{wf})]$ (MPa/s)	$[m(p_i) - m(p_{wf})]/q_{sc}$
0.5	10.406292×10^{12}	3.468760×10^7
1.0	10.798580×10^{12}	3.599527×10^7
1.5	11.026704×10^{12}	3.675568×10^7
2.0	11.188736×10^{12}	3.729579×10^7
2.5	11.314524×10^{12}	3.771508×10^7
3.0	11.418992×10^{12}	3.806331×10^7

When we plot the $[m(p_i) - m(p_{wf})]/q_{sc}$ against q_{sc} for each value of t (Fig. E7/3.1), all the data sets lie on parallel lines.

The general equation of these lines is

$$\frac{m(p_i) - m(p_{wf})}{q_{sc}} \left[\frac{\text{MPa/s}}{\text{Nm}^3/\text{d}} \right] = A + F q_{sc} \; [\text{Nm}^3/\text{day}] \; .$$

Applying a least-squares fit to the data for each t, we get:

t (h)	A $\left(\dfrac{\text{MPa/s}}{\text{Nm}^3/\text{d}} \right)$	F $\left(\dfrac{\text{MPa/s}}{(\text{Nm}^3/\text{d})^2} \right)$
0.5	3.17136×10^7	9.9135
1.0	3.30093×10^7	9.9494
1.5	3.37815×10^7	9.9139
2.0	3.43098×10^7	9.9494
2.5	3.47409×10^7	9.9139
3.0	5.50773×10^7	9.9496

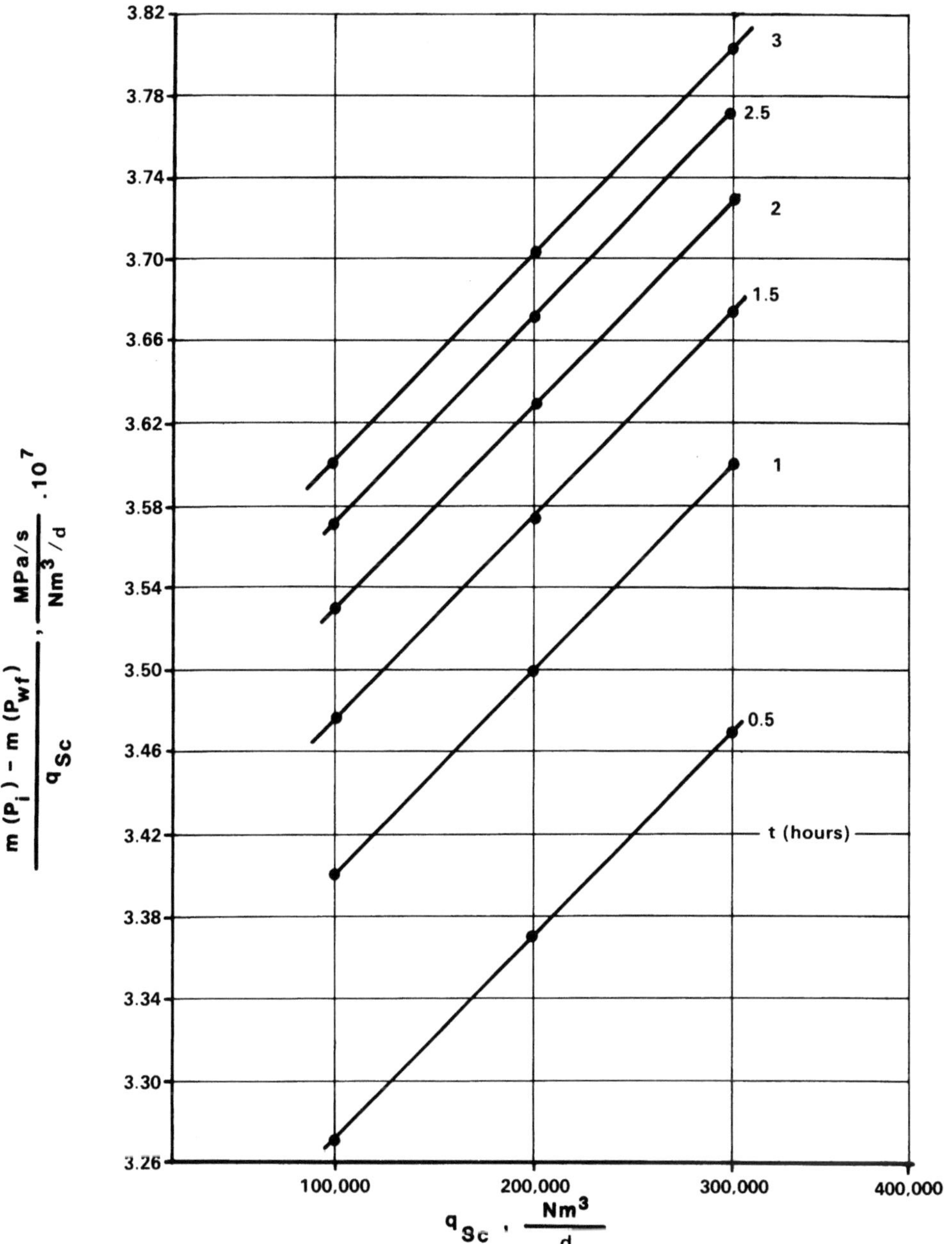

Fig. E7/3.1

The average of the slopes is

$$\bar{F} = 9.9317 \, \frac{\text{MPa/s}}{(\text{Nm}^3/\text{d})^2}$$

The line corresponding to pseudo-steady state flow will have the same slope as these transient data, and will pass through the single data point measured at the end of the extended flow period.

This point is

$$p_{wf} = 145.7 \, \text{kg/cm}^2 \, \text{abs} ,$$

at

$$q_{sc} = 300\,000 \, \text{Nm}^3/\text{day} ,$$

Therefore,

$$\frac{2.132 \times 10^{11} \, (205 - 145.7)}{300\,000} = A + 9.931 \times 300\,000 \, ,$$

so that,

$$A = 3.916 \times 10^{7} \, \frac{\text{MPa/s}}{\text{Nm}^3/\text{d}} \, .$$

In terms of $m(p)$, the pseudo-steady state flow equation is therefore,

$$m(p_{\text{wf}}) = 2.745 \times 10^{13} - 3.916 \times 10^{7} \, q_{\text{sc}} - 9.932 \, q_{\text{sc}}^2 \, , \tag{7/3.2}$$

with $m(p)$ in MPa/s and q_{sc} in Nm^3/day.

In Eq. (7/3.2), note that

$$2.745 \times 10^{13} \, \text{MPa/s} \equiv 205 \, \text{kg/cm}^2 = p_{\text{i}} \, .$$

In practical metric units, Eq. (7/3.2) becomes

$$p_{\text{wf}} (\text{kg/cm}^2) = 205 - 1.837 \times 10^{-4} q_{\text{sc}} \, (\text{Nm}^3/\text{day}) - 4.659 \times 10^{-11} q_{\text{sc}}^2 \, (\text{Nm}^3/\text{day})^2 \, . \tag{7/3.3}$$

Equation (7/3.3) is only valid in the pressure range 140–$210 \, \text{kg/cm}^2$, where the relationship between $m(p)$ and p described by Eq. (7/3.1) applies.

We should therefore use Eq. (7/3.2), in $m(p)$, to calculate the AOF.

Given that at $p_{\text{sc}} = 0.1013 \, \text{MPa}$, $m(p) = 8.212 \times 10^{8} \, \text{MPa/s}$ (see Ex. 7/1), we have

$$9.932 \, (\text{AOF})^2 + 3.916 \times 10^{7} \, (\text{AOF}) - 2.745 \times 10^{13} = 0$$

from which we conclude that,

$$\text{AOF} = 607\,000 \, \text{Nm}^3/\text{day} \, .$$

For the calculation of k, S and D, we need to interpret any two of the three drawdown tests for which we have data.

From Eq. (7/3.1),

$$\Delta m(p) \, [\text{Pa/s}] = 2.132 \times 10^{17} \Delta p \, [\text{kg/cm}^2] \, .$$

From Table 7.2, we have, in SI units,

$$b = -\frac{dm(p_{\text{wf}})}{d \log t} = 128.81 \, \frac{q_{\text{sc}} T}{kh} \, ,$$

from which

$$b = -2.132 \times 10^{17} \, \frac{dp_{\text{wf}}(\text{kg/cm}^2)}{d \log t}$$

$$= 128.81 \times \frac{q_{\text{sc}}(\text{Nm}^3/\text{day})}{86\,400} \, \frac{1}{k \, (\text{md}) \times 9.869233 \times 10^{-16}} \, \frac{T(\text{K})}{h(\text{m})}$$

and:

$$-\frac{dp_{\text{wf}} \, (\text{kg/cm}^2)}{d \log t} = 7.085 \times 10^{-6} \, \frac{q_{\text{sc}} \, (\text{Nm}^3/\text{day}) \, T \, (\text{K})}{k \, (\text{md}) \, h \, (\text{m})}$$

Referring to the plots of p_{wf} versus $\log t$ in Fig. E7/3.2, we can make two estimates of k (which should be the same, of course), remembering that for this example we have $h = 12 \, \text{m}$ and $T = 71.1 \,^\circ\text{C} = 344.3 \, \text{K}$:

q_{sc} (Nm^3/day)	$- dp_{\text{wf}}/d \log t$ $(\text{kg cm}^{-2}/\text{cycle})$	k (md)
100\,000	2.023	10.0
300\,000	6.316	9.7

These two permeability estimates are very close.

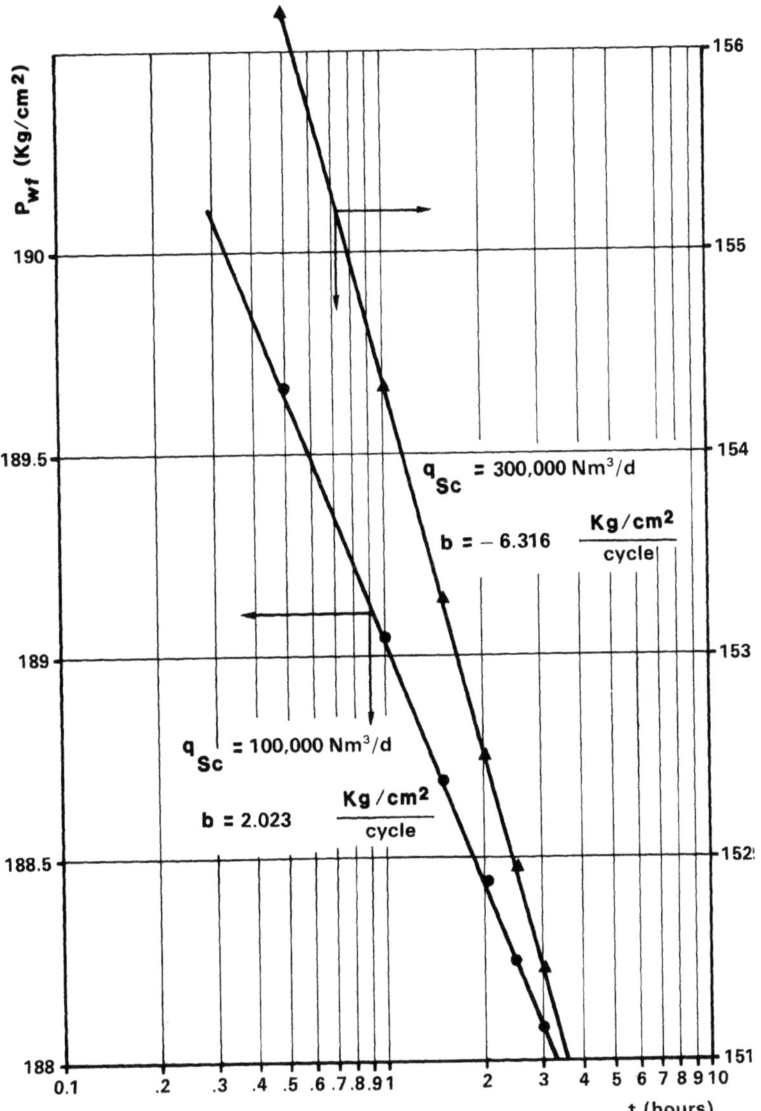

Fig. E7/3.2

We can now calculate S and D with Eq. (7.47b), using the practical metric form presented in Table 7.2:

$$S + Dq_{sc} = 1.151 \left[\frac{m(p_i) - m(p_{1\,h})}{b} - \log \frac{k}{\phi \mu_g c_g r_w^2} + 3.107 \right].$$

Since $m(p)$ and the coefficient b have the same units, their ratio is independent of the units system in use. In this example, we will work in kg/cm².

In Ex. 7.1 we had the following average values in the pressure range of interest:

$$\mu_g = 1.8 \times 10^{-5} \text{ Pa s} = 0.018 \text{ cP}$$

$$c_g = 5.28 \times 10^{-8} \text{ Pa}^{-1} = 5.18 \times 10^{-3} \text{ cm}^2/\text{kg} .$$

The relevant flowing test data is

q_{sc}	(Nm^3/day)	100 000	300 000
p_i	(kg/cm^2)	205	205
p_{1h}	(kg/cm^2)	189.05	154.35
b	$\left(\dfrac{\text{kg/cm}^2}{\text{cycle}}\right)$	2.023	6.316
k	(md)	10.0	9.7

which gives us the following two equations:

$$S + 1 \times 10^5 D = 3.81146 \, ,$$

$$S + 3 \times 10^5 D = 3.98205 \, .$$

These can be solved to give

$$S = 3.726 \, ,$$

$$D = 8.530 \times 10^{-7} \, (\text{Nm}^3/\text{day})^{-1} \, .$$

The results are summarised as follows:

1. *Flow equation*

$$p_{wf} = 205 - 1.837 \times 10^{-4} q_{sc} - 4.659 \times 10^{-11} q_{sc}^2$$

with

p_{wf} in kg/cm^2 (abs)

q_{sc} in Nm^3/day

valid in the range $140 < p_{wf} < 210$.

$$\text{AOF} = 600 000 \, \text{Nm}^3/\text{day}$$

2. *Average permeability to gas:*   10 md

3. *Skin factor:* $S = 3.73$

4. *Non-Darcy coefficient:* $D = 8.53 \times 10^{-7} \, (\text{Nm}^3/\text{day})^{-1}$

$$\Diamond \, \Diamond \, \Diamond$$

Exercise 7.4

Two short duration production tests, each followed by a shut in, were performed in another well in the reservoir described in Ex. 7.3. A high resolution gauge was used to record the bottom-hole pressures of the buildups, but no data were acquired during the flowing periods.

The sequence of events was as follows:

1st flowing test   – Initial static pressure: 203 kg/cm^2 abs

  – Flow rate: 200 000 Nm^3/day for 3 h

  – Well shut in for a 9 h pressure buildup

2nd flowing test   – Initial static pressure: 203 kg/cm^2 abs

  – Flow rate: 400 000 Nm^3/day for 3 h

  – Well shut in for a 9 h pressure buildup

The following table lists the pressures recorded during the first 4 h of each buildup (transient flow periods):

q_{sc} (Nm³/day)	200 000	400 000
Time since shut in (h)	p_{ws} (kg/cm²)	p_{ws} (kg/cm²)
0	180.53	136.15
0.5	201.36	199.73
1	201.83	200.67
1.5	202.08	201.15
2	202.23	201.46
2.5	202.34	201.67
3	202.42	201.83
4	202.53	202.06

The reservoir temperature was 71.1 °C, and the produced gas had the same properties as in Ex. 7.1.

The well was completed with a $9\frac{5}{8}''$ casing.

Formation thickness at the well was 14 m, and porosity 20%.

The gas properties pertaining to Ex. 7.3, which was another well in the same reservoir, are applicable to the current example. The relevant data are:

For $136 \text{ kg/cm}^2 < p < 203 \text{ kg/cm}^2$

$$m(p) \text{ [Pa/s]} = 2.132 \times 10^{17} \, p \text{ [kg/cm}^2 \text{ abs]} - 1.6254 \times 10^{19} \, ,$$

$$\mu_g = 0.018 \text{ cP} \, ,$$

$$c_g = 5.18 \times 10^{-3} \text{ cm}^2/\text{kg} \, .$$

Calculate the following parameters from the two buildups:

- average permeability in the drainage area of the well,
- skin factor S,
- non-Darcy coefficient D.

Solution

Since there is a conveniently linear relationship between $m(p)$ and p in the pressure range of interest, it is quite easy to calculate the slope of the buildup curve. We have

$$b = -\frac{\mathrm{d}\, m(p_{ws})}{\mathrm{d} \log \dfrac{t+\Delta t}{\Delta t}} = -2.132 \times 10^{17} \frac{\mathrm{d}\, p_{ws}}{\mathrm{d} \log \dfrac{t+\Delta t}{\Delta t}},$$

where t, the constant rate flowing time, was 3 h before both buildups.

p_{ws} is plotted against $(t + \Delta t)/\Delta t$ on semi-log axes in Fig. E7/4.1. There is a clear linear trend to both sets of data, indicating that both tests were entirely in the transient flow regime.

If we make least-squares fits to each data set we get

Buildup after $q_{sc} = 200\,000 \; Nm^3/day$

$$b = 1.94753 \, \frac{\text{kg/cm}^2}{\text{cycle}} = 4.15213 \times 10^{17} \, \frac{\text{Pa/s}}{\text{cycle}}$$

Buildup after $q_{sc} = 400\,000 \; Nm^3/day$

$$b = 3.86473 \, \frac{\text{kg/cm}^2}{\text{cycle}} = 8.2396 \times 10^{17} \, \frac{\text{Pa/s}}{\text{cycle}}$$

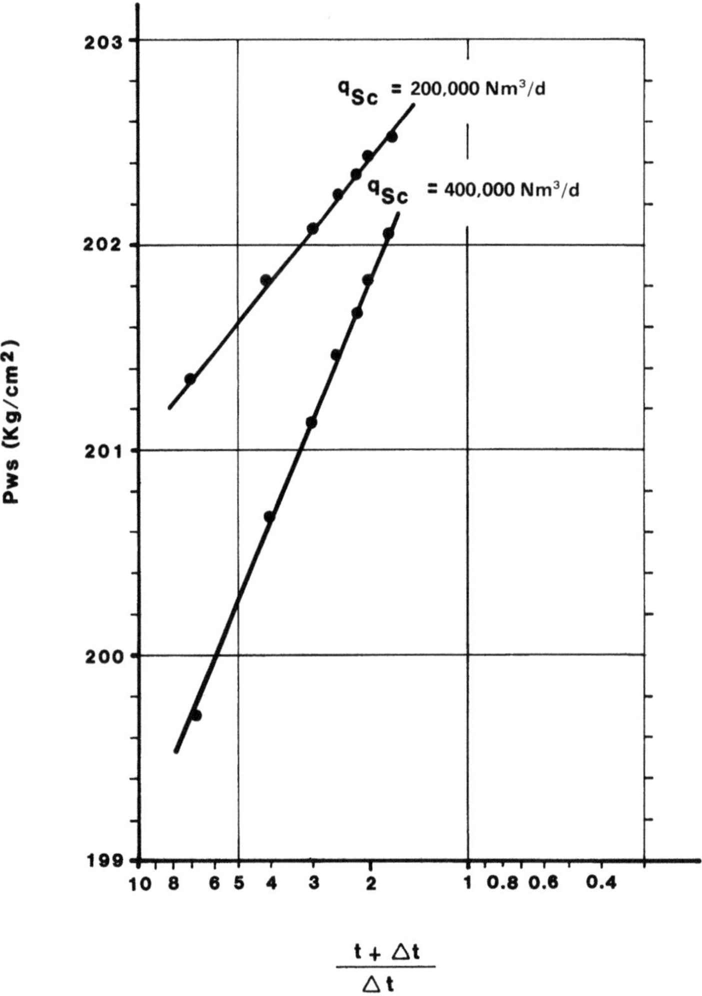

Fig. E7/4.1

Expressing all parameters in SI units (Table 7.2), we have

$$k = 128.81 \frac{200\,000}{86\,400} \frac{344.3}{14 \times 4.15213 \times 10^{17}} = 1.76605 \times 10^{-14}\,\text{m}^2 = 17.9\,\text{md}$$

$$k = 128.81 \frac{400\,000}{86\,400} \frac{344.3}{14 \times 8.2396 \times 10^{17}} = 1.77991 \times 10^{-14}\,\text{m}^2 = 18.0\,\text{md}$$

There is close agreement between the two interpretations.

Because of the direct proportionality that exists between $m(p_{ws})$ and p_{ws}, we can work in kg/cm² for b, $m(p_{ws})$ (1 h) and $m(p_{wf})$.

Referring to the equations in Table 7.2, we have:

For the test at $q_{sc} = 200\,000\ Nm^3/day$

$$S + 2 \times 10^5 D = 1.151 \left[\frac{201.83 - 180.53}{1.94753} - \log \frac{17.9}{0.2 \times 1.8 \times 10^{-2} \times 5.18 \times 10^{-3} \times 0.1222^2} + 3.107 \right]$$

$$= 7.1770\,,$$

where

$$p_{wf} = 180.53\ \text{kg/cm}^2$$

$$p_{1\,h} = 201.83\ \text{kg/cm}^2$$

For the test at $q_{sc} = 400\,000$ Nm^3/day

$$S + 4 \times 10^5 D = 1.151 \left[\frac{200.67 - 136.15}{3.86473} - \log \frac{18.0}{0.2 \times 1.8 \times 10^{-2} \times 5.18 \times 10^{-3} \times 0.1222^2} + 3.107 \right]$$

$$= 13.8013$$

where

$$p_{wf} = 136.15 \text{ kg/cm}^2$$

$$p_{1h} = 200.67 \text{ kg/cm}^2$$

We have two equations with two unknowns

$$\begin{cases} S + 2 \times 10^5 D = 7.1770 \\ S + 4 \times 10^5 D = 13.8013 \end{cases}$$

from which we can derive

$$S = 0.55$$

$$D = 3.31 \times 10^{-5} \text{ (Nm}^3\text{/day)}^{-1}$$

To summarise the results, we have

- average permeability in the drainage area of the well: $k = 18$ md
- skin factor: $S = 0.55$
- non-Darcy flow coefficient: $D = 3.31 \times 10^{-5}$ (Nm^3/day)$^{-1}$

Note that we could make a reasonable estimate of the deliverability equation for the well

$$m(\bar{p}) - m(p_{wf}) = A q_{sc} + B q_{sc}^2 , \tag{7.43}$$

using the transient results and other data in the theoretical pseudo-steady state flow Eq. (7.26), which, in SI units is

$$m(\bar{p}) - m(p_{wf}) = 55.955 \frac{q_{sc} T}{kh} \left(\ln \frac{A}{C_A r_w^2} + 0.809 + 2S + 2D q_{sc} \right) . \tag{7.26}$$

In order to use Eq. (7.26), we need to input the drainage area size and geometry, which can in fact be easily deduced from the well spacing and pattern.

Suppose the drainage area is a square with sides of 800 m. Then

$$A = 6.4 \times 10^5 \text{ m}^2 ,$$

$$C_A = 30.88 \quad \text{(Fig. 5.7)} .$$

In the present case, therefore, we will have the following equation:

$$[m(\bar{p}) - m(p_{wf})] \left(\frac{\text{Pa}}{\text{s}} \right) = [\bar{p} - p_{wf}] \left(\frac{\text{kg}}{\text{cm}^2} \right) \times 2.132 \times 10^7$$

$$= 55.955 \frac{q_{sc}(\text{Nm}^3/\text{day})}{86\,400} \frac{344.3}{18(\text{md}) \times 9.869233 \times 10^{-16} \times 14(\text{m})}$$

$$\times \left(\ln \frac{6.4 \times 10^5}{30.88 \times 0.1222^2} + 0.809 + 2 \times 0.55 + 2 \times 3.31 \times 10^{-5} q_{sc}(\text{Nm}^3/\text{day}) \right),$$

which reduces to

$$[\bar{p} - p_{wf}] \text{ (kg/cm}^2) = 4.205 \times 10^{-6} q_{sc} \text{ (Nm}^3\text{/day) } [16.052 + 6.62 \times 10^{-5} q_{sc} \text{ (Nm}^3\text{/day)}]$$

or:

$$\bar{p} - p_{wf} = 6.750 \times 10^{-5} q_{sc} + 2.784 \times 10^{-10} q_{sc}^2 \tag{7/4.1}$$

This is valid in the range 136 kg/cm^2 < p < 203 kg/cm^2 where we found there was a linear relationship between $m(p)$ and p.

By means of Eq. (7/4.1) we can calculate the flowing pressures for the two flow rates, assuming pseudo-steady state flow and $\bar{p} = 203$ kg/cm^2

$p_{wf} = 178.36$ kg/cm^2 for $q_{sc} = 200\,000$ Nm^3/day

$p_{wf} = 131.46$ kg/cm^2 for $q_{sc} = 400\,000$ Nm^3/day

These pressures are consistent with the measured transient data.

Calculating the deliverability equation from S and D, as we have done here, has the advantage over the flow-after-flow test method (described in Sect. 7.8.2) that it can be applied at any time subsequent to a flowing test, and therefore for any value of $m(p)$.

On the other hand, the estimates of S and D obtained from buildup data are highly sensitive to pressure measurement errors: for this reason, gauges of high resolution and accuracy are essential for this kind of test.

8 Downhole Measurements for the Monitoring of Reservoir Behaviour*

8.1 Introduction

The well testing techniques described in Chaps. 6 and 7 provide very useful information about a number of aspects of the reservoir: the overall productive behaviour of the hydrocarbon-bearing intervals open to flow in each well; near-wellbore conditions (skin); and, by means of interference or pulse tests, the *average* permeability and porosity–compressibility product between wells.

However, with the exception of those cases where it is possible to perform vertical interference tests between layers in the same well (Sect. 6.12), conventional well tests are unable to provide detail regarding the vertical distribution of permeability over the producing interval.

Well and reservoir behaviour is nevertheless strongly affected by formation heterogeneity, and, as was explained in some depth in Chap. 3, a hydrocarbon-bearing zone may well consist of several adjacent sedimentary units (layers) with widely varying petrophysical properties.

Some typical reservoir problems confronting the engineer are:

- by how much has the oil/water (or gas/oil) contact advanced at a given time, as a result of hydrocarbon production?
- has the water (or gas) front advanced uniformly (remaining essentially horizontal), or has it advanced preferentially along layers of high permeability? In the latter case, where are the oil/water (or gas/liquid) contacts in different parts of the reservoir?

As for individual well behaviour, the obvious question to ask when there is an increase in produced water/oil ratio (WOR) or gas/oil ratio (GOR) is – "where exactly is the water or free gas (i.e. gas not in solution in the oil) entering the wellbore?"

When there is bottom water drive, we might expect the water to be entering at the bottom-most perforations; but it is quite possible for water to be produced through perforations higher up the well, if they happen to be in a high permeability stratum which offers a preferential path for the water, leading to early break-through.

In the latter case, there is always the risk that some oil will be bypassed by the water because of its non-uniform advance, and will not be recovered.

Water can also flow vertically *behind* the casing (in a cement channel, for instance), entering the wellbore via perforations which are perhaps a considerable distance from the true source of the water in reservoir terms.

---

* The author wishes to acknowledge the valuable contribution of additional material made by Peter Westaway during the translation of this chapter

The workover procedure needed to eliminate unwanted water entry into the well and to maintain the production of dry oil, depends on the cause of the problem, but in all cases it is essential to locate the point or points of water entry.

These comments can also be applied to gas reservoirs under water drive, and, with appropriate modifications, to the entry of free gas from a gas cap.

A number of methods are available to investigate the downhole condition of a producing well, and to locate the oil/water or gas/liquid contacts:

A. *In open hole*: Selective formation testing (RFT = Repeat Formation Tester).
B. *In cased hole*: B1 – production logs, for the identification of the fluids present and the profiling of their relative flow rates over the producing interval (sometimes limited to qualitative or semi-quantitative results).
B2 – nuclear logs, for the identification of fluid contacts (close to the wellbore) and estimation of hydrocarbon saturation.

The following sections outline the applications of these methods and the information they can provide the reservoir engineer.

Please refer to the list of references at the end of the chapter if you wish to study these topics in greater depth.

8.2 Selective Formation Testing (RFT)

The downhole device used for selective formation testing is commonly referred to generically as the RFT (this is actually the name employed by one of the service companies for its own equipment[7] – the Repeat Formation Tester). The basic design is presented schematically in Fig. 8.1. Its principal application is in open hole, although cased hole versions (with a built-in perforator) are available. The RFT performs the following functions:

– measurement of the local pore pressure in the formation. In open hole, any number of these measurements can be made at different depths in a single run in the well. The cased hole version is, however, limited to a small number of "shots".
– retrieval of samples of formation fluid (typically a maximum of two per run).

The RFT is run in-hole on a multi-conductor cable. Depth control is assured firstly by measurement of cable length spooled into the well (as for any logging survey), with precise adjustment being made by means of a simultaneously recorded natural gamma-ray or spontaneous potential log, which can be correlated against other open hole logs.

The tool is stopped at the first measurement depth (specified by the operator). Triggered by the logging engineer, a back-up arm then opens out and presses against one side of the hole, pushing the rubber packer assembly (shown in Fig. 8.1) against the opposite side. At the same time, a hollow metal probe in the middle of the packer pushes forward, through the mud cake and into (or, at least, hard against) the rock face.

The pore fluid is now in communication, through the probe, with the measurement and sampling system within the tool. The packer seal ensures that wellbore fluid is isolated from the probe, while the mud cake, of course, provides the

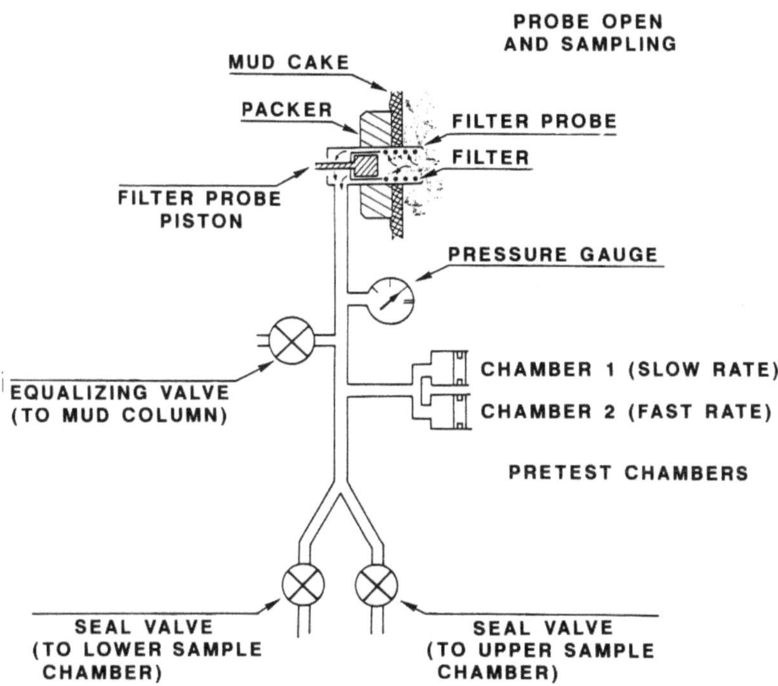

Fig. 8.1. *Schematic of the basic components of a repeat formation tester. From Ref. 7, reprinted courtesy of Schlumberger*

hydraulic seal over the rest of the hole wall. The probe contains a filter, and this is automatically cleaned after each stop by means of a small piston.

Formation fluid is drawn into each of two "pretest" chambers (*No. 1* and *No. 2* in Fig. 8.1) by means of two pistons which are retracted at different speeds. The second piston does not start moving until the first has stopped, thereby achieving what amounts to two successive drawdown tests, at different flow rates (in a ratio approximately 1 : 5). The actual volume of fluid withdrawn in each test is only about 10 cm^3. When the piston of the second pretest chamber reaches the end of its travel and the chamber is full, there will follow a period of pressure buildup.

The pressure in the flow stream is measured using a strain gauge (resolution ~ 1 kPa, accuracy ~ 100 kPa) or quartz gauge (resolution ~ 0.1 kPa, accuracy ~ 10 kPa). The pressure data is transmitted to surface via the cable, and recorded in real time in analogue (print) and digital (tape) form. Data from the two drawdowns and the buildup are then available for analysis.

At the end of the pretest, the engineer can either retract the backup arm, probe and packer and move on to another station depth, or keep the tool set in place and take a fluid sample.

For formation fluid sampling, the RFT can be fitted with two chambers, each of a volume between 5 and 50 l for normal applications. The actual chamber sizes selected will depend on the fluid recovery expected within a reasonable time.

Unfortunately, the samples are sometimes contaminated with mud filtrate (which is, of course, the first fluid to enter the probe unless the two pretests have been sufficient to deplete the local filtrate content of the pores). In addition, if there is an imperfect packer seal, the drilling mud itself may mix with the sample.

However, enough pore fluid may be retrieved to allow its nature (gas, oil or water) to be discerned, and some measurements to be made.

With a good sample recovery, it is possible to determine R_s (Sect. 2.3.2) of the reservoir oil.

But the most important application of the RFT is in establishing the pressure profile point-by-point over the reservoir interval, using the pretest buildup data (analysed in a manner similar to the conventional buildup tests described in Chap. 6).

In a virgin reservoir, or a developed reservoir which is in hydrostatic equilibrium, where the pressure gradients characterising the different fluids are clearly distinguishable, it is possible (Sect. 1.5) to determine the depths of the gas/oil and oil/water contacts from a plot of formation pressure versus depth, as shown in Fig. 8.2.

The pressure profile measured in a newly-drilled well, in a reservoir which has been on production for some time, can become significantly different from this. Intervals of uneven depletion will be observed, with pressures higher or lower than the mean reservoir pressure. These correspond to strata which have a productivity per unit thickness which is less than or greater than the average for the reservoir. One such profile, shown in Fig. 8.3, was measured in an in-fill well prior to completion.

It will be up to the reservoir engineer to determine the causes of these anomalies: strata in poor vertical communication with (or completely isolated from) the rest of

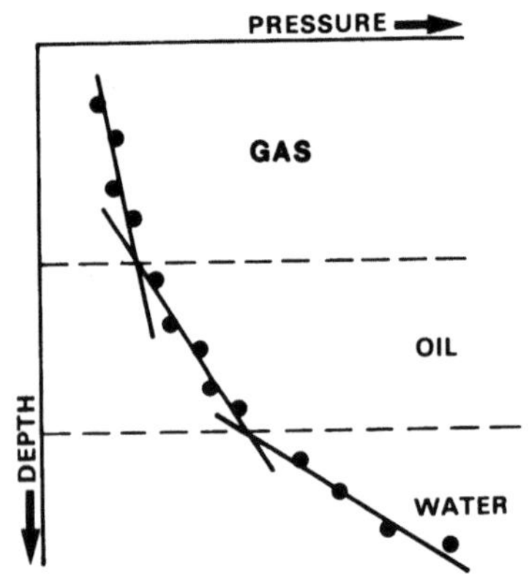

Fig. 8.2. *Pressure-depth profile in a virgin reservoir containing an oil section with gas cap and underlying aquifer*

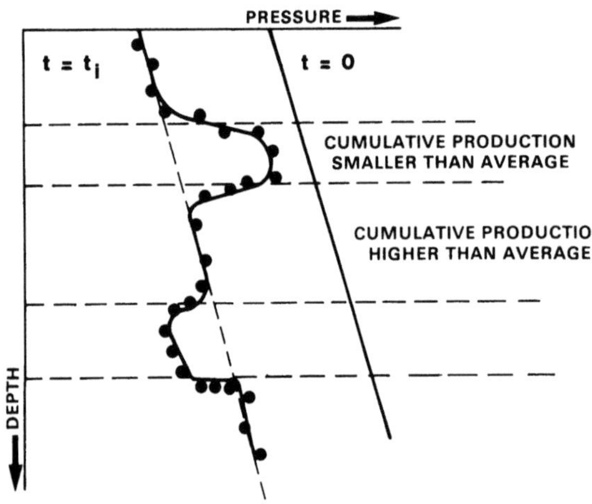

Fig. 8.3. *Pressure profile measured after drilling an in-fill well in a stratified reservoir exhibiting non-uniform depletion*

the reservoir, preferential support from the aquifer to certain strata, etc.), and to propose remedial action (workover, selective reperforation, etc.) to improve the productivity of the interval.

This pressure profile should, of course, also be matched by the field simulation model. Any discrepancies will indicate that modification to the reservoir description is required.

8.3 Production Logs

Production logging tools (generic PLT) are designed to measure the following parameters with respect to depth, in producing or injection wells:[4]

– local fluid velocity
– local fluid density, pressure gradient, or dielectric constant
– water fraction (holdup)
– temperature and pressure
– inside diameter of hole or casing

By analysing the profile of these parameters logged across the interval of interest, it is possible to determine the amount and nature of the fluid or fluids entering the wellbore at every depth. In this way, for example, it is possible to identify perforated sections which are not producing; where water or free gas entry is occurring; thief zones (taking fluid in from the wellbore – quite common in developed reservoirs under shut in well conditions, but also possible with the well on production); a leaking packer; a split liner; channelling of fluid behind casing through the cement sheath; etc.

There follows a brief review of the main PLT sensors, and a discussion of the interpretation of the PL log data. The PLT is usually run on a small diameter steel-armoured monoconductor cable. All sensor measurements are transmitted simultaneously up a single wire inside the cable. Power and control signals travel down the same wire.

The last few years have seen the increasing use of "slick line" or downhole memory PLTs. Here, the data is stored in an integral memory system which is downloaded after the tool has been brought back to surface, after which it can be presented and analysed in the normal way. The equipment is powered by a built-in battery pack, and since it requires no power from surface, nor complex electronics to transmit measurements to surface, it is very compact, and can be run on a thin metal slick line without the need for an internal conductor wire.

The PLT can be run in cased or open hole ("barefoot") completions. Open hole completions present certain operational problems when there is a large difference in diameter between the production tubing (through which the tool has to be run) and the open section.

8.3.1 The Measurement of Fluid Velocity in the Wellbore

Local wellbore fluid velocity is almost always measured by means of spinner flowmeters. The measurement is the *rotational frequency* of the spinner (rps, or revolutions per second), which is proportional to the fluid speed.

There are various designs to cope with different applications and operating conditions. They can be grouped into two broad categories:

a) Continuous Profiling

a.1: Fullbore Spinner Flowmeter (FBS) or Folding Flowmeter (Fig. 8.4A)
a.2: Continuous Flowmeter (CFS) (Fig. 8.4B).

The first category consists of large spinners which can be collapsed or folded in to a small diameter to pass through the tubing (Fig. 8.4A, left), and opened out to their full size to record in the casing (Fig. 8.4A, right). The second category consists of smaller, non-collapsible spinners. Both groups are in fact "continuous" in the sense that they can be logged continuously along the well, although the term tends to be attributed to the CFS type.

Because of frictional and viscous drag effects inherent in any spinner system, there is a lower limit to the fluid speeds that can be measured. This is typically ~ 5–10 ft/min, which, in a 7″ casing, for instance, corresponds to a flow rate of 250–500 bbls/day.

The log is recorded with the tool moving at a constant speed across the interval of interest. The added effect of the tool velocity must be removed algebraically to obtain the true fluid speed. This is commonly achieved by calibrating the flowmeter response in situ, which involves additionally logging up and down at different speeds,[5] each constant, over part or all of the interval.

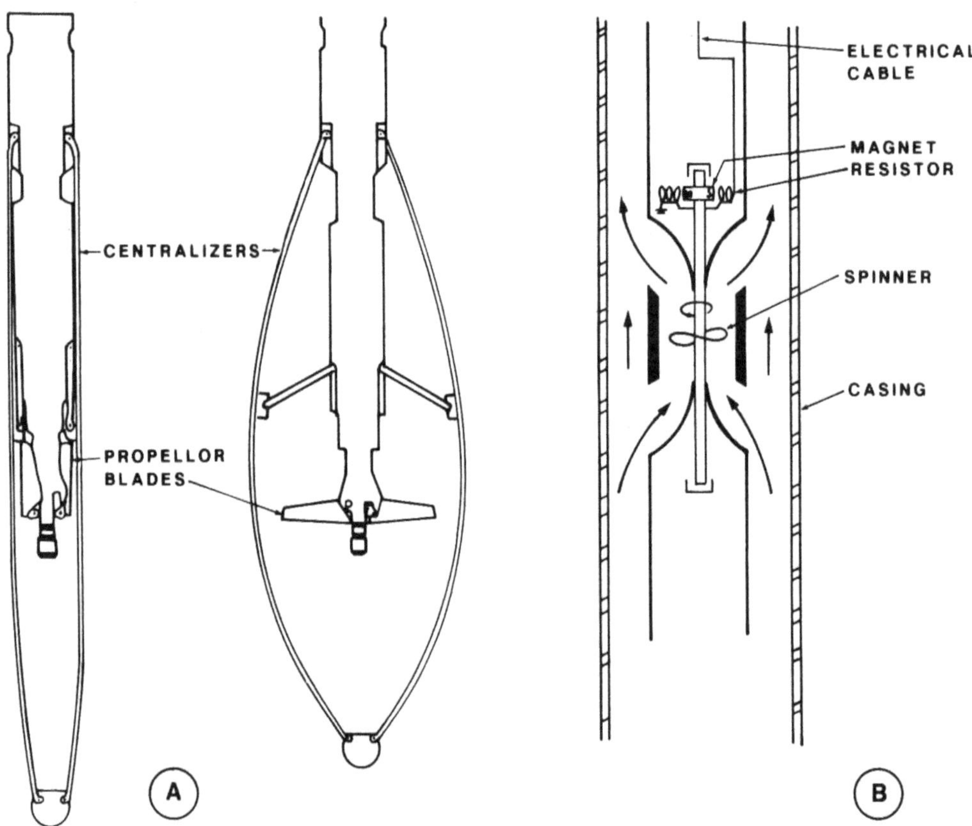

Fig. 8.4. A *Fullbore spinner flowmeter (FBS);* **B** *Continuous flowmeter (CFS)*

b) Stationary Measurements (Point-by-Point Profiling)

This type of flowmeter overcomes the low fluid velocity limitation of the fullbore and continuous tools by means of a flow converging element, which forces the fluid into the hollow body of the flowmeter. This represents a significant reduction in diameter, and therefore an increase of fluid speed, sufficient to rotate a spinner mounted in this internal flow stream. In this way, wellbore fluid velocities as low as a few ft/min (say 100 bbls/day in a 7″ casing) can be measured with reasonable precision.

Two types of converging flowmeter have been developed:

b.1: Petal Basket Flowmeter (PFS) or Umbrella Flowmeter, Fig. 8.5A.
b.2: Packer Flowmeter (PFM), Fig. 8.5B.

The petal basket type of flowmeter is equipped with a conical flow converger made out of overlapping metal petals supported on bow-spring arms, while the umbrella type uses a rubberised umbrella. The packer flowmeter has an inflatable rubber packer to achieve the same result. These converger systems can be opened and closed using power sent down the cable from the logging unit.

Not surprisingly, these instruments are not intended for continuous logging. Readings (rps) are taken instead with the tool stationary and the converger fully opened. As soon as a satisfactory measurement has been obtained, the basket or

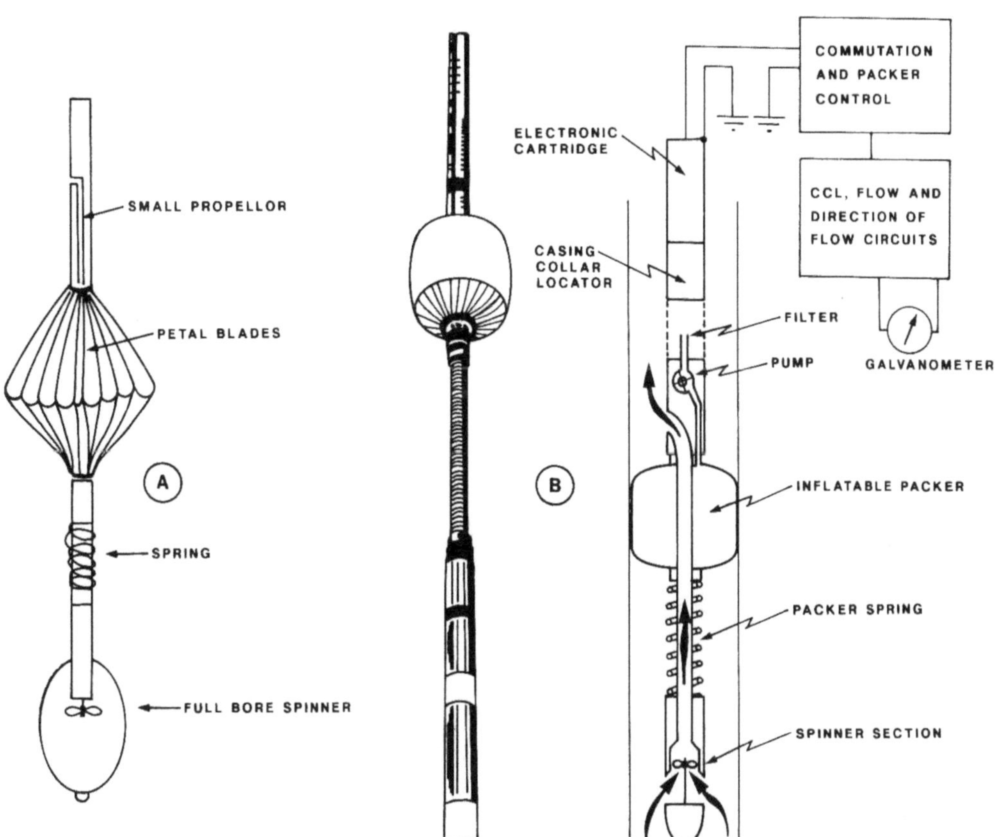

Fig. 8.5. A *Petal Basket Flowmeter (PFS))*; **B** *Packer Flowmeter (PFM)*

umbrella is collapsed, or the packer deflated. The tool can then be repositioned at
the next station and the converger redeployed.

The flow profile therefore consists of a series of point readings. At moderate
flow rates (e.g. ~ 3000 bbls/day in 7″ casing), fluid begins to leak around the edges
of the converger and the readings become progressively less accurate. At higher
rates, there is a risk that the whole tool will be pushed up-hole.

For very low fluid velocities, or very viscous fluids in which a spinner would
have difficulty in turning, a totally different measuring technique is employed.

The tracer ejector tool (TET) shown in Fig. 8.6 contains an ejector sub which is
triggered from surface to squirt out a measured dose of a radioactive isotope (the

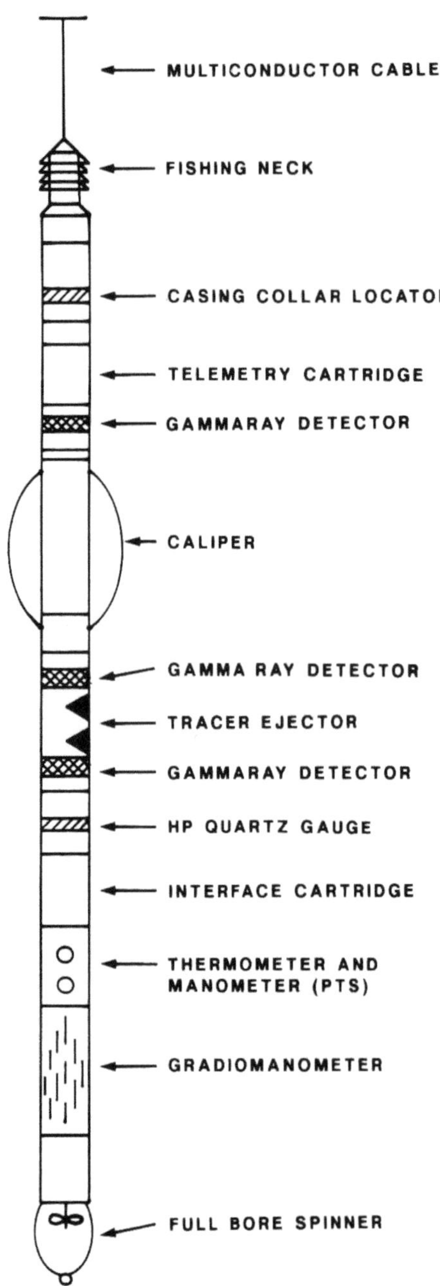

Fig. 8.6. *The tracer ejector tool (TET)*

tracer) which is miscible with the wellbore fluid. Two natural gamma ray scintillation detectors are located in the tool string some distance from the ejector. The passage of the small tracer slug, carried by the moving wellbore fluid, past each detector is registered as a momentary peak in each count-rate level, a short time apart.

With the tool stationary, if the time between the two gamma-ray peaks is Δt, and the distance between the detectors is L, then the fluid velocity past the tool is simply $L/\Delta t$. Conversion to volumetric flow rate can then be made by taking into account the inside diameter of the casing and the diameter of the tool housing.

Stationary readings are taken at a number of depths to build a point-by-point flow profile. This is referred to as a "velocity shot" survey.

The detectors must of course be located *below* the ejector sub if downflow is expected, or *above* for upflow. The TET is most commonly used in injection wells.

A second application of the TET is monitoring zonal injectivity in water injection wells. Here, the radioactive tracer travels with the injection water into the rock strata.

A number of continuous gamma ray surveys is then run at regular intervals across the zone of interest. The resulting log profiles show the distribution of the tracer with time, and will contrast zones of good injectivity (high gamma ray activity) from bad (low to no activity). This technique, called the "timed run" method, may also detect the movement of injected fluid behind the casing through a cement channel, sometimes towards a permeable zone that may be some distance from the perforated section.[5]

8.3.2 Measurement of Wellbore Fluid Density

As well as serving to identify the presence of the individual gas, oil or water phases, a measurement of the wellbore fluid density can be interpreted in terms of the percentages of the phases present in the event that the flow stream is a multi-phase mixture.[5]

Using the density measurement alone, the method is limited to two phases (e.g. oil and water), and the density of each pure phase must of course be known for the calculation to be possible. The presence of a third phase requires an additional measurement such as the fluid capacitance (see Sect. 8.3.3) or photoelectric absorption coefficient (see Sect. 8.3.4).

8.3.2.1   The Gradiomanometer (GM)

As we have already seen, the difference in pressure between two points a vertical distance Δz apart in a fluid of density ρ_f is:

$$\Delta p = \rho_f g \Delta z \ . \tag{8.1}$$

This is the principle of the gradiomanometer, a differential pressure sensor which measures the pressure gradient over 2 ft (61 cm) of the wellbore, by means of two metal bellows (Fig. 8.7).

Equation (8.1) is only valid in static fluid in a vertical well. In an inclined well, a correction must be made for the fact that Δz (vertical) is less than 2 ft, using the appropriate angle of deviation.

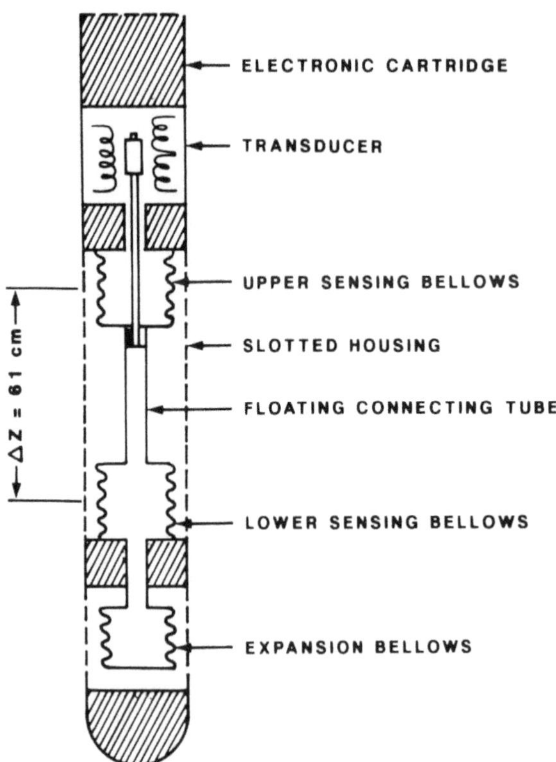

Fig. 8.7. *The gradiomanometer (GM)*

In a moving fluid, the raw log measurement includes an additional pressure drop caused by frictional drag along the casing wall and the tool housing. At very high fluid speeds, there may also be a further pressure drop from kinetic energy losses[5] (this is caused on a local scale by sudden changes in pipe diameter, such as when the gradiomanometer enters the tubing).

These must be subtracted from the log reading (in general, they are negligible for low to moderate fluid speeds – for example, less than ~ 1500 m³/day in 7″ casing); the hole deviation correction is then applied, and the resultant (hydrostatic) pressure gradient corresponds to the fluid density.

The gradiomanometer log is scaled directly as fluid density (g/cm³), although it may need correcting as described already before it can be used for interpretation. It has a measurement range of 0–1.60 g/cm³, with a resolution of 0.05 g/cm³ and an absolute accuracy of 3% if properly calibrated (g/cm³ is the conventional logging unit).

An amplified curve ($\times 5$) is also presented on the log to permit precise reading of small changes in density.

New generation gradiomanometers (PTS) utilise a strain-type (e.g. quartz diaphragm) differential pressure sensor, but otherwise operate on the same principles, and are subject to the same friction, kinetic and deviation effects (see Sect. 8.3.5).

Note that because the gradiomanometer depends on the hydrostatic pressure gradient to derive fluid density, *it cannot be used in a horizontal well.* In fact, the measurement is of little use at deviations higher than about 60°.

8.3.2.2   Nuclear Fluid Densimeter (NFD)

The nuclear fluid densimeter measurement is based on the fact that the gamma ray absorption coefficient of a medium is directly proportional to its density (see Sect. 8.3.4). Therefore, by measuring the absorption of gamma rays passing through the wellbore, the fluid density can be derived.

The tool consists of a small Cs^{137} 0.1 Curie (Ci) gamma ray source which emits radiation into the wellbore fluid. The count-rate registered at a Geiger–Muller or scintillation detector a short distance from the source is a measure of the degree of absorption suffered by the gamma radiation, and can therefore be related directly to the fluid density.

There are two basic tool designs:

- the "focused" configuration, in which the wellbore fluid passes through a small chamber between the source and detector, entering and exiting via ports at each end. The gamma radiation is focused along the axis of the tool so that most absorption occurs within this constantly changing fluid sample.
- the "non-focused" configuration, in which the gamma radiation is directed along the outside of the tool towards the detector 2 ft (61 cm) away, so that it is absorbed in the surrounding wellbore fluid.

The denser the fluid, the stronger the attenuation and the lower the count-rate at the detector.

The NFD can be used over a range $0.2-1.2 \ g/cm^3$, with a resolution of $0.01 \ g/cm^3$ and an absolute accuracy of $0.02 \ g/cm^3$.

The advantage of this measurement is that it is not susceptible to deviation, friction or kinetic effects. It can therefore be used in highly deviated and horizontal wells.

Disadvantages include the risks associated with a radioactive source in a producing well, and radioactive statistics which make the reading noisy and require slow logging speeds and filtering. The non-focused tool must obviously be very well centralised to avoid being affected by the steel casing; this is not a problem with the focused tool. There may also be some influence from the natural background radiation emitted from rock strata (e.g. shales), and from scale deposits on the casing wall, which tend to be highly gamma-active.

In addition, the focused NFD measures a small sample of fluid, which may or may not be truly representative of the average wellbore mixture outside the tool.

8.3.3   Measurement of the Fluid Capacitance

The fluid capacitance tool, or holdup meter (HUM), is configured as a capacitance cell, with the wellbore fluid as the "electrolyte" between its "plates". The measuring section consists of a hollow chamber between the tool housing and a central mandrel, whose surfaces form the two plates of what amounts to a cylindrical capacitor.

As with the NFD, fluid can flow through ports into and out of the chamber. An oscillator circuit continuously measures the capacitance of the cell, which varies with the dielectric properties of the wellbore fluid passing through the chamber. The oscillator frequency signal (Hz) is transmitted to surface and recorded against depth.

The dielectric constant ε_m of a fluid mixture is a function of the dielectric constants of the constituent fluids. For a producing well we have:

$$\varepsilon_w \approx 80 \text{ (depending on salinity)},$$

$$\varepsilon_o = 4,$$

$$\varepsilon_g = 2.$$

There is a strong contrast between water and hydrocarbons, but oil and gas are effectively indistinguishable. It is therefore possible to estimate the fraction of water present at any depth in a water/hydrocarbon (oil and/or gas) mixture by profiling its dielectric constant, knowing the values of ε for the pure constituents.

In practice, the raw frequency signal is used directly for interpretation, and it is common to establish the pure phase frequencies by calibrating the tool in water and oil (or gas) whenever possible (e.g. in the segregated fluid columns with the well shut in).

The capacitance measurement is not affected by hole deviation, friction or kinetic effects, and will function in highly deviated and horizontal wells. The relationship between frequency and water fraction is, however, not linear, and this should be considered as a semi-quantitative measurement.

Like the focused NFD, it measures a small sample of fluid, which may or may not be truly representative of the average wellbore mixture outside the tool.

8.3.4 Measurement of the Photoelectric Properties of the Wellbore Fluid

A recently developed PLT sensor, the X-ray fluid analyser (XFT), measures the X-ray absorption coefficient, which is a function of the relative proportions of oil, water and gas present locally in the wellbore fluid.

Photons in general (and X-rays in particular) interact with matter in three ways:

- *Compton scatter*: the photon collides with an orbiting electron in an atom, imparting some of its energy to the electron, which is ejected from its orbit. Repeated Compton scatter therefore results in the degradation and deflection of photon energy. It occurs with a high probability for radiation between about 0.4 and 4 MeV. The absorption coefficient is proportional to the density of the target medium.
- *Photo-electric effect*: the photon collides with an orbiting electron, but this time is completely absorbed. The photon energy (E) triggers the ejection of an electron from its orbit. The photoelectric (PE) interaction therefore results in the annihilation of photon energy. The absorption coefficient is high for low energy photons (<0.1 MeV), and increases with the atomic number Z of the target atoms, according to the rule

$$\text{PE} = \frac{Z^{3.6}}{E^3}. \tag{8.2}$$

- *Pair production*: this is an interaction with the atomic nucleus, and results in the annihilation of the incident photon energy, and the production of an electron–positron pair, each with an energy of 0.51 MeV. The photon energy must be at least 1.02 MeV for this to occur, and interaction probability becomes appreciable above about 10 MeV.

The density ρ of the medium is derived from the measurement of its coefficient of absorption for Compton scatter, by counting gamma rays in the appropriate energy range (see Sect. 8.3.2.2).

Additionally, the mean atomic number Z of the medium is derived from the absorption coefficient for the photoelectric effect, obtained by measuring low energy gamma rays (X-rays). From this, an estimate of the composition of the medium can be made.

In the case of wellbore fluids in a producing well we have:

- Water: NaCl-saturated $\rho_w = 1.190 \ \text{g/cm}^3$ PE = 1.32

               fresh $\rho_w = 1.000 \ \text{g/cm}^3$ PE = 0.37

- Oil (stock tank) $\rho_o = 0.800\text{–}0.900 \ \text{g/cm}^3$ PE = 0.129

- Methane (at 35 MPa) $\rho_g = 0.240 \ \text{g/cm}^3$ PE = 0.085

If the measured PE of an oil/water mixture containing a fraction f_w of water is $(\text{PE})_m$, then

$$\rho_m(\text{PE})_m = \rho_w(\text{PE})_w f_w + \rho_o(\text{PE})_o (1 - f_w) \tag{8.3a}$$

so that

$$f_w = \frac{\rho_m(\text{PE})_m - \rho_o(\text{PE})_o}{\rho_w(\text{PE})_w - \rho_o(\text{PE})_o} . \tag{8.3b}$$

The local fraction of water present in the wellbore fluid mixture can therefore be calculated from the simultaneous measurement of $(\text{PE})_m$ and ρ_m, using a value of $(\text{PE})_w$ derived from the water salinity.

The X-ray Fluid Analyser Tool (XFT) is shown schematically in Fig. 8.8.

The X-rays, with energies below 100 keV, are produced by bombarding a tungsten target with high energy electrons (Brehmstrahlung effect).

A metal spacer prevents any X-radiation from travelling directly to the short spacing (SS) and long spacing (LS) detectors; the X-rays are therefore constrained to pass through the wellbore fluid.

The absorption coefficient for Compton scatter is determined from the ratio of the count rates (cps) of scattered X-ray photons recorded at the two detectors. The average fluid density ρ_f is then derived from this.

The photons arriving at the LS detector are also analysed in terms of their incident energy, by means of two energy "windows": LS_1 for the low end of the spectrum, and LS_2 for the high end.

The photoelectric absorption coefficient $(\text{PE})_m$ is determined from the ratio of cps in LS_1 to cps in LS_2.

The two measurements ρ_m and $(\text{PE})_m$ are recorded on surface against depth, and can be interpreted as a continuous profile of the fraction of water in the wellbore fluid.

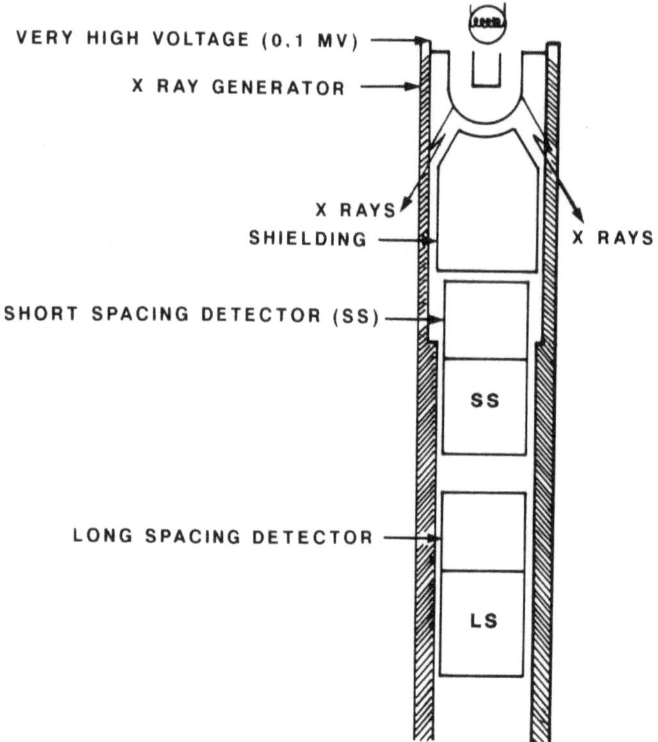

Fig. 8.8. *The X-ray Fluid Analyser Tool (XFT)*

8.3.5 The Measurement of Temperature and Pressure

Pressure and temperature are usually measured simultaneously, often with a single tool. One example is the pressure temperature sonde (PTS), which records the following data:

- pressure
- pressure gradient derived from the difference in pressure over a distance of 2 ft (61 cm) along the tool
- temperature
- temperature gradient calculated by differentiating over a 4 ft (122 cm) interval

In the PTS, the pressure sensor is the strain gauge type (accuracy 0.1% of full scale, although other systems may use a quartz gauge (accuracy 0.025% of full scale), or capacitance gauge (accuracy 0.05% of full scale).

The differential pressure sensor used in the PTS for the pressure gradient is a quartz diaphragm. This measurement is peculiar to the PTS and one or two other tools, and is not a feature of the majority of pressure gauges.

Because it can record the pressure gradient, the PTS is able to function as a gradiomanometer for fluid density. It is more compact and robust than the bellows type tool (GM) introduced in Sect. 8.3.2.1, and has no moving parts.

The measurement range is 0–2 g/cm^3, with an absolute accuracy of 2% of full scale ($\sim$0.04 g/cm^3).

The thermometer is a DIN-rated platinum wire sensor, with a rapid response, an accuracy of 0.5% of full scale ($\pm 1\,°C$), and a measuring range of $-25\,°C$ to $+190\,°C$.

By mounting two such sensors on either side of the measuring point, it is possible to measure the differential temperature directly (rather than by mathematical differentiation of a single profile), with an accuracy of $\pm 0.1\,°C$.

8.3.6 Auxiliary Measurements

In addition to the electronic cartridge necessary to handle the interrogation of the various sensors, and transmission of the data to surface, certain other sensors are commonly run as part of the PLT string:

a) *Casing collar locator (CCL)*. The depths of the joints (collars) between each section of casing are measured quite precisely when the casing is run into the hole during completion of the well. During a production logging survey, these collars are detected by means of the CCL – a simple coil and magnet system which responds to the variation in magnetic flux caused by the changing metal thickness at each joint.

On the log, the collar signals are recorded as spikes on an otherwise straight (if slightly noisy) base line. By matching the collar signals to the casing record, the depth of the log can be precisely adjusted. In practice, production logs are usually depth matched to a reference CCL log which was run shortly after completion and has already been adjusted to the casing record.

b) *Gamma ray log (GR)*. The gamma ray tool is a standard component of the PLT string, and is an alternative, and, in many circumstances, a preferable means of depth control. The GR log, introduced in Sect. 3.2b, is a record of the contrasting natural gamma ray emission from different lithologies, notably shales (usually high activity) and clean strata (lower activity). Since this gamma ray emission can travel through casing, a GR log can be recorded with each PLT pass, and can be correlated against a reference open-hole GR profile run before completion.

c) *Caliper log*. In order to convert a measurement of fluid velocity to a volumetric flow rate, we need to have a profile of the hole diameter. This is measured by means of a caliper tool consisting of three or more arms which open under spring pressure against the hole wall. Their radial extension is recorded and transmitted to surface. One such tool is the Through-Tubing Caliper (TCS), which consists of three bow-spring arms and measures a single average diameter. More versatile caliper tools exist with two independent pairs of arms, measuring two perpendicular diameters (very useful in oval or irregular holes).

The caliper tool is an essential part of the PLT string in open hole completions, where diameter may vary considerably. In cased hole, it is often omitted (and the nominal casing ID assumed for interpretation purposes), unless significant variations in diameter are expected (e.g. in the event of scale deposits or damaged casing).

In the case of the slick line tools using downhole memory storage, because of the limited data capacity, only the gamma ray is usually run.

8.3.7 The Recording of Production Logs

All the sensors of a production logging tool are recorded simultaneously on each logging pass. A full PL survey therefore requires only a single run in the hole. The PLT string will consist of a combination of some or all of the sensors described previously, depending on the objectives of the survey, and operating constraints.

Figure 8.9 is a schematic of two possible sensor combinations that might be run in a high (A) and low (B) flow rate well. (B) is now an obsolescent tool configuration, and these days it would be more usual to include a petal basket flowmeter in tool string (A) for low rates.

The slick line PLTs are usually run with the following sensor combination: flowmeter, pressure, temperature and gamma ray, although other sensors can in principle be included (gradiomanometer, caliper) at the expense of increased tool length and memory requirement.

Figure 8.10 is an example[5] of a PLT log run in an oil well, with an open hole section below casing. This is a fairly high rate well, producing with a large water cut.

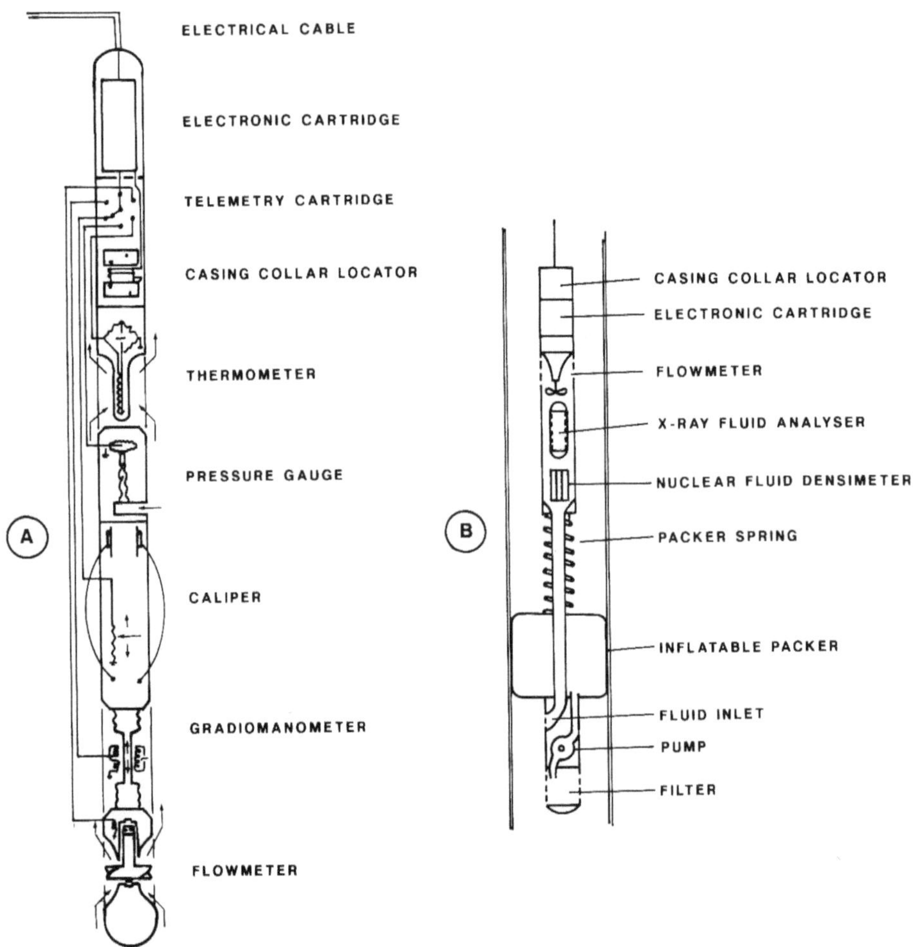

Fig. 8.9A, B. *Schematics of production logging tool combinations for logging high* **A** *and low* **B** *flow rate wells*

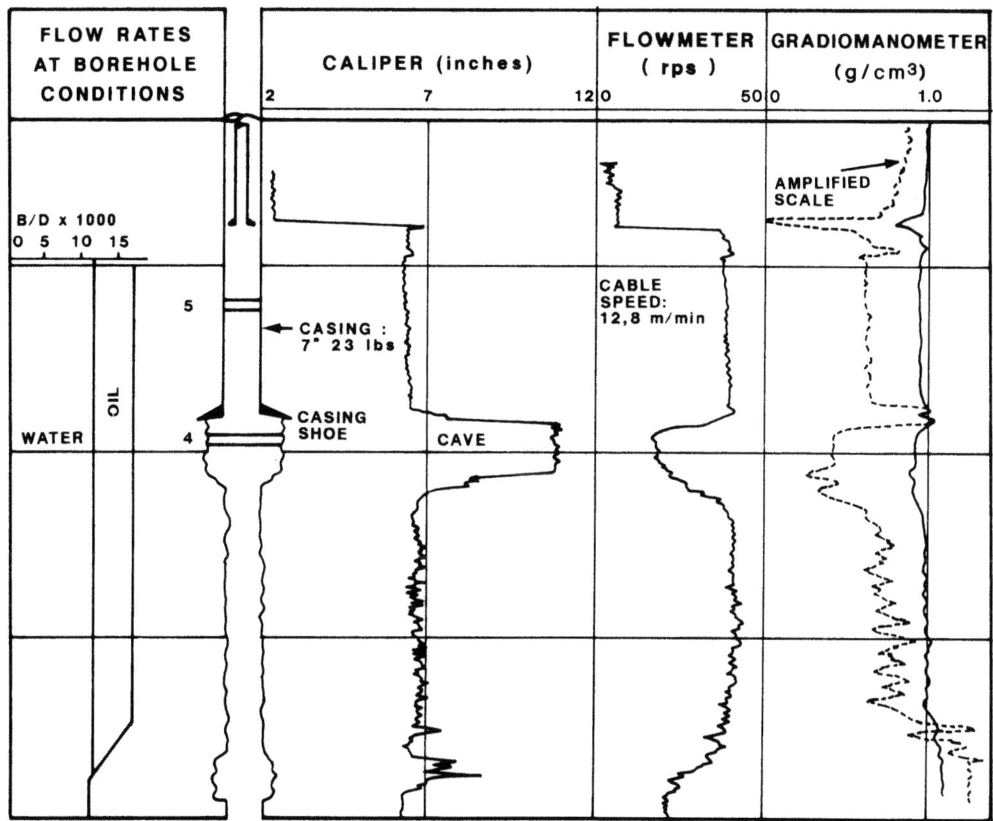

Fig. 8.10. *Example of a suite of production logging measurements (flowmeter, gradiomanometer, caliper) in an oil well producing from an open section below casing. From Ref. 5, reprinted courtesy of Schlumberger*

8.4 Criteria for the Interpretation of Production Logs

For diagnostic purposes, production logs are run to meet the following three objectives:

a) To evaluate the efficiency of a well completion (perforations, packers, valves, etc.), or of subsequent treatment (workover, stimulation, etc.)
b) To check on the frontal advance of reservoir fluids
c) To investigate anomalous production behaviour (sudden or gradual decline in productivity, the onset of water or free gas production, etc.).

The range of possible situations that could be encountered is so vast that it would be impossible to cover all of them in a single chapter, and they would be worthy of a book devoted entirely to the topic. Space permits only a few of the more common situations to be included, and in the following subsections we will see how the PLT is used to diagnose the causes of anomalous well behaviour and to suggest possible remedial action.

8.4.1   Production Log Diagnosis in Oil Wells

As soon as the well has been completed and put on production, it is advisable to run a PLT survey: flowmeter (FBS, CFS or PFM, depending on the flow rate), pressure, temperature, fluid density (GM, PTS or NFD) and/or fluid analyser (HUM, XFT).

If the oil is above its bubble point in the reservoir, the interpretation of the flowmeter profile (Fig. 8.11) is straightforward, and the productivity of the layers open to flow can be compared with what is expected from their permeability and thickness. This will reveal, for instance, if certain layers are underproducing because the perforations have not cleaned up effectively, or there are not enough shots per metre, or because the near-wellbore zone is heavily damaged by mud filtrate.

If the well is producing water with the oil, the point or points of water entry can be distinguished by the local increase in wellbore fluid density,[6] or by means of the HUM or XFT.

If the produced $\mathrm{GOR} \gg R_s$, the entry of free gas will be characterised by a significant increase in fluid velocity accompanied by a reduction in density (Fig. 8.12).

Because of buoyancy forces resulting from their different densities[6] $(\rho_w > \rho_o > \rho_g)$, the water, oil and gas phases move upwards with different velocities. Consequently, the percentages of free gas and water (their "holdups") calculated from the local fluid density or XFT cannot be interpreted directly in terms of *flow rates*, and serve only as qualitative indications of their presence.

Quantitative interpretation requires knowledge of the different phase velocities. Since these cannot be measured individually, recourse is taken to empirical correlations to obtain an estimate of the "slip velocity" (i.e. the relative velocity) of one phase (the lighter) in another (the heavier). This approach will often provide

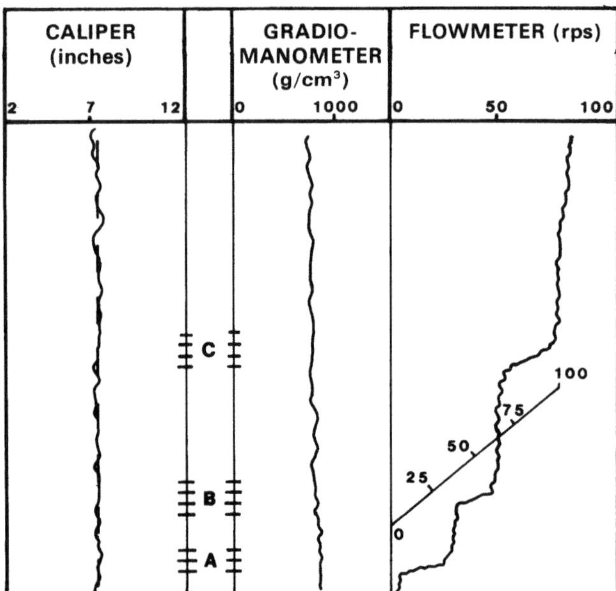

Fig. 8.11. *Determination of the percentage productivity from each perforated interval in a well producing from an undersaturated oil reservoir*

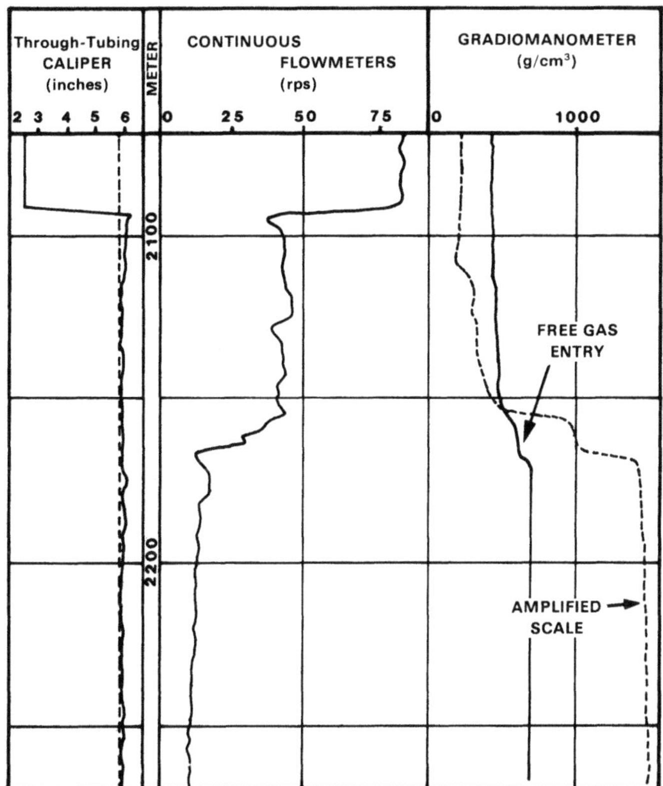

Fig. 8.12. *Localising the point of entry of free gas into an oil well. From Ref. 5, reprinted courtesy of Schlumberger*

satisfactory results where only two phases are involved; three phases require more specialised treatment.

The well should be surveyed regularly by PLT during its producing life, to monitor any changes in layer productivity. In addition, any significant variations in surface production (fluctuations in flow rate, changes in stock tank oil gravity, appearance of water or free gas, etc.) should be investigated as soon as they occur.

A PLT survey might identify, for example, a split casing allowing water influx from a zone above the reservoir interval, or, worse, allowing leakage of oil upwards through permeable formations, with catastrophic consequences should it reach the surface in an inhabited area.

8.4.2 Production Log Diagnosis in Gas Wells

In gas wells, production logs will be used to determine the layer productivities, perforation efficiency, etc., and to localise any points of water entry.

Interpretation is based on the combination of flowmeter and fluid density, in the same way as for oil wells. Since $\rho_w \gg \rho_g$, the sensitivity to the entry of water is enhanced in gas wells.

In wells producing from gas condensate reservoirs, a problem sometimes occurs when the reservoir pressure falls below the dew point. If the wellbore velocity of the gas stream is too low, it will not be able to transport its load of liquid condensate droplets to surface. The consequent liquid fallback results in the development of a sump of oil at the bottom of the well. In high productivity wells, where the pressure drawdown ($p_{ws} - p_{wf}$) is small, the sump may build up to such an extent that it impedes production from the lowermost intervals.

The presence of this sump can be clearly distinguished on the gradiomanometer log; in many cases it is sufficient to put the well on production at a high rate for a short time to displace the liquid.

Gas entries can almost always be identified from the temperature log (PTS).

Under isoenthalpic expansion, most gases become cooler: this is the well-known Joule–Thomson effect, described by

$$\mu = \left(\frac{\partial T}{\partial p}\right)_H = \frac{RT^2}{C_p p}\left(\frac{\partial z}{\partial T}\right)_p, \tag{8.4}$$

where μ is the Joule–Thomson coefficient, H is the specific enthalpy, C_p is the specific heat of the gas at constant pressure, and the other coefficients have the usual definitions.

For moderate pressures ($p_{pr} < 10$, Sect. 2.3.1.1), $(\partial z/\partial T)_p$ is positive, so that expansion will lead to cooling.

On the other hand, as has been noted by Vaghi et al.,[14] at very high pressures, such as those found in the Malossa field in Italy, $(\partial z/\partial T)_p$ is negative, so that *warming* will occur when the gas expands.

In either case, a temperature log run in a flowing gas well should show the gas entries[3] (Fig. 8.13), and will identify any open intervals which are not producing as expected.

Since wellbore heating or cooling is transmitted by conduction into the local formation, these effects may still be observed in a shut in well if the log is recorded soon after closure.

8.4.3 Production Log Diagnosis in Water Injection Wells

The principal concern in injection wells is how uniformly the injection water is distributed amongst the open intervals (the injectivity profile).

It is also of prime importance to ascertain whether water is channelling behind the casing as a result of poor cement placement, or damage. If injection pressure is increased (to obtain higher rates), the cement bond to the casing may be broken down or the cement damaged. If injection pressure exceeds the parting pressure of the formation, a fracture or fractures will be propagated out from the well.

The consequences of channelling of water behind casing may be serious: (1) a perhaps significant volume of injection water is not able to contribute to the pressure maintenance and frontal sweep for which it was intended; (2) injection water may reach shallow porous intervals, contaminating water-bearing strata which, now being overpressured, immediately erupt on surface.

The flowmeter and the temperature log are the most commonly used for injectivity profiling.

The spinner flowmeter usually gives excellent results, since this is a monophasic environment.

Because the injected water is cooler than the rock, the temperature log gives useful information, even with the well shut in after injection.

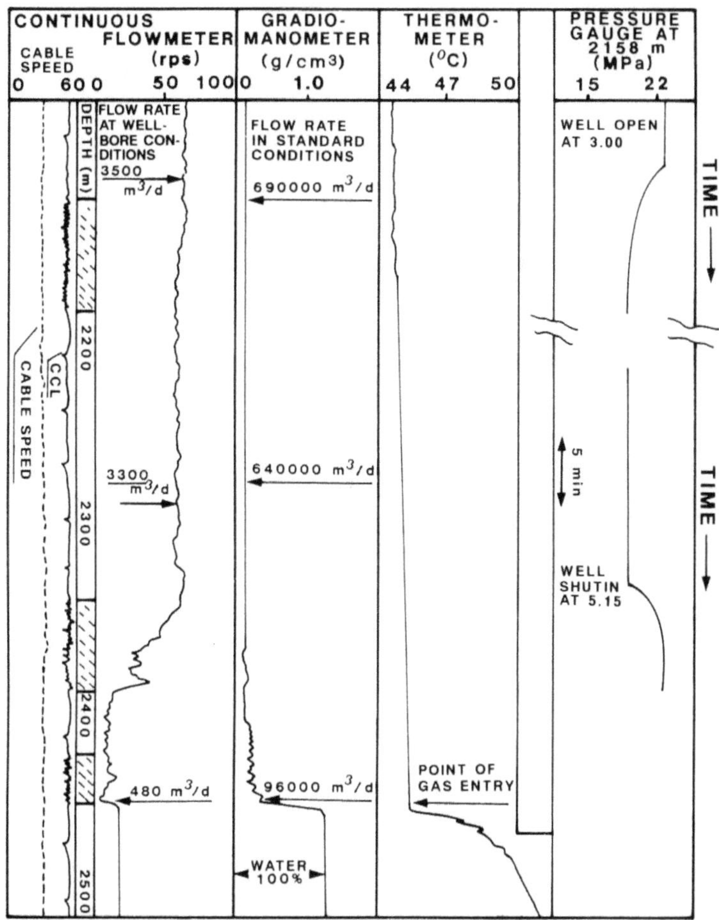

Fig. 8.13. *Production logs in a gas well. Note the cooling where the gas expands on entry into the wellbore*

The radioactive tracer ejector tool (TET) is often effective in locating cement channels behind casing. One of the most commonly used tracers is NaI, containing the isotope I^{131}, which emits gamma rays.

Following ejection of the tracer, gamma ray profiles are recorded at regular intervals across the section of interest (Fig. 8.14). In this way, the advance of the tracer can be monitored, highlighting zones taking water (c-l-m-q in Fig. 8.14), and, of course, any movement of water behind the casing (c-f-j-n-v, and l-p in Fig. 8.14).

Wherever channelling is detected, a cement squeeze (injection of cement into the problem section, through existing or newly shot perforations) will probably be required as remedial treatment.

8.5 Nuclear Logs

8.5.1 Overview

With the aid of production logs we can usually determine with a fair degree of precision where water, oil and free gas *enter the wellbore*. We can therefore identify

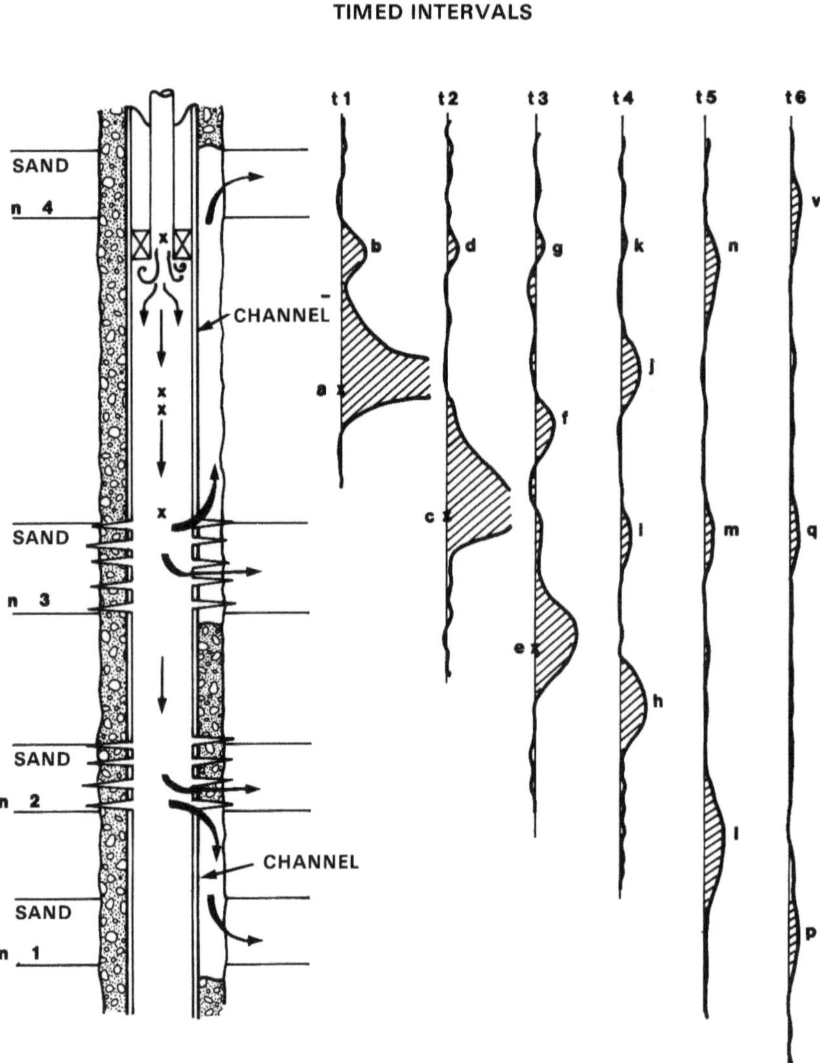

Fig. 8.14. *A succession of GR profiles run after injection of radioactive tracer in a poorly cemented interval – the "timed run method". From Ref. 5, reprinted courtesy of Schlumberger*

the highest *open interval* into which the gas cap has expanded, or the lowest *open interval* into which the aquifer, or injection front, has encroached.

It is good practice not to perforate ("shoot") an oil-bearing section over its entire thickness when gas/oil or oil/water contacts are present, but to keep a "safe" distance from them so as to avoid – for a certain time at least – gas or water production as a result of gas cap or aquifer expansion, or coning (to be dealt with in Chap. 12; Vol. 2).

The same can be said for gas reservoirs, where only the upper part of the interval will be perforated to avoid early influx of water.

However, until the eventual appearance of gas or water in the production stream, the reservoir engineer would be effectively "in the dark" about the regular (or anomalous) movement of the fluid contacts. Fortunately, logging techniques exist which are able to "see" into the near-wellbore rock, through the casing and cement, without the need to establish any sort of hydraulic communication: these are the *nuclear logs*.

Only the nuclear family of logs provides information about lithology, porosity, fluid type and saturation in both open hole (during the drilling phase), *and when the well has been completed with cemented casing.*

Electrical logs, so important in open hole, are shorted out by the steel casing (an infinite conductor) and can therefore provide no information about the formation. Acoustic logs are also adversely affected by the presence of casing and cement sheath. The elastic wave travelling through the casing is usually faster than in the formation and therefore arrives first at the receivers. A formation signal (albeit a much weaker one than in open hole) can be obtained by analysing the entire acoustic wave train, although this depends on there being a good cement bond to both casing and rock face so that acoustic energy can be transmitted. In cased hole, acoustic logs are in fact run expressly to examine the casing and cement quality.

In order to understand nuclear logging measurements, we will first review some of the fundamentals of radioactivity and the interaction of nuclear particles with matter.[11,12]

8.5.2 Some Fundamentals of Radioactivity

Natural radioactivity is a characteristic of certain number of elemental nuclei, which emit particles and/or radiation following spontaneous decay to a nucleus of lower atomic number.

Two types of particle are emitted:

- *α-particles* – helium nuclei (He_2^4), with an atomic number of 4 and an electrical charge of 2. They have a relatively weak penetrating power through matter, and are therefore of no use for well logging applications.
- *β-particles* – high velocity electrons or positrons (negatively or positively charged). Although they have a greater penetrating power than α-particles, they are still too weak for any logging application.

Radioactive emissions in the form of waves are referred to as γ-rays (or γ-photons). These are similar to visible light or X-rays, but are of higher energy. Energy is related to wavelength by Planck's law:

$$E = hv = h\frac{c}{\lambda},\tag{8.5}$$

where

E = energy of the radiation (J),
h = Planck's constant ($= 6.627 \times 10^{-34}$ J s),
c = velocity of light in a vacuum ($= 2.998 \times 10^8$ m/s),
v = frequency of radiation (s^{-1}),
λ = wavelength of radiation (m).

Particle and wave energies are conventionally measured in electron-volts (eV), where

1 eV = energy acquired by an electron when submitted to a potential gradient of 1 V/cm.

The eV is too small a unit for practical use in a logging context; the MeV is a more convenient multiple:

$$1 \text{ MeV} = 10^6 \text{ eV} = 1.6 \times 10^{-13} \text{ J} .$$

For example, applying Eq. (8.5) to a photon energy of 1 MeV, we have

$$\lambda = 1.242 \times 10^{-12} \text{ m} = 0.0124 \text{ Å} ,$$

$$\nu = 2.414 \times 10^{20} \text{ s}^{-1} .$$

Radioactive emission is a statistical phenomenon; in other words, if N is the number of particles or photons emitted from a certain quantity of radioactive matter, dN/dt is not rigorously constant, but varies statistically about a mean value.

Since emission is accompanied by a transformation of the source nucleus into a new element, which may or may not be radioactive itself, the decay process eventually exhausts itself after a certain time.

Radioactive decay obeys the law

$$\frac{1}{N} \frac{dN}{dt} = -\beta , \tag{8.6a}$$

so that

$$N = N_0 e^{-\beta t} , \tag{8.6b}$$

where:

N_0 = number of emitting nuclei at time $t = 0$,
N = number of emitting nuclei at time t,
β = decay constant (s^{-1}).

It is customary to characterise the longevity of a radioactive substance by its *half-life*. This is a measure of the rate at which it decays – to be more specific, the time required for the number of radioactive nuclei to decrease by one half. If we substitute $N/N_0 = 0.5$ in Eq. (8.6b), we get

$$\text{half-life} = \frac{0.693}{\beta} . \tag{8.6c}$$

The γ-ray sources most commonly used in logging tools contain capsules of Co^{60} ($E = 1.17$–1.33 MeV) or Cs^{137} ($E = 0.662$ MeV) at a strength of 2–4 Ci.

The *neutron* is a particle with a mass of 1 and an electrical charge of zero. The absence of charge gives the neutron an appreciable penetrating power, and several important cased hole logging techniques have been developed based on the interaction of neutrons with matter.

Atomic nuclei do not emit neutrons spontaneously. Instead, emission is induced by bombarding source nuclei with other particles or ions.

Continuous neutron emission for logging purposes can be achieved by means of a source containing beryllium intimately mixed with an α-emitter. The following reaction occurs:

$$\text{Be}_9^4 + \text{He}_2^4 \rightarrow \text{C}_6^{12} + \text{n}_0^1 ,$$

Ra_{88}^{226}, Pu_{94}^{239}, or Am_{95}^{241} are in common use as the α-emitters.

The Ra–Be source emits neutrons with energies in the range 1–13 MeV (peak value 4.5 MeV). It has a half-life of 1620 yr, and is usually supplied with an activity level of 300 mCi (1 Ci = 3.7×10^{10} emissions per second.)

It is, however, not a "clean" source, the neutron emission being heavily contaminated by γ-rays (about 10 000 γ-photons for every neutron), and it is rarely used.

The Pu–Be source emits neutrons at a mean energy of 4.5 MeV, with an activity of 5 Ci, almost free of photons. It has a half-life of 24 300 yr.

Am–Be is the most widely used combination. The neutron emission, at a mean energy of 4.35 MeV, is relatively strong (16 Ci) and free of photons. Its half-life is 458 yr.

Pulsed neutron emission: to generate fast neutrons, with energies as high as 14 MeV, requires a different approach, based on the reaction between deuterium and tritium:

$$H_1^2 + H_1^3 \rightarrow He_2^4 + n_0^1 \quad (14 \text{ MeV}) .$$

A linear accelerator operating at 125–200 kV is supplied with ionised deuterium. The high voltage source consists of a bank of capacitors charged in parallel, and discharged in series, or a radio-frequency quadrupole (RFQ).

The ions, now travelling at high velocity, are focused onto a titanium target containing adsorbed heavy water (H_2^3O). The deuterium and tritium interact, releasing high energy neutrons.

This sort of neutron emission is not continuous – the accelerator is pulsed repeatedly for, typically, 100 µs every 1000 µs. Various interactions between the neutrons and the surrounding medium are monitored between the pulses, by counting the resulting γ-ray emissions, or epithermal and thermal neutron populations. New generation pulsed neutron tools also look at the γ-ray emissions from very high energy neutron interactions which occur during the neutron burst itself.

8.5.3 The Interaction of Gamma Rays with Matter

The three interactions between photons and matter – Compton scatter, the photo-electric effect, and pair production – were described in detail in Sect. 8.3.4. Of these, only Compton scatter (in which a γ-photon collides with one of the outer electrons of an atom, and suffers a loss of energy and a change of direction), and the photoelectric effect (in which a γ-photon collides with one of the outer electrons and is annihilated), are used in nuclear logging.

The absorption coefficient for Compton scatter of γ-photons is directly proportional to the density of the absorbing medium.

So by measuring the absorption coefficient, the bulk density ρ_b can be logged across a formation interval.

Figure 8.15 is a schematic of the logging sonde used for this purpose[10] in open hole. The tool is run in to the bottom of the well, whereupon the pad section (shown shaded in Fig. 8.15) is pressed against the hole wall by means of a back-up arm which is opened from the surface. The pad contains the γ-ray source (1.5 Ci Cs[137]) and two scintillation detectors which count the returning scattered γ-photons. A heavy metal shield between the source and the detectors prevents direct communication.

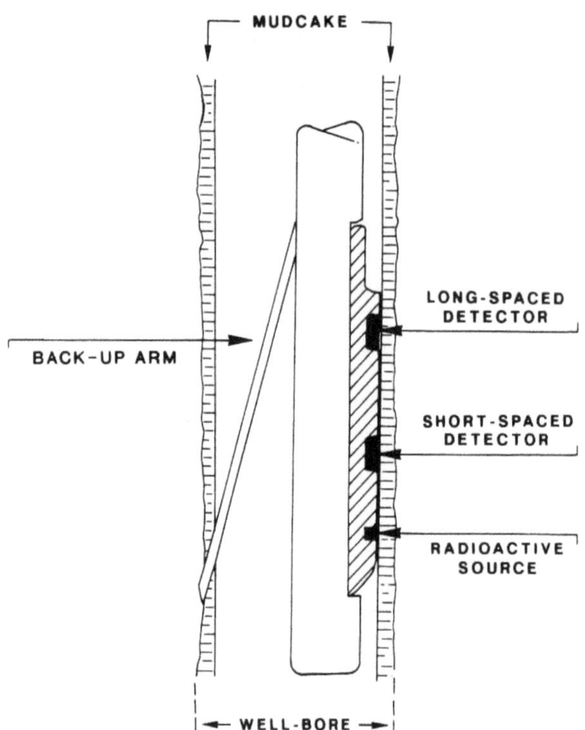

Fig. 8.15. *Schematic of the logging tool used to measure rock density (Formation Density Compensated, FDC). From Ref. 10, reprinted courtesy of Schlumberger*

The log is recorded by pulling the tool slowly up through the open hole section with the pad pressed against the wall.

There is a *decrease* in count-rates at the two detectors as the density of the rock opposite the pad increases, and vice versa. Although the basic measurement could be made with a single detector, the use of a second one facilitates correction of the reading for the presence of mud cake or mud-filled rugosity between the surface of the pad and the rock face. Hence, the name Formation Density *Compensated* (FDC).

The FDC is primarily an open hole tool, but useful readings can be obtained in cased hole provided the correction (for casing and cement sheath rather than mud cake) is not excessive.

Detector count-rates are transmitted to surface via the multi-conductor logging cable, and are functioned for log presentation as a density curve in g/cm^3 (Fig. 8.16). The magnitude of the mud cake correction that has been applied is also displayed.

The radius of investigation of the FDC reading is about 10 cm.

It is usual to record the natural γ-ray activity of the formation (*Gamma Ray Log, GR* – measured with a different detector further up the tool) and the hole diameter (*Caliper* – measured with the pad/backup arm assembly) along with the formation density (Fig. 8.16).

The GR log shows up shale strata, and can be used as an estimator of the shaliness of a formation, because shales generally exhibit a natural gamma ray level which is higher than most other sedimentary rocks.

The Litho-Density Tool (LDT) is an improved version of the FDC which also measures the photoelectric absorption coefficient (described in Sect. 8.3.4) by selectively counting the low energy gamma rays arriving at the long-spacing

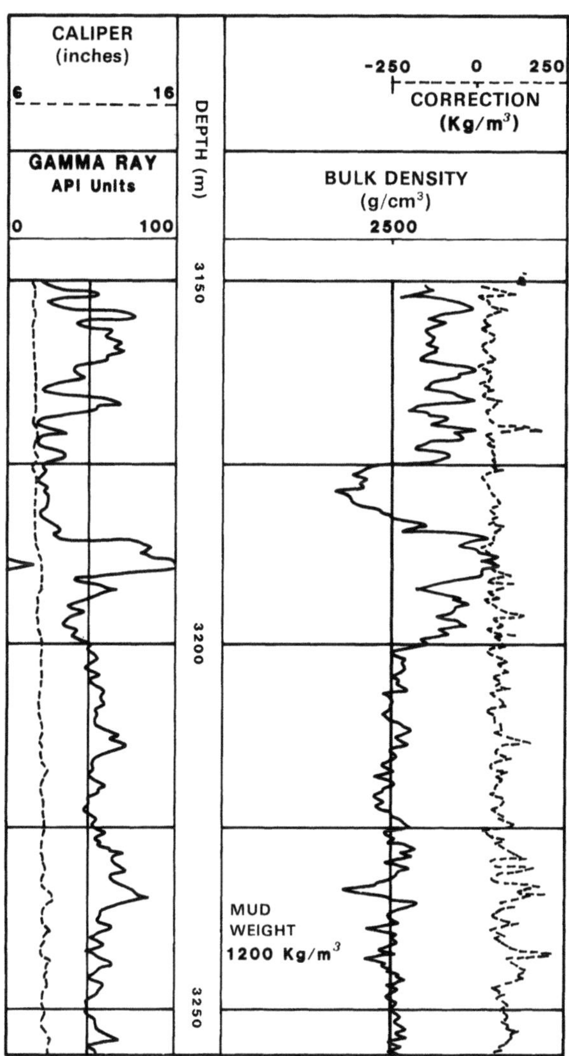

Fig. 8.16. *Example of a combination FDC–GR log. From Ref. 10, reprinted courtesy of Schlumberger*

detector.[11] This provides lithological information based on the aggregate atomic number of the rock, and is useful for mineralogical analysis when combined with the bulk density measurement.

8.5.4 The Interactions of Neutrons with Matter

The type of interaction that occurs when a neutron collides with an atom depends principally on its incident energy. Some result in the annihilation of the neutron, others in a slowing down process which eventually leaves the neutron at a minimum "thermal energy" (0.025 eV at 25 °C).

We will first review the possible interactions between neutrons and matter, starting from the initial energy of emission (14 MeV) from the accelerator in the logging tool, and working down the energy scale as shown in Fig. 8.17.

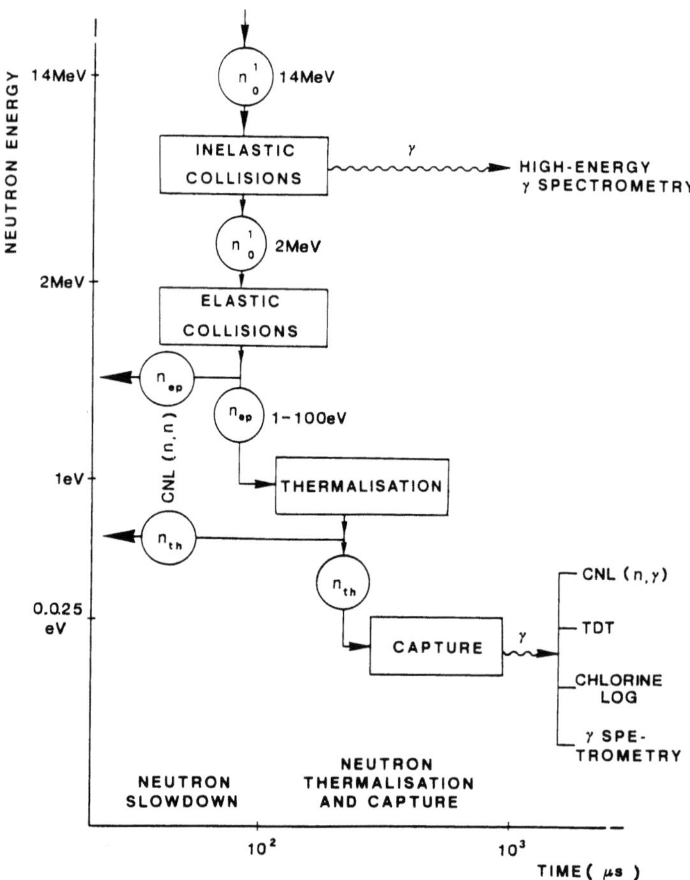

Fig. 8.17. *Interactions of neutrons with matter as a function of neutron energy. From Ref. 11, reprinted with permission of Elf-Aquitaine*

8.5.4.1   Inelastic Collisions

Inelastic interactions between neutrons and matter are predominant at neutron energies higher than about 2 MeV. The neutron imparts some or all of its energy to the target nucleus, which is momentarily excited to a higher energy level.

The neutron, if not annihilated, continues with a reduced energy and, perhaps, a change of direction.

The nucleus, meanwhile, returns to its ground state, emitting the excess energy in the form of a gamma photon. Each element can exhibit *several different, well-defined, possible energies of emission.*

By analysing the "induced" (as opposed to "natural") gamma ray spectrum resulting from inelastic neutron interactions with an absorbing medium, we can identify the elements present from their characteristic energy peaks.[8]

The common elements giving measurable gamma yields in a practical logging context are: C, O, Si, Ca, Fe and S.

This is the principle of the *Gamma Ray Spectrometry Tool (GST)*[15] and the *Multi-Channel Spectroscopy Instrument (MSI-CO)*, among others.

Most importantly, the presence of oil in the near-wellbore region can be ascertained, and an estimate of its saturation made, from the relative proportions of carbon and oxygen present in the spectrum. Hence the generic name for this type of survey – the Carbon/Oxygen (CO) or Carbon/Oxygen Ratio (COR) log.[19]

Since the inelastic interactions only occur when the neutrons have high energy, the gamma ray spectrum is acquired during the neutron burst and for perhaps 20 µs afterwards.

8.5.4.2 Elastic Collisions

When neutron energies fall below about 2 MeV, interactions with matter become predominantly elastic in nature. At each collision, some of the neutron's energy is transmitted to the atomic nucleus. However, the collision is purely ballistic in nature, and no radiation is emitted. Elastic interactions follow the law:

$$\frac{E_2}{E_1} = \frac{1}{2}[(1 + a) + (1 - a)\cos\theta]\,, \tag{8.7a}$$

where

$$a = \left(\frac{A - 1}{A + 1}\right)^2\,, \tag{8.7b}$$

E_1 = incident energy of neutron
E_2 = rebound energy of neutron
θ = change in direction of path of neutron,
A = mass of the target atom

Equation (8.7) says that, for a given value of "a" E_2/E_1 can vary from 1 to $(1 + a)/2$ as θ varies from 0 to π. θ itself varies statistically from collision to collision.

It also follows that, for a given θ, E_2/E_1 is a function of "a".

"a" has its minimum value (almost zero), for the *hydrogen* atom, which is the lightest of all the atoms ($A = 1.008$, $a = 1.59 \times 10^{-5}$, $E_2/E_1 \approx 0$ for $\theta = \pi$).

The heavier the atom, the larger "a" becomes ($a = 0.778$ for oxygen, 0.867 for silicon, etc.). The limit would be $a = 1$ for an atomic mass of infinity. Since for larger "a", the ratio E_2/E_1 increases, it follows that *the maximum energy loss* (i.e. minimum E_2/E_1) *results from a collision with a hydrogen atom*.

After a number of elastic collisions, the neutron will fall into what is called the *epithermal energy* range (100–0.1 eV), and is now referred to as an "epithermal neutron".

After further elastic collisions, the neutron's energy will have been decreased to a level corresponding to the *thermal energy* of the surrounding atoms at the prevailing temperature [at 25 °C, for example, this is 0.025 eV, equivalent to a (vibratory) velocity of 2200 m/s]. In this thermal state, it is referred to as a *thermal neutron*.

Table 8.1 lists the *average* number of elastic collisions required with different types of atom to reduce the energy of a 2 MeV neutron to 0.025 eV. Note that the heavier the atom, the more the collisions needed.

The following important conclusion can be drawn from these observations: the more hydrogen there is present in a medium, the shorter the path travelled by a neutron before being "thermalised". In other words, *the neutrons reach thermal energy closer to the neutron source the higher the concentration of hydrogen per unit volume in the surrounding medium*.

Table 8.1. *Number of elastic collisions required to reduce the energy of a 2 MeV neutron to 0.025 eV*

Element	Number of collisions
Hydrogen	18
Carbon	114
Oxygen	150
Silicon	257
Chlorine	329
Calcium	368

The concentration of hydrogen per unit volume is conventionally expressed in terms of the *hydrogen index* I_H, which is normalised as 1.0 for fresh water at 15 °C. A rock of porosity 0.25, with $S_w = 1.0$, therefore represents an $I_H = 0.25$.

For hydrocarbons,

$$I_H = \frac{9n}{12 + n} \frac{\rho_o}{1000} \tag{8.8}$$

where "n" is the number of hydrogen atoms in the empirical molecular formula $(CH_n)_m$ of the hydrocarbon.

Taking hexane as an example: for C_6H_{14} we have $n = 14/6 = 2.333$; $\rho = 664.1 \text{ kg/m}^3$, so that $I_H = 0.973$. This is very close to the index for water.

Gas, on the other hand, has a density of a few hundred kg/m^3 or less at reservoir conditions, and consequently its $I_H \ll 1$.

The marked contrast between the hydrogen indices of gas on the one hand, and oil and water on the other, is used as a means of identifying gas-bearing intervals in an oil reservoir.

8.5.4.3 Thermal Neutron Capture

Once a neutron has been thermalised ($E = 0.025$ eV), it will eventually undergo another type of interaction with an atomic nucleus – *thermal neutron capture*. The neutron is totally absorbed by the nucleus, which is momentarily excited to a higher energy state. It returns to ground state by emitting a γ-photon.

The probability that a thermal neutron will be captured is different for each element. This is quantified as the "capture cross section", measured in *barns* (1 barn = 10^{-28} m^2).

Figure 8.18 is a graphical comparison of the "microscopic" thermal neutron capture cross-sections of the principal elements found in reservoir rock and fluid;[12] note how *chlorine* stands out from the others. The probability of capture by chlorine is therefore much higher than the other elements in the Fig. 8.18. This fact is the basis of a pulsed neutron logging technique called the thermal neutron decay (TDT) log, which responds to the presence of saline water.

The γ-photon emitted when the capturing nucleus returns to its ground state has an energy which is characteristic of the element involved.[12] Each element may exhibit *several different, well-defined, energies of emission* following thermal neutron capture (Fig. 8.19).

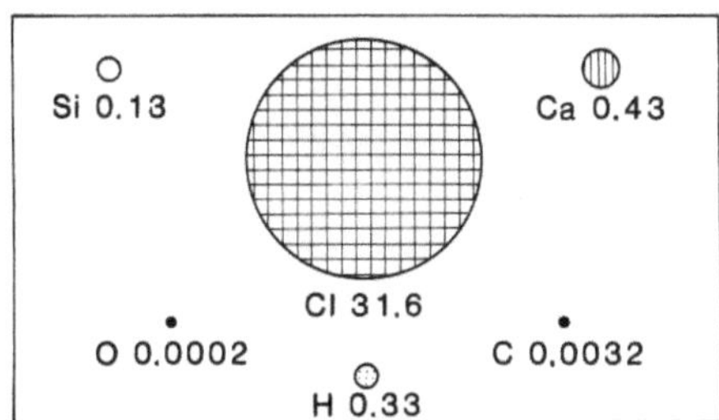

Fig. 8.18. *A comparison of the microscopic thermal neutron capture cross sections of some of the priciple elements found in reservoir rock. From Ref. 11, reprinted with permission of Elf Aquitaine*

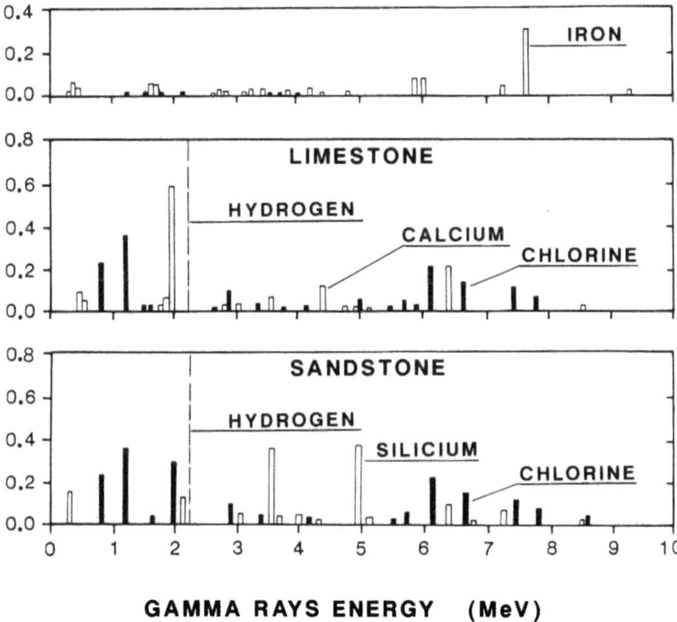

Fig. 8.19. *Gamma ray spectra resulting from thermal neutron capture by steel casing (iron), a carbonate rock and a sandstone. From Ref. 11, reprinted with permission of Elf Aquitaine*

Some of the elements present in the rock and fluid – notably Cl, H, Si, Ca, Fe, and S – can therefore be identified by spectroscopic analysis of capture gamma ray emission.

Although some of the same elements can also be distinguished in the inelastic gamma ray spectrum (Sect. 8.5.4.1), different emission energies are involved and their spectra are not the same in the two processes.

Since the capture interactions only occur when the neutrons have reached thermal energy, the gamma ray spectrum is acquired some time (~ 500 µs) after the neutron burst.

Capture gamma ray spectrometry is the basis of the *Chlorine Log*, used to determine the presence of saline water in the near-wellbore formation.

The "Carbon/Oxygen" logging tools introduced in Sect. 8.5.4.1 also acquire and analyse the capture spectrum during the latter part of each neutron pulse cycle. This adds information about salinity and lithology to the inelastic C/O (hydrocarbon saturation) measurement.

By counting the epithermal neutron population, or the gamma rays emitted after thermal neutron capture, and using one or two detectors at fixed distances from the source, it is possible to log the hydrogen index I_H of the roughly spherical volume of formation comprised between the source and the detector(s).

Since elements other than hydrogen contribute (to a lesser extent) to the slowing down process, the log reading will not be exactly equal to the true I_H, and must be corrected for lithological and other effects.

The larger the value of I_H, the less extensive will be the thermalised neutron cloud around the source (they will have been slowed down more effectively). Consequently, the count-rate of epithermal neutrons will be lower at the detectors (there will be fewer of them since more have been thermalised). Similarly, the capture gamma ray count-rate will decrease because the thermal neutron population being captured will be predominantly close to the source (i.e. farther from the detector) than for a low I_H medium.

Because of the relationship between I_H and porosity (they are equal if the pore fluid is oil or water), it is possible to measure the porosity of the formation *provided there is no gas present. In a gas-bearing interval, the log porosity is less than the true porosity* because of the lower I_H of gas.

The now obsolete *Gamma Ray – Neutron Tool* (NL or GNT) was a single-detector porosity logging device. The log, presented in Neutron API units, was interpreted in terms of porosity by means of a cross plot calibration.

Its successor for both open and cased hole applications was the *Compensated Neutron Tool (CNT)*. This two-detector tool, which is still in current use, provides a log scaled directly in porosity based on the ratio of the two count-rates (Fig. 8.20). Various corrections[9] are applied in real time so that the effects of the borehole and formation fluid salinity (and casing) are already absent from the log. The correctness of the porosity curve then depends only on the following assumptions:

- the pore fluid is water (or oil of $I_H = 1$); $S_g = 0$
- the formation lithology is limestone (the tool response is calibrated on surface so as to read the correct porosity in limestone).

Charts[9] are available to correct for non-limestone lithologies.

8.5.4.4   Thermal Neutron Decay

As we have seen, the thermal neutron cloud that forms around a pulsed neutron source is depleted by the capture process, with the emission of gamma photons.

This results in the relatively gradual decay of the thermal neutron population over a period of $\sim 500-1000$ μs. The rate of decay depends on the overall capture cross section of the mixture of atoms interacting with the neutrons (large cross section = rapid decay).

Because the capture cross section of chlorine (31.6 barns) is so much larger than that of any other common reservoir rock or fluid element (Fig. 8.18), a measurement of the rate of decay (or relaxation) of thermal neutrons will be primarily influenced by the amount of chlorine (i.e. saline water) present in the environment.

This is the principle of the *Thermal Decay Time (TDT) log.*

Measurement of the thermal neutron decay time requires a *pulsed* source of high energy neutrons (Sect. 8.5.2), so that the decay can be monitored between pulses. The pulse length is typically $\sim 80-100$ μs, repeated every 1000 μs.

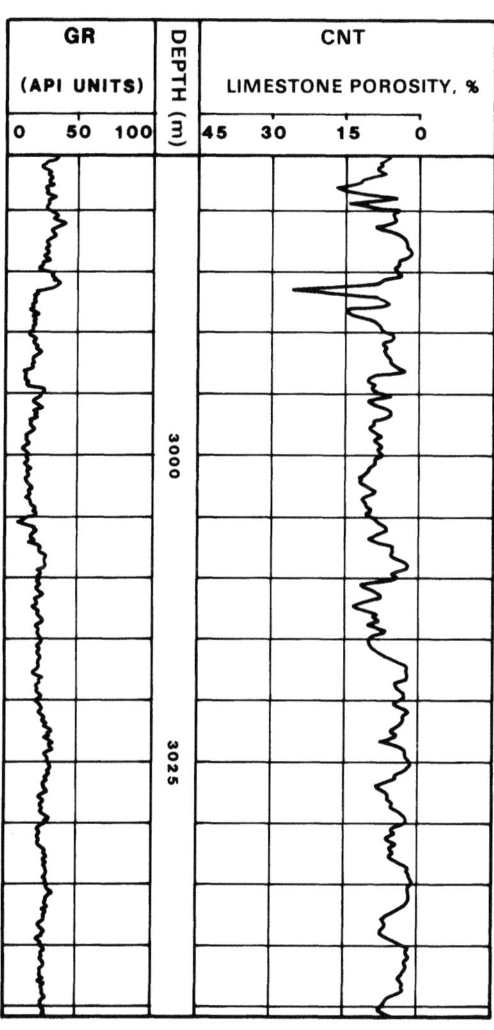

Fig. 8.20. *Example of a combination GR + CNT log*

The buildup and decay of the thermal neutron population after a pulse are shown schematically[1,2] in Fig. 8.21. The decay is essentially an exponential function of time.

Suppose N_1 is the average number of thermal neutrons counted in a "window" of duration T centred at time t_1 µs after the peak of the curve; N_2 is the average number in a second window, also of duration T, centred at t_2 µs from the peak. Then

$$N_1 = N_0\, e^{-v\,\Sigma_t t_1}, \tag{8.9a}$$

$$N_2 = N_0\, e^{-v\,\Sigma_t t_2}, \tag{8.9b}$$

where Σ_t is the total ("macroscopic") capture cross section of the formation (the sum of microscopic capture cross sections of the elements present weighted by their relative volumetric percentages); and v is the average velocity of the thermal neutrons ($v \sim 2200$ m/s).

From Eq. (8.9) we have

$$\Sigma_t = \frac{1}{(t_2 - t_1)v} \ln \frac{N_1}{N_2}. \tag{8.10}$$

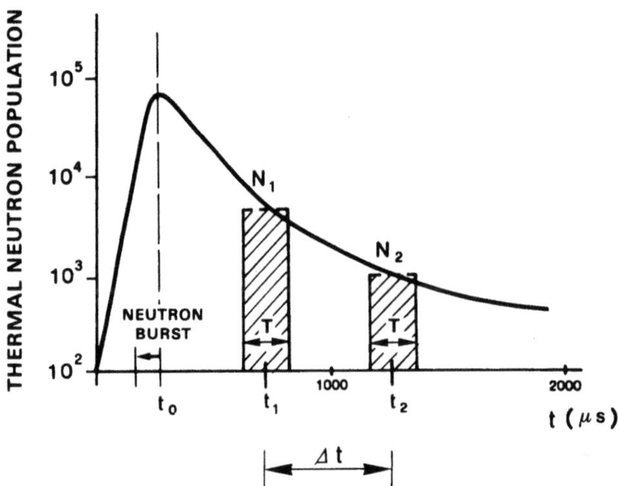

Fig. 8.21. *Decay of thermal neutrons following the bombardment of matter with a high energy neutron burst. Two detector counting windows are used for the calculation of Σ_t [see Eq. (8.10)]*

For log presentation, Σ_t is conveniently expressed in "capture units" (cu):

$$1\ \mathrm{cu} = 10^{-3}\ \mathrm{cm}^{-1}\ .$$

With time in μs, so that $v = 0.22\ \mathrm{cm/\mu s}$, we have

$$\Sigma_t(\mathrm{cu}) = \frac{10\,500}{\Delta t}\log\frac{N_1}{N_2} \tag{8.11a}$$

after substituting

$$(t_2 - t_1) = \Delta t \quad (\mu s)\ . \tag{8.11b}$$

Therefore, by counting the thermal neutrons in two equal time periods of duration T, spaced Δt apart, it is possible to determine the Σ_t of the formation.

The *decay time* (τ) itself is defined as the time required for the counts to decline to 0.37 ($= 1/e$) of their initial value (starting anywhere on the decay slope because it is an exponential). (In Fig. 8.21, this would be the value of Δt for which $N_1/N_2 = e = 2.718$).

Referring to Eq. (8.11a), it follows that $\Sigma_t = 4550/\tau$. You may find the τ curve displayed as well as Σ on early TDT logs.

For technical reasons, it is more efficient to count the *gamma rays emitted when the neutrons are captured* than the neutrons themselves. Since the number of captures occurring at any time is proportional to the number of thermal neutrons still in existence, the gamma emission will decay at the same rate. TDT tools therefore use gamma ray detectors.

This is a very simplistic treatment of the operating principle behind thermal decay logging. Modern tools have two detectors, and more sophisticated methods are employed to estimate Σ_t:[16,17,18] more counting windows, of the same or different widths, are employed, and the basic timings of the pulse cycles may vary depending on the magnitude of Σ_t being measured.

8.5.5   Neutron Logging Tools

Neutron logging tools (NL, GNT, CNT, TDT, GST, Chlorine Log) consist of a neutron generator (pulsed or continuous), and one or two detectors of gamma rays, or epithermal or thermal neutrons.

The gamma ray detectors are usually scintillation counters consisting of a sodium iodide crystal activated by thallium (NaI–Tl), coupled to a photomultiplier. For spectrometry, this system is adequate in general logging applications, but specialised high resolution work requires a germanium crystal activated by lithium[11] (Ge–Li), which needs to be kept at a very low operating temperature $(-196\,°C)$.

Epithermal neutron detectors use activated boron or lithium fluoride crystals, which have a relatively large interaction cross section to neutrons in the epithermal energy range. The neutrons interact to produce α-particles, which are counted with a proportional counter:

$$B_5^{10} + n_0^1 \rightarrow Li_3^7 + \alpha_2^4 .$$

Thermal neutrons can be counted indirectly by means of the gamma rays emitted when they are captured in the formation; or directly, using a detector

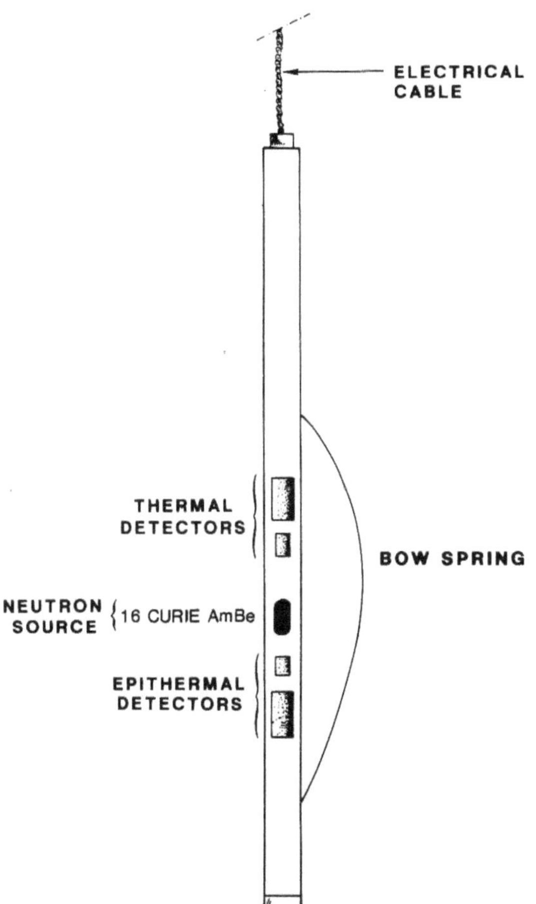

Fig. 8.22. *Schematic of the CNT logging sonde. From Ref. 10, reprinted courtesy of Schlumberger*

containing He^3. The He^3 interacts with thermalised neutrons to emit α-particles, which are counted with a proportional counter:

The source-detector configuration varies with the logging application.

As an example, Fig. 8.22 shows the Schlumberger CNT sonde for open and cased hole compensated neutron porosity logging.

The CNT sonde is run off centre of the bore hole, pressed against the hole wall by means of a bow-spring. The neutron source is of the continuous emission type (Am–Be). Two thermal neutron detectors are placed above the source, and two epithermal neutron detectors below.

The usefulness of having two detector spacings will become apparent in the following sections.

The simultaneous measurement of thermal and epithermal neutrons offers a number of advantages in terms of response. In particular, the epithermal measurement suffers less interference from the effects of chlorine and other strong neutron absorbers, which distort the I_H measurement by affecting the thermal neutron distribution.

Count-rates are transmitted to surface via the logging cable (multi- or mono-conductor, depending on the log and whether in open or cased hole).

In the well site logging unit, the downhole data are processed, various environmental corrections applied where necessary, and the results presented in analogue form on film, and stored in digital form on magnetic tape.

In the case of a high energy pulsed neutron source, the timings of the generator and detector windows in the latest generation logging tools are controlled from the surface unit. In earlier tools, these were either fixed or self-adjusting.

8.6 Identification of Gas-Bearing Intervals

8.6.1 Newly Drilled or Recently Completed Wells

In a newly drilled well, or one that has just been completed, a zone containing a high saturation of mud filtrate $[S_\mathrm{w} \sim (1 - S_\mathrm{or})]$ usually extends some distance out from the wellbore (Chap. 3).

In this environment, the dual-detector CNT is very effective at distinguishing gas-bearing intervals.

The radius of investigation of each detector is approximately equal to the source-detector spacing. Consequently, the short-spacing (SS) detector is influenced by the invaded zone to a greater degree than the long-spacing (LS) detector, which will read some of the uninvaded reservoir as well (except in the case of very deep invasion).

In a gas-bearing formation, then, the region investigated by the LS detector will have a lower hydrogen index $(I_\mathrm{H})_\mathrm{LS}$ than that investigated by the SS, $(I_\mathrm{H})_\mathrm{SS}$, because of the low I_H of gas (Fig. 8.23).

However, the two detectors read slightly differently even opposite an uninvaded formation, simply because of their different spacings.

To compensate for this, the two responses are normalised to read the same I_H opposite an uninvaded interval, such as a shale, by adjusting the count-rate scales on the log so that the curves overlay.

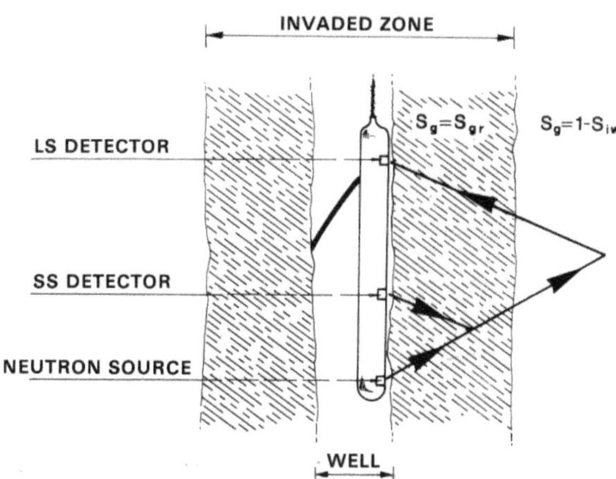

Fig. 8.23. *The principle of the dual-spacing CNT for the detection of gas-bearing strata*

Since the I_H of oil and of liquid hydrocarbon are very similar (Sect. 8.5.4.2), the two curves should still overlay – $(I_H)_{SS} = (I_H)_{LS}$ – opposite an oil-bearing formation, even if it has been invaded.

On the other hand, opposite a gas-bearing formation,

$$(I_H)_{LS} < (I_H)_{SS} \tag{8.12}$$

because of the low hydrogen index of the gas.

In other words, in the presence of an invaded zone, the apparent porosity "seen" by the long-spacing detector will be lower than that seen by the short-spacing detector. In terms of count-rates, the LS detector will read higher than the SS (Fig. 8.24).

Note also in Fig. 8.24 that the count-rates of *both* detectors have increased above the gas/oil contact, relative to the readings below it. This suggests that gas is affecting the response of the SS detector to some extent (but not as strongly as the LS).

8.6.2 Producing Wells

If vertical permeability is favourable, once the well has been completed the mud filtrate will gradually migrate under gravitational forces to the base of the stratum. The eventual dissipation of invaded zone renders the dual-spacing CNL technique described in the previous section ineffective at identifying gas.

One solution to this problem is to run a cased hole CNL porosity log and to compare it with the CNL porosity previously recorded in open hole (or, in fact, with any reliable open hole porosity measurement). Any gas in the pores will have a strong effect on the cased hole log (low apparent porosity) since it is very probably seen by both detectors.

A less commonly used alternative is to record a combination FDC–CNL log in casing.

The FDC measures the formation bulk density ρ_b, while the CNL measures the hydrogen index, I_H, which is equal to ϕ provided $(I_H)_f = 1$. The log is therefore a recording of the two curves, ρ_b and ϕ.

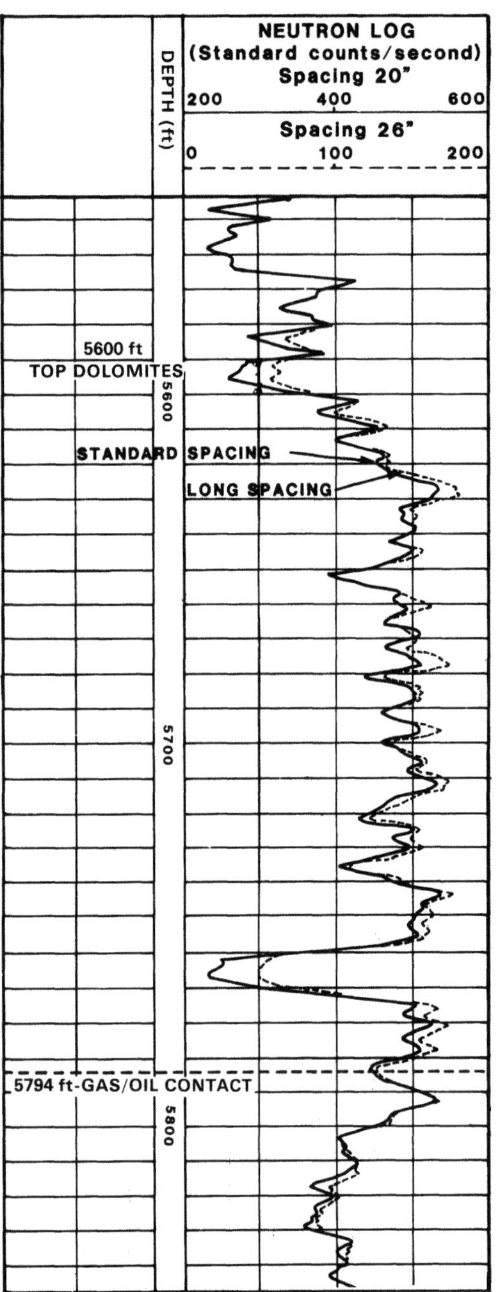

Fig. 8.24. *Example of a dual-spacing CNL, showing evidence of a gas/oil contact*

Using the subscript "f" for pore fluid, and "r" for rock, we have for the FDC:

$$(\rho_b)_{FDC} = \phi \rho_f + (1 - \phi) \rho_r \tag{8.13a}$$

so that

$$\phi = \frac{\rho_r - (\rho_b)_{FDC}}{\rho_r - \rho_f} . \tag{8.13b}$$

Looking at Eq. (8.13a), for a given ϕ, the ρ_b curve will decrease in the presence of gas, relative to its readings in oil or water-bearing strata. Furthermore, if ϕ is calculated from $(\rho_b)_{FDC}$ [Eq. (8.13b)] using too large a value for ρ_f (i.e. ρ_o or ρ_w when it should be ρ_g), it will be *higher* than the true porosity.

For the CNL, if we ignore the apparent hydrogen index of the rock,

$$(I_H)_{CNL} = \phi(I_H)_f \ . \tag{8.14}$$

The CNL porosity curve is computed and displayed assuming $(I_H)_f = 1.0$. Consequently when gas is present, since $(I_H)_g < [(I_H)_w$ and $(I_H)_o]$ ϕ_{CNL} will read less than the true porosity.

Therefore, a conventional FDC–CNL interpretation performed assuming oil or water in the pores will, in a gas-bearing interval, produce $\phi_{FDC} > \phi_{true}$ and $\phi_{CNL} < \phi_{true}$, and consequently $\phi_{CNL} \ll \phi_{FDC}$.

Water and oil-bearing formations can be distinguished from those which are gas-bearing by plotting log readings on a Cartesian cross-plot of ϕ_{FDC} against ϕ_{CNL} (Fig. 8.25). *All gas-bearing points will lie above the line* $\phi_{FDC} = \phi_{CNL}$.

In fact, the gas-bearing sections stand out clearly on the FDC–CNL log itself. This is demonstrated in Fig. 8.26. The FDC log is customarily presented as ρ_b rather than ϕ_{FDC}, but the same rules apply. There is a distinct separation of the

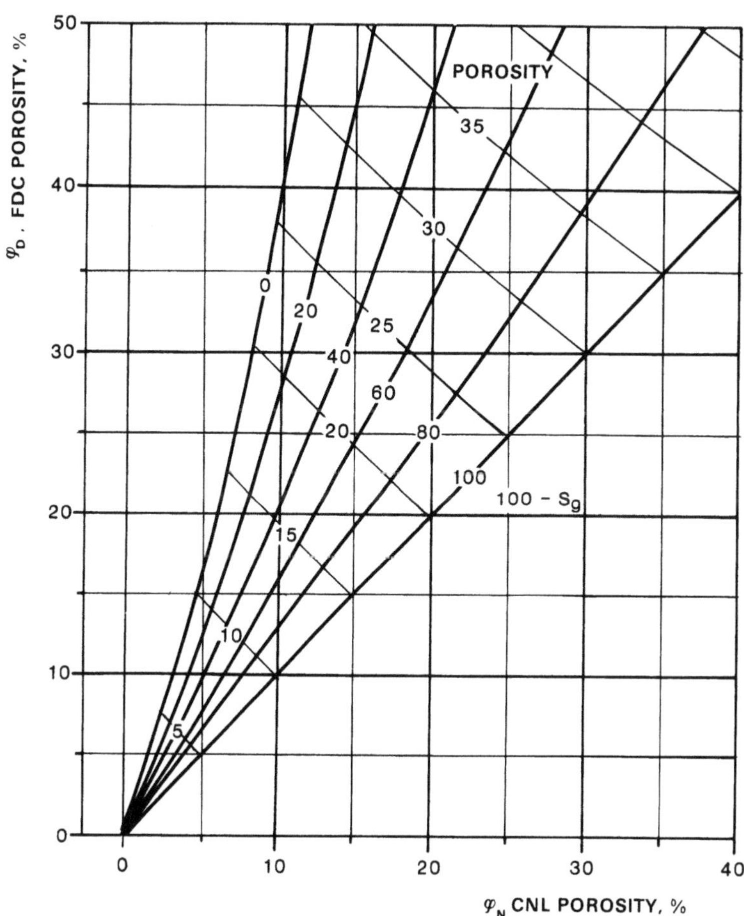

Fig. 8.25. *Chart for the determination of gas saturation from* ϕ_{CNL} *and* ϕ_{FDC} *for the depth range 3000–4000 m. (Adapted from Ref. 10.)*

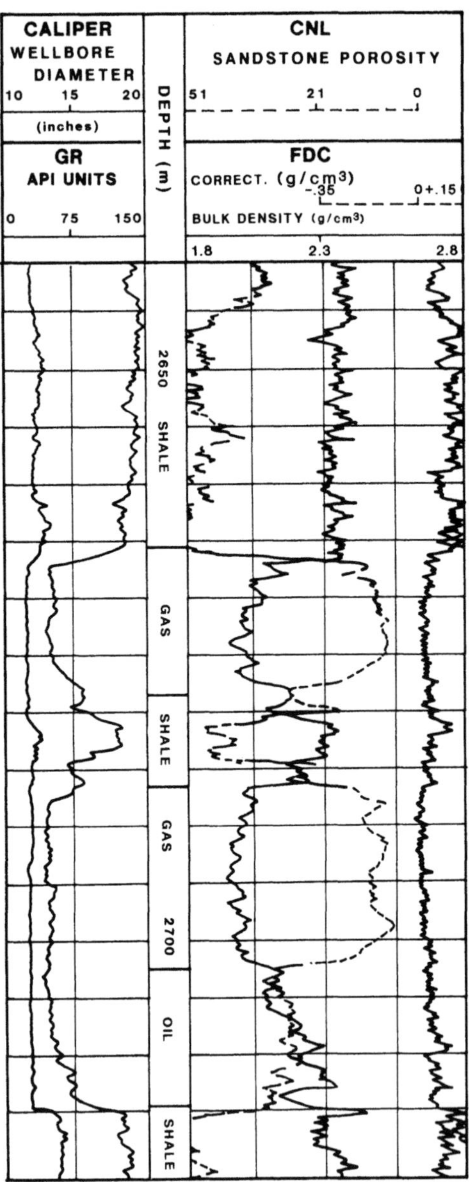

Fig. 8.26. *Identification of gas zones in a producing well using the combination FDC–CNL log*

two curves where there is gas, since ϕ_{CNL} decreases and ϕ_{FDC} increases (ρ_b decreases), whereas they overlay in the oil-bearing interval. Note that they separate in the opposite direction in the shales.

8.7   Identification of Water Flooded Zones ($S_w > S_{iw}$)

8.7.1   Chlorine Log

The chlorine log[13] has now been largely superseded by the Thermal Decay Time (TDT) log. It is based on the simultaneous recording of a conventional NL and an NL where only gamma rays of energy greater than 7 MeV are counted.

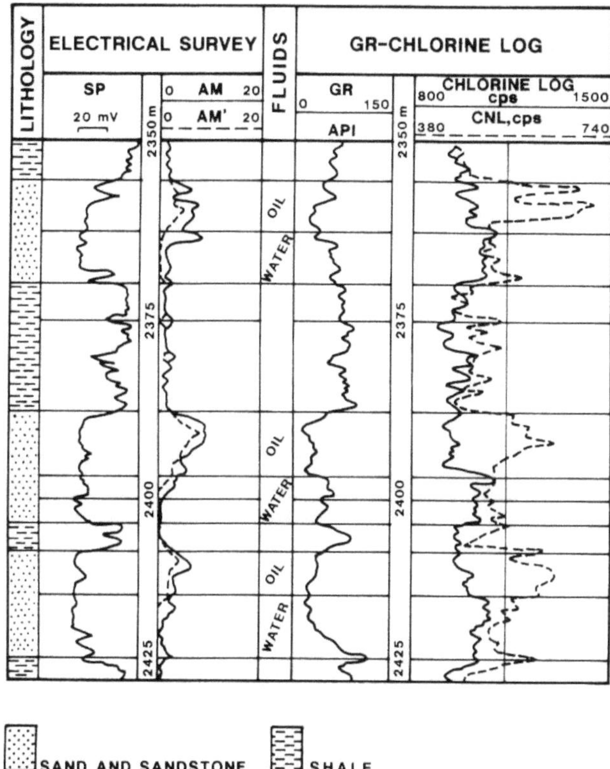

Fig. 8.27. *Example of a chlorine log run in cased hole in combination with a GR. Also shown is an early open hole electrical log (ES) consisting of the SP, AM and AM'. From Ref. 11, reprinted with permission of Elf Aquitaine*

Because chlorine has a high thermal neutron capture cross section (Sect. 8.5.4.3), and emits capture gamma rays with energies 7.42 and 7.77 MeV, the Chlorine Log is sensitive to the chlorine content (i.e. *saline water*) in the formation.

To compensate for their different response levels, the two NL measurements are normalised opposite a zone which is known to be water-bearing ($S_w = 1.0$), by adjusting the scale of one or the other until the two curves overlay.

Over the rest of the log, wherever $S_w < 1$ the Chlorine Log curve will read a lower count-rate than the NL because of the reduced Cl content. The chlorine curve will separate to the left, showing a higher apparent porosity, as in Fig. 8.27. Any such separation of the two logs can be assumed to indicate the presence of hydrocarbon provided the water salinity remains the same as in the normalisation zone: in this way it is possible to distinguish zones which are still oil- and gas-bearing from those which have been water flooded.

8.7.2 Thermal Decay Time Log (TDT)

The basic operating principles of the TDT, and the measurement of the macroscopic thermal neutron capture cross section Σ_t, have been introduced in Sect. 8.5.4.4.

In order to estimate S_w from Σ_t, we have to know the salinity of the formation water. We will simplify the explanation by assuming that we are dealing with a "clean" lithology – there are no laminar or dispersed shales present (Σ_{sh} is very large).

The measured Σ_t is made up of the sum of the volumetrically weighted cross sections of the various constituents, which are conventionally categorised in terms of water (w), hydrocarbon (h), and rock matrix (r):

$$\Sigma_{TDT} = \phi S_w \Sigma_w + \phi(1 - S_w)\Sigma_h + (1 - \phi)\Sigma_r . \tag{8.15}$$

Σ_w is an increasing function of the water salinity[9] (Fig. 8.28). For example, $\Sigma_w = 20$ cu for distilled water, and 58 cu for a (NaCl) salinity of 100 g/l.

Σ_h varies between about 22 and 16 cu for liquid hydrocarbons and can be derived from a knowledge of the oil gravity and solution GOR (Σ_h decreases as the GOR increases). For gases it is typically in the range 5–10 cu, depending on the gas density, and therefore on the reservoir pressure and temperature. Charts have been published[10] to estimate these values.

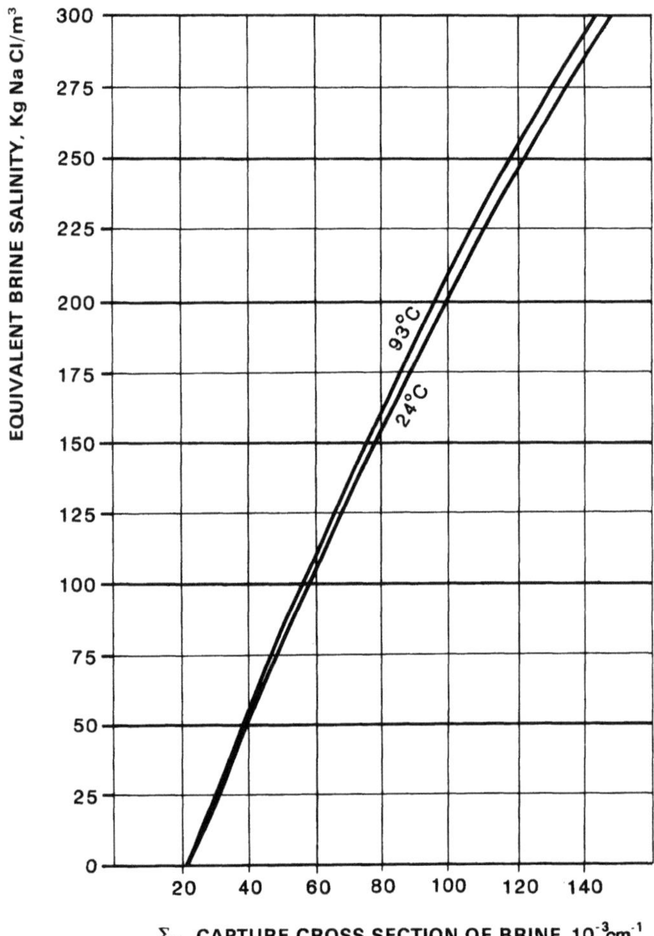

Fig. 8.28. *The capture cross section of water, Σ_w, as a function of (NaCl equivalent) salinity and temperature. (Adapted from Ref. 10.)*

For clean and moderately clean reservoir rock, Σ_r lies between about 8 and 12 cu, depending on lithology.

In order for the TDT to distinguish oil from water with reasonable certainty, *the water salinity must be high enough to give a useful contrast between the fluid capture cross sections.* For practical purposes, this means at least 50 g/l (corresponding to a Σ_w of 40 cu) in an oil reservoir ($\Sigma_h \sim 21$ cu), and slightly less in a gas reservoir ($\Sigma_h \sim 10$ cu).

The porosity ϕ is best determined from the open hole FDC–CNL logs if available, or from a cased hole porosity log. We can then calculate S_w from

$$S_w = \frac{\Sigma_{TDT} - \Sigma_r - \phi(\Sigma_h - \Sigma_r)}{\phi(\Sigma_w - \Sigma_h)} . \tag{8.16}$$

Equation (8.15) can be applied numerically, or in a convenient graphical or "cross-plot" approach, described here.

Rearranging Eq. (8.15) as follows:

$$\Sigma_{TDT} - \Sigma_r = \phi(\Sigma_w - \Sigma_r) - \phi(1 - S_w)(\Sigma_w - \Sigma_h) \tag{8.17}$$

note that all water-bearing points ($S_w = 1$) lie on the straight line:

$$\Sigma_{TDT} - \Sigma_r = \phi(\Sigma_w - \Sigma_r) . \tag{8.18}$$

If the water salinity and the rock type are known, precise estimates of Σ_w and Σ_r can be obtained. Since at $\phi = 0$, $\Sigma_{TDT} = \Sigma_r$, and at $\phi = 1$, $\Sigma_{TDT} = \Sigma_w$, the line corresponding to $S_w = 1$ can be drawn on a Cartesian plot of Σ_{TDT} versus ϕ (Fig. 8.29).

Equation (8.17) can also be used to define the straight line representing $S_w = 0$, which should of course read $\Sigma_{TDT} = \Sigma_r$ at $\phi = 0$; $\Sigma_{TDT} = \Sigma_h$ at $\phi = 1$.

Lines corresponding to intermediate values of S_w can be constructed in a similar manner (Fig. 8.29).

Data points are now read from the TDT log and marked along with the corresponding porosity values (from open or cased hole porosity logs) on the cross-plot. Water-bearing points should fall on (or very close to) the line for $S_w = 1$. Undepleted oil sections will plot down towards (but not on) the $S_w = 0$ line. Where

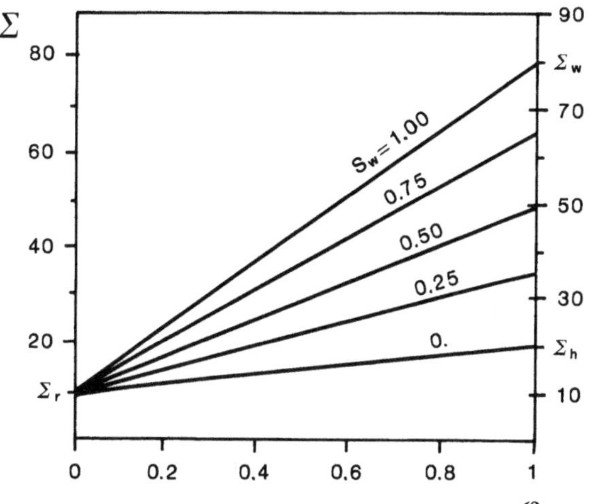

Fig. 8.29. *Graphical interpretation of the TDT log in clean lithology*

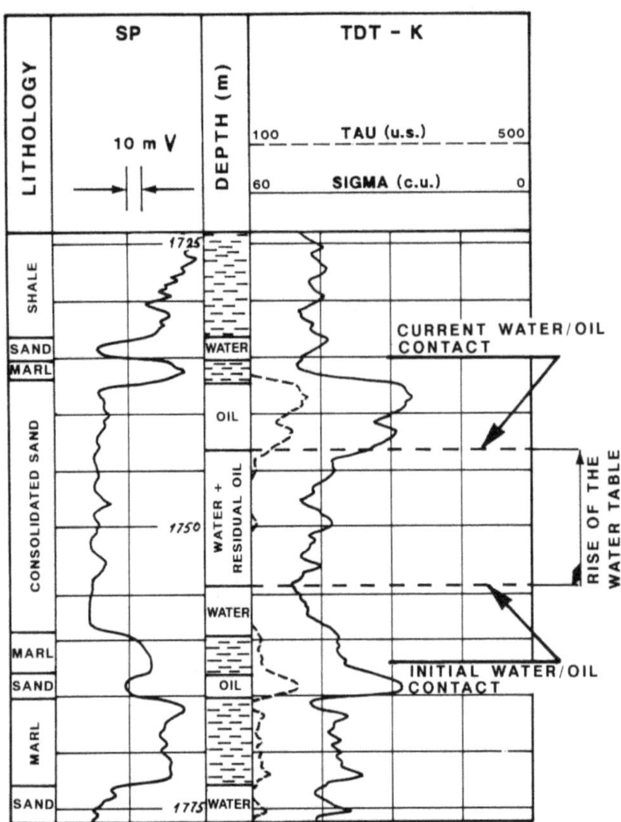

Fig. 8.30. *Example of a TDT log in a producing well, showing the rise of the oil/water contact as a result of aquifer expansion following production of oil*

water influx has occurred, the data will plot at intermediate saturations $[S_w \sim (1 - S_{or})]$, provided there is no change in water salinity.

Figure 8.30 is an example of a TDT log, showing both the capture cross section (sigma) and decay time (tau) curves. The open hole SP curve has been traced on to highlight the shaley intervals.

The TDT is a small diameter logging tool, run on monoconductor cable, and suitable for through-tubing work. It is extremely effective at monitoring water contact movement behind casing, particularly where breakthrough of water (perhaps unexpected) has occurred at the well, along highly permeable strata. If the layer has not been perforated, this water will, of course, not be detected by a production log, nor will it be produced to surface in this well.

The TDT is, however, *only applicable where the water salinity is sufficiently high, and quantitative interpretation is only possible when Σ_w is known and constant.*

8.7.3 The Carbon/Oxygen Ratio, COR

The relatively recent technique of induced gamma ray spectrometry, introduced in Sects. 8.5.4.1 and 8.5.4.3, analyses the gamma ray emission resulting from the bombardment of the formation with high energy (14 MeV) neutrons, which undergo inelastic and capture interactions with atomic nuclei in the near-wellbore region.

The standard log presentation of the GST (Gamma Ray Spectrometry Tool)[8] consists of the following curves, based on the ratios of elemental percentage yields in the spectra:

COR: carbon/oxygen ratio – the ratio of C/O, derived from the inelastic gamma ray spectrum

SIR: salinity indicator ratio – ratio of Cl/H, derived from the capture gamma ray spectrum

PIR: porosity indicator ratio – ratio H/(Si + Ca), from the capture gamma ray spectrum

LIR: lithology indicator ratio – ratio Si/(Si + Ca), from the capture gamma ray spectrum

IIR: iron indicator ratio – ratio Fe/(Si + Ca), from the capture gamma ray spectrum

The C/O ratio responds to changing S_w, *but is not affected by salinity. The GST can therefore be used in fresh formation water environments, or where the salinity is variable or unknown* (e.g. in the presence of injection water breakthrough).[19]

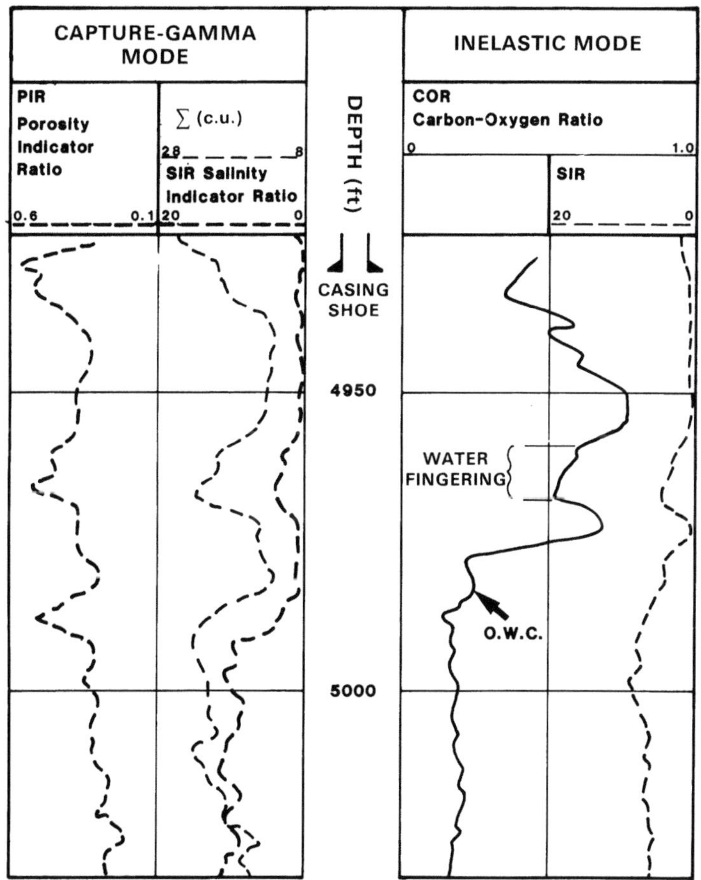

Fig. 8.31. *Example of an induced gamma ray spectrometry log (GST), obtained from inelastic and thermal neutron capture interactions, run in the open hole section of an observation well to identify oil and water zones and estimate saturations. COR, SIR and Σ respond to water saturation, PIR to porosity. From Ref. 8, reprinted courtesy of Schlumberger*

Some of the ratio curves are presented[8] in Fig. 8.31. This particular example shows a major oil/water contact with a zone of water fingering above it. The large increase in COR just above the OWC is caused by a change of borehole fluid from water to oil.

The GST delivers a large amount of information: the other measurements can be used to evaluate water salinity, porosity and lithology, and a simultaneous Σ is also recorded.

It is, however, a large diameter tool, run on multiconductor cable, and requires a minimum tubing diameter of $4\frac{1}{2}''$, a fact which severely limits its applicability. In addition, the inelastic readings are very strongly influenced by the presence of the wellbore, and, where formation water salinity is favourable, the TDT is preferred for saturation monitoring.

A new generation tool[20], the RST, has recently been introduced. Running on single-conductor cable, it is available in two sizes – $1\frac{11}{16}''$ and $2\frac{1}{2}''$ – for through-tubing work. The use of two detectors, and improved detector technology, have greatly improved the quality of the tool response.

References

1. Clavier C, Hoyle W, Meunier D (1971) Quantitative interpretation of thermal decay time logs. Part I. Fundamentals and techniques Part II. Interpretation example, interpretation accuracy, and time lapse technique. J Petrol Tech (June): 743–763; Trans AIME 251
2. Dewan JT, Johnstone CW, Jacobson LA, Wall WB, Alger R (1971) Thermal neutron decay time logging using dual detection. SPWLA, 14th Annu Log Symp Trans
3. Kunz KS, Tixier MP (1955) Temperature surveys in gas producing wells. J. Petrol Tech (July): 111–119; Trans AIME 204
4. Meunier D, Tixier MP, Bonnet JL (1971) The production combination tool – a new system for production monitoring. J. Petrol Tech (May): 603–613
5. Schlumberger Ltd (1973) Production log interpretation. Schlumberger Ltd, Houston
6. Schlumberger Ltd (1974) Fluid conversions in production log interpretation. Schlumberger Ltd, Houston
7. Schlumberger Ltd (1981) RFT – essentials of pressure test interpretation. Schlumberger ATL, Montrouge, France
8. Schlumberger Ltd (1983) Gamma ray spectrometry. London
9. Schlumberger Ltd (1985) Log interpretation charts. Schlumberger Well Services, Houston
10. Schlumberger Ltd (1987) Log interpretation principles/application. Schlumberger Educational Services, Houston
11. Serra O (1979) Diagraphies différées – bases de l'interprétation, Tome 1: acquisition des données diagraphiques. Bull des Centres de Recherche Exploration-Production Elf-Aquitaine, Mémoire 1, Pau
12. Serra O (1985) Diagraphies différées – bases de l'interprétation, Tome 2: interprétation des données diagraphiques. Bull des Centres de Recherche Exploration-Production Elf-Aquitaine, Mémoire 7, Pau
13. Tanner HL (1987) Evaluation of low-resistivity cased-off reserves with the shale-compensated chlorine log. SPE Formation Evaluation (Sept): 284–288
14. Vaghi GC, Torricelli L, Pulga M, Giacca D, Chierici GL, Bilgeri D (1980) Production in the very deep Malossa field, Italy. Proc 10th World Petrol Congr, vol. 3. Hayden & Son, London, pp. 371–384
15. Westaway P, Hertzog R, Plasek RE (1983) Neutron-induced gamma ray spectroscopy for reservoir analysis, Soc Petrol Eng J (June): 553–564; Trans. AIME 275.
16. Olesen J-R, Mahdavi M, Steinman DK, Yver J-P (1987) Dual-burst thermal decay time logging overview and examples. Pap. SPE 15716, presented at the 5th SPE Middle East Oil Show, Manama, Bahrain, March 7–10, 1987

17. Randall RR, Oliver DW, Hopkinson EC (1986) PDK-100: a new generation pulsed neutron logging system. Pap presented at the 10th Eur Formation Evaluation Symp, Aberdeen, April 22–25, 1986
18. Schultz WE, Smith HD Jr, Verbout JL, Bridges JR (1983) Experimental basis for a new borehole corrected pulsed neutron capture logging system (TMD), Paper presented at the SPWLA 24th Annual Logging Symp, June 27–30, 1983
19. Woodhouse R, Kerr SA (1988) The evaluation of oil saturation through casing using carbon/oxygen logs. Pap. SPE 17610, presented at the SPE Int Meet on Petroleum Engineering, Tianjin, China, Nov 1–4, 1988
20. Scott HD, Stoller C, Roscoe BA, Plasek RE, Adolph RA (1991) A new compensated through-tubing carbon/oxygen tool for use in flowing wells, Paper M presented at the SPWLA Annual Logging Symposium, June 16–19, 1991

9 The Influx of Water into the Reservoir

9.1 Introduction

Chapter 1 explained how all hydrocarbon reservoirs, at the time of their formation, must have been in contact with an aquifer; the very existence of a hydrocarbon accumulation depended, of course, on the migration of oil or gas through permeable strata saturated with water.

If, over geological time, subsequent faulting or diagenesis happened to create a localised zone of zero permeability somewhere between the reservoir and the aquifer, hydraulic continuity would be interrupted. This might have occurred close to the reservoir or some distance from it, and, consequently, the volume V_{aq} of aquifer remaining in communication could be anything from zero to a very large value.

In any case, *there is no such thing as an infinitely large aquifer*. The term "infinite aquifer" is however a commonly used reservoir engineering term; its meaning will be explained shortly.

The decrease in pressure which results from the production of oil or gas from the reservoir, propagates with a finite velocity into the aquifer if it is in hydraulic communication. The reduced pressure allows the water to expand.

As long as the pressure disturbance has not reached the external limits of the aquifer, it is said to be "infinite acting" – it is in this context that the term *infinite aquifer* is used.

The expansion of aquifer water induced by the declining pressure over a period of time t since the onset of production, represents a volume $W_e(t)$. The water/hydrocarbon contact will be displaced upwards from its original position by a corresponding amount – this is "water drive".

The calculation of $W_e(t)$ *from the production history of the reservoir* is straightforward.

As a simple example, we can consider a dry gas reservoir. G is the initial volume of gas in place, $G_p(t)$ is the volume of gas produced in time t, and B_{gi} and $B_g(p)$ are the gas volume factors (Sect. 2.3.1.1) at initial pressure p_i and $p(t)$, respectively.

If we ignore any change in the porosity of the reservoir rock with pressure, we have by material balance:

$$GB_{gi} = [G - G_p(t)]B_g(p) + W_e . \tag{9.1}$$

Equation (9.1) is in fact saying that the pore volume V_p above the initial gas/water contact remains constant with time (Fig. 9.1).

$W_e(t)$ can of course only increase with t. In reality, successive estimations of $W_e(t)$ calculated with Eq. (9.1) at various times in the producing life of the reservoir often display a quite irregular trend (Fig. 9.2), especially in the early stages. This is primarily the result of errors in the evaluation of the initial reserves G (calculated

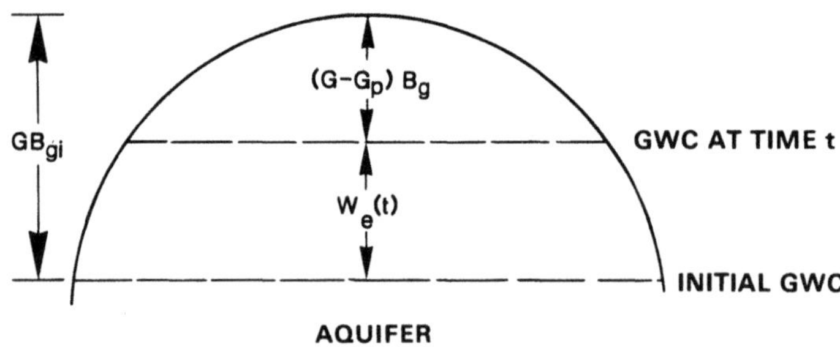

Fig. 9.1. *Representation of the terms in Eq. (9.1) for material balance in a gas reservoir under water drive*

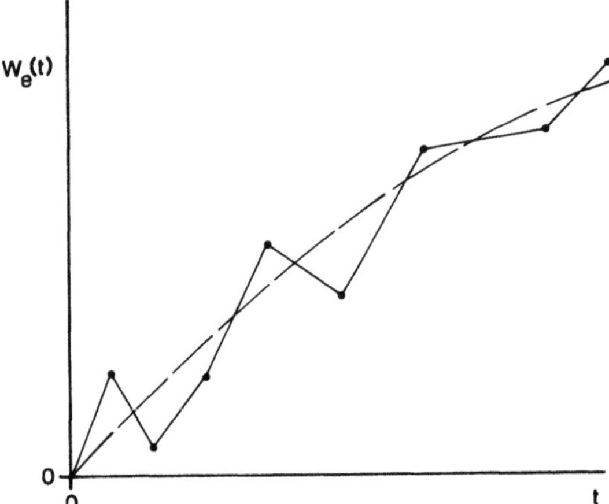

Fig. 9.2. *Water influx* $W_e(t)$ *calculated by material balance* (points), *and average trend* (dashed line)

volumetrically as described in Chap. 4), and the pressures p_i and $p(t)$, on which $B_g(p)$ depends.

The actual trend more closely follows an average curve through the data, as shown in Fig. 9.2.

Prediction of $W_e(t)$ *is far more difficult* when production constraints are imposed. Nevertheless, it is absolutely essential to know $W_e(t)$ if we are to evaluate the future productivity and pressure behaviour in any hydrocarbon reservoir in communication with an aquifer.

If the geometry and lateral extent of the aquifer are known, and its thickness, porosity and permeability are known and constant, $W_e(t)$ can be calculated from $p(t)$ quite easily using the theory introduced in Chaps. 5 and 6 for wells producing from reservoirs containing fluids of small but constant compressibility.

The reservoir itself can, in fact, be represented as a very large well which is draining the surrounding aquifer.

Unfortunately, the aquifer geometry is frequently known only roughly, through a study of the basin geology. Also, its thickness, porosity and permeability cannot be measured directly, and are far from constant. Since wells are rarely drilled intentionally into the aquifer, information about it can only be obtained from "dry" wells (unsuccessful exploration or development wells), which are usually in the immediate vicinity of the reservoir and therefore provide only a local picture.

This chapter will introduce a number of methods which have been developed to estimate $W_e(t)$. In the period from the mid-1930s to the early 1970s, computational capabilities were such that the reservoir–aquifer system could only be modelled using a concentrated parameter set, with a single reservoir cell of simple (circular or linear) geometry, in contact with an aquifer of finite or infinite extent. These highly simplistic models, not surprisingly, achieved mixed success.

With the advent of powerful high-speed computers in the mid-1960s, it became possible to study reservoirs using distributed parameters. The reservoir–aquifer system was modelled as a large number of discrete intercommunicating grid blocks with, where necessary, different parameters associated with different blocks. The actual geometry and distribution of petrophysical characteristics could thus be reproduced quite accurately, in addition to the positions of wells and their production/injection rates.

However, even with this degree of sophistication, computations still suffer from uncertainty in the estimation of the properties of some of the grid blocks, especially in the aquifer.

In the following pages, we will look at the principal "single cell model" (SCM) methods, using concentrated parameters, for the estimation of $W_e(t)$. The grid block model, using distributed parameters, will be dealt with in Chap. 13; Vol. 2.

9.2 Empirical Equations for the Calculation of $W_e(t)$

9.2.1 The Schilthuis Equation (1936)

The first published equation for the calculation of $W_e(t)$ was proposed by Schilthuis:[13]

$$W_e(t_n) = C \int_0^{t_n} (p_i - p)\mathrm{d}t \ . \tag{9.2}$$

Here, p_i is the initial reservoir pressure, $p(t)$ the pressure at time t, and C is a constant which can be evaluated from the water influx history $W_e(t)$ of the reservoir.

Differentiating Eq. (9.2) with respect to t, and designating e_w as the rate of water encroachment across the initial hydrocarbon/water contact, we have:

$$e_w(t) = \frac{\mathrm{d}W_e(t)}{\mathrm{d}t} = C(p_i - p) \ , \tag{9.3}$$

If we compare Eq. (9.3) with Eq. (5.53) (for steady state radial flow), it is apparent that Eq. (9.2) describes water encroachment into the reservoir *from an aquifer with a constant pressure p_i at its outer boundary, ignoring any transient effects caused by the variation of the pressure $p(t)$ within the reservoir.*

From this comparison we can write:

$$C = \frac{2\pi k h}{\mu} \ \frac{1}{\ln(r_e/r_o)} \ , \tag{9.4}$$

where r_e and r_o are, respectively, the external radius of the aquifer and the radius of the reservoir.

Referring to Fig. 9.3, the integral in Eq. (9.2) is equal to the area delimited by the straight lines ($t = t_n = $ const.), ($p = p_i = $ const.) and the curve $[p = p(t)]$. It is most simply performed as a summation.

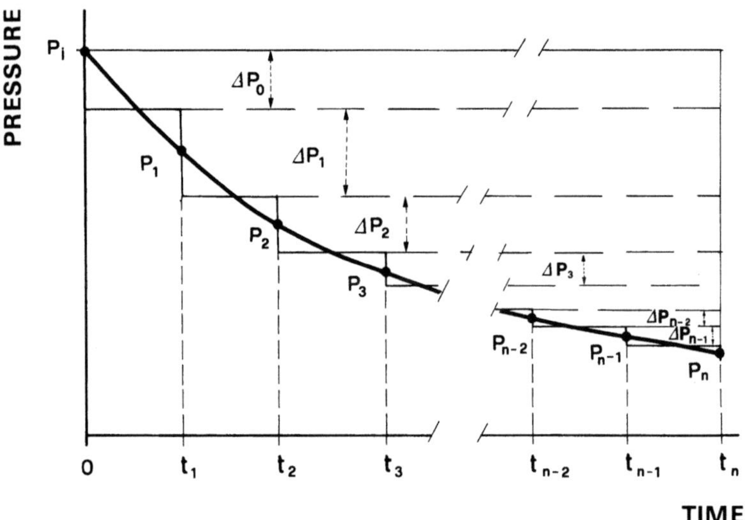

Fig. 9.3. *Schematic for the calculation of the integral in Eq. (9.2)*

If we approximate the curve $p(t)$ by a series of steps as shown, then:

$$W_e(t_n) = C[\Delta p_0(t_n - 0) + \Delta p_1(t_n - t_1) + \Delta p_2(t_n - t_2)$$

$$+ \cdots + \Delta p_{n-2}(t_n - t_{n-2}) + \Delta p_{n-1}(t_n - t_{n-1})]$$

$$= C \sum_{j=0}^{n-1} \Delta p_j(t_n - t_j) \,. \tag{9.5}$$

The pressure increments are:

$$\Delta p_0 = \frac{p_i - p_1}{2} \,,$$

$$\Delta p_1 = \frac{p_i - p_1}{2} + \frac{p_1 - p_2}{2} = \frac{p_i - p_2}{2} \,,$$

$$\Delta p_2 = \frac{p_1 - p_2}{2} + \frac{p_2 - p_3}{2} = \frac{p_1 - p_3}{2} \,,$$

$$\Delta p_3 = \frac{p_2 - p_3}{2} + \frac{p_3 - p_4}{2} = \frac{p_2 - p_4}{2}$$

$$\cdots \cdots \cdots \cdots$$

$$\Delta p_{n-1} = \frac{p_{n-2} - p_{n-1}}{2} + \frac{p_{n-1} - p_n}{2} = \frac{p_{n-2} - p_n}{2}$$

or, more generally:

$$\Delta p_j = \frac{p_{j-1} - p_{j+1}}{2} \,. \tag{9.6}$$

Combining Eqs. (9.5) and (9.6), we get:

$$W_e(t_n) = C \sum_{j=0}^{n-1} \frac{p_{j-1} - p_{j+1}}{2} (t_n - t_j) \,, \tag{9.7}$$

with

$$p_{-1} = p_i .$$

9.2.2 Hurst's Equation (1943)

In 1943, Hurst[10] proposed an equation which accounted in an approximate manner for the fact that at least part of the aquifer flow towards the reservoir was transient:

$$W_e(t_n) = C_1 \int_0^{t_n} \frac{p_i - p}{\log(C_2 t)} \, dt . \tag{9.8}$$

Therefore:

$$e_w = C_1 \frac{p_i - p}{\log(C_2 t)} . \tag{9.9}$$

In order to appreciate the physical significance of Eqs. (9.8) and (9.9), we can compare them with the equation from Chap. 5 for the wellbore flowing pressure in constant rate transient radial flow:

$$p_i - p_{wf} = \frac{q\mu}{4\pi kh} \ln \frac{4kt}{\gamma \phi \mu c_t r_w^2} , \tag{5.32a}$$

which is valid for:

$$\frac{kt}{\phi \mu c_t r_w^2} > 25 . \tag{5.34b}$$

Replacing q by e_w for the case of aquifer influx, Eq. (5.32a) becomes:

$$e_w = \frac{4\pi kh}{\mu_w} \frac{p_i - p}{2.302 \log \dfrac{4kt}{\gamma \phi \mu_w c_t r_o^2}} . \tag{9.10}$$

If we make:

$$C_1 = 5.458 \frac{kh}{\mu_w} \tag{9.11a}$$

and

$$C_2 = 2.246 \frac{k}{\phi \mu_w c_t r_o^2} , \tag{9.11b}$$

then Eqs. (9.10) and (9.9) are identical.

Therefore, the term $\log(C_2 t)$ in the denominator of Eq. (9.9) models, in an approximate manner, the transient flow condition in the aquifer which results from the continuous variation of $p(t)$.

The weakness of Eq. (9.8) lies in its range of validity, as expressed by Eq. (5.34b) (in which r_w should be replaced by r_o).

Take, for example, an aquifer where:

$$k = 100 \, \text{md} = 9.87 \times 10^{-14} \, \text{m}^2$$

$$\phi = 0.20 \, ,$$

$$\mu_\text{w} = 0.6 \, \text{cP} = 6 \times 10^{-4} \, \text{Pa} \, \text{s} \, ,$$

$$c_\text{t} = c_\text{w} + c_\text{f} = 9.5 \times 10^{-5} \, \text{cm}^2 \, \text{kg}^{-1} = 9.69 \times 10^{-10} \, \text{Pa}^{-1} \, ,$$

$$r_\text{o} = 1 \, \text{km} = 1000 \, \text{m} \, .$$

Equation (5.34b) tells us that Eq. (5.32a) and, consequently, Eqs. (9.8) and (9.9) are valid for:

$$t > 29.45 \times 10^6 \, \text{s} \quad \text{(which is about 11 months)} \, .$$

This means that Eq. (9.8) is ill-suited to describe the early transient period. For this we would really need to use the exponential integral function $ei(x)$ (Sect. 5.6.1.1).

The integral in Eq. (9.8) can be evaluated graphically in the same way as indicated in Sect. 9.2.1, with the additional constraint that:

$$\log(C_2 t) > 0 \, . \tag{9.12}$$

Note that Eqs. (9.8) and (9.9) allow calculation of the coefficients C_1 and C_2 based on the historical variation of $W_\text{e}(t)$ for the reservoir. Unfortunately, owing to the frequently erratic nature of $W_\text{e}(t)$ (Fig. 9.2), resulting from errors in the measurement of $p(t)$, it is usually necessary to resort to linear regression techniques to obtain stable values of the two coefficients. Once C_1 and C_2 have been established, predictions can be made about the future behaviour of $W_\text{e}(t)$, assuming a projected trend in pressure $p(t)$.

9.3   A Global Equation for the Calculation of $W_\text{e}(t)$

In 1959, Chierici et al.[4] proposed an approximate treatment for aquifers of limited volume V_aq, and high permeability. Presented again in 1978 by Dake[7], this method is based on the hypothesis that the pressure at any point in the aquifer is, at any instant, the same as the pressure p at the boundary between the reservoir and the aquifer.

From Sect. 3.4.3, the pore compressibility c_p in the aquifer is expressed as:

$$\frac{1}{V_\text{p}} \left(\frac{\partial V_\text{p}}{\partial p} \right)_{\bar{\sigma}} = c_\text{p} - c_\text{r} = \frac{c_\text{f} + \phi c_\text{r}}{1 - \phi} \cong c_\text{p} \, . \tag{9.13}$$

If c_w is the isothermal compressibility of the water (Sect. 2.4), then we have:

$$W_\text{e} = \phi V_\text{aq}(c_\text{w} + c_\text{p})(p_\text{i} - p) = W_{ei}\left(1 - \frac{p}{p_\text{i}}\right) , \tag{9.14a}$$

where:

$$W_{ei} = \phi V_\text{aq}(c_\text{w} + c_\text{p}) p_\text{i} = \phi V_\text{aq} c_\text{t} p_\text{i} \, , \tag{9.14b}$$

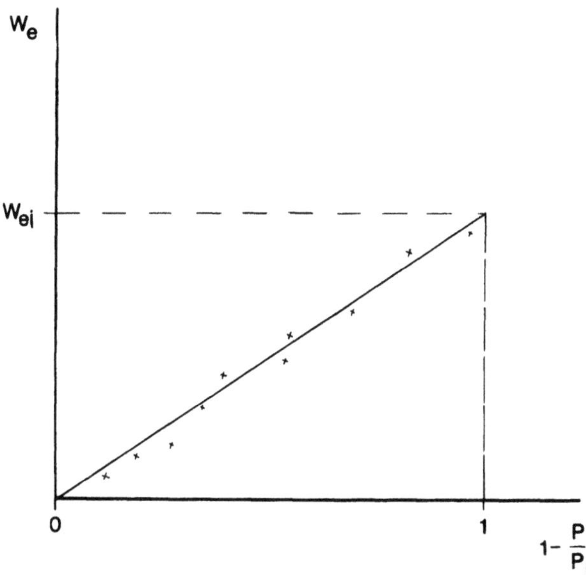

Fig. 9.4. *Graphical solution to Eq. (9.14a) for the estimation of* W_{ei}

W_{ei} represents the maximum volume of water that the aquifer could theoretically supply to the reservoir if its pressure were reduced to zero over a sufficiently large period of time.

Note that Eq. (9.14) does not contain an integral; W_{ei} can, in fact, be found graphically by plotting values of W_e against $(1 - p/p_i)$ as shown in Fig. 9.4.

This method does not allow for any transient flow in the aquifer, and its use is therefore restricted to aquifers of limited extent and high permeability.

9.4 Calculation of W_e Using the Solution to the Diffusivity Equation for Fluids of Constant Compressibility

$p_D(t_D)$ forms of the solutions to the radial diffusivity equation for fluids of small and constant compressibility, *with constant flow rate at the inner boundary*, were described in Chaps. 5 and 6. These can be applied rigorously to $W_e(t)$ in all the flow regimes (transient, late transient and pseudo-steady state) which occur in aquifers.

This approach will be described in Sect. 9.5. We will first look at an alternative solution, presented by van Everdingen and Hurst[14] in 1949, which is based on the assumption of *constant pressure at the inner boundary* (i.e. at the aquifer/reservoir contact). This was applicable to aquifers of finite size (i.e. closed), or infinite size.

There follows a detailed description of the case of radial geometry. Linear,[12] elliptical,[5] spherical[2] and other geometries are dealt with in the published literature.

The radial flow of a fluid of small and constant compressibility is described in dimensionless terms by:

$$\frac{1}{r_D} \frac{\partial}{\partial r_D}\left(r_D \frac{\partial p_D}{\partial r_D}\right) = \frac{\partial p_D}{\partial t_D} . \tag{5.15}$$

In the present context, we replace the "well" referred to in Chaps. 5 and 6 by the outer radius of the reservoir, r_o (Fig. 9.5).

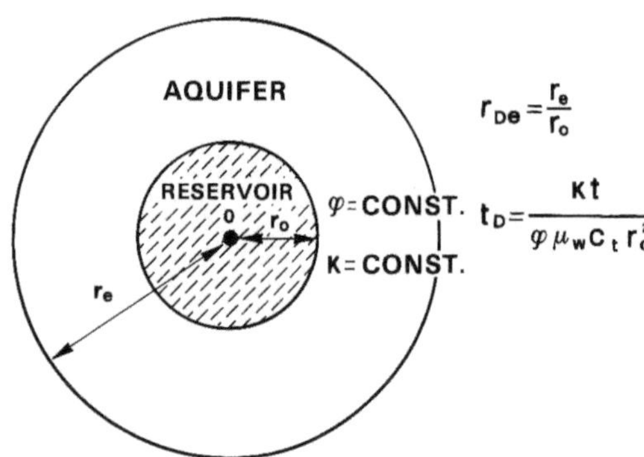

Fig. 9.5. *Schematic of a cylindrical reservoir surrounded by a uniform concentric aquifer of finite extent, as used by van Everdingen and Hurst in their solution of the diffusivity equation for the case of radial geometry*

In order to solve for the influx of water from the aquifer, we need to make appropriate modifications to Eqs. (5.14), replacing the wellbore radius r_w by the reservoir outer radius r_o:

$$r_D = \frac{r}{r_o}, \tag{5.14a$'$}$$

$$t_D = \frac{k}{\phi \mu_w c_t r_o^2} t \tag{5.14b$'$}$$

Suppose that water enters the reservoir across its boundary with the aquifer at a flow rate $e_w(t)$, following a drop in pressure Δp at time $t = 0$ at this boundary. We define a new dimensionless variable:

$$q_D = \frac{\mu_w}{2\pi k h} \frac{e_w}{\Delta p}. \tag{9.15}$$

where Δp is constant.

Therefore:

$$
\begin{aligned}
W_e(t) &= \int_0^t e_w \, dt = \int_0^t \frac{2\pi k h}{\mu_w} \Delta p \, q_D(t_D) \, dt \\
&= \int_0^{t_D} \frac{2\pi k h}{\mu_w} \Delta p \, q_D(t_D) \frac{\phi \mu_w c_t r_o^2}{k} \, dt_D \\
&= 2\pi r_o^2 h \phi c_t \, \Delta p \int_0^{t_D} q_D(t_D) \, dt_D ,
\end{aligned}
\tag{9.16}
$$

which can be expressed as:

$$W_e(t) = C \, \Delta p \, Q_D(t_D) , \tag{9.17a}$$

where:

$$C = 2\pi r_o^2 h \phi c_t . \tag{9.17b}$$

In the case of a reservoir/aquifer system in the form of the segment of a circle of angle θ (radians):

$$C = \theta r_o^2 h \phi c_t \ . \tag{9.17c}$$

Note that C represents twice the volume of water that would be expelled at equilibrium ($t \to \infty$) from an aquifer of the same size as the reservoir, following a unit pressure decrease Δp.

van Everdingen and Hurst[14] demonstrated that the following relationship exists between $p_D(t_D)$ and $Q_D(t_D)$:

$$\int_0^{t_D} \frac{dp_D(\tau)}{d\tau} Q_D(t_D - \tau) d\tau = \int_0^{t_D} p_D(t_D - \tau) \frac{dQ_D(\tau)}{d\tau} d\tau = t_D \ . \tag{9.18}$$

In other words, the convolution integral of $p_D(t_D)$ with $Q_D(t_D)$, calculated between zero and t_D, is equal to t_D.

van Everdingen and Hurst published values of $Q_D(t_D)$ and $p_D(t_D)$ for infinite and bounded radial aquifers (r_{De} between 1.5 and 10), and for infinite and bounded linear aquifers.

They were calculated subject to the classical conditions already mentioned in Chap. 5: the aquifer is of uniform thickness; porosity and permeability are uniform and constant; initial pressure is the same throughout the aquifer and equal to the initial pressure p_i of the reservoir.

These are the most significant limitations to the applicability of any reservoir engineering method for evaluating $W_e(t)$ from $Q_D(t_D)$.

van Everdingen and Hurst[14] presented values of $Q_D(t_D)$ in tabular and graphical form (Fig. 9.6).

For aquifers of limited extent, $Q_D(t_D)$ not surprisingly tends toward a constant value:

$$Q_D(\text{max}) = \tfrac{1}{2}(r_{De}^2 - 1) \ , \tag{9.19}$$

for which we have, via Eq. (9.17):

$$W_e(\text{max}) = \phi V_{aq} c_t \Delta p \ , \tag{9.20a}$$

where:

$$V_{aq} = \pi(r_e^2 - r_o^2)h \ . \tag{9.20b}$$

Note that Eq. (9.20a) is equivalent to Eq. (9.14b) which, as we have seen, corresponds to the equilibrium situation attained after a certain dimensionless time. In real terms, this equilibrium is reached more rapidly the larger the value of the constant:

$$\frac{t_D}{t} = \frac{k}{\phi \mu_w c_t r_o^2} \ , \tag{9.21}$$

Various authors have presented these $Q_D(t_D)$ functions as polynomials or exponentials for computational purposes.

For example, Table 9.1 is the listing of a computer subroutine for the calculation of $Q_D(t_D)$ for circular aquifers of infinite and finite extent ($r_{De} < 200$), and linear aquifers, accurate to within 1% of published values.

As seen in Sect. 5.7 the principle of superposition (Duhamel's theorem) was valid for the diffusivity equation.

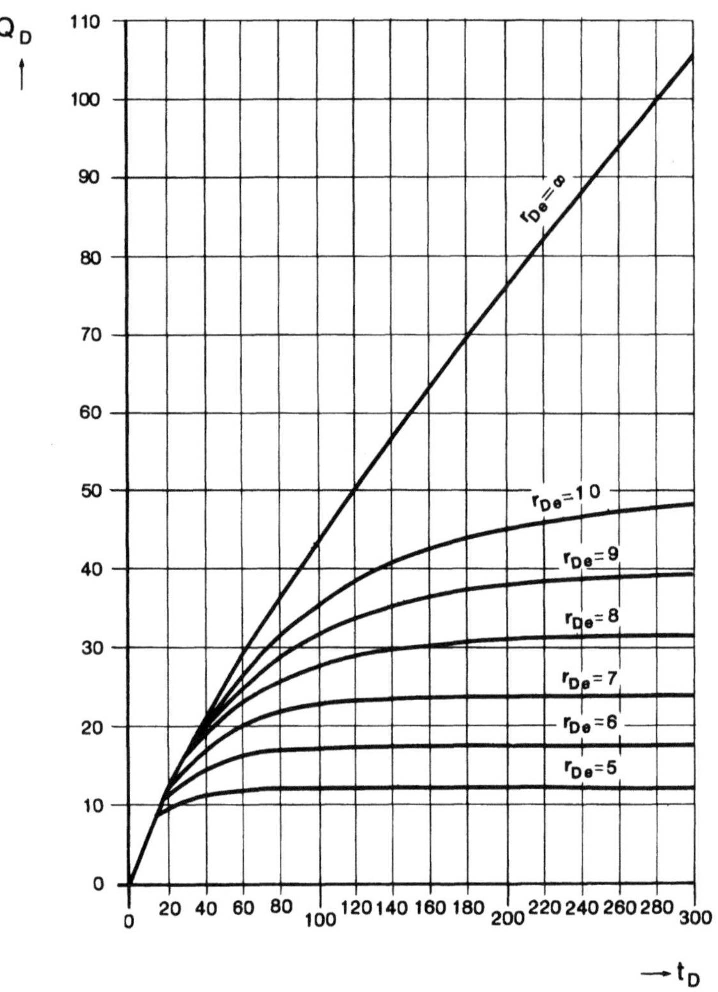

Fig. 9.6. *The function $Q_D(t_D)$ for a circular aquifer with $r_{De} \geq 5$. From Ref. 14, 1949, Society of Petroleum Engineers of AIME, reprinted with permission of the SPE*

We can therefore calculate the cumulative influx $W_e(t_n)$ of water into the reservoir at time t_n by approximating the aquifer pressure decline as a series of steps, starting at the initial value p_i at $t_0 = 0$, then:

p_1 at time t_1

p_2 at time t_2

. . . . . .

p_n at time t_n

and superposing the effects of each pressure drop (Fig. 9.3):

$$\Delta p_0 = \frac{p_i - p_1}{2} \quad \text{over the time interval } (t_n - 0)$$

$$\Delta p_1 = \frac{p_i - p_2}{2} \quad \text{over the time interval } (t_n - t_1)$$

Table 9.1. *Program[3] for the calculation of $Q_D(t_D)$ for linear aquifers, and circular aquifers of infinite and finite extent* $(r_{De} < 200)$

```
      SUBROUTINE QTD(Q,TD,RDE)
C                THIS SUBROUTINE IS DESIGNED TO CALCULATE THE Q(TD)
C                FUNCTION FOR THE CONSTANT TERMINAL PRESSURE CASE,
C                AS DEFINED BY VAN EVERDINGEN-HURST IN J. PET. TECH.
C                (DECEMBER 1949).
C                FOR SELECTING AQUIFER SIZE AND GEOMETRY USE:
C
C      RDE=0.    INFINITE LINEAR AQUIFER
C      RDE=1.    FINITE LINEAR AQUIFER
C      RDE=R     FINITE RADIAL AQUIFER WITH DIMENSIONLESS RADIUS
C                BETWEEN 2. AND 200.
C      RDE=1001. INFINITE RADIAL AQUIFER, TD LESS THAN 1.0E+11
      IF (RDE-1.) 203,302,306
  302 IF(TD.LE.0.25) GOTO 203
      IF(TD.GT.2.5) GOTO 305
      Q=0.25119+1.5522*TD-1.2888*TD**2+0.48855*TD**3-0.06944*TD**4
      GOTO 20
  305 Q=1.
      GOTO 20
  306 IF(TD.GT.0.01) GOTO 204
  203 Q=2.*SQRT(TD/3.1415)
      GOTO 20
  204 IF(RDE.GT.1000.) GOTO 206
      X=ALOG(RDE)
      TMIN=(-7.9152033+8.854302*X-2.2820631*X**2+0.23883683*X**3+
     &      0.110437671/X**5)
      IF(TMIN.GE.TD) GOTO 206
      A1=EXP(-1.1375115+1.8705656*X+0.033155554*X**2-
     &      0.0028802629*X**3-0.43012557/X)
      A2=EXP(-5.8976609+3.1917553*X-0.30111549*X**2+
     &      0.0096058316*X**3)
      B1=EXP(9.2970758-0.0067698009*X-0.19062197*X**2+
     &      0.010848388*X**3-9.7094342*(X**(1./3.)))
      B2=EXP(5.2786904-6.2218634*X+1.1091244*X**2-0.091446385*X**3)
      IF(B1*TD.LE.40.) GOTO 207
      IF(B2*TD.LE.40.) GOTO 213
      Q=(RDE**2-1.)/2.
      GOTO 20
  213 Q=(RDE**2-1.)/2.-2.*A2/EXP(B2*TD)
      GOTO 20
  207 IF(B2*TD.LE.40.) GOTO 209
      Q=(RDE**2-1.)/2.-2.*A1/EXP(B1*TD)
      GOTO 20
  209 Q=(RDE**2-1.)/2.-2.*(A1/EXP(B1*TD)+A2/EXP(B2*TD))
      GOTO 20
  206 YY=ALOG(TD)
      IF(TD.GT.100.) GOTO 402
      Q=EXP(.45222+.63124*YY+.018504*YY**2+.00011697*YY**3)
      GOTO 20
  402 IF(TD.GT.1.0E+5) GOTO 404
      Q=EXP(.34712+.66544*YY+.018592*YY**2-.00045818*YY**3)
      GOTO 20
  404 IF(TD.GT.1.0E+11) GOTO 406
      Q=EXP(-.24344+.81681*YY+.0054309*YY**2-.000069851*YY**3)
   20 RETURN
  406 WRITE(6,410)
  410 FORMAT('TD ABOVE 1.E+11')
      END
```

$$\Delta p_2 = \frac{p_1 - p_3}{2} \quad \text{over the time interval } (t_n - t_2)$$

. . . .

$$\Delta p_{n-1} = \frac{p_{n-2} - p_n}{2} \quad \text{over the time interval } (t_n - t_{n-1})$$

This results in the following summation:

$$W_e(t_n) = C \sum_{j=0}^{n-1} \Delta p_j Q_D(t_{D,n} - t_{D,j}) \,, \tag{9.22a}$$

which can also be expressed as:

$$W_e(t_n) = \frac{C}{2} \sum_{j=0}^{n-1} (p_{j-1} - p_{j+i}) Q_D(t_{D,n} - t_{D,j}) \,, \tag{9.22b}$$

with

$$p_{-1} = p_0 = p_i$$

The calculation is simplified by considering equal time intervals Δt_D. We can then substitute:

$$t_{D,j} = j \, \Delta t_D \,,$$

$$t_{D,n} = n \, \Delta t_D \,.$$

Equation (9.22a) then reduces to the form:

$$W_e(t_n) = C \sum_{j=0}^{n-1} \Delta p_j \, Q_D[(n - j)\Delta t_D] \,. \tag{9.23}$$

In the limit as the time step Δt_D approaches zero, Eq. (9.22a) becomes the integral:

$$W_e(t_n) = C \int_0^{t_{Dn}} \frac{d\Delta p}{d\tau} Q_D(t_{D,n} - \tau) d\tau \,, \tag{9.24}$$

which is none other than *the integral for the convolution of $\Delta p(t_D)$ with $Q_D(t_D)$.*

As a consequence of the properties of the convolution integral, Eq. (9.24) can be rearranged as:

$$W_e(t_n) = C \int_0^{t_{Dn}} \Delta p(t_{D,n} - \tau) \frac{dQ_D(\tau)}{d\tau} d\tau \,. \tag{9.25}$$

The calculation of $W_e(t)$ was first formulated in this way by Hurst[10] in 1943.

Numerical deconvolution techniques, widely used in seismic stratigraphy, can be used here to calculate $Q_D(t_D)$ from the curves of $W_e(t)$ and $[p_i - p(t)]$ obtained from the recorded history of the reservoir, even where data are noisy as a result of measurement errors.

This technique is certainly in use in at least one major oil company. Unfortunately, no work has been published on the topic, so it will not be discussed further here.

However, a method of deconvolution used to calculate aquifer "influence functions," starting from an equation very similar to Eq. (9.24), will be described in Sect. 9.5.

van Everdingen and Hurst's work was, in its time, highly acclaimed in reservoir engineering circles, even though it was published a full 2 years after the definitive treatise[1] on heat diffusion by Carslaw and Jäger, which presented the same solutions in a different context. It would appear that the petroleum industry was, on the whole, unaware of this earlier work.

With the passage of time, it has become apparent that the quality of results obtained using van Everdingen and Hurst's equation is very much a matter of chance, despite its rigorous mathematical treatment of the underlying physics. As a consequence, it has been the cause of many errors and misconceptions.

The reasons for this are quite simple. We need the following information in order to use van Everdingen and Hurst's equation:

- aquifer geometry (radial, linear, etc.),
  dimensions of the aquifer relative to those of the reservoir (r_{De} in the case of a circular aquifer),
- values of the aquifer thickness h; porosity ϕ; permeability k; and compressibility c_p (all assumed to be constant over the entire aquifer),
- values of the viscosity μ_w and compressibility c_w of the aquifer water.

Now, in reality, with the exception of μ_w and c_w, few if any of these parameters are known accurately before the reservoir is put on production. This is, however, precisely the time when the reservoir engineer is asked to design the development programme for the field, and the potential for error is, consequently, at a maximum.

Only after a sufficient period of production can early comparisons be made between the estimate of $W_e(t)$ calculated from material balance (Chap. 10), and that derived from a theoretical model such as the one described in this section. As more data becomes available, the key terms r_{De}, C and t_D/t can be "recalibrated" and the updated values used to predict (with an improved chance of success) the future ingression of water into the reservoir.

9.5 Aquifer Influence Functions – Their Calculation from Reservoir History and Their Use in Predicting $W_e(t)$

9.5.1 Calculation of $W_e(t)$ from the Constant Rate $p_D(t_D)$ Solution

In Sect. 9.4 it was explained that the solutions to the diffusivity equation for the case of constant rate influx across the internal boundary of the aquifer could be used to calculate $W_e(t)$. This provides an alternative to van Everdingen and Hurst's solution[14] $Q_D(t_D)$ for constant pressure at the inner boundary.

As a precursor to the concept of aquifer "influence functions", we will first examine this approach in some detail.

We have the dimensionless functions:

$$t_D = \frac{k}{\phi \mu_w c_t r_0^2} t \,, \tag{5.14b'}$$

$$p_D = \frac{2\pi k h}{e_w \mu_w}(p_i - p) \,, \tag{5.14c'}$$

p_D is defined only for $e_w = $ constant, so as to take into account the case of constant rate influx across the inner boundary of the aquifer.

We also have the appropriate solution $p_D = p_D(t_D)$ for the geometrical configuration of the aquifer/reservoir system in question.

In the case of a well, with a drainage area A_w, this is [from Eq. (6.20)]:

$$p_D(t_D) = 2\pi \frac{r_o^2}{A_w} t_D + \frac{1}{2}\ln t_D - \frac{1}{2}p_{D(MBH)}\left(\frac{r_o^2}{A_w} t_D\right) + 0.405 , \tag{9.26a}$$

where:

$$t_{DA} = \frac{r_o^2}{A_w} t_D \tag{9.26b}$$

and $p_{D(MBH)}$ is the Matthews, Brons and Hazebroek function (Sect. 6.4).

Equation (9.26) is valid for all flow regimes, from transient, through late transient, to pseudo-steady state. It was, however, derived for a *line source*, with sufficiently *large* t_D, and it can only be applied to the aquifer problem if certain conditions are met.

As seen in Sect. 5.7, any $p_D(t_D)$ solution to the diffusivity equation is amenable to the principle of superposition.

Suppose we wish to calculate $W_e(t)$ for a reservoir whose initial pressure p_i and pressure history $p(t)$ are known.

The function $e_w(t)$ can be approximated by a step-function consisting of equal steps $\Delta e_w = $ constant (Fig. 9.7). From Eq. (5.14c'), we then have:

At $t = t_1$

$$p_i - p_1 = \frac{\mu_w}{2\pi k h} \Delta e_{w,0} p_D(t_{D,1}) , \tag{9.27a}$$

from which:

$$\Delta e_{w,0} = \frac{2\pi k h}{\mu_w} \frac{p_i - p_1}{p_D(t_{D,1})} . \tag{9.27b}$$

At $t = t_2$

$$p_i - p_2 = \frac{\mu_w}{2\pi k h} [\Delta e_{w,0} p_D(t_{D,2}) + \Delta e_{w,1} p_D(t_{D,2} - t_{D,1})] , \tag{9.28a}$$

from which:

$$\Delta e_{w,1} = \frac{[2\pi k h/\mu_w](p_i - p_2) - \Delta e_{w,0} p_D(t_{D,2})}{p_D(t_{D,2} - t_{D,1})} \tag{9.28b}$$

At $t = t_3$

$$p_i - p_3 = \frac{\mu_w}{2\pi k h} [\Delta e_{w,0} p_D(t_{D,3}) + \Delta e_{w,1} p_D(t_{D,3} - t_{D,1})$$

$$+ \Delta e_{w,2} p_D(t_{D,3} - t_{D,2})] , \tag{9.29a}$$

from which:

$$\Delta e_{w,2} = \frac{[2\pi k h/\mu_w](p_i - p_3) - \Delta e_{w,0} p_D(t_{D,3}) - \Delta e_{w,1} p_D(t_{D,3} - t_{D,1})}{p_D(t_{D,3} - t_{D,2})} \tag{9.29b}$$

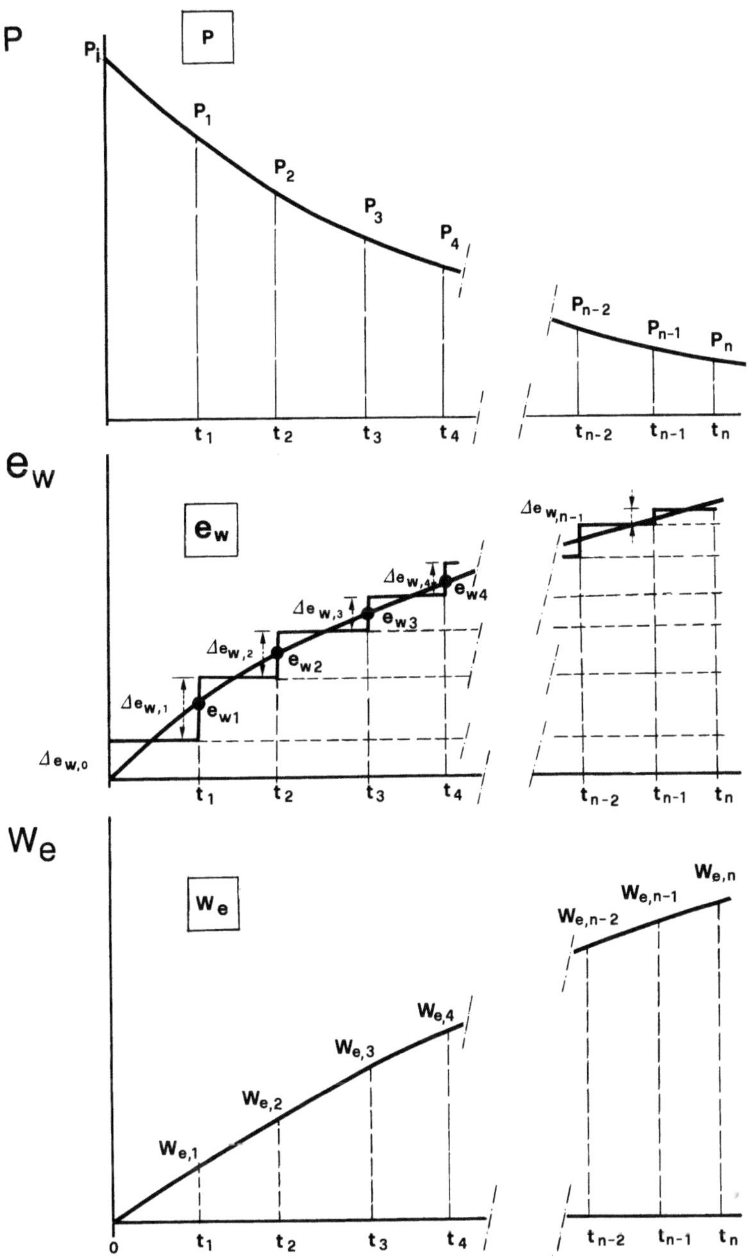

Fig. 9.7. *Influx rate $e_{w,j}$ and cumulative volume $W_{e,j}$ of aquifer water versus time, calculated using the $p_D(t_D)$ function Eq. (9.31), for a reservoir whose pressure history is known*

At $t = t_n$

$$p_i - p_n = \frac{\mu_w}{2\pi k h} \sum_{j=0}^{n-1} \Delta e_{w,j}\, p_D(t_{D,n} - t_{D,j}) \tag{9.30a}$$

and:

$$\Delta e_{w,n-1} = \frac{\dfrac{2\pi k h}{\mu_w}(p_i - p_n) - \displaystyle\sum_{j=0}^{n-2} \Delta e_{w,j}\, p_D(t_{D,n} - t_{D,j})}{p_D(t_{D,n} - t_{D,n-1})} . \tag{9.30b}$$

The cumulative water influx is now obtained by the summation:

$$W_e(t_n) = \sum_{j=0}^{n-1} \Delta e_{w,j}\,(t_n - t_j) \tag{9.30c}$$

and the problem is completely solved.

The calculation is simplified by imposing equal time intervals $t_j - t_{j-1} = $ constant $= \Delta t$, so that:

$$t_j = j\,\Delta t$$

$$t_n = n\,\Delta t$$

Equations (9.30) then reduce to:

$$p_i - p_n = \frac{\mu_w}{2\pi kh} \sum_{j=0}^{n-1} \Delta e_{w,j}\, p_D[(n-j)\,\Delta t_D]\,, \tag{9.31a}$$

$$\Delta e_{w,n-1} = \frac{\dfrac{2\pi kh}{\mu_w}(p_i - p_n) - \displaystyle\sum_{j=0}^{n-2} \Delta e_{w,j}\, p_D[(n-j)\,\Delta t_D]}{p_D(\Delta t_D)}\,, \tag{9.31b}$$

$$W_e(t_n) = \sum_{j=0}^{n-1} \Delta e_{w,j}\, [(n-j)\,\Delta t]\,. \tag{9.31c}$$

As the time interval Δt tends to zero, both Eqs. (9.30a) and (9.31a) are transformed into convolution integrals:

$$p_i - p_n = \frac{\mu_w}{2\pi kh} \int_0^{t_{Dn}} \frac{de_w}{d\tau}\, p_D(t_{D,n} - \tau)d\tau$$

$$= \frac{\mu_w}{2\pi kh} \int_0^{t_{Dn}} e_w(t_{D,n} - \tau)\frac{dp_D(\tau)}{d\tau}d\tau\,. \tag{9.32}$$

No $Q_D(t_D)$ function is needed to calculate $W_e(t)$ from Eq. (9.30) (or (9.31)). They simply use well-known functions of the form $p_D(t_D)$ which describe the declining pressure at the inner boundary of the aquifer as a result of a constant rate transfer of water to the reservoir.

9.5.2 Aquifer "Influence Functions"

When we have a record of the reservoir history in terms of $p(t)$ and $W_e(t)$, from Eqs. (9.32) we can, by deconvolution, calculate the $p_D(t_D)$ function out as far as the value of t_D which corresponds to total production time to date.

Hutchinson and Sikora[11] and Hicks et al.[9] published this technique within a few months of one another in 1959. It was subsequently presented in a more rational form by Coats et al.[6] in 1964.

These authors all availed themselves of an innovative mathematical formalism which is of great practical significance. It involved replacing $p_D(t_D)$ by the *aquifer influence function F(t)*, defined as:

$$F(t) = \frac{p_i - p(t)}{\Delta e_w}\,, \tag{9.33a}$$

so that, rearranging:

$$p(t) = p_\mathrm{i} - \Delta e_\mathrm{w} F(t) \; . \tag{9.33b}$$

$F(t)$ is the pressure drop $[p_\mathrm{i} - p(t)]$ after time t at the internal boundary of the aquifer associated with a unit increase $\Delta e_\mathrm{w} = 1$ in the water influx rate into the reservoir. *This function takes into account the geometry of the aquifer as well as the global distribution of petrophysical parameters and heterogeneity within it.*

Note that $F(t)$ is defined in terms of *real time* – not dimensionless, as was the case for $p_\mathrm{D}(t_\mathrm{D})$; in fact, there are *no* explicit constants in Eq. (9.33) (compare Eq. (5.14c′), which contains $\mu_\mathrm{w}/2\pi k h$). The constants are implicit in $F(t)$ itself.

Like $p_\mathrm{D}(t_\mathrm{D})$, $F(t)$ is a solution to the diffusivity equation, and the principle of superposition can be applied to it.

Unlike $p_\mathrm{D}(t_\mathrm{D})$ however, $F(t)$ cannot be calculated beforehand from the geometry and petrophysical parameters; it must be deduced from the reservoir history $W_\mathrm{e}(t)$ [and $e_\mathrm{w}(t)$] up to the current time t_n. Once established, $F(t)$ can be extrapolated forwards in time to predict future reservoir behaviour. The rules for performing this extrapolation are outlined next.

In order to calculate $F(t)$, the reservoir history, total duration t_n, is subdivided into equal time steps Δt. Then:

$$t_0 = 0$$

$$t_1 = \Delta t$$

$$t_2 = 2\,\Delta t$$

$$\dots\dots$$

$$t_j = j\,\Delta t$$

$$\dots\dots$$

$$t_n = n\,\Delta t$$

For convenience, we define:

$$F_j = F(j\,\Delta t) \tag{9.34}$$

so that Eq. (9.31) can be written as (see Fig. 9.7):

$$p_\mathrm{i} - p_n = \sum_{j=0}^{n-1} \Delta e_{\mathrm{w},j}\, F_{n-j} \tag{9.35a}$$

or, from the properties of the convolution integral:

$$p_\mathrm{i} - p_n = \sum_{j=0}^{n-1} e_{\mathrm{w},j}\,(F_{n-j} - F_{n-j-1}) \; . \tag{9.35b}$$

$F(t)$ can now be calculated directly from Eq. (9.35).
For example, using Eq. (9.35a) we have:

$$p_\mathrm{i} - p_1 = \Delta e_{\mathrm{w},0}\, F_n$$

$$p_\mathrm{i} - p_2 = \Delta e_{\mathrm{w},0}\, F_n + \Delta e_{\mathrm{w},1}\, F_{n-1}$$

$$p_\mathrm{i} - p_3 = \Delta e_{\mathrm{w},0}\, F_n + \Delta e_{\mathrm{w},1}\, F_{n-1} + \Delta e_{\mathrm{w},2}\, F_{n-2} \; .$$

.... etc. The $F(t)$ step function is then:

$$F_n = \frac{p_i - p_1}{\varDelta e_{w,0}}$$

$$F_{n-1} = \frac{p_1 - p_2}{\varDelta e_{w,1}}$$

$$F_{n-2} = \frac{p_2 - p_3}{\varDelta e_{w,2}}$$

$$\cdots\cdots$$

$$F_j = \frac{p_{n-j} - p_{n-j+1}}{\varDelta e_{w,n-j}}$$

$$\cdots\cdots$$

$$F_1 = \frac{p_{n-1} - p_n}{\varDelta e_{w,n-1}}$$

In reality, errors in the measurement of p, and in the evaluation of initial reserves on which the accuracy of $W_e(t)$ [and therefore $e_w(t)$] depend, often result in such severe fluctuations in the values of $F_1, F_2, \ldots F_n$ that an average (smoothed) $F(t)$ curve cannot be identified.

For this reason, $F(t)$ is best calculated by a computerised linear programming technique, as described by Coats et al.[6] There follows a brief outline of this approach.

As Coats demonstrated, the $F(t)$ function has to satisfy a number of constraints when $e_w > 0$, over the entire field from $t = 0$ to $t = t_n$:

$$F(t) \geq 0 \tag{9.36a}$$

$$\frac{dF(t)}{dt} > 0 \tag{9.36b}$$

$$\frac{d^2F(t)}{dt^2} < 0 \tag{9.36c}$$

In other words, $F(t)$ must be positive, increasing with t, and concave downwards. More generally:

$$\frac{d^{(2k+1)}F(t)}{dt^{(2k+1)}} > 0 \tag{9.36d}$$

$$\frac{d^{(2k)}F(t)}{dt^{(2k)}} < 0 \tag{9.36e}$$

For $e_w < 0$, the inequalities in Eqs. (9.36) are reversed.

The objective function to be minimised is the sum of the absolute values of the differences between measured $(p_i - p)$ and those calculated from Eqs. (9.35a) assuming $F(t)$ is known, subject to the constraints cited in Eqs. (9.36):

$$\varDelta = \sum_{k=1}^{n} \left| p_i - p_k - \sum_{j=0}^{k-1} \varDelta e_{w,j} F_{k-j} \right| . \tag{9.37}$$

An optimum function $F(t)$ producing a minimum error $\varDelta$, subject to the constraints listed in Eqs. (9.36), is calculated by a linear programming routine. The

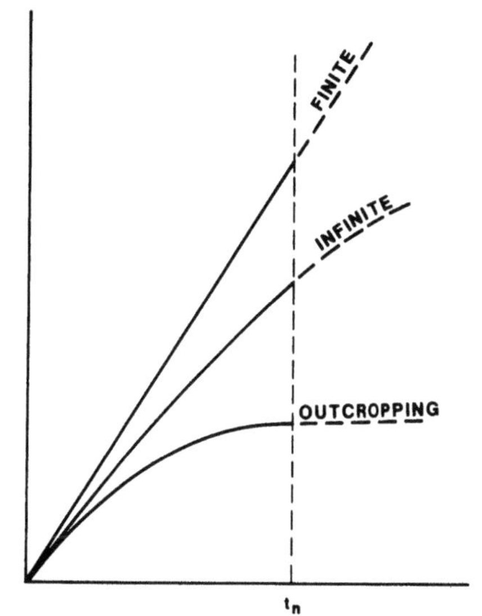

Fig. 9.8. *Behaviour of the aquifer influence function $F(t)$ for aquifers of finite and infinite extent, and for an aquifer with a constant pressure outer boundary*

maximum value of t for which $F(t)$ can be estimated is, of course, limited to t_n, the total time that the reservoir has been on production.

The manner in which $F(t)$ is extrapolated forward in time beyond t_n depends on the behaviour of the optimised $F(t)$ as t approaches t_n.

If it is *linear* over this final period (on a Cartesian plot), it means the aquifer is of finite extent (Fig. 9.8) since, by analogy with $p_D(t_D)$, pseudo-steady state flow conditions have already been attained.

In this case, for $t > t_n$, $F(t)$ can simply be extrapolated forward as a continuation of the linear trend.

When the final part of the $F(t)$ curve is *horizontal* (Fig. 9.8), such that $F(t) = $ constant, it means the aquifer is being fed from an external source (probably the surface), with a sufficient rate to maintain a constant pressure at its outer boundary. For $t > t_n$, the horizontal trend is simply extrapolated forwards.

Curvature in $F(t)$ as t approaches t_n is taken as an indication that pseudo-steady state flow conditions have not yet been reached and the aquifer is still infinite acting.

Now, in transient flow, we know that $p_D(t_D)$ is a straight line when plotted semi-logarithmically: i.e. as $p_D = f(\log t_D)$.

$F(t)$ should therefore be plotted against $\log t$, so that the semi-log trend can be extrapolated beyond $t = t_n$.

There is, of course, an element of risk in this, since we do not know just how long the aquifer will remain infinite acting, nor when the pressure disturbance will reach the outer boundaries (closed or constant pressure), causing $F(t)$ to deviate from the semi-log trend.

9.6 The Fetkovich Method for Finite Aquifers

In the early 1970s, Fetkovich[8] published a method for calculating water influx based on the concept of the *aquifer productivity index J_w*, defined as:

$$e_w(t) = J_w[\bar{p}_{aq}(t) - p(t)] \,, \tag{9.38}$$

where $\bar{p}_{aq}(t)$ is the *average pressure* of the aquifer at time t, and $p(t)$ is the pressure at its inner boundary with the reservoir, again at time t.

Fetkovich then made the somewhat questionable assumption that $J_w =$ constant throughout the entire life of the reservoir.

This amounts to saying that the flow of water towards the reservoir is *permanently* in either pseudo-steady state, in which case (Table 5.1):

$$J_w = \frac{kh}{C\mu_w} \frac{1}{\ln(r_e/r_o) - \frac{3}{4} + S},\tag{9.39}$$

or in steady state, in which case:

$$J_w = \frac{kh}{C\mu_w} \frac{1}{\ln(r_e/r_o) - \frac{1}{2} + S},\tag{9.40}$$

but the flow is never transient.

Note that in Eqs. (9.39) and (9.40), S represents a skin at the water/hydrocarbon contact (caused, for instance, by the presence of oxidised oil, or tarmat, etc.), and the constant C has the following values.

$C = 1/2\pi$ in SI units

$C = 19.03$ in practical metric units

$C = 141.2$ in oilfield units.

We saw in Sect. 5.4 that, during the transient period:

$$J_w(t) = \frac{e_w(t)}{p_i - p(t)} = \frac{2\pi kh}{\mu_w} \frac{1}{p_D(t_D)},\tag{9.41}$$

where p_i is the *initial pressure* in the aquifer, and

$$p_D(t_D) = \frac{1}{2} ei\left(\frac{\phi\mu_w c_t r_o^2}{4kt}\right) + S,\tag{9.42a}$$

when

$$\frac{kt}{\phi\mu_w c_t r_o^2} < 25\tag{9.42b}$$

and

$$p_D(t_D) = \frac{1}{2}\left(\ln\frac{kt}{\phi\mu_w c_t r_o^2} + 0.809\right) + S,\tag{9.43a}$$

when

$$\frac{kt}{\phi\mu_w c_t r_o^2} > 25,\tag{9.43b}$$

ei in Eq. (9.42a) is the exponential integral, which was presented in graphical form in Fig. 5.5.

Since the average reservoir radius r_o is very large, the transient behaviour of $p_D(t_D)$ will in most cases be described by Eq. (9.42a).

It is obvious from this that the Fetkovich method should only be used for finite aquifers with high permeability, where the duration of the transient period will be very short.

With this limitation in mind, Fetkovich's method is nonetheless an improvement over the "global" approach outlined in Sect. 9.3.

Within these constraints, the theory is developed as follows. From:

$$W_e = W_{ei}\left(1 - \frac{\bar{p}_{aq}}{p_i}\right),$$
(9.14a)

with

$$W_{ei} = \phi V_{aq} c_t p_i ,$$
(9.14b)

we have:

$$e_w = \frac{dW_e}{dt} = -\frac{W_{ei}}{p_i}\frac{d\bar{p}_{aq}}{dt} .$$
(9.44)

Combining Eqs. (9.38) and (9.44), we get:

$$-\frac{W_{ei}}{p_i}\frac{d\bar{p}_{aq}}{dt} = J_w(\bar{p}_{aq} - p) .$$
(9.45a)

Rearranging:

$$\frac{d\bar{p}_{aq}}{\bar{p}_{aq} - p} = -\frac{J_w p_i}{W_{ei}}dt ,$$
(9.45b)

so that

$$\ln(\bar{p}_{aq} - p) = -\frac{J_w p_i}{W_{ei}}t + C .$$
(9.45c)

At $t = 0$,

$$C = \ln(p_i - p) ,$$
(9.45d)

giving

$$\ln\frac{\bar{p}_{aq} - p}{p_i - p} = -\frac{J_w p_i}{W_{ei}}t ,$$
(9.46a)

which can be written as:

$$\bar{p}_{aq} - p = (p_i - p)e^{-(J_w p_i/W_{ei})t} .$$
(9.46b)

Finally, combining Eqs. (9.38) and (9.46b), we obtain:

$$e_w(t) = J_w(p_i - p)e^{-(J_w p_i/W_{ei})t} .$$
(9.47)

If a pressure step $(p_i - p) = $ constant is imposed at $t = 0$ at the inner boundary of the aquifer:

$$W_e(t) = \int_0^t e_w\,dt = J_w(p_i - p)\frac{W_{ei}}{J_w p_i}(1 - e^{-(J_w p_i/W_{ei})t})$$

$$= \frac{W_{ei}}{p_i}(p_i - p)(1 - e^{-(J_w p_i/W_{ei})t}) ,$$
(9.48)

which describes the behaviour of $W_e(t)$ for $\Delta p = $ constant. Note that as $t \to \infty$, Eq. (9.48) reduces to Eq. (9.14a).

In order to calculate $W_e(t)$ for a given reservoir pressure decline $p(t)$, the time is subdivided into equal intervals $\Delta t = $ constant. The pressure record is then

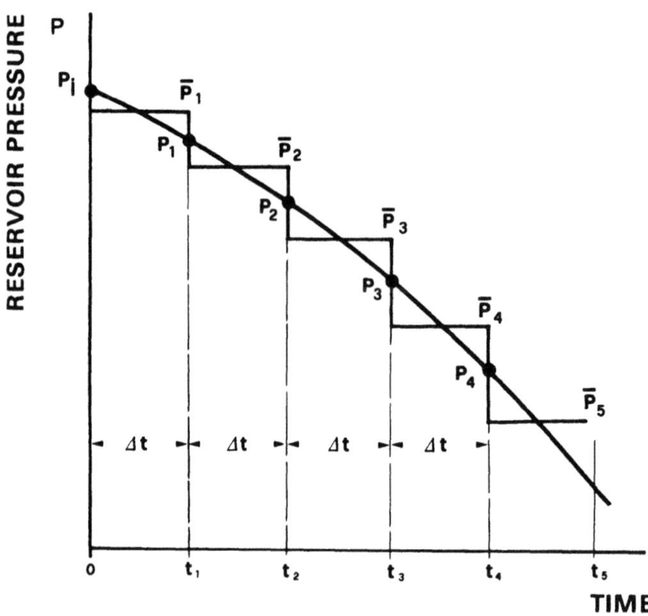

Fig. 9.9. *Step discretisation of the curve of pressure p(t) for the calculation of water influx by the Fetkovich method (Eq. 9.53)*

approximated as a series of steps, using the average pressure $\bar{p}_j$ in each interval j, as shown in Fig. 9.9:

$$\bar{p}_j = \frac{p_{j-1} + p_j}{2} . \tag{9.49}$$

We will simplify the equation by defining:

$$\frac{W_{ei}}{p_i}(1 - e^{-(J_w p_i/W_{ei})\Delta t}) = A = \text{const.} \tag{9.50}$$

so that

$$\Delta W_{e,1} = A(p_i - \bar{p}_1) \tag{9.51a}$$

$$\Delta W_{e,2} = A(\bar{p}_{aq,1} - \bar{p}_2) \tag{9.51b}$$

$$\cdots \cdots$$

$$\Delta W_{e,n} = A(\bar{p}_{aq,n-1} - \bar{p}_n) \tag{9.51c}$$

where:

$$\bar{p}_{aq,1} = p_i\left(1 - \frac{\Delta W_{e,1}}{W_{ei}}\right), \tag{9.52a}$$

$$\cdots \cdots$$

$$\bar{p}_{aq,n-1} = p_i\left(1 - \frac{\displaystyle\sum_{j=1}^{n-1} \Delta W_{e,j}}{W_{ei}}\right), \tag{9.52b}$$

We can now express the cumulative water influx as a summation:

$$W_{e,n} = \sum_{j=1}^{n} \Delta W_{e,j} = A \left\{ \sum_{j=1}^{n} \left[p_i \left(1 - \frac{\sum_{k=0}^{j-1} \Delta W_{e,k}}{W_{ei}} \right) - \bar{p}_j \right] \right\} \tag{9.53}$$

with $\Delta W_{e,0} = 0$.

Equation (9.53) contains three constants, one of which (p_i) is known, while the others (A and W_{ei}) must be estimated from the extent, thickness and properties of the aquifer.

These constants must satisfy two conditions:

$$W_{ei} > 0 \tag{9.54a}$$

$$0 < \frac{Ap_i}{W_{ei}} < 1 \tag{9.54b}$$

If the water influx history $W_e(t)$ of the reservoir is known, the constants A and W_{ei} can be calculated by linear programming techniques analogous to those described in Sect. 9.5.2. In a more simplistic approach, non-linear regression analysis may be used. It is, however, difficult to respect the constraints imposed in Eqs. (9.54) by this method.

In the case of a reservoir not yet on production, or with a very short production history, A and W_{ei} will have to be estimated using Eqs. (9.50), (9.14b), and (9.39), (9.40). Obviously, maximum use should be made of whatever information is available about the basin geology (to evaluate the lateral extent of the aquifer), and of petrophysical data from any wells which intercept the aquifer (to evaluate J_w).

9.7 Final Considerations

Over the last twenty years, interest in the various single cell model (SCM) methods described in the previous sections – where the reservoir is a single cell surrounded by an aquifer, all described by a single set of parameters – has given way to the vastly more powerful computer-based numerical modelling techniques (finite difference and finite element – see Chap. 13) using grid blocks and distributed parameters.

This increased complexity does not, of course, mean that the results will be any better! The reservoir engineer now has to define the *real* geometrical configuration of the aquifer (it is not necessarily circular or linear, etc.), and the distribution of thickness and petrophysical parameters over *every block* in the aquifer model.

Single parameter models do, however, find wide use where a simplistic approach is adequate, and a detailed knowledge of the distribution of pressure and saturation in the reservoir is not required. A typical case would be a medium-small gas reservoir of good permeability in contact with an aquifer, in which it would be reasonable to assume a uniform pressure throughout the reservoir, owing to the high gas mobility and the tendency for the water–gas contact to advance as a horizontal front.

The methods described in this chapter are well suited to this sort of situation, and can be used to estimate the cumulative volume of water influx $W_e(t)$. There are two distinct cases:

a)   Reservoir Not Yet on Production

Since there will be very little information available about the aquifer, a relatively simplistic approach can be taken to calculate $W_e(t)$: there is usually nothing to be gained in accuracy by using the more complex van Everdingen and Hurst method (Sect. 9.4).

The Hurst equation (Sect. 9.2.2) takes account to some extent of the transient phase in the aquifer, and the constants C_1 and C_2 can be estimated to a first approximation using Eq. (9.11).

As soon as some production data becomes available, these constants can be reevaluated so as to match the actual $W_e(t)$.

b)   Producing Reservoir

When a suitably long production history is available, it is recommended to use the aquifer influence functions $F(t)$ described in Sect. 9.5.2, extrapolating as explained to predict future behaviour.

Alternatively, the $Q_D(t_D)$ function (Sect. 9.4) can be derived by deconvolution of Eq. (9.24). Forward extrapolation of $Q_D(t_D)$ is then achieved by comparison with published curves.

References

1. Carslaw HS, Jäger JC (1947) Conduction of heat in solids. Clarendon Press, Oxford
2. Chatas AT (1966) Unsteady spherical flow in petroleum reservoirs. Soc Petrol Eng J (June): 102–114; Trans AIME 237
3. Chierici GL, Ciucci GM (1969) Application de programmes de production à débit cyclique aux gisements à poussée d'eau. Influence sur les limites d'indétermination dans le calcul des reserves. Rev Inst Fr Pétrole (March) 24(3): 337–350
4. Chierici GL, Long G (1959) Compressibilité et masse spécifique des eaux de gisement dans les conditions des gisements. Application à quelques problèmes de reservoir engineering. Proc Fifth World Petrol Congr, New York, 1–5 June 1959 Fifth World Petroleum Congresses, New York, pp 187–210
5. Coats KH, Tek MR, Katz DL (1959) Unsteady state liquid flow through porous media having elliptic boundaries. Trans AIME 216: 460–464
6. Coats KH, Rapoport LA, McCord JR, Drews WP (1964) Determination of aquifer influence functions from field data. J Petrol Tech (Dec): 1417–1424; Trans AIME 231
7. Dake LP (1978) Fundamentals of reservoir engineering. Elsevier, Amsterdam
8. Fetkovich MJ (1971) A simplified approach to water influx calculations – finite aquifer systems. J Petrol Tech (July): 814–828
9. Hicks AL, Weber AG, Ledbetter RL (1959) Computing techniques for water-drive analysis. Trans AIME 216: 400–402
10. Hurst W (1943) Water influx into a reservoir and its application to the equation of volumetric balance. Trans AIME 151: 57–72
11. Hutchinson TS, Sikora VJ (1959) A generalized water-drive analysis. Trans AIME 216: 169–178
12. Nabor GW, Barham RH (1964) Linear aquifer behaviour. J Petrol Tech (May): 561–563, Trans AIME 231
13. Schilthuis RJ (1936) Active oil and reservoir energy. Trans AIME 118: 33–52
14. van Everdingen AF, Hurst W (1949) The application of the Laplace transformation to flow problems in reservoirs. Trans AIME 186: 305–324

EXERCISES

Exercise 9.1

A layer of coarse sandstone is 80 m thick and in the form of an anticline, bounded laterally by sealing faults.

The area within the faulted boundary is 42 km^2. The layer is oil-bearing at the crest of the structure, over an area of 1.6 km^2. The initial reservoir pressure was 250 kg cm^{-2}, and temperature 90 °C.

Aquifer water salinity is 100 kg/m^3.
The formation porosity is 25%, with pore compressibility $c_p = 2 \times 10^{-4}$ cm^2 kg^{-1}.

The reservoir has been on production for 4 yr; the average reservoir pressure, measured at yearly intervals, was as follows:

t (years)	$\bar{p}$ (kg cm^{-2})
0	250
1	242
2	230
3	214
4	194

Calculate the maximum volume of water which could have entered the reservoir as a function of time, using the global equation from Sect. 9.3.

Solution

From Eq. (9.14), we have:

$$W_e = W_{ei}\left(1 - \frac{p}{p_i}\right), \tag{9.14a}$$

where

$$W_{ei} = \phi V_{aq} c_t p_i . \tag{9.14b}$$

and

$$c_t = c_w + c_p$$

For water with a salinity of 100 kg m^{-3}, we can read from a correlation chart, at 90 °C:

$$c_w = 3.7 \, 10^{-5} \text{ cm}^2 \text{ kg}^{-1}$$

Therefore

$$c_t = 2.37 \, 10^{-4} \text{ cm}^2 \text{ kg}^{-1}$$

Then:

$$V_{aq} = (42 - 1.6) \times 10^6 \times 80 = 3.232 \times 10^9 \text{ m}^3$$

and therefore:

$$W_{ei} = 0.25 \times 3.232 \times 10^9 \times 2.37 \times 10^{-4} \times 250 = 47.874 \times 10^6 \text{ m}^3$$

The *maximum* volume of water that can have entered the reservoir from the aquifer, assuming an effectively infinite aquifer permeability, is therefore:

Year	Pressure $(\mathrm{kg\,cm^{-2}})$	W_e $(\mathrm{m^3 \times 10^6})$
0	250	zero
1	242	1.532
2	230	3.830
3	214	6.894
4	194	10.724

For comparison, the initial reservoir hydrocarbon volume was 25×10^6 m^3.

◇ ◇ ◇

Exercise 9.2

For the reservoir of Ex. 9.1, calculate the volume $W_e(t)$ of water influx at $t = 1, 2$ and 3 yr, under the following conditions:

Case A: aquifer permeability $k = 20$ md

Case B: $k = 100$ md

Case C: $k = 500$ md

Solution

If we assume the reservoir and aquifer to be circular, then:

- area of reservoir: $\qquad\qquad\qquad A_o = 1.6$ km^2
- total area of reservoir + aquifer: $\quad A_e = 42$ km^2
- reservoir radius: $\qquad\qquad\qquad r_o = \sqrt{(A_o/\pi)} = 713.65$ m
- external radius of aquifer: $\qquad r_e = \sqrt{(A_e/\pi)} = 3656.4$ m

$$r_{\mathrm{De}} = \frac{r_e}{r_o} = 5.12$$

The water viscosity (from a correlation chart) is:

$$\mu_w = 0.38 \text{ cP}$$

In practical metric units, we have:

$$t_{\mathrm{D}} = \frac{kt}{\phi \mu_w c_t r_o^2} \qquad\qquad\qquad (5.14b')$$

$$= \frac{k(\mathrm{md}) \times 9.869233 \times 10^{-16} t(\mathrm{yr}) \times 86\,400 \times 365.25}{\phi \mu_w(\mathrm{cP}) \times 10^{-3} [c_t(\mathrm{cm^2\,kg^{-1}})/98\,066.5] r_o^2(\mathrm{m^2})}$$

$$= 0.26636 \, \frac{k(\mathrm{md}) \times t(\mathrm{yr})}{\phi \mu_w(\mathrm{cP}) \times c_t(\mathrm{cm^2\,kg^{-1}}) \times r_o^2(\mathrm{m^2})}$$

Therefore:

k (md)	t_{D}
20	$5.327 \times t(\mathrm{yr})$
100	$26.636 \times t(\mathrm{yr})$
500	$133.18 \times t(\mathrm{yr})$

Furthermore:

$$C = 2\pi r_o^2 h \phi c_t$$

$$= 2\pi \times 713.65^2 \times 80 \times 0.25 \times 2.37 \times 10^{-4}$$

$$= 15\,168 \, \frac{\text{m}^3}{\text{kg cm}^{-2}} \tag{9.17b}$$

$p(t)$ can now be discretised as shown in Fig. E9/2.1. The step values are:

$$\Delta p_0 = \frac{p_i - p_1}{2} = \frac{250 - 242}{2} = 4 \, \text{kg cm}^{-2} \quad \text{at } t = 0$$

$$\Delta p_1 = \frac{p_i - p_2}{2} = \frac{250 - 230}{2} = 10 \, \text{kg cm}^{-2} \quad \text{at } t = t_1$$

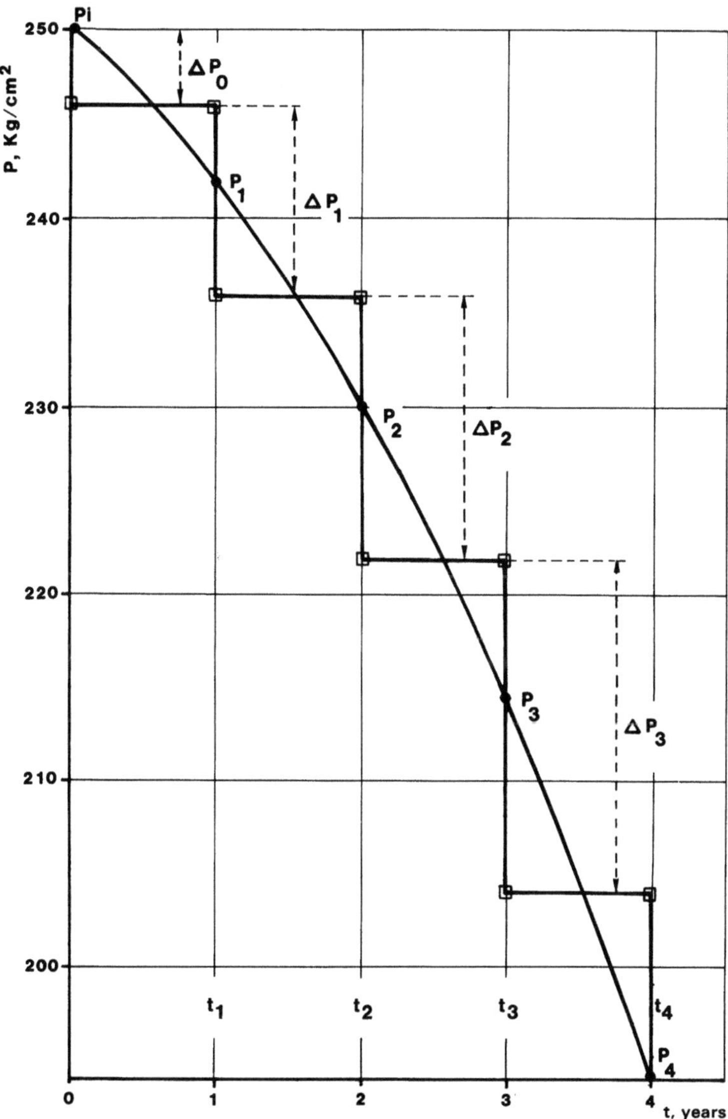

Fig. E9/2.1

$$\Delta p_2 = \frac{p_1 - p_3}{2} = \frac{242 - 214}{2} = 14 \, \text{kg cm}^{-2} \quad \text{at } t = t_2$$

$$\Delta p_3 = \frac{p_2 - p_4}{2} = \frac{230 - 194}{2} = 18 \, \text{kg cm}^{-2} \quad \text{at } t = t_3$$

If we divide the time into equal intervals (1 yr) the convolution integral takes the form of Eq. (9.23):

$$W_e(t_n) = C \sum_{j=0}^{n-1} \Delta p_j Q_D[(n-j)\Delta t_D] \, . \tag{9.23}$$

The calculations can be presented as a series of tables, as follows:

$k = 20 \, \text{md}$

j	$j\Delta t_D$	$Q_D(j\Delta t_D)$	Δp_j	$\sum_{j=0}^{n-1} \Delta p_j Q_D[(n-j)\Delta t_D]$	W_e $(\text{m}^3 \times 10^6)$
0	0	0	4	0	0
1	5.327	4.691	10	18.764	0.285
2	10.654	7.217	14	75.778	1.149
3	15.981	8.874	18	173.340	2.629
4	21.308	9.953	–	314.028	4.763

$k = 100 \, \text{md}$

j	$j\Delta t_D$	$Q_D(j\Delta t_D)$	Δp_j	$\sum_{j=0}^{n-1} \Delta p_j Q_D[(n-j)\Delta t_D]$	W_e $(\text{m}^3 \times 10^6)$
0	0	0	4	0	0
1	26.636	10.657	10	42.628	0.647
2	53.272	11.829	14	153.886	2.334
3	79.908	11.980	18	315.408	4.784
4	106.544	12.0	–	525.232	7.967

$k = 500 \, \text{md}$

j	$j\Delta t_D$	$Q_D(j\Delta t_D)$	Δp_j	$\sum_{j=0}^{n-1} \Delta p_j Q_D[(n-j)\Delta t_D]$	W_e $(\text{m}^3 \times 10^6)$
0	0	0	4	0	0
1	133.18	12	10	48	0.728
2	266.36	12	14	168	2.548
3	399.54	12	18	336	5.096
4	532.72	12	–	552	8.372

The values of $W_e(t)$ calculated for each of the aquifer permeabilities are presented graphically in Fig. E9/2.2, together with the values derived using the step approximation to $p(t)$ with the global equation as in Ex. 9.1.

There are two important points to note:

1. With a finer discretisation of $p(t)$ (smaller Δt intervals), $W_e(t)$ calculated from the global equation would be almost identical to the $W_e(t)$ calculated for $k = 500 \, \text{md}$.
2. As k increases, $W_e(t)$ approaches the curve calculated with the global equation.

◇ ◇ ◇

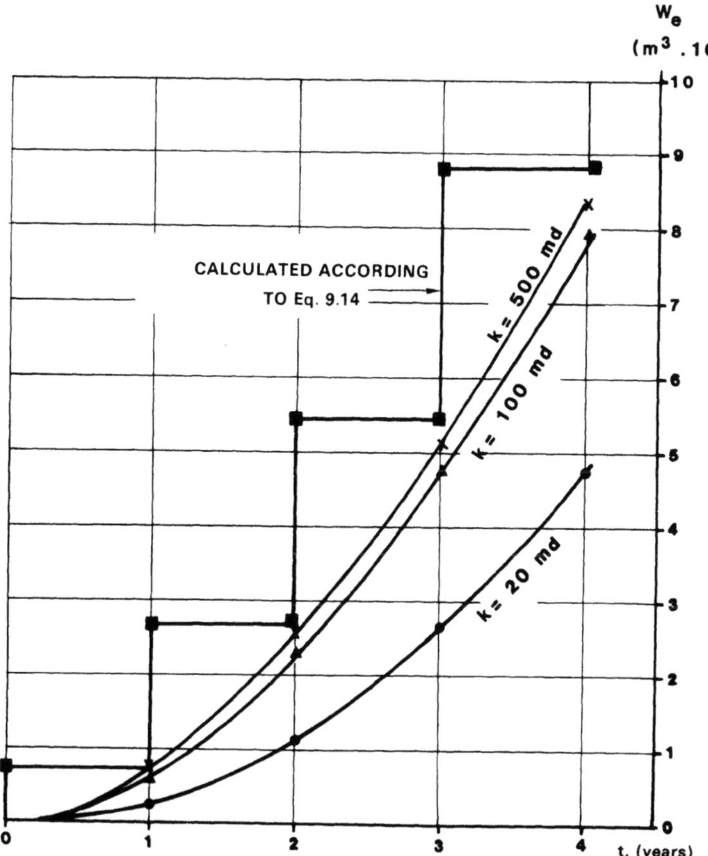

Fig. E9/2.2

Exercise 9.3

The reservoir is the same as in Exs. 9.1 and 9.2. Calculate the water influx as a function of time by the Fetkovich method (Sect. 9.6) for the same three permeabilities $k = 20, 100$ and 500 md.

Compare the results with those calculated in Ex. 9.2 by the van Everdingen and Hurst method.

Solution

From Eq. (5.46), ignoring the skin factor S, we have:

$$\bar{p} - p_{wf} = \frac{q\mu_w}{2\pi kh}\left(\ln\frac{r_e}{r_w} - \frac{3}{4}\right). \tag{5.46}$$

If we express $q = e_w$ in $\text{m}^3\,\text{yr}^{-1}$ and the other parameters in practical metric units, then:

$$98\,066.5\,(\bar{p} - \bar{p}_{wf})[\text{kg cm}^{-2}] = \frac{e_w(\text{m}^3\,\text{yr}^{-1})}{365.25 \times 86\,400}\;\frac{\mu_w(\text{cP}) \times 10^{-3}}{2\pi h(\text{m})}\;\frac{\ln(r_e/r_w) - 3/4}{k(\text{md}) \times 9.869233 \times 10^{-16}}$$

so that

$$J_w\left(\frac{\text{m}^3\,\text{yr}^{-1}}{\text{kg cm}^{-2}}\right) = \frac{e_w(\text{m}^3\,\text{yr}^{-1})}{(p - p_{wf})\text{kg cm}^{-2}}$$

$$= 19.19\,\frac{k(\text{md})h(\text{m})}{\mu_w(\text{cP})}\;\frac{1}{\ln(r_e/r_w) - 3/4}$$

The following data apply in this particular:

$$h = 80\ \text{m}$$

$$\mu_w = 0.38\ \text{cP}$$

$$r_e = 3656.4\ \text{m}$$

$$r_w = r_o = 713.65\ \text{m}$$

Therefore:

$$J_w\left(\frac{\text{m}^3\,\text{yr}^{-1}}{\text{kg cm}^{-2}}\right) = 4570.95 \times k(\text{md})$$

Case A: $k = 20$ md

$$J_w = 4570.95 \times 20 = 91\,419.1\ \frac{\text{m}^3\,\text{yr}^{-1}}{\text{kg cm}^{-2}}$$

$$W_{ei} = 47.874 \times 10^6\ \text{m}^3$$

$$p_i = 250\ \text{kg cm}^{-2}$$

$$A = \frac{W_{ei}}{p_i}(1 - e^{-(J_w p_i / W_{ei})\Delta t}) \tag{9.50}$$

If we stipulate that:

$$\Delta t = 1\ \text{yr}$$

then

$$A = 72\,692.28\ \frac{\text{m}^3}{\text{kg cm}^{-2}}$$

The pressure history can be discretised in the same way as for the previous exercise (Fig. E9/2.1). We will then get the following results:

t (yr)	$\bar{p}_{aq}$ (kg cm^{-2})	$\bar{p}_j$ (kg cm^{-2})	$\Delta W_{e,j}$ (m$^3 \times 10^6$)	$W_e(t)$ (m$^3 \times 10^6$)
0	250	–	0	0
1	248.48	246	0.291	0.291
2	243.74	236	0.907	1.198
3	235.49	222	1.580	2.778
4	223.53	204	2.289	5.067

Case B: $k = 100$ md

$$J_w = 4570.95 \times 100 = 457\,095\ \frac{\text{m}^3\,\text{yr}^{-1}}{\text{kg cm}^{-2}}$$

and

$$A = 173\,896\ \frac{\text{m}^3}{\text{kg cm}^{-2}}$$

t (yr)	$\bar{p}_{aq}$ (kg cm^{-2})	$\bar{p}_j$ (kg cm^{-2})	$\Delta W_{e,j}$ (m$^3 \times 10^6$)	$W_e(t)$ (m$^3 \times 10^6$)
0	250	–	0	0
1	246.37	246	0.696	0.696
2	236.95	236	1.803	2.499
3	223.37	222	2.600	5.099
4	205.78	204	3.368	8.467

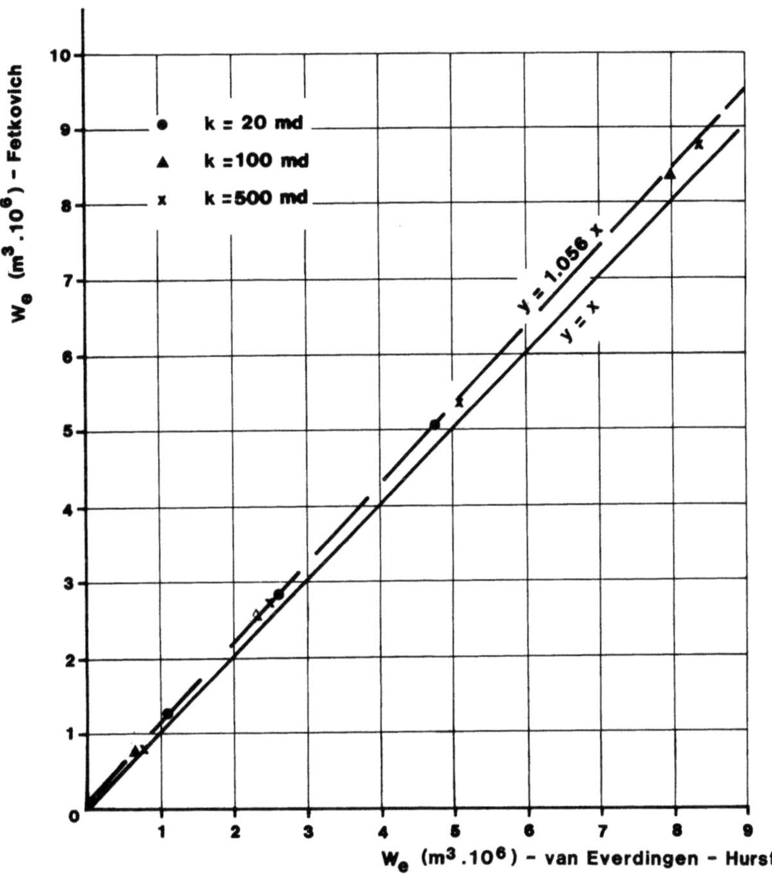

Fig. E9/3.1

Case C: $k = 500$ md

$$J_{\mathrm{w}} = 4570.95 \times 500 = 2\,285\,475 \,\frac{\mathrm{m^3\,yr^{-1}}}{\mathrm{kg\,cm^{-2}}}$$

and

$$A = 1\,91\,495 \,\frac{\mathrm{m^3}}{\mathrm{kg\,cm^{-2}}}$$

t (yr)	$\bar{p}_{\mathrm{aq}}$ (kg cm^{-2})	$\bar{p}_j$ (kg cm^{-2})	$\Delta W_{\mathrm{e},j}$ (m$^3 \times 10^6$)	$W_{\mathrm{e}}(t)$ (m$^3 \times 10^6$)
0	250	–	0	0
1	246.00	246	0.766	0.766
2	236.00	236	1.915	2.681
3	222.00	222	2.681	5.362
4	204.00	204	3.447	8.809

Fig. E9/3.1 compares $W_{\mathrm{e}}(t)$ calculated by the Fetkovich and van Everdingen/Hurst methods. They are in excellent agreement, with the Fetkovich results (y-axis) 5.6% higher than those from van Everdingen and Hurst (x-axis) for all permeabilities.

10 The Material Balance Equation

10.1 Introduction

The material balance equation was first formulated by Schilthuis[14] and introduced in 1936. Until well into the 1950s it was the only means of determining the production drive mechanism in oil and gas reservoirs, and of predicting their behaviour.

The equation, in its various forms, represents a very simple and intuitive statement of a physical fact – the sum:

> (volume of reservoir occupied by hydrocarbons and interstitial water) +
> (volume of water influx from the aquifer and/or injection from surface) +
> (reduction in pore volume caused by the compaction of the reservoir rock
> induced by reduction in reservoir pressure)

remains constant throughout the life of the reservoir, and is equal to the initial interconnected pore volume of the reservoir (Fig. 10.1), hence the reference to "material balance".

The material balance equation describes the *whole* reservoir using global average terms (mean pressure, total oil, gas and water volumes present above the *initial* oil/water or gas/water contacts). It was the culmination of long series of theoretical and experimental studies which began in 1914, when the Oil Division of the U.S. Bureau of Mines (USBM) was founded in Bartlesville, Oklahoma. This marks the time when "reservoir engineering" emerged from within petroleum engineering as a recognised science in its own right.

It is worth giving a brief overview of the evolution of reservoir engineering from those early days until now.[1]

By 1914, industrial exploitation of oil reserves had already been underway for 55 years (the beginning is usually attributed to the drilling of the famous Titusville well in Pennsylvania by one "Colonel" Drake in 1859).

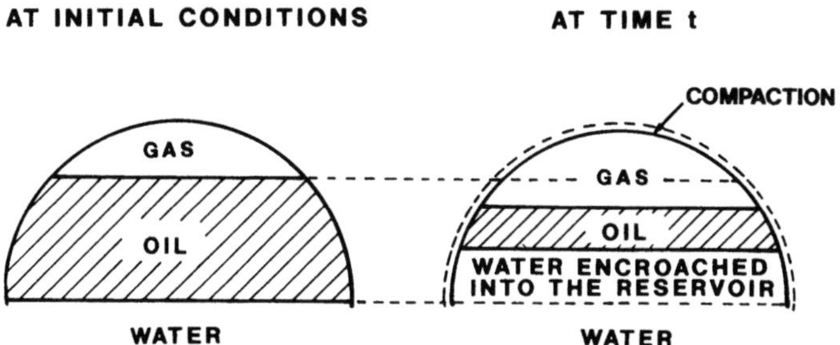

Fig. 10.1. *Graphic representation of the material balance equation*

Between 1859 and the early 1900s, as more and more wells were drilled, the reservoirs were produced in a wholly haphazard and undisciplined fashion. This led inevitably to the premature watering out of many wells, if not entire fields, excessive gas production, well collapse, and numerous other catastrophes of economic significance.

The founding of the Bartlesville laboratory was only a first step by the U.S. government towards gaining an understanding of the mechanisms of hydrocarbon production so as to be able to control subsequent reservoir behaviour. A second and more important step came with the decision in 1924 by the then president Calvin Coolidge to set up a Conservation Board at federal government level. The Board's mandate was to "make recommendations concerning the duty of federal government to regulate the methods used by the petroleum industry, with the aim of preventing the damaging of oil and gas reservoirs".

1927 saw yet another significant step with the setting up of the Development and Production Engineering Division of the American Petroleum Institute (API).

Up to the late 1920s, a major part of the research undertaken by the USBM and the API was concentrated on the thermodynamic behaviour of reservoir fluids, and the development of methods and equipment for the study of reservoir rock properties. The importance of dissolved gas in the production of oil was realised (at long last!), and the fundamental concepts concerning the displacement of oil by water in the reservoir were laid down.

From the early 1930s the petroleum industry (complying somewhat begrudgingly with President Coolridge's commercially restrictive initiative), began to come to grips on a large scale with the problems of reservoir management. There followed the first electrical wireline logs, the first notions of permeability, relative permeability and capillary pressure and the equipment to measure them, and an appreciation of the importance of the phase diagram to describe the behaviour of reservoir fluids. In particular, the phenomenon of retrograde condensation was both discovered and explained.

Other scientists, mostly physicists of European origin, developed the equations describing the mono- and polyphasic flow of fluids through porous media.[7, 11] This culminated with the work of van Everdingen and Hurst[16] in 1949, and Muskat's second book.[12] By this time, Schilthuis had published his work on the material balance equation.

Unfortunately, as has already been mentioned, the equation is only applicable to the reservoir *as a whole*, using globally averaged parameters. It cannot, therefore, be applied to individual well behaviour to describe the variation of flow rate, pressure, gas/oil, water/oil and water/gas ratios, nor the vertical and areal movement of oil/water or gas/oil fronts, etc., as each well is produced through time.

We should remember, however, that up to the late 1950s the limited computational capabilities of the time would not have permitted engineers to undertake anything more complex anyway. For example, as we saw in Chap. 9, to perform a water influx calculation, the reservoir had to be approximated by a cylindrical (Fig. 10.2) or linear geometry, with all the inherent errors that this might entail.

It was in the early 1960s that high-speed third generation computers and numerical techniques appeared, and now it became possible to simulate the movement of fluids in the reservoir and the aquifer, while retaining the true geometry of the system. The areal and vertical movements of fluid contacts could thus be modelled and predicted.

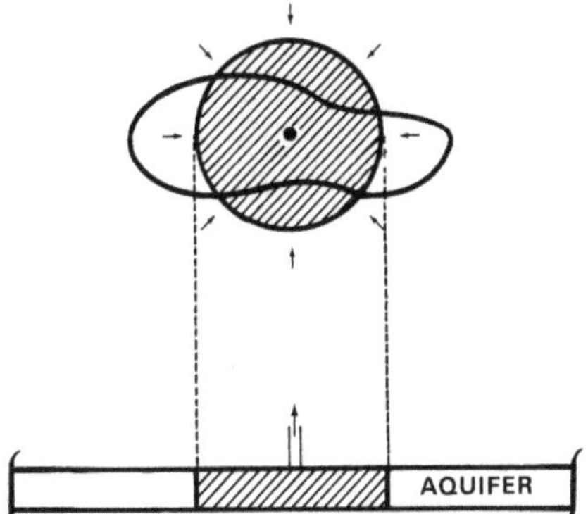

Fig. 10.2. *Schematic of a reservoir + aquifer represented as a single cell with radial geometry. From Ref. 1, reprinted with permission of Multi-Science Publishing Co. Ltd.*

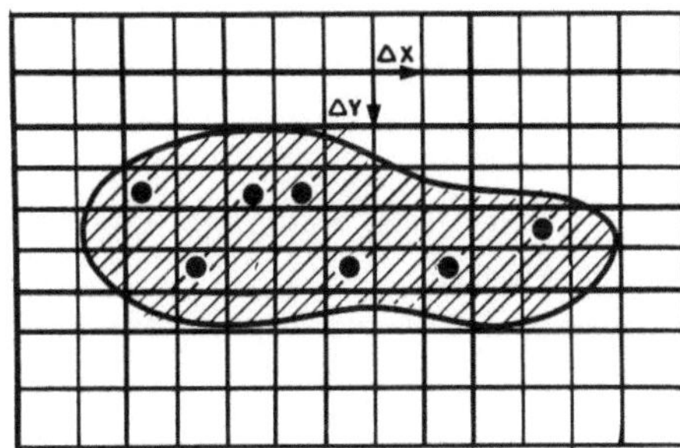

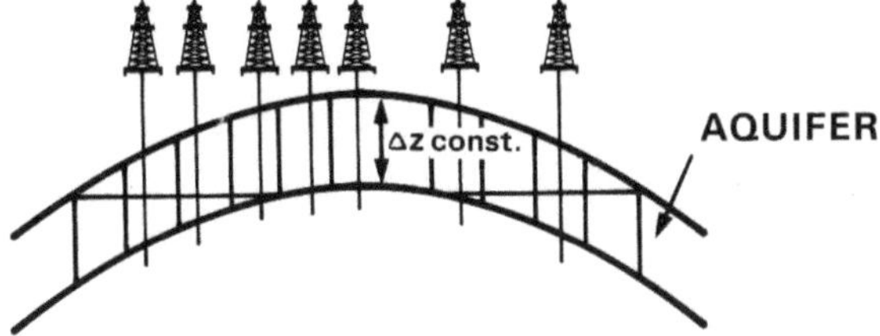

Fig. 10.3. *Two-dimensional (x–y) model of a reservoir + aquifer using grid blocks. From Ref. 1, reprinted with permission of Multi-Science Publishing Co. Ltd.*

Chapter 13 (Vol. 2) is devoted entirely to this technique of "numerical reservoir simulation".

To demonstrate the significance of this approach, Figs. 10.3 and 10.4 show how a reservoir which, in the context of the material balance equation, is simulated as a single "cell", can be subdivided into a number of small cells (grid blocks) both areally and, in the case of a thick or multi-zoned reservoir, vertically (Fig. 10.4). A large number of grid blocks can be specified. Oil or gas flows between cells and is extracted from those located at the wells.

Numerical simulation allows us to calculate, for each cell, the variation with time of its average pressure and fluid saturations, and, where there is a well, the relative flow rates. A correctly adjusted model, capable of matching past production data, can therefore be used to predict these important parameters according to any given production programme.

Obviously, judicious use of a numerical simulation will provide information which is far more detailed and reliable than the results of simplistic material balance calculations. Consequently, use of the material balance equation is rather limited at present, and as far as oil reservoirs are concerned, its main application is for a global evaluation of the past behaviour of the reservoir – in particular the

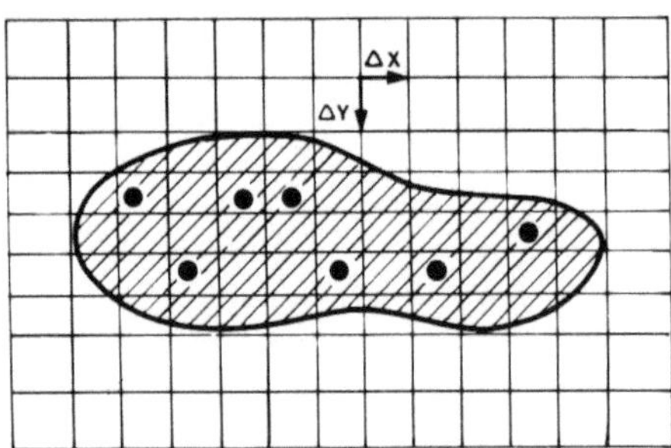

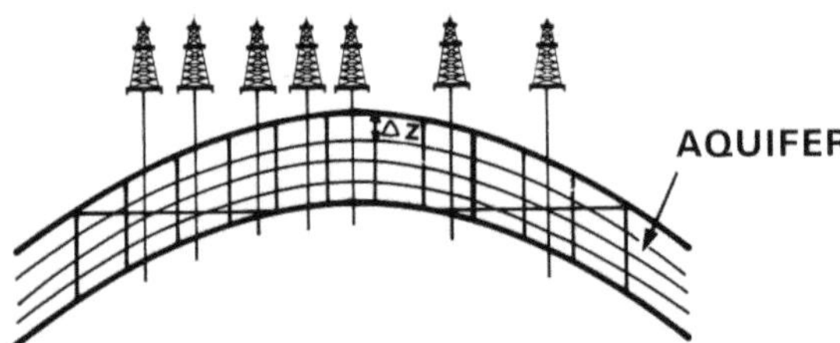

Fig. 10.4. *Three-dimensional (x–y–z) model of a reservoir + aquifer using grid blocks. From Ref. 1, reprinted with permission of Multi-Science Publishing Co. Ltd.*

analysis of the contribution made by the various drive mechanisms (water drive, dissolved gas drive, gas cap expansion, rock compaction).

Developed oil reservoirs are generally characterised by quite significant areal variations in pressure caused by the different production rates from each well. This poses an immediate computational problem in the material balance equation: what do we use for the average reservoir pressure $p(t)$ at any specified time? This will be the first topic to be dealt with in the following pages.

In developed gas reservoirs, on the other hand, because of the low gas viscosity, the reservoir pressure tends to remain fairly uniform throughout (except in the case of very low permeability). Consequently, the material balance equation is more widely applicable, especially in smaller reservoirs, to both the analysis of the drive mechanism and semi-quantitative prediction of future behaviour.

Details of the techniques used will be provided in the following sections.

10.2 Average Reservoir Pressure

The material balance equation is formulated in terms of the average *static* reservoir pressure $\bar{p}(t)$ at time t, and the corresponding thermodynamic properties $B_g(\bar{p})$, $B_o(\bar{p})$, $B_w(\bar{p})$, $R_s(\bar{p})$, $c_t(\bar{p})$, etc.

We therefore need to know the value of $\bar{p}(t)$ at any chosen time. Direct measurement would require closing in all the wells and noting the bottom-hole pressure long enough afterwards for the localised pressure gradients caused by fluid movement to have balanced out throughout the reservoir.

This is obviously not a widely acceptable approach because lost production represents lost revenue (except in extreme cases such as the closure of a field because production has exceeded the agreed quota, or a war, etc.).

A means must therefore be found to *calculate* the average static reservoir pressure, still using measurements made in individual wells, but not entailing a total shut-down.

It is generally accepted that the average static pressure of the reservoir is equal to the weighted mean of the average pressures $\bar{p}_j(t)$ of the drainage areas of the

n wells in the field. Each term $\bar{p}_j(t)$ is weighted by the pore volume $V_{p,j}$ of its drainage area, and is referenced to a common datum depth.

The problem can therefore be broken down into two tasks:

a) determination of the size A_j and shape of the drainage area of each of the n wells,
b) determination of the static pressure $\bar{p}_j(t)$ at time t in each drainage area.

We have already covered the first task in Sect. 5.6.2 in the context of pseudo-steady state flow in radial geometry.

For any geometry, we assume that the drainage area A_j of a well producing at a rate $q_{j,\mathrm{sc}}$ from a pay thickness h_j is proportional to $q_{j,\mathrm{sc}}/h_j$, the production per unit thickness:

$$\frac{A_j}{q_{j,\mathrm{sc}}/h_j} = K \tag{10.1}$$

If A is the total surface area of the reservoir, and there are n wells, then:

$$A_j = A \frac{q_{j,\mathrm{sc}}/h_j}{\displaystyle\sum_{j=1}^{n} q_{j,\mathrm{sc}}/h_j} \tag{10.2}$$

Once the value of A_j for each drainage area has been calculated, the reservoir area can be subdivided into regions of appropriate size, with the shapes approximated so as to give total coverage with no overlap.

In order to evaluate the average static pressure $\bar{p}_j(t)$ in each of these drainage areas, we need data from pressure buildup tests performed in each well, fairly close together in time.

The evaluation of this well-test data is described for oil wells in Sect. 6.6. Similar methods can be applied to gas wells.

In the following recap of the method, a simplification will be introduced. In Fig. 10.5, the straight line $p_{\mathrm{ws}} = f\{\ln[(t + \Delta t)/\Delta t]\}$ is extrapolated to $\ln[(t + \Delta t)/\Delta t] = 0$ (equivalent to $\Delta t = \infty$) to obtain the pressure p^*. The average

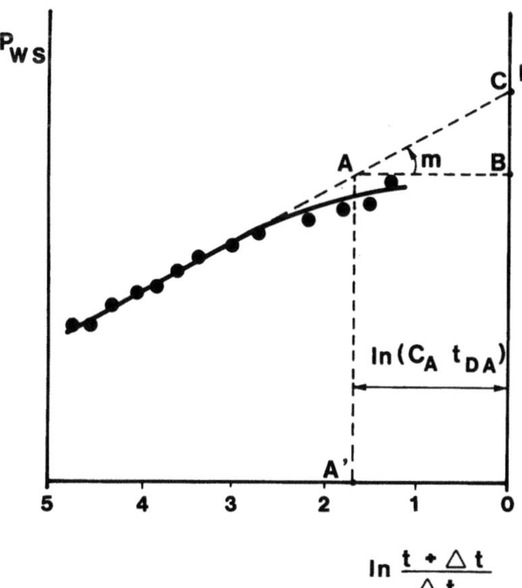

Fig. 10.5. *Calculation of the average drainage area pressure $\bar{p}$ from a buildup test, in a well that was producing in pseudo-steady state prior to shut in. From Ref. 4, reprinted with permission of Elsevier Science Publishers and Prof. Laurie P. Dake*

drainage area pressure can be estimated from this using the relationship:

$$\bar{p} = p^* - \frac{q_{sc} B_o \mu_o}{4\pi k_o h} \, p_{D(MBH)}(t_{DA}) \,, \tag{6.50}$$

where:

$$t_{DA} = \frac{k_o}{\phi \mu_o c_t A} t \tag{6.16}$$

and $p_{D(MBH)}(t_{DA})$ is the value read at time t_{DA} from the Matthews, Brons and Hazebroek[8] curves (Figs. 6.5–6.11) for the appropriate drainage area geometry. It is therefore essential to know the size A_j and shape of the drainage area of each of the n wells in the field.

Equation (6.50) can usually be simplified, because the well will probably have been producing in pseudo-steady state flow prior to shut in. We can write:

$$\bar{p} = p^* - \frac{q_{sc} B_o \mu_o}{4\pi k_o h} \ln(C_A t_{DA}) \,, \tag{6.51}$$

where C_A is the Dietz shape factor for drainage area in question, read from Fig. 5.7.
Since:

$$m = -\frac{dp_{ws}}{d \ln \dfrac{t + \Delta t}{\Delta t}} = \frac{q_{sc} B_o \mu_o}{4\pi k_o h} \,, \tag{6.43}$$

we can substitute into Eq. (6.51) to get:

$$\bar{p} = p^* - m \ln(C_A t_{DA}) \tag{10.3}$$

Figure 10.5 shows a very elegant graphic solution to the problem, based on Eq. (10.3).

$\bar{p}$ can be obtained[5] by simply reading the pressure p_{ws} from the extrapolated straight line corresponding to an x-axis value equal to $\ln(C_A t_{DA})$. Referring to Fig. 10.5:

$$\bar{p} = p^* - \overline{CB} = p^* - \overline{AB} \tan B\hat{A}C$$
$$= p^* - m \ln(C_A t_{DA}) \,. \tag{10.4}$$

This value of $\bar{p}$ should now be corrected to the reference datum depth by adding or subtracting the hydrostatic pressure difference (corresponding to a column of reservoir fluid) between the gauge depth and the datum.

Once the $\bar{p}_j(t)$ for each of the n wells has been calculated in this way, we can compute the average reservoir pressure from:

$$\bar{p}(t) = \frac{\displaystyle\sum_{j=1}^{n} \bar{p}_j(t) \, V_{p,j}}{\displaystyle\sum_{j=1}^{n} V_{p,j}} \,, \tag{10.5}$$

where $V_{p,j}$ is the connected pore volume of the drainage area of the jth well.

10.3 Dry Gas Reservoirs

The very simple case of an isolated (or "closed") dry gas reservoir sealed off from any support from an aquifer will be treated first. We will then look at the more complex case of the dry gas reservoir receiving aquifer support in the form of an influx of water as the production of gas brings the reservoir pressure down.

10.3.1 Closed Reservoir

10.3.1.1 Material Balance Equation

This is the most straightforward application of the material balance equation to the study and prediction of reservoir behaviour.

We use the following nomenclature:

G: the initial volume of gas in place in the reservoir (m^3), expressed under standard conditions of 0.1013 MPa and 288.2 K.

$G_p(t)$: the cumulative volume of gas produced up to time t, (m^3), also expressed under standard conditions.

p_i: the initial reservoir pressure (MPa) at datum depth.

$\bar{p}(t)$: the average reservoir pressure at time t, (MPa).

B_{gi}: volume factor of the gas (see Sect. 2.3.1.1) at initial reservoir pressure p_i and reservoir temperature T_R.

B_g: volume factor of the gas at pressure $\bar{p}(t)$ and temperature T_R.

z_i: real gas deviation factor (Sect. 2.3.1.1) at initial reservoir pressure p_i and reservoir temperature T_R.

$\bar{z}$: real gas deviation factor at pressure $\bar{p}(t)$ and temperature T_R

T_R: reservoir temperature (K).

We ignore the reduction in pore volume owing to compaction caused by a reduction in reservoir pressure. We then have:

– initial volume of gas in the reservoir: GB_{gi}
– volume of gas remaining in the reservoir at time t: $(G - G_p)B_g$

Since the pore volume available to the gas is constant if there is no water influx and we ignore compaction, we can write:

$$G B_{gi} = (G - G_p) B_g \, ,\tag{10.6a}$$

which, referring to Eq. (2.8b), becomes:

$$\frac{\bar{p}}{\bar{z}} = \frac{p_i}{z_i}\left(1 - \frac{G_p}{G}\right).\tag{10.6b}$$

This can be simplified to:

$$\frac{\bar{p}}{\bar{z}} = \frac{p_i}{z_i} - K\,G_p \, ,\tag{10.6c}$$

where:

$$K = \frac{p_i}{z_i\,G}.\tag{10.6d}$$

Equations (10.6a)–(10.6c) represent three different formulations of the material balance equation for a closed dry gas reservoir.

Equation (10.6c) is of particular interest, because it states that in a closed dry gas reservoir, there is a linear relationship between $\bar{p}/\bar{z}$ and the cumulative volume of produced gas G_p (Fig. 10.6).

If the trend is extrapolated to $\bar{p}/\bar{z} = 0$, we have an estimate of the initial gas in place, G, since when $\bar{p}/\bar{z} = 0$:

$$1 - \frac{G_p}{G} = 0, \qquad G_p = G \; . \tag{10.7}$$

It is important to realise at this stage that the linear relationship between G_p and $\bar{p}/\bar{z}$ expressed in Eq. (10.6c) *is a necessary but not sufficient condition for the reservoir to be of the "closed" type* (a point overlooked by several authors[6]).

It has been demonstrated by Chierici et al.,[2] and verified with data from a large number of real gas reservoirs, that this sort of linear relationship can also frequently be exhibited when the reservoir is in contact with an aquifer, particularly a fairly small one.

In such a case, the estimate of G obtained from the extrapolation made in Eq. (10.7) will be wrong, and too large. Many values of G could in fact be postulated, each of which, when combined with an appropriate set of aquifer parameters, would reproduce the observed data, within the limits of measurement error in p and G_p (Fig. 10.7).

In systems analysis terminology, the *reservoir + aquifer system is a black box whose internal structure* (in particular: G) *cannot be uniquely determined from external observation of its behaviour*[2] (variation of p with G_p).

Therefore, before using Eq. (10.7) to estimate G, *it is essential to confirm from geological evidence that the reservoir is indeed of the "closed" type.* If this is not the case, although the estimate of G obtained in this way provides an upper limit, it can be as much as three times larger than the true initial volume of gas in place.

Where the reservoir is of the closed type, Eqs. (10.6) can be used directly to obtain an estimate of $\bar{p}/\bar{z}$ (and therefore $\bar{p}$) for any postulated future value of G_p.

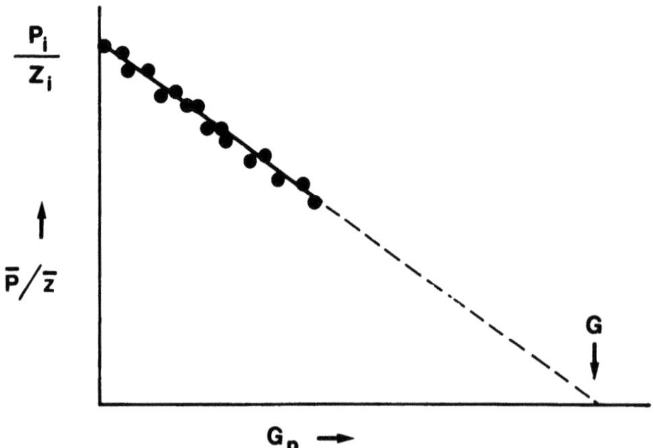

Fig. 10.6. *Variation of $\bar{p}/\bar{z}$ as a function of G_p in a closed dry gas reservoir*

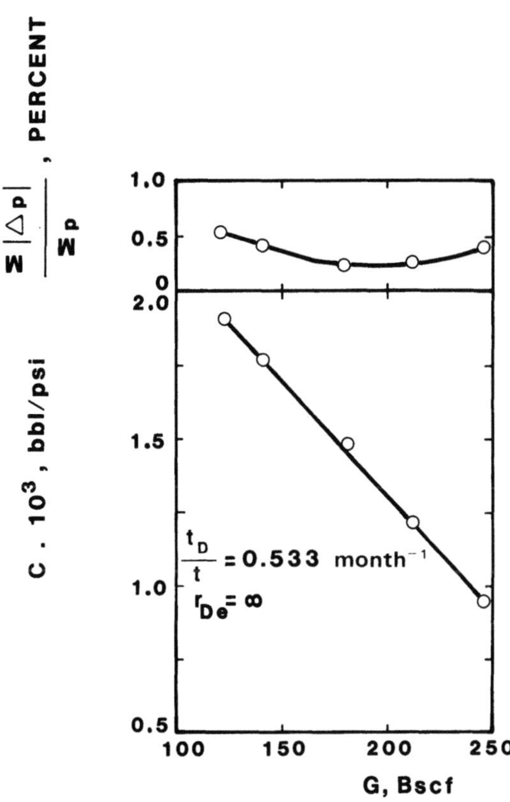

Fig. 10.7. *Gas reservoir under water drive, in the case where the initial resources G cannot be uniquely determined from the material balance calculation.*

Relationship between G and the aquifer constant C, which makes possible the simulation of the production history of the reservoir, to within measurement error on the pressure $\bar{p}$. From Ref. 2, 1967, Society of Petroleum Engineers of AIME, reprinted with permission of the SPE

In the case of a virgin reservoir, since no production data are yet available to establish the straight line trend represented by K in Eq. (10.6c), Eq. (10.6b) must be used. This requires knowledge of the initial volume of gas in place, G.

If, on the other hand, the reservoir has been producing long enough for a useful number of data points $(\bar{p}/\bar{z}, G_p)$ to be available, the most probable value of K can be computed by least squares. $\bar{p}/\bar{z}$ can then be calculated directly for any future value of G_p.

10.3.1.2 Prediction of the Gas Production Rate as a Function of Time

A highly significant factor in the development of a reservoir is the prediction of flow rate $q_g(t)$ and cumulative gas production $G_p(t)$ that the reservoir will be able to deliver at any future time t. This is because the gas is sold to the distributors on long-term contracts, which fix the annual volumetric quota and the maximum delivery rate (and also the minimum rate if it is a "take or pay" contract). The supplier faces "shortfall" penalties if these terms are not met.

Since by definition there is no water influx into a "closed" reservoir, calculations of gas deliverability do not need to take account of well shutdowns because they have flooded out, nor loss of productivity from individual wells through water "coning" (described at length in Chap. 12, Vol. 2).

The only constraints on the flow rate are:

– tubing head flowing pressure p_{tf} must be greater than the inlet pressure to the pipeline (or the minimum inlet pressure of the compressor),

– the gas flow rate must be less than the critical rate for "sanding" (dislodging of sand grains and particle "fines" and their transport into the production stream) in poorly consolidated sandstone reservoirs.

The maximum production that can be obtained from the reservoir is obviously the sum of the maxima of all the wells in the field.

For each well, the calculations are performed as follows:

1. Decide on the length of the time step Δt to be used for the calculations (typically 3 or 6 months),
2. Calculate the reservoir pressure $\bar{p}_j$ at time t_j corresponding to the start of the time step Δt specified in 1. If $t_j = 0$, then obviously $\bar{p}_j = p_i$; if $t_j > 0$, use Eq. (10.6c) with $G_p = G_{p,j}$, the cumulative produced volume of gas up to time t_j,
3. Starting with the minimum allowed tubing head flowing pressure $p_{tf,\,min}$ and assuming an approximate flow rate $q_{g,j}^{k=1}$, calculate the corresponding minimum bottom-hole flowing pressure $p_{wf,\,min}^{k=1}$ that can be applied to the reservoir. Cullender and Smith's equation,[3] integrated by Young's method,[17] provides a very simple means of estimating bottom-hole pressure from tubing head pressure in a flowing gas well. We neglect any inertial effects, so that:

$$p_{wf}^2 = e^s p_{tf}^2 + \left(\frac{3.907 \times 10^{-3} f}{d^5} q_{g,j}^k \, \bar{T} \bar{z} \right)^2 (e^s - 1), \qquad (10.8a)$$

where:

$$s = 0.0684 \frac{\gamma_g D}{\bar{T} \bar{z}} \qquad (10.8b)$$

and:

$$f = \frac{5.387 \times 10^{-3}}{d^{0.224}} \qquad (10.8c)$$

with:

   d: inside diameter of tubing (cm)
   D: depth of well (m)
   e: base of natural logarithms = 2.718281
   f: frictional coefficient (dimensionless)
   p_{wf}:   bottom-hole flowing pressure (MPa)
   p_{tf}:   tubing head flowing pressure (MPa)
   $q_{g,j}^k$:   estimate of gas flow rate $q_{g,j}$ at time t_j, at kth iteration (m³/day under standard conditions)
   $\bar{T}$:   average well temperature $(T_{tf} + T_R)/2$ (K)
   $\bar{z}$:   gas deviation factor at $p = (p_{tf} + p_{wf})/2$ and $T = \bar{T}$
   γ_g: specific gravity of the gas at standard conditions (air = 1.00)

Note that for every trial value of $q_{g,j}^k$ further iterations may be required because $\bar{z}$ itself depends on p_{wf}.

The friction factor f calculated by Eq. (10.8c) is one-quarter of the value determined by Moody[10] for fully turbulent flow.

4. We now calculate the maximum flow rate $q_{g,j,\,max}^{k=1}$ that each well can deliver at a reservoir pressure $\bar{p}_j$ and bottom-hole flowing pressure $p_{wf,\,min}^{k=1}$, using the

deliverability equation (7.43) with the A and B coefficients determined from field data for each well. If:

$$\left| \frac{q_{g,j,\max}^{k=1} - q_{g,j}^{k=1}}{q_{g,j,\max}^{k=1}} \right| \leqslant \varepsilon \tag{10.9a}$$

with ε sufficiently small (typically $\varepsilon = 0.01$), the calculation of $q_{g,\max}$ can be terminated. Failing this, steps 3 and 4 must be repeated, this time assuming:

$$q_{g,j}^{k=2} = \frac{q_{g,j,\max}^{k=1} + q_{g,j}^{k=1}}{2} \tag{10.10}$$

and iterating until convergence is attained:

$$\left| \frac{q_{g,j,\max}^{k} - q_{g,j}^{k}}{q_{g,j,\max}^{k}} \right| \leqslant \varepsilon \tag{10.9b}$$

The resulting value of $q_{g,j,\max}$ satisfies the conditions of minimum tubing head pressure and static reservoir pressure.

5. Check that the flow rate $q_{g,j,\max}$ obtained in step 4 is lower than the critical rate for sanding. If not, then use the critical sanding rate for $q_{g,j,\max}$.
6. Calculate the maximum total deliverability from the reservoir at time t_j as the sum of all the maximum well rates. Check that this total is greater than or equal to the highest daily delivery rate demanded by the contract in the time period t_j to $t_j + \Delta t$.

   If not, it will be necessary to perforate more intervals in existing wells, to drill more wells or, if feasible, modify the terms of the supply contract.
7. Using the average total gas production rate $q_{g,av}$ specified in the contract for the period t_j to $t_j \Delta t$ (which, being less than $q_{g,\max}$, can certainly be attained without problem), calculate the volume of gas

$$\Delta G_{p,j} = q_{g,av} \Delta t \tag{10.11}$$

that will be produced over this period, and then the cumulative volume of gas produced from the reservoir up to time t_{j+1}, which is simply: $G_{p,j+1} = G_{p,j} + \Delta G_{p,j}$.

8. Repeat the procedure starting from step 2, this time with $t_j = t_{j+1}$, then t_{j+2}, and so on in steps of Δt until the required evaluation period has been covered.

   Figure 10.8 is a flow diagram of the calculation procedure. It is best performed on a programmable calculator or computer.

10.3.2 Reservoir in Contact with an Aquifer

10.3.2.1 Material Balance Equation

In the case of a gas reservoir in contact with an aquifer, the material balance equation must obviously take into account the reduced volume occupied by the hydrocarbon following the influx of a volume $W_e(t)$ of water.

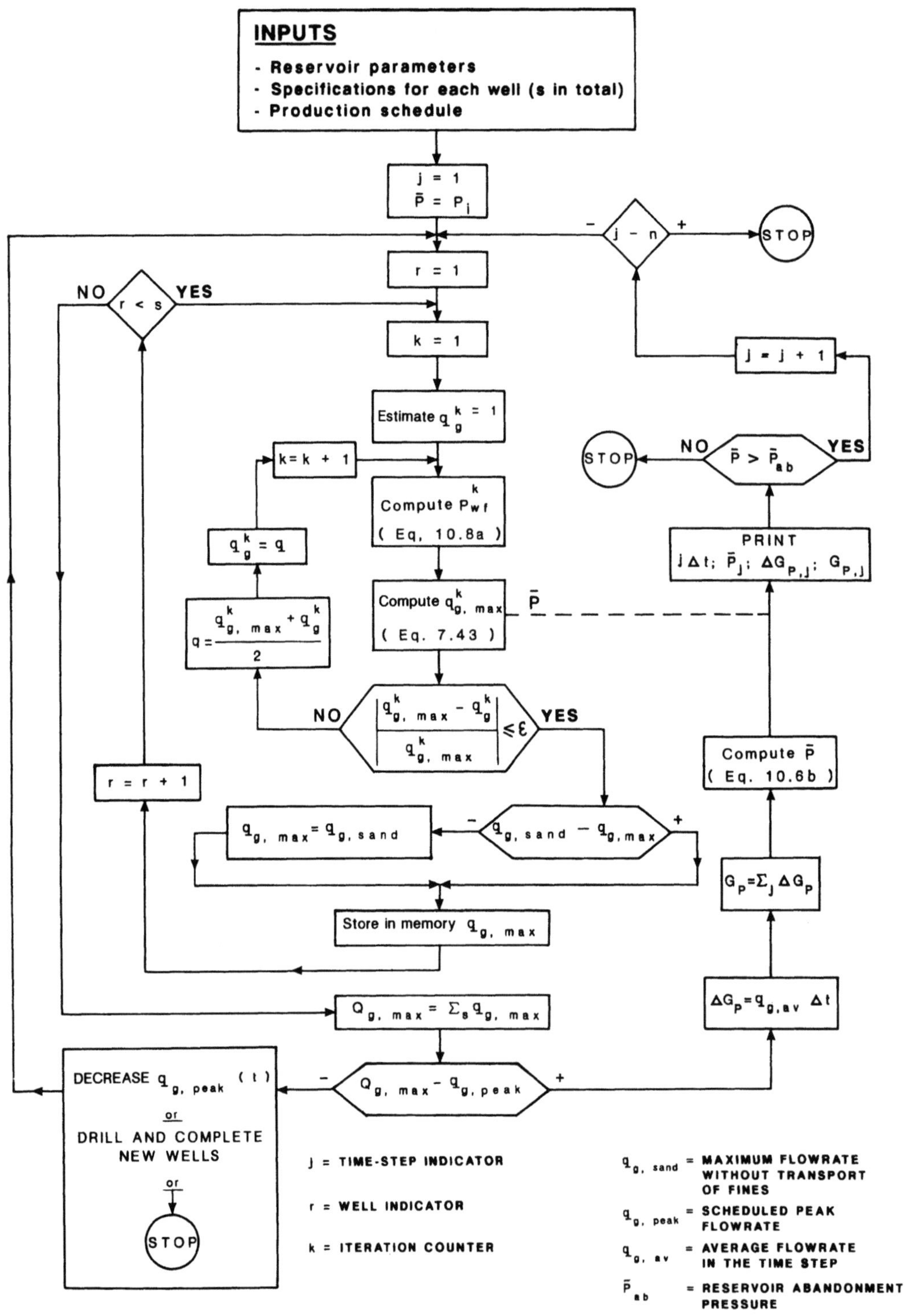

Fig. 10.8. *Flow chart for the calculation of the gas flow rate from a "closed" type reservoir*

If again we ignore the reduction in pore volume which follows a decline in reservoir pressure, we will have:

(volume occupied by gas in the reservoir at time t) =
(volume occupied by gas initially) −
(volume of water that has entered the reservoir by time t)

which can be expressed as:

$$(G - G_p)B_g = GB_{gi} - W_e \,. \tag{10.12}$$

Here we should recall the important point made in Chap. 9 that $W_e(t)$ *is a function of the entire pressure history $p(t)$ of the reservoir, from the start of production at $t = 0$ to the current time t under investigation.*

To handle this, we will need to apply the convolution integral Eq. (9.24) for the van Everdingen-Hurst method, or the summation Eq. (9.53) for the Fetkovich method.

A further point is that $\bar{p}(t)$ itself is a function of $W_e(t)$. This can be appreciated very easily by rewriting Eq. (10.12) with Eq. (2.8b) substituted for B_g:

$$(G - G_p)\frac{\bar{z}}{\bar{p}} = G\frac{z_i}{p_i} - \frac{W_e}{3.515 \times 10^{-4}\, T_R} \,, \tag{10.13a}$$

so that:

$$\frac{\bar{p}}{\bar{z}} = \frac{p_i}{z_i}\, \frac{1 - \dfrac{G_p}{G}}{1 - \dfrac{W_e}{3.515 \times 10^{-4}\, \dfrac{z_i}{p_i}\, T_R\, G}} \tag{10.13b}$$

If we now define the volume of gas in the reservoir under initial conditions as:

$$V_{g,o} = G\, B_{gi} = 3.515 \times 10^{-4}\, \frac{z_i}{p_i}\, T_R\, G \,, \tag{10.14}$$

we get:

$$\frac{\bar{p}}{\bar{z}} = \frac{p_i}{z_i}\, \frac{1 - \dfrac{G_p}{G}}{1 - \dfrac{W_e}{V_{g,o}}} \,. \tag{10.15}$$

The analysis of the pressure response $\bar{p}(t)$ of a gas reservoir to a specified production schedule $G_p(t)$ therefore involves the simultaneous solution of Eq. (10.12) for material balance, and one of the water influx equations for $W_e(t)$ presented in Chap. 9.

The most efficient method of solution involves iteration on the pressure p, and this will be described in Sect. 10.3.2.3.

Two cases must be considered for the calculation of $W_e(t)$:

1. Virgin reservoir – no production data $\bar{p}(t) = f[G_p(t)]$ are available.

Either the van Everdingen-Hurst (see Sect. 9.4) or the Fetkovich (Sect. 9.6) method must be used to calculate $W_e(t)$. Since we have no historical data to calibrate against, we must assume values for the aquifer parameters θ, h, ϕ, k, c_t and r_{De} with all the inaccuracies that this will probably introduce.

Recently, there has been a tendency to treat the problem probabilistically, attributing a likely range of values to each of the aquifer parameters (rather than a single value). Computations are then performed for all the possible combinations of parameter values.

The result in this approach is obviously not a single curve relating $\bar{p}(t)$ to the production schedule $G_p(t)$, but a *family of curves* within a band. The "true" reservoir behaviour lies somewhere in this band.

2. The reservoir has been on production, and we have the initial pressure p_i, the deviation factor $z(p)$, and a record of $(\bar{p}, G_p)$ over the production history.

The most commonly used methods to estimate $W_e(t)$ in this situation are those of Fetkovich (Sect. 9.6) and the influence function (Sect. 9.5).

The historical data are used to calculate the most likely values of $W_{ei}(t)$ and A, or of the $F(t)$ function, using the procedure already described. The influence function $F(t)$ is then extrapolated forward in time using the criteria provided in Sect. 9.5.

10.3.2.2 Prediction of the Gas Production Rate as a Function of Time

To be able to predict the incremental production ΔG_p that will be attained in each time interval Δt over the future life of the reservoir (or to confirm it if a supply contract already exists) is essential for the same reasons as were introduced in Sect. 10.3.1.2, even when the reservoir benefits from water drive.

The same procedure is used as for the closed type reservoir described in Sect. 10.3.1.2, with two extra conditions:

a) At each time step it is necessary to check for wells which have begun to produce water because of the advance of the water front;
b) The production rate from each well must not exceed the critical value for water coning from an underlying aquifer.

The critical coning rate in point (b) can be calculated by a number of methods, covered in Chap. 12.

Point (a) is checked in the following manner.

Construct a diagram (similar to Fig. 4.2 in Sect. 4.3.2) of the hydrocarbon-bearing area versus the height above the initial water/gas contact (WGCI).

Mark each well on the diagram, drawing in each case the interval open to production (Fig. 10.9).

If we know the volume of water $W_{e,j}$ which has entered the reservoir at time t_j, we can calculate the bulk volume of reservoir rock that has been invaded:

$$V_{R,j} = \frac{W_{e,j}}{\phi(1 - S_{iw} - S_{gr})} . \tag{10.16}$$

The height above the WGCI which corresponds to this volume of rock can be calculated quite simply. Now, assuming the fluid contact to be horizontal, the WGC at time t_j can now be drawn on the diagram.

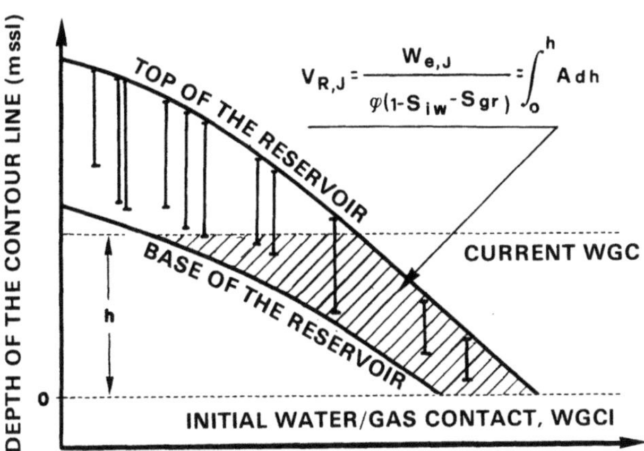

$$V_{R,J} = \frac{W_{e,J}}{\varphi(1-S_{iw}-S_{gr})} = \int_{o}^{h} A\,dh$$

Fig. 10.9. *Reservoir under water drive. Graphic method for forecasting which wells will not cut water at time t, and will therefore still be able to produce gas*

The disposition of the producing intervals should now be examined relative to the WGC:

- any well in which the top of the open interval is below the WGC is shut in and the interval is plugged off. Any higher gas-bearing intervals can be completed to restore production;
- any well where the WGC lies part way up the producing interval is a candidate for a workover – most likely a cement plug to seal off the watered-out section of perforations plus a safety margin above the WGC.

In this latter case, if the production from the remaining perforations is not sufficient to even pay for the cost of the workover, the interval should be closed off completely. If there is another gas-bearing interval higher up with good potential deliverability, this could be completed to sustain production from the well.

10.3.2.3 *Prediction of the Reservoir Pressure Behaviour*

Once we have calculated the incremental amount of gas $\Delta G_{p,j}$ that will be produced in the time interval $\Delta t = t_{j+1} - t_j$ [Eq. (10.11)], we need to estimate the average reservoir pressure $\bar{p}_{j+1}$ at t_{j+1}, the end of the interval.

To calculate $W_e(t_{j+1})$, we will use the Fetkovich method, for which we will assume we have been able to estimate W_{ei} and A (Sect. 9.6) from the production history (a degree of "guesstimation" may be required if the reservoir has only a short history).

Similar treatments can be developed for the van Everdingen-Hurst and influence function approaches.

Let t_j be the start time for the prediction of the reservoir behaviour. In the special case of a virgin reservoir, of course, $t_j = 0$.

We have the following data at our disposal: G, p_i, the function $B_g(p)$, and (produced reservoirs only) the production history $\bar{p}(t) = f[G_p(t)]$ from $t = 0$ up to $t = t_j$.

At time t_j we have already calculated the average total gas production rate $q_{g,av}(t_j)$, as described in Sect. 10.3.2.2.

The calculation procedure can be summarised as follows:

1. From the production history, read the values of:

$$\bar{p}_{j-1} \text{ at time } t_{j-1} = t_j - \Delta t \,,$$

$$G_{\mathrm{p},j-1} \text{ at time } t_{j-1} = t_j - \Delta t \,.$$

2. Calculate the value of:

$$G_{\mathrm{p},j+1} = G_{\mathrm{p},j} + \Delta G_{\mathrm{p},j} = G_{\mathrm{p},j} + q_{\mathrm{g,av}}(t_j)\Delta t \,.$$

3. For time:

$$t_{j+1} = t_j + \Delta t \,.$$

calculate a first estimate, $\bar{p}_{j+1}^{k=1}$, of the average reservoir pressure $\bar{p}_{j+1}$.
For reservoirs that have already been on production, we assume:

$$\bar{p}_{j+1}^{k=1} = \bar{p}_j + (\bar{p}_j - \bar{p}_{j-1})\frac{G_{\mathrm{p},j+1} - G_{\mathrm{p},j}}{G_{\mathrm{p},j} - G_{\mathrm{p},j-1}} \,. \tag{10.17a}$$

For reservoirs that have not yet begun production $(t_j = 0)$, we assume:

$$\bar{p}_1^{k=1} = p_{\mathrm{i}}\left(1 - \frac{G_{\mathrm{p},1}}{G}\right) \tag{10.17b}$$

with

$$t_1 = \Delta t$$

4. Using Eq. (9.53) with $n = j + 1$, calculate $W_{\mathrm{e},j+1}^{k=1}$
5. Calculate $B_{\mathrm{g},j+1}^{k=1}$ at $\bar{p}_{j+1}^{k=1}$, and then:

$$(G - G_{\mathrm{p},j+1})B_{\mathrm{g},j+1}^{k=1} \,.$$

6. If:

$$\left|\frac{(G - G_{\mathrm{p},j+1})B_{\mathrm{g},j+1}^{k=1} + W_{\mathrm{e},j+1}^{k=1}}{GB_{\mathrm{gi}}} - 1\right| \leq \varepsilon \,, \tag{10.18}$$

where ε is sufficiently small (typically 0.01), we can consider that material balance has been satisfied, and:

$$\bar{p}_{j+1}^{k=1} = \bar{p}_{j+1} \,.$$

7. If the left-hand term in Eq. (10.18) is not within the tolerance ε, steps 3 to 6 must be reiterated until convergence is achieved. Starting with $k = 2$ we will have:

$$p_{j+1}^{k=2} = p_{j+1}^{k=1}\frac{(G - G_{\mathrm{p},j+1})B_{\mathrm{g},j+1}^{k=1} + W_{\mathrm{e},j+1}^{k=1}}{GB_{\mathrm{gi}}} \tag{10.19}$$

and so on for $k = 3, 4, \ldots$ until Eq. (10.18) is satisfied.

The flow diagram for this procedure is presented in Fig. 10.10. Again, this is better suited to a programmable calculator or computer than to hand calculation.

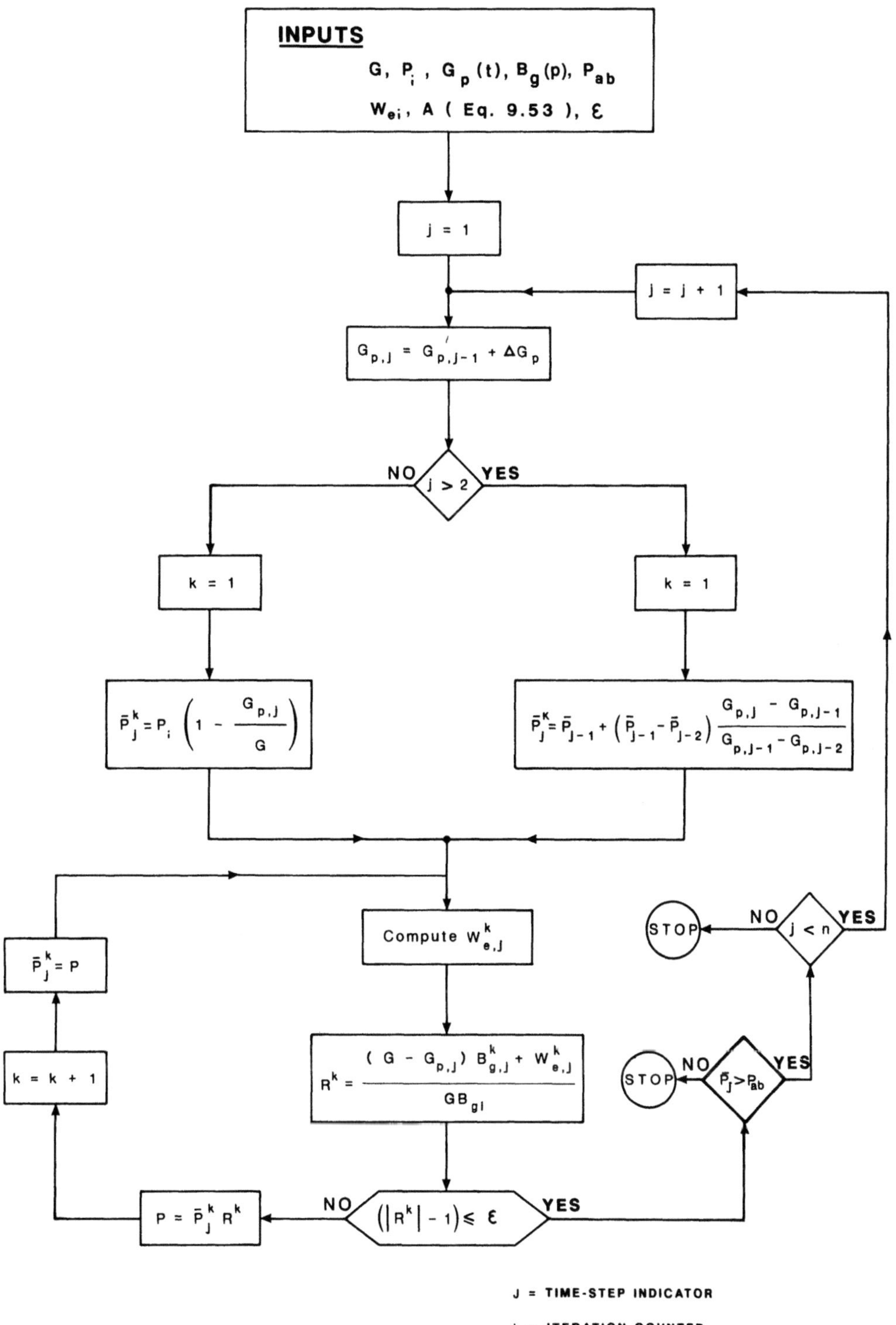

Fig. 10.10. *Flow chart for the calculation of reservoir pressure as a function of a specified production schedule, in a gas reservoir under water drive*

10.4 Gas Condensate Reservoirs

10.4.1 Introduction

The material balance equation for a retrograde gas condensate reservoir is the same as for a dry gas reservoir, whether it is of the closed type or under water drive.

However, there are some very important differences in volumetric and phase behaviour between these two fluids which must be taken into account.

The composition of a dry gas remains unchanged throughout the entire life of the reservoir. By contrast, the gas condensate composition changes continuously as a result of the phenomenon of retrograde condensation (see Sect. 2.2), which occurs once the production of gas has brought the reservoir pressure below the dew point.

In particular, as the pressure falls through the dew point (p_{dp} in Fig. 10.11) towards inversion pressure (p_{inv}), progressively more of the heavier hydrocarbon components in the reservoir fluid (gas) condense, or "drop out", as liquid droplets. The liquid volume which builds up in this manner (Fig. 10.11) tends to be trapped in the pore spaces like a residual oil saturation S_{or}.

If the pressure declines further – from inversion pressure towards reservoir abandonment pressure (p_{ab} in Fig. 10.11) – an increasing proportion of this liquid will revaporise, thereby decreasing its volume.

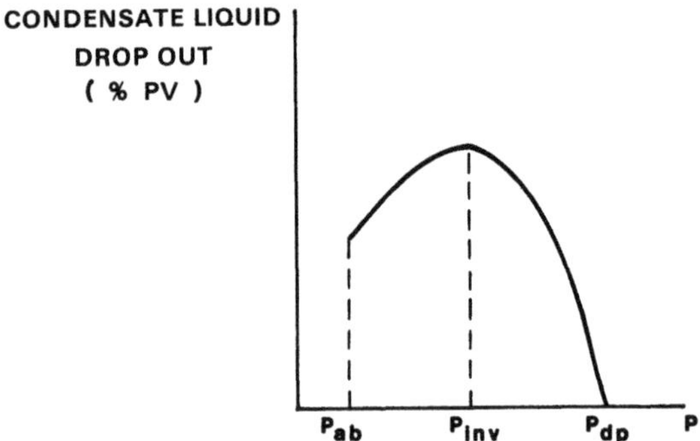

Fig. 10.11. *Retrograde condensate reservoir. Typical variation in the volume of condensate liquid drop-out in the reservoir with changing pressure*

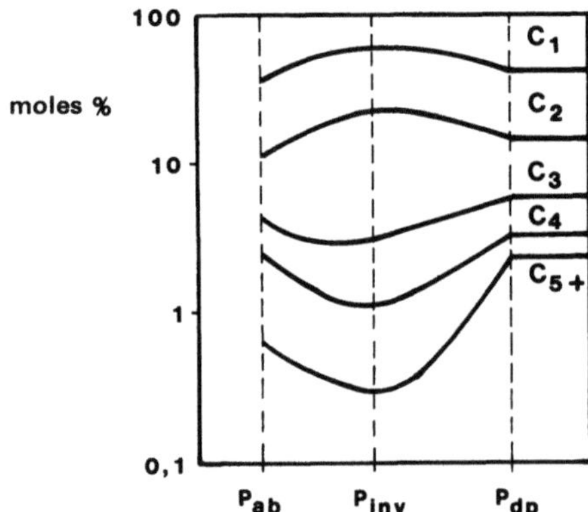

Fig. 10.12. *Retrograde condensate reservoir. Typical variation in the composition of produced gas as the reservoir pressure changes*

Consequently, two important characteristics of retrograde condensate reservoir behaviour are:

1. The composition of the produced gas varies continuously with time (Fig. 10.12)
2. The volume factor B_g (and therefore the deviation factor z) of the *biphasic* fluid in the reservoir will vary, not only as a function of pressure, but also of the overall composition of this fluid.

By the time reservoir abandonment pressure has been reached, when the field would normally be shut down, there still remains a certain quantity of liquid drop-out trapped in the rock. This could represent a quite significant volume of oil, and therefore of lost earnings from the field.

The only effective method of preventing this loss of revenue is to maintain the reservoir pressure above the dew point throughout the entire life of the field. There are two techniques in common use and, for completeness, they will be described briefly in the next section.

10.4.2 Pressure Maintenance by Water Injection or Gas Recycling

The simplest way to keep the reservoir pressure above the dew point is to inject water into the producing strata through peripheral wells, at a rate equal to or greater than the rate of gas production, measured under *reservoir* conditions.

Cost advantages make this approach very attractive: the additional investment in drilling, completing and equipping injection wells is not excessive; gas production can commence (and revenue earned!) as soon as the pressure maintenance has been started up.

However, it does suffer from one serious disadvantage: a fraction of the gas, corresponding to its residual saturation S_{gr} (Sect. 3.5.2.4), will remain trapped in the pores that have been swept by water, as well as in any low permeability regions of the reservoir, and this means lost production.

Therefore, although the gas that is produced retains all the hydrocarbon components that were present initially (there is *no* variation in composition with time, since retrograde condensation is not allowed to happen), a proportion of the gas initially in place is not recovered.

This method of pressure maintenance should therefore only be employed if economically viable: on the asset side, the increased value of the production stream due to the heavy components that would not otherwise have been recovered; on the debit side, the cost of the injection wells, and the lost value of the unrecovered gas trapped in the reservoir.

The second method of pressure maintenance, illustrated schematically in Fig. 10.13, is that of gas recycling.

Produced gas is first treated for the removal of all C_3 and heavier hydrocarbon components, which are then sold commercially. The resulting dry gas from the plant, plus perhaps gas produced from other fields, is now compressed and reinjected into the reservoir, usually through peripheral wells. In very large reservoirs, distributed injection of gas through alternating lines of injection and production wells is sometimes used.

The injected dry gas drives the reservoir condensate gas towards the producing wells. By balancing injected and produced volumes (measured at reservoir

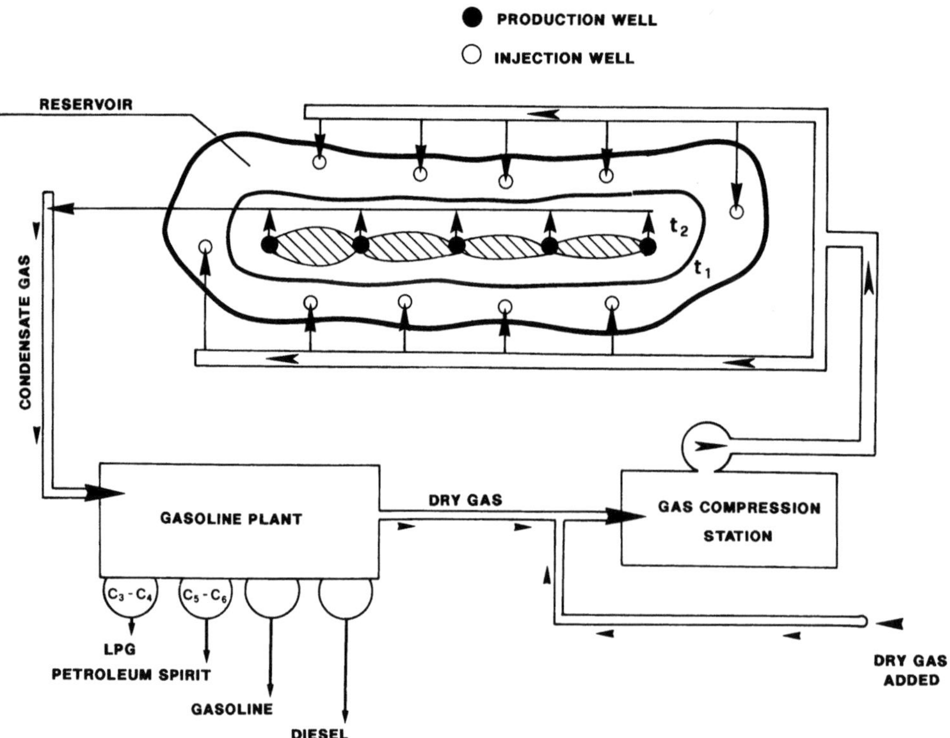

Fig. 10.13. *Schematic of the gas recycling process in a condensate gas reservoir*

conditions), the reservoir pressure can be maintained above the dew point, so that no liquid drop-out occurs.

When the production becomes predominantly dry gas (indicating that the reservoir between the injectors and producers has been swept), injection is terminated. There now follows a phase of production (and sale!) of the dry gas that was injected, plus whatever residual condensate gas that can be recovered.

The phenomenon of retrograde condensation will not occur on any significant scale during this phase, because the reservoir fluid composition has been changed from that of a condensate (curve ACB in the phase diagram Fig. 10.14) to that of a dry gas (curve A′C′B′).

Volume for volume, gas recycling is invariably a more costly investment than water injection. The surface facilities required to process and compress gas for injection represent a bigger outlay than pumping equipment for water, and running costs are higher because the compressor turbine burns more gas as fuel.

On the positive side, gas recycling offers the possibility of producing all of the gas present in the reservoir. Because the injected gas is completely miscible with the original reservoir gas, there is no residual saturation of lost condensate gas in the regions which have been swept.

However, the time delay before any produced gas is available for sale, which recycling introduces, is a serious disadvantage. Consequently, all economic factors must be evaluated carefully to determine whether gas recycling is the best approach to take for a particular field.

When examining the economics of both gas and water injection, the first step should be to evaluate the quantity (and commercial value) of the heavy

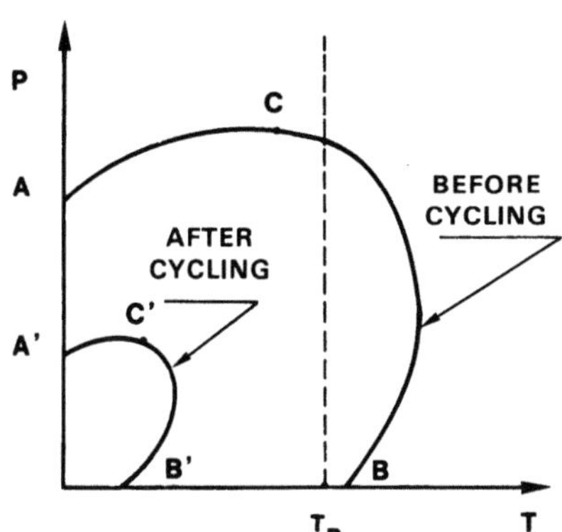

Fig. 10.14. *Phase diagram for the reservoir fluid before and after gas recycling*

hydrocarbons that would be recovered by simply producing the reservoir *without* pressure maintenance, allowing it to fall below the dew point.

For this, it is essential to know how the composition of the gas stream will change with cumulative volume produced.

This subject is dealt with separately for closed and water drive reservoirs in the following sections.

10.4.3 Closed Reservoir

The methods already described in Sect. 10.3.1 for dry gas reservoirs are equally applicable to condensates for the calculation of single well and field production rates, and the prediction of the average reservoir pressure $\bar{p}(t)$ versus cumulative production $G_p(t)$. The linear relationship between $\bar{p}/\bar{z}$ and $G_p(t)$ in Eq. (10.6c) is also valid, but two problems must be dealt with here in the context of closed condensate reservoirs:

– determination of how the produced gas composition will vary with $G_p(t)$ and $\bar{p}(t)$,
– determination of the volumetric behaviour of the biphasic hydrocarbon mixture present in the reservoir as a function of $\bar{p}$ when $\bar{p} < p_{dp}$.

These can both be resolved by means of PVT laboratory measurements on fluid samples (Sect. 2.3).

A sample of reservoir fluid, obtained by recombination of produced gas and liquid (according to the gas/liquid ratio measured at the field separator during a production test) is fed into a constant volume cell at reservoir temperature T_R.

The cell pressure is initially higher than that of the reservoir (and therefore the dew point), and is reduced in a large number of carefully controlled flash liberation steps (Sect. 2.3.2.1), down to abandonment pressure, by bleeding off successive volumes of gas $\Delta G_{p,j}$. At each step, the pressure $\bar{p}$, ΔG_p, and the volume of liquid present in the cell, are measured precisely. The composition of the removed gas is also analysed.

Data from these measurements are used to establish diagrams like Figs. 10.11, 10.12 and 10.15 for the volumetric and phase behaviour of the reservoir fluid.

In particular, the experimental data in Figs. 10.12 and 10.15 provide the means of estimating the cumulative volume of each hydrocarbon component that will be produced by straightforward expansion of the reservoir fluid.

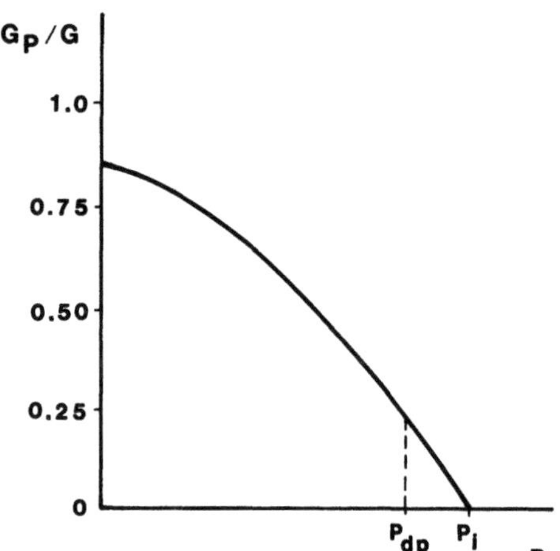

Fig. 10.15. *Proportion of the gas initially in place that is produced from a closed type condensate reservoir as a function of reservoir pressure*

10.4.4 Reservoir in Contact with an Aquifer

In the case of a condensate reservoir in communication with an aquifer, it is not simply a matter of using experimental data for the compositional variations of the produced gas, and volumetric behaviour of the reservoir fluids (gas + liquid hydrocarbons) with respect to $\bar{p}(t)$.

This is because the pore volume available to the hydrocarbons at any given time (or, rather, pressure) depends on the amount of water influx $W_e(t)$ from the aquifer.

$W_e(t)$ is itself a function of $\bar{p}(t)$, and is therefore dependent on the production schedule $q_g(t)$ that has been imposed. Note that, theoretically, it would be possible to produce the field at such a low gas flow rate that the reservoir pressure at abandonment is barely less than the initial pressure.

In any case, predicting $W_e(t)$ will involve the same uncertainties as were outlined in Chap. 9. Consequently, specialised calculation methods are required to evaluate the behaviour of a condensate reservoir under water drive.

The most common approach makes use of an equation of state (EOS) to predict the fluid PVT properties, and simulates the producing life of the field as a series of steps. During a step, the reservoir pressure *is assumed to remain constant*, while a certain mass of gas is removed. At the end of the step, the remaining reservoir gas is allowed to expand to fill the new volume not yet infiltrated by aquifer water, and the new pressure is calculated for the next step.

10.4.4.1 *Calculation of the Volumetric and Phase Behaviour of a Condensate Gas Using an EOS*

A large number of equations of state for hydrocarbon systems under reservoir conditions have been published.[18] In this section we see how to use one of them – the Soave–Redlich–Kwong[15] (SRK) equation.

We use $v = V/N$ to represent the molar volume of the reservoir fluid (where V is the total volume occupied by N kg mol of the fluid), and z_i ($i = 1, 2, \ldots, n$) for the mole fraction of its ith hydrocarbon component C_i. The SRK equation can then be expressed as:

$$p = \frac{R T_{\mathrm{R}}}{v - B} - \frac{A}{v(v + B) T_{\mathrm{R}}^{0.5}} \, ,$$ (10.20)

where:

R = the universal gas constant = 8.312×10^{-3} MN m K^{-1}

T_{R} = reservoir temperature (K)

and

$$A = \sum_{i=1}^{n} \sum_{j=1}^{n} z_i z_j a_{ij} \, ,$$ (10.21a)

$$B = \sum_{i=1}^{n} z_i b_i \, .$$ (10.21b)

a_{ij} is a numerical coefficient which has a specific value for each hydrocarbon pair C_i–C_j, while b_i has a value for each individual component C_i.

More precisely:

$$a_{ij} = a_{\mathrm{c},ij} \sqrt{\alpha_i} \sqrt{\alpha_j} \, ,$$ (10.22)

where

$$a_{\mathrm{c},ij} = (1 - \delta_{ij}) \sqrt{a_{\mathrm{c}i}} \sqrt{a_{\mathrm{c}j}} \, ,$$ (10.23a)

δ_{ij} = empirical coefficient of interaction between hydrocarbons
$\qquad$ C_i and C_j (10.23b)

$$a_{\mathrm{c}i} = 0.4275 \frac{R^2 T_{\mathrm{c},i}^{2.5}}{p_{\mathrm{c},i}} \, ,$$ (10.23c)

and:

$$\sqrt{\alpha_i} = T_{\mathrm{r},i}^{0.25} [1 + \chi(1 - T_{\mathrm{r},i}^{0.5})] \, ,$$ (10.24a)

$$\chi = 0.480 + 1.574 \omega_i - 0.176 \omega_i^2 \, ,$$ (10.24b)

ω_i = acentric coefficient for component C_i (after Pitzer) (10.24c)

$$b_i = 8.664 \times 10^{-2} \frac{R T_{\mathrm{c},i}}{p_{\mathrm{c},i}} \, ,$$ (10.25)

$p_{\mathrm{c},i}$ and $T_{\mathrm{c},i}$ are the critical pressure and temperature of the ith hydrocarbon component C_i, and $T_{\mathrm{r},i}$ is the reduced temperature: $T_{\mathrm{r},i} = T_{\mathrm{R}}/T_{\mathrm{c},i}$.

This group of equations may appear rather complex. In fact, the computation of the A and B coefficients in Eq. (10.21) for a mixture of hydrocarbons with known mole fractions (z_i for $i = 1, 2, \ldots, n$) can be performed on a computer. This requires a table of the following parameters for each component (readily available in textbooks[19]):

$p_{\mathrm{c},i}$ = critical pressure (MPa) ,
$T_{\mathrm{c},i}$ = critical temperature (K) ,
ω_i = Pitzer acentric coefficient (dimensionless) .

Tables of interaction coefficients δ_{ij} between components C_i and C_j are also available.[19]

In the course of the simulation, the δ_{ij} coefficients will in fact be adjusted as part of the process of matching computed behaviour to experimental data from constant volume PVT measurements.

We now define:

$$A^* = A \frac{p}{R^2 T_R^{2.5}} \,, \tag{10.26a}$$

$$B^* = B \frac{p}{R T_R} \,. \tag{10.26b}$$

For $N = 1$ we have:

$$z = \frac{pv}{R T_R} \,, \tag{10.26c}$$

and, by substitution, we can rewrite Eq. (10.20) as follows:

$$z^3 - z^2 + [A^* - B^*(1 + B^*)]z - AB = 0 \,. \tag{10.27}$$

This can be simplified further if we define two more coefficients:

$$A^* - B^*(1 + B^*) = C \,, \tag{10.28a}$$

$$AB = D \,, \tag{10.28b}$$

so that Eq. (10.27) becomes:

$$z^3 - z^2 + Cz - D = 0 \,. \tag{10.29}$$

Equation (10.29) is a third order polynomial in z, the deviation factor. Consequently, it has three roots – real (positive or negative) and/or imaginary.

We will consider the function:

$$f(z) = z^3 - z^2 + Cz - D \,. \tag{10.30}$$

Referring to Fig. 10.16, $f(z)$ will have three zero values in the interval LM provided it has a maximum and a minimum.

The values of z at which the maximum and minimum occur are calculated by solving $f'(z) = 0$:

$$f'(z) = 3z^2 - 2z + C = 0 \tag{10.31a}$$

from which:

$$z(\max, \min) = \frac{1 \pm \sqrt{1 - 3C}}{3} \,. \tag{10.31b}$$

Equation (10.31b) has two real positive roots (z must be positive!) if:

$$0 < C < \tfrac{1}{3} \,. \tag{10.32a}$$

In other words:

$$0 < [A^* - B^*(1 + B^*)] < \tfrac{1}{3} \,. \tag{10.32b}$$

If Eq. (10.29) has three real positive roots, the smallest corresponds to the liquid state, the largest to the gaseous state, and the middle value to the biphasic state.

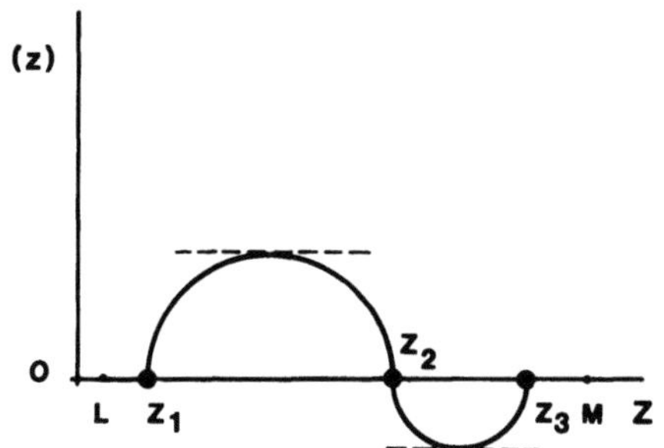

Fig. 10.16. *The function f(z) of Eq. (10.30) in real space*

If Eq. (10.32) does not hold true, there is only one root of Eq. (10.29) which is real and positive, and therefore which has any physical significance. The other two are either negative or complex.

Equation (10.27) is particularly useful because it allows us to calculate the deviation factor of a natural hydrocarbon system *even when it is biphasic*.

At this point we will consider a biphasic hydrocarbon system, in which x_i $(i = 1, 2, \ldots, n)$ is the mole fraction of the ith liquid component, and y_i the mole fraction of the ith gas component.

We know from thermodynamics that such a system is in equilibrium when, *for each component*, the fugacities of the liquid and gas phases are equal:

$$f_{iL} = f_{iV}, \quad i = 1, 2, \ldots, n. \tag{10.33}$$

The *coefficient of fugacity* Ψ is defined as:

$$
\begin{aligned}
\psi_{iL} &= \frac{f_{iL}}{x_i p} \\
\psi_{iV} &= \frac{f_{iV}}{y_i p}
\end{aligned}
\quad i = 1, 2, \ldots, n. \tag{10.34}
$$

We can calculate the coefficient of fugacity for each phase from its deviation factor (z_L for liquid, z_V for vapour) using the following equation:

$$\ln \psi_i = b_i^*(z - 1) - \ln(z - B^*) - A^* \frac{2a_i^* - b_i^*}{2\sqrt{2}B^*} \ln \frac{z + B^*(\sqrt{2} + 1)}{z - B^*(\sqrt{2} - 1)}, \tag{10.35a}$$

where:

$$a_i^* = \frac{\sum_{j=1}^{n} a_{ij} z_j}{A}, \tag{10.35b}$$

and:

$$b_i^* = \frac{b_i}{B}. \tag{10.35c}$$

A^* and B^* have already been defined in Eq. (10.26).

At *equilibrium*, we define the *equilibrium constant* K_i for the ith hydrocarbon component in the system, at a specified temperature and pressure, as:

$$K_i = \frac{y_i}{x_i} = \frac{\psi_{iL}}{\psi_{iV}}. \tag{10.36}$$

To a first approximation we can usually assume:

$$K_i^{(1)} = \frac{1}{p_{r,i}} \exp\left[5.3727(1 + \omega_i)\left(1 - \frac{1}{T_{r,i}}\right)\right].$$
(10.37)

We will now go through the calculation of the vapour-liquid equilibrium for a hydrocarbon system of known mole composition z_i $(i = 1, 2, \ldots, n)$ at a specified temperature T_R and pressure p.

We define:

N: total number of moles in the system,
L: the number of moles in the liquid phase at equilibrium,
V: the number of moles in the vapour phase at equilibrium.

so that:

$$N = L + V.$$
(10.38)

For the ith component, we can write a mass balance equation between the total system (in which it has a mole fraction z_i) and the liquid (x_i) and vapour (y_i) fractions:

$$Nz_i = Lx_i + Vy_i.$$
(10.39)

Substituting from Eqs. (10.36) and (10.38), we get:

$$Nz_i = (N - V)x_i + VK_ix_i = (N - V)\frac{y_i}{K_i} + Vy_i,$$
(10.40)

which, when rearranged, gives:

$$x_i = \frac{Nz_i}{V(K_i - 1) + N},$$
(10.41a)

and:

$$y_i = \frac{NK_iz_i}{V(K_i - 1) + N},$$
(10.41b)

Now, by definition, we know that:

$$\sum_{i=1}^{n} x_i = 1,$$
(10.42a)

and:

$$\sum_{i=1}^{n} y_i = 1.$$
(10.42b)

Substituting for x_i and y_i from Eqs. (10.41a) and (10.41b) and equating, we obtain:

$$\sum_{i=1}^{n} \frac{(K_i - 1)z_i}{(V/N)(K_i - 1) + 1} = 0.$$
(10.43)

If we know the composition z_i of the system, and have a first approximation, $K_i^{(1)}$, of the equilibrium constants, Eq. (10.43) can be solved by successive approximation for the unknown term V/N.

The following is a practical procedure for doing this:

1. Calculate first approximations for $K_i^{(1)}$ using Eq. (10.37), and solve the summation in Eq. (10.43) for V/N.
2. Once V/N (and hence V) is known, calculate the liquid and gas phase compositions using Eq. (10.41a) for x_i (liquid) and Eq. (10.41b) for y_i (gas).
3. With x_i for each of the n liquid phase components, calculate the EOS coefficients from Eqs. (10.21)–(10.25), the deviation factor z_L from Eqs. (10.26)–(10.29), and the fugacity coefficient ψ_{iL} from Eq. (10.35).
4. With y_i, perform the calculations indicated in 3 for the n vapour phase components, to obtain the fugacity coefficient ψ_{iV}.
5. Now calculate a second approximation of the equilibrium constant $K_i^{(2)}$ for each component C_i, from Eq. (10.36):

$$K_i^{(2)} = \frac{\psi_{iL}}{\psi_{iV}}, \tag{10.44}$$

If:

$$\sum_{i=1}^{n} |K_i^{(2)} - K_i^{(1)}| < \varepsilon . \tag{10.45}$$

with ε sufficiently small, the calculations can be terminated.

If this is not the case, repeat the calculation procedure, starting from step 1 with $K_i^{(2)}$, through to step 5, and so on until Eq. (10.45) is satisfied.

At the conclusion of these calculations, we will have the following equilibrium values:

- L, the number of moles of liquid phase, its composition x_i, and deviation factor z_L,
- V, the number of moles of vapour phase, its composition y_i, and deviation factor z_V.

The calculations can of course be performed on a computer.

10.4.4.2 Prediction of Reservoir Behaviour

In order to predict the behaviour of the reservoir, *we must first of all know the composition z_i of the monophasic hydrocarbon system above its dew point.*
In fact, when the pressure has fallen below the dew point and the system is biphasic, it is not possible to assess the volume of liquid nor its composition, nor therefore to derive the overall composition of the reservoir fluid.
For the same reason, even if the reservoir has been on production for some time and data have been accumulated on the variation of pressure and produced gas composition with total produced volume, *all calculations concerning the phase behaviour of the reservoir fluid must be run from the very start of production.*
From the initial monophasic reservoir fluid composition $z_i^{(0)}$, we calculate the deviation factor $z_{V,0}$ by the method described in Sect. 10.4.4.1. From this we can obtain the initial number of moles N_0 present in the reservoir:

$$N_0 = \frac{p_i V_H}{z_{V,0} R T_R}, \tag{10.46}$$

where:

p_i is the initial reservoir pressure, in MPa
V_H is the volume occupied by the condensate gas in the virgin reservoir, in m^3 (Eq. 4.12).

The actual production schedule is simulated by a series of steps of equal duration Δt.

We will assume that the initial reservoir pressure is above the dew point $(p_i > p_{dp})$.

Expressed under standard conditions, G is the initial volume of gas in the reservoir, and $\Delta G_{p,1}$ the volume produced during the first time step, from $t = 0$ to $t = t_1 = 1\,\Delta t$.

The number of moles remaining in the reservoir at time t_1 will be:

$$N_1 = N_0\left(1 - \frac{\Delta G_{p,1}}{G}\right). \tag{10.47a}$$

The calculation procedure is as follows:

1. Make a first approximation $p_1^{(1)}$ as to the reservoir pressure at time t_1.
2. Calculate V/N from Eq. (10.43) by following the procedure in Sect. 10.4.4.1. If $V/N = 1$, then $p_1^{(1)} > p_{dp}$ and the system is still monophasic.
3. Calculate the deviation factor $z_1^{(1)}$ of the reservoir fluid (monophasic if $p_1^{(1)} > p_{dp}$, biphasic otherwise), again following the procedure in Sect. 10.4.4.1. Now calculate $B_g(p_1^{(1)})$ from Eq. (2.8b).
4. Now use the procedure outlined in Sect. 10.3.2.3 for a dry gas to check if the estimate of the reservoir pressure $p_1^{(1)}$ at t_1 satisfies the material balance equation, including, of course, the water influx term W_e.
5. If so, continue to the next time step Δt, "remove" a further volume of gas $\Delta G_{p,2}$, and repeat the calculations in steps 1–5. Then go to step 7.
6. If material balance is not satisfied in step 4, try a different estimate, $p_1^{(2)}$, for the reservoir pressure at time t_1, and repeat steps 2–4. Reiterate with different values of the pressure if necessary until balance is achieved, then proceed with step 5.
7. Perform the calculations described above for successive time steps, withdrawing a volume of gas $\Delta G_{p,i}$ each time, until at a certain time step j you find that the reservoir pressure p_j is less than the dew point pressure p_{dp}, so that $V/N < 1$.

    Up to this moment, the composition of the produced gas will have remained constant – the same, in fact, as the initial reservoir fluid composition z_i – because there has been no retrograde condensation in the reservoir.

    The number of moles of gas in the reservoir at time $t_j = j\,\Delta t$ will be:

$$N_j = N_0\left(1 - \frac{\sum_{i=1}^{j} \Delta G_{p,i}}{G}\right). \tag{10.47b}$$

8. Calculate V_j and L_j (see step 2 above) and get the composition $x_i(t_j)$ and $y_i(t_j)$ of the liquid and vapour phases from Eqs. (10.41). We must keep in mind from this point onwards that any liquid phase will be trapped in the reservoir and only the vapour phase can be produced.

9. For the volume of gas $\Delta G_{\mathrm{p},j+1}$ produced during the next time interval $\Delta t = t_{j+1} - t_j$, calculate the corresponding number of moles ΔV_{j+1} withdrawn from the reservoir.

Since each kg mol of gas takes up a volume of 23.645 m^3 under standard conditions (0.1013 MPa, 288.2 K), this will be:

$$\Delta V_{j+1} = \frac{\Delta G_{\mathrm{p},j+1}}{23.645} \, . \tag{10.48}$$

At time t_{j+1}, therefore, the following quantities of each phase will remain in the reservoir:

L_j moles of liquid of composition $x_i(t_j)$
$V_{j+1} = V_j - \Delta V_{j+1}$ moles of vapour of composition $y_i(t_j)$.

10. The *global* (liquid + vapour) composition $z_i(t_{j+1})$ of the system at time t_{j+1} is computed from mass balance. We have:

$$N_{j+1} = L_j + V_{j+1} \, , \tag{10.49a}$$

$$z_i(t_{j+1}) = \frac{L_j x_i(t_j) + V_{j+1} y_i(t_j)}{N_{j+1}} \, . \tag{10.49b}$$

11. Decide on a first estimate $p_{j+1}^{(1)}$ of the reservoir pressure p_{j+1} at time t_{j+1} and go through steps 2–10 (skipping steps 5 and 7, of course, and replacing the indices "1" and "j" by "$j + 1$" where necessary). Proceed in this way for successive time intervals Δt until the predetermined reservoir abandonment pressure is reached.

These calculations would hardly be feasible by hand, and should be programmed for a computer.

Experience has shown that the equilibrium constant K_i is not very sensitive to small changes in pressure. We can therefore save some computing time by only recalculating V/N (step 2) for the first two or three iterations; any subsequent (minor) adjustment to the pressure will not have a significant effect on the ratio V/N.

10.5 Oil Reservoirs

10.5.1 General Material Balance Equation for Oil Reservoirs

We will consider the completely general case of an oil reservoir with a gas cap, and in contact with an aquifer, shown schematically in Fig. 10.17. We can assume, since the oil and gas are in equilibrium, that the oil is at its bubble-point pressure throughout the oil-bearing section.

Starting from the volumetric material balance equation for a reservoir like this one, we can derive the equations for an oil reservoir without gas cap (i.e. undersaturated oil), with or without aquifer contact, as special cases.

We define:

N: the initial volume of oil in the reservoir, measured at standard conditions (0.1013 MPa, 288.2 K), and therefore without dissolved gas (m^3)

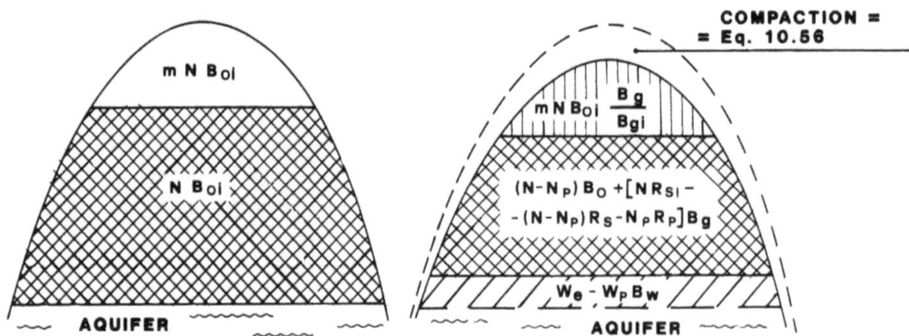

Fig. 10.17. *Schematic representation of the volumetric material balance equation for an oil reservoir with gas cap and aquifer*

m:	the ratio of the volume of the gas cap to the volume of the oil, measured at initial reservoir conditions (dimensionless)
N_p:	the cumulative volume of oil produced at time t, measured under standard conditions (m^3)
G_p:	the cumulative volume of gas produced at time t, measured under standard conditions (m^3)
$R_p = G_p/N_p$:	the *average* produced gas/oil ratio for the time period from 0 to t (dimensionless)
W_p:	the cumulative volume of water produced at time t, measured under standard conditions (m^3).

Bearing in mind what was stated in Sect. 10.1, we will have:

(reservoir volume initially occupied by hydrocarbons) = (reservoir volume occupied by hydrocarbons at time t) + (cumulative net volume of water influx from the aquifer) + (reduction in pore volume available to the hydrocarbons as a result of rock compaction and the expansion of the interstitial water in the hydrocarbon section, caused by reduction in reservoir pressure from p_i to p).

The terms in this equation will be explained in detail.

If B_{oi} is the volume factor of the oil at initial pressure p_i of the virgin reservoir, the initial volume of oil in place will be NB_{oi}, and the initial volume of the gas cap will be mNB_{oi}.

Therefore:

Reservoir volume initially occupied by the hydrocarbons $= (1 + m)NB_{oi}$

$$(10.50)$$

The volume of hydrocarbon remaining in the reservoir at time t is the sum of three terms:

a) the volume of the oil remaining $= (N - N_p)B_o$. $\qquad\qquad$ (10.51a)

b) the volume of the gas in the original gas cap $= mNB_{oi}B_g/B_{gi}$. $\qquad$ (10.51b)

c) the volume of gas liberated from the oil as a result of the decreased pressure (but not produced to surface), expressed at reservoir conditions

$$= \{[NR_{si} - (N - N_p)R_s] - N_p R_p\} B_g \, .$$ (10.51c)

In Eqs. (10.50), (10.51), B_o and R_s are the oil volume factor and gas solubility determined in the lab by multi-stage flashing to standard condition (Sect. 2.3.2.1) reservoir oil samples extracted from the PVT cell at each pressure step of the differential liberation.

In Eq. (10.51b), the term B_g/B_{gi} represents the expansion of the gas cap which has resulted from the reduction in reservoir pressure from p_i to p.

In Eq. (10.51c), NR_{si} is the volume of gas dissolved in the initial total oil volume N, while $(N - N_p)R_s$ is the volume of gas in solution at time t in the remaining oil volume $(N - N_p)$, and $N_p R_p$ is the cumulative volume of gas produced by time t.

The summation described in Eqs. (10.51) is therefore:

Reservoir volume occupied by hydrocarbons at time t

$$= (N - N_p)B_o + mN B_{oi} \frac{B_g}{B_{gi}}$$

$$+ \{[NR_{si} - (N - N_p)R_s] - N_p R_p\} B_g \, .$$ (10.52)

The *net* volume of water influx from the aquifer is the actual volume of water influx W_e minus the volume W_p of any water produced with the oil, plus the volume W_i of any water injected into the reservoir. W_p and W_i are measured at standard conditions. This is expressed as:

Cumulative net volume of water influx from the aquifer at time t

$$= W_e - B_w W_p + B_w W_i \, .$$ (10.53)

Finally, the variations in the volume actually available to the hydrocarbons at any given time: these are due to the reduction in pore volume and the expansion of the interstitial water which are a consequence of the decrease in pressure $\Delta p = p_i - p$. These are accounted for in the following way.

The initial reservoir volume occupied by hydrocarbons (oil + gas) is $(1 + m)N B_{oi}$. The pore volume is therefore:

$$V_p = \frac{(1 + m)N B_{oi}}{1 - \bar{S}_{wi}} \, ,$$ (10.54a)

and the volume of interstitial water it contains is:

$$V_w = (1 + m)N B_{oi} \frac{\bar{S}_{wi}}{1 - \bar{S}_{wi}} \, ,$$ (10.54b)

where $\bar{S}_{wi}$ is the average water saturation in the virgin reservoir.

With pore compressibility c_p (Sect. 3.4.3) and water compressibility c_w (Sect. 2.4), a reduction Δp in the pressure will cause an expansion

$$\Delta V_w = c_w V_w \Delta p$$ (10.55a)

in the interstitial water volume, and a reduction

$$\Delta V_{\mathrm{p}} = (c_{\mathrm{p}} - c_{\mathrm{r}}) V_{\mathrm{p}} \Delta p = \frac{c_{\mathrm{f}} + \phi c_{\mathrm{r}}}{1 - \phi} V_{\mathrm{p}} \Delta p \cong c_{\mathrm{p}} V_{\mathrm{p}} \Delta p \tag{10.55b}$$

in the pore volume.

The combined effect will be:

Reduction in volume available to the hydrocarbons $= \Delta V_{\mathrm{w}} + \Delta V_{\mathrm{p}}$

$$= (1 + m) N B_{\mathrm{oi}} \frac{\bar{S}_{\mathrm{wi}} c_{\mathrm{w}} + c_{\mathrm{p}}}{1 - \bar{S}_{\mathrm{wi}}} \Delta p \;. \tag{10.56}$$

Bringing together Eqs. (10.50)–(10.56), we obtain the overall material balance equation:

$$\begin{aligned}
(1 + m) N B_{\mathrm{oi}} = {} & (N - N_{\mathrm{p}}) B_{\mathrm{o}} + m N B_{\mathrm{oi}} \frac{B_{\mathrm{g}}}{B_{\mathrm{gi}}} \\
& + \left\{ [N R_{\mathrm{si}} - (N - N_{\mathrm{p}}) R_{\mathrm{s}}] - N_{\mathrm{p}} R_{\mathrm{p}} \right\} B_{\mathrm{g}} \\
& + \left\{ W_{\mathrm{e}} - B_{\mathrm{w}} W_{\mathrm{p}} + B_{\mathrm{w}} W_{\mathrm{i}} \right\} \\
& + \left\{ (1 + m) N B_{\mathrm{oi}} \frac{\bar{S}_{\mathrm{wi}} c_{\mathrm{w}} + c_{\mathrm{p}}}{1 - \bar{S}_{\mathrm{wi}}} \Delta p \right\} .
\end{aligned} \tag{10.57}$$

In Eq. (10.57), the left-hand term represents the initial volume of reservoir occupied by the oil with its dissolved gas, plus the gas cap. The right-hand term is the volume of as yet unproduced oil in the reservoir at time t, plus the gas cap (which has expanded), the volume of solution gas liberated from the oil but not produced with it, and the water that has entered the reservoir (from the aquifer, and by injection if relevant) minus the water produced. The volume of reservoir that was initially available to the hydrocarbons has decreased by time t, because of the rock compaction and expansion of interstitial water induced by the falling pressure.

An alternative, and very useful, way of stating the material balance relationship is from the viewpoint of the volume occupied by the cumulative volume of fluids produced up to time t, expressed at the reservoir conditions prevailing at time t.

The cumulative volume of oil produced at time t is N_{p} (at standard conditions). Under the current reservoir conditions, this volume becomes $N_{\mathrm{p}} B_{\mathrm{o}}$. This includes a volume $R_{\mathrm{s}} N_{\mathrm{p}}$ of solution gas which was dissolved in the oil.

The cumulative volume of gas produced at time t is $N_{\mathrm{p}} R_{\mathrm{p}}$ (at standard conditions). Of this, a volume $R_{\mathrm{s}} N_{\mathrm{p}}$ of solution gas was liberated from the produced oil N_{p}, while the remaining volume $(R_{\mathrm{p}} - R_{\mathrm{s}}) N_{\mathrm{p}}$ was produced as gas from the reservoir – either from the gas cap, or from gas that had been in solution in the unproduced oil. The volume of this "free" gas at reservoir conditions at time t is $(R_{\mathrm{p}} - R_{\mathrm{s}}) N_{\mathrm{p}} B_{\mathrm{g}}$.

The volume of water produced at time t, expressed at standard conditions, is W_{p}. Under reservoir conditions this will occupy a volume $W_{\mathrm{p}} B_{\mathrm{w}}$.

The cumulative volume of fluids produced up to time t, expressed at the reservoir conditions prevailing at that time

$$= N_{\mathrm{p}} [B_{\mathrm{o}} + (R_{\mathrm{p}} - R_{\mathrm{s}}) B_{\mathrm{g}}] + W_{\mathrm{p}} B_{\mathrm{w}} \;. \tag{10.58}$$

We can rearrange Eq. (10.57) so that these terms appear on the left-hand side, giving:

$$N_p[B_o + (R_p - R_s)B_g] + W_p B_w = \{N(B_o - B_{oi}) + N(R_{si} - R_s)B_g\}$$

$$+ \left\{mNB_{oi}\left(\frac{B_g}{B_{gi}} - 1\right)\right\} + \{W_e + B_w W_i\}$$

$$+ \left\{(1 + m)NB_{oi}\frac{\bar{S}_{wi}c_w + c_p}{1 - \bar{S}_{wi}}\Delta p\right\}. \qquad (10.59)$$

The physical significance of each of the terms as they appear in Eq. (10.59) will be explained in the next section.

10.5.2   Drive Indices[13]

If we divide the right-hand side of Eq. (10.59) by the term on the left-hand side, we can define:

$$\text{DDI} = \frac{N(B_o - B_{oi}) + N(R_{si} - R_s)B_g}{N_p[B_o + (R_p - R_s)B_g] + W_p B_w}, \qquad (10.60\text{a})$$

$$\text{SDI} = \frac{mNB_{oi}((B_g/B_{gi}) - 1)}{N_p[B_o + (R_p - R_s)B_g] + W_p B_w}, \qquad (10.60\text{b})$$

$$\text{WDI} = \frac{W_e + B_w W_i}{N_p[B_o + (R_p - R_s)B_g] + W_p B_w}, \qquad (10.60\text{c})$$

$$\text{CDI} = \frac{(1 + m)NB_{oi}[(\bar{S}_{wi}c_w + c_p)/(1 - \bar{S}_{wi})]\Delta p}{N_p[B_o + (R_p - R_s)B_g] + W_p B_w}. \qquad (10.60\text{d})$$

These new terms are, of course, related by:

$$\text{DDI} + \text{SDI} + \text{WDI} + \text{CDI} = 1. \qquad (10.61)$$

We will now see what their physical significance is.

The numerator in Eq. (10.60a) is the sum of $N(B_o - B_{oi})$ and $N(R_{si} - R_s)B_g$. $N(B_o - B_{oi})$ is the change in the initial volume, N, of the oil from time $t = 0$ (pressure p_i, volume factor B_{oi}) to time t [pressure $p(t)$, volume factor B_o], following a reduction in reservoir pressure from p_i to $p(t)$. $N(R_{si} - R_s)$ is the change in the total volume of gas dissolved in the oil initially, in going from initial conditions (p_i, R_{si}) to $[p(t), R_s]$ at time t. Therefore, the full term $N(R_{si} - R_s)B_g$ represents the volume occupied, under reservoir conditions prevailing at time t, by the dissolved gas liberated from the oil as a result of the pressure decrease from p_i to $p(t)$.

The numerator in Eq. (10.60a), then, is the global volume change undergone by the initial oil and its dissolved gas because of the reduction in reservoir pressure from p_i to $p(t)$.

Equation (10.60a) is the ratio of the expansion of the initial oil volume and the volume occupied under reservoir conditions by all the produced fluids. It constitutes an indicator of *that fraction of the total energy available from the combined reservoir production mechanisms, which is provided by the expansion of the oil and its dissolved gas alone.* This indicator is called the *depletion drive index* or *DDI*.

Now we will take a close look at the numerator in Eq. (10.60b). It can be written as follows:

$$mNB_{oi}\left(\frac{B_g}{B_{gi}} - 1\right) = \frac{mNB_{oi}}{B_{gi}} B_g - mNB_{oi} , \tag{10.62}$$

where mNB_{oi} is the initial volume of gas in the gas cap ($p = p_i$, $B_g = B_{gi}$), so that mNB_{oi}/B_{gi} is its volume under standard conditions.

It follows that $(mNB_{oi}/B_{gi})B_g$ represents the volume at time t [pressure $= p(t)$, gas volume factor $= B_g$] of the gas initially present in the gas cap. The numerator in Eq. (10.60b) is therefore the overall expansion of the gas cap from time $t = 0$ to time t.

Equation (10.60b) defines the ratio between this gas cap expansion and the volume occupied, under reservoir conditions, by the totality of the produced fluids. This is an indicator of *the fraction of the total energy available from the combined reservoir production mechanisms, which is provided by the expansion of the gas cap.* It is referred to as the *segregation drive index* or *SDI*. This assumes that the gas cap, while expanding, always remains perfectly segregated from the oil.

It should now be an easy matter to understand the remaining two indicators defined in Eqs. (10.60c) and (10.60d).

WDI, the water drive index, is the fraction of the total energy available from the combined reservoir production mechanisms, which is provided by the water drive from the aquifer, plus any water that has been injected.

CDI is the compaction drive index. This is the fraction of the total energy available from the combined reservoir production mechanisms, which is provided by the compaction of the reservoir rock and the expansion of the interstitial water as a result of the decrease in reservoir pressure from p_i to $p(t)$ at time t.

CDI is usually negligible in a light oil reservoir or where there is a gas cap; it can, however, be significant in undersaturated reservoirs, particularly if they are at shallow depth and in poorly compacted rock[9] (high c_p), with no water drive (WDI $= 0$).

The evaluation of these drive indices from the production history of the reservoir, and of their evolution with time, can provide a useful insight into the prevalent drive mechanism and the kind of recovery factor $E_{R,O}$ that can be expected. For a given reservoir, the overall recovery factor is in fact the weighted mean of the recovery factors for the specific drive processes:

$$E_{R,O} = DDI\,(E_R)_{DDI} + SDI\,(E_R)_{SDI} + WDI\,(E_R)_{WDI} + CDI \tag{10.63}$$

the recovery factor for compaction drive being 1.

It is straightforward to calculate values for DDI, SDI and CDI from the production history of the reservoir and the thermodynamic properties of the oil and gas, using Eqs. (10.60a–d).

Except in very special cases, however, we cannot estimate $W_e(t)$ from the production history. WDI therefore is usually taken as the difference between 1.0 and the sum of the other indices:

$$WDI = 1 - (DDI + SDI + CDI) , \tag{10.64}$$

which then enables us to calculate W_e from Eq. (10.60c).

10.5.3 Special Cases

The drive mechanisms described in Sect. 10.5.2 are not necessarily all present in a reservoir. We will next examine the most important cases where only some of the drives are present.

10.5.3.1 Undersaturated Oil Reservoir at Pressures Above Bubble Point

We will consider a reservoir containing undersaturated oil, which means that its initial pressure p_i is higher than the bubble point p_{bp} (Sect. 2.3.2). A gas cap cannot exist in equilibrium with the oil in such a situation.

If we only look at the early part of the production history where $p(t) > p_{bp}$, we have:

$$R_s = R_p = R_{si} \qquad (10.65a)$$

and, in addition [Eq. (2.15)]:

$$B_o - B_{oi} = c_o B_{oi} \, \Delta p \; . \qquad (10.65b)$$

In view of Eqs. (10.65) and the absence of a gas cap, the volumetric balance Eq. (10.59) can be simplified to:

$$N_p B_o + W_p B_w = N B_{oi} c_o \, \Delta p + N B_{oi} \frac{\bar{S}_{wi} c_w + c_p}{1 - \bar{S}_{wi}} \, \Delta p + W_e + B_w W_i \; . \qquad (10.66)$$

Since $\bar{S}_{oi} = 1 - \bar{S}_{wi}$, we can write Eq. (10.66) as:

$$N_p B_o + W_p B_w = N B_{oi} c_{o,e} \, \Delta p + W_e + B_w W_i \; , \qquad (10.67a)$$

where:

$$c_{o,e} = \frac{\bar{S}_{oi} c_o + \bar{S}_{wi} c_w + c_p}{1 - \bar{S}_{wi}} \; . \qquad (10.68)$$

$c_{o,e}$ is the *equivalent oil compressibility*, or the volume of oil produced per m^3 of oil initially in the reservoir, per unit decrease in pressure ($\Delta p = 1$).

In the particular case where the reservoir is not in communication with an aquifer, and no water is being injected into it, ($W_e = 0$, $W_i = 0$, $W_p = 0$), Eq. (10.67a) reduces to:

$$N_p B_o = N B_{oi} c_{o,e} \, \Delta p \; . \qquad (10.67b)$$

$c_{o,e}$ is usually very small (between 1×10^{-2} and 1×10^{-4} MPa^{-1}), so in the absence of any form of water drive, the pressure declines rapidly (Fig. 10.18). To sustain well productivity, this must be prevented by pressure maintenance through water injection. This will be described in Chap. 12 (Vol. 2).

Whether pressure support is due to water injection or to natural water drive from an aquifer, the water/oil ratio (WOR) will increase steadily (Fig. 10.19) once water has broken through at the nearest producing well. The produced gas/oil ratio (GOR), on the other hand, will remain constant so long as $p > p_{bp}$.

There also exist a number of reservoirs, particularly in shallow non-consolidated rock (e.g. the Bachaquero, Tia Juana and Lagunillas fields on the

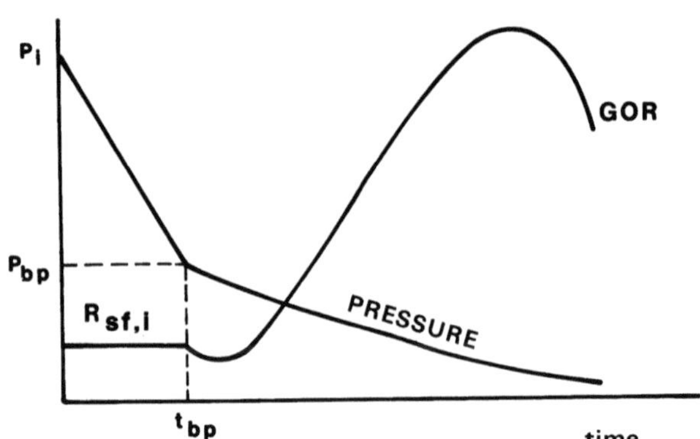

Fig. 10.18. *Behaviour of the pressure and produced GOR in an initially undersaturated oil reservoir with no aquifer support, and no water injection*

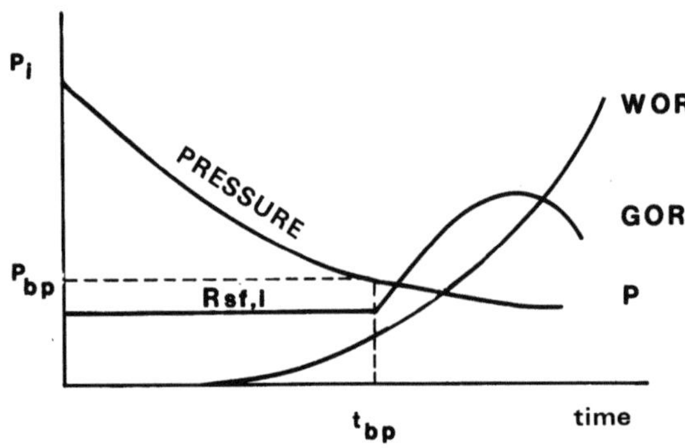

Fig. 10.19. *Behaviour of the pressure, produced GOR and WOR in an initially undersaturated oil reservoir under aquifer or injection water drive*

Bolivar Coast of Venezuela,[4] and the Wilmington group of fields on the Californian coast), but also in certain carbonates (such as the biomicrites found in the Ekofisk field in the Norwegian sector of the North Sea), where c_p, and therefore $c_{o,e}$ [Eq. (10.68)], is exceptionally large. In some of these reservoirs,[9] it is rock compaction that provides the major part of the drive energy for production.

10.5.3.2   Saturated Oil Reservoirs, Without Gas Cap or Aquifer

In a saturated oil reservoir that has neither gas cap, aquifer support nor water injection, we have $m = 0$, $W_e = 0$, $W_i = 0$ and $W_p = 0$.

Because the compressibility of the gas liberated since the onset of production ($p = p_{bp}$) is very much larger than $c_{o,e}$ as defined in Eq. (10.68), the reservoir rock compaction term [Eq. (10.56)] is relatively insignificant and can be ignored.

The volumetric material balance equation [Eq. (10.59)] for a saturated oil reservoir without gas cap, aquifer or water injection therefore reduces to:

$$N_p[B_o + (R_p - R_s)B_g] = N[(B_o - B_{oi}) + (R_{si} - R_s)B_g] \tag{10.69}$$

after eliminating all the zero or negligible terms.

The typical behaviour of the pressure and GOR below the saturation pressure is shown in Fig. 10.18.

The GOR rises rapidly as the pressure declines. This is due to the formation of a high saturation of free gas throughout the reservoir, and the favourable relative permeability to gas that ensues (Sect. 3.5.2.5).

The GOR will eventually reach a maximum and go into decline. Practically speaking, from now on almost all the gas liberated from the oil is produced to surface and makes no contribution to the drive process. Some of this liberated gas might migrate vertically to form what is known as a secondary gas cap: the extent to which this happens, if at all, depends on the vertical permeability and the rate of decline of the reservoir pressure.

On the whole, the solution gas drive mechanism described above is at best an inefficient one, since the major part of the potential drive energy (in the form of the gas dissolved in the oil) is lost as the liberated solution gas moves towards the wells and is produced to surface, without making any contribution to sweeping the oil itself.

Better recovery can be achieved in practice by replacing the solution gas drive by water drive. This requires drilling injection wells in the periphery of the field and/or amongst the producers, and will be described more fully in Chap. 12 (Vol. 2).

10.5.3.3 *Oil Reservoir with a Gas Cap, Not in Communication with an Aquifer*

In a reservoir with a gas cap in equilibrium with the oil column, the oil should be saturated with gas. However, there are many cases where equilibrium conditions are found to exist for only a few metres, or tens of metres, below the gas/oil contact: deeper than this, the oil is observed to be undersaturated. Non-equilibrium cases like this can only be handled using 3-D numerical simulators capable of modelling the variations in the oil PVT properties throughout the pay zone.

In the much simpler equilibrium case, the material balance equation [Eq. (10.59)] can be written as follows, remembering that $W_e = 0$, $W_i = 0$ and $W_p = 0$, and ignoring the equivalent oil compressibility $c_{o,e}$ as in the previous section:

$$N_p[B_o + (R_p - R_s)B_g] = NB_{oi}\left[\frac{(B_o - B_{oi}) + (R_{si} - R_s)B_g}{B_{oi}} + m\left(\frac{B_g}{B_{gi}} - 1\right)\right]. \tag{10.70}$$

If the production rate is sufficiently low, and the vertical permeability good enough, to allow a *frontal* advance of the gas cap (gravity stabilisation is covered in Chap. 11, Vol. 2), and, in addition, the open intervals in the wells are located far enough below the gas/oil contact to prevent downward *coning* of the gas (also dealt with in Chap. 12, Vol. 2), the gas cap expansion [the term $m[(B_g/B_{gi}) - 1]$ in Eq. (10.70)] can prove even more efficient as a recovery mechanism than water drive.

Ideally, the gravity segregation of the gas/oil system should occur without the development of gas *fingering* within the oil. If these gas fingers extend from the gas cap to the wells, so that gas is produced to surface, drive energy is lost.

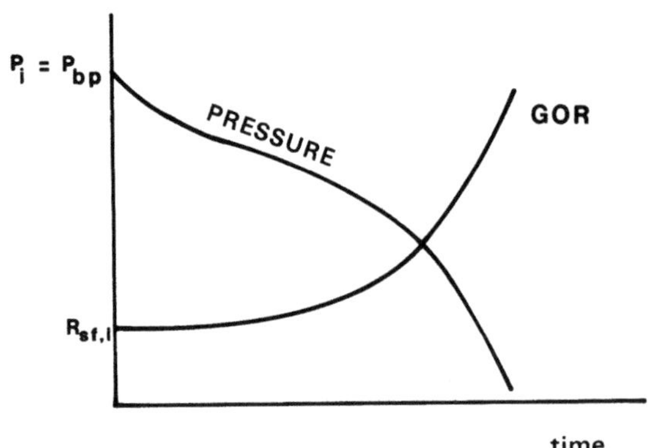

Fig. 10.20. *Behaviour of the pressure and produced GOR in an oil reservoir with a gas cap, producing under gravitational segregation drive*

Figure 10.20 shows the typical behaviour of an oil reservoir with a gas cap producing under gravitational segregation.

References

1. Chierici GL (1985) Petroleum reservoir engineering in the year 2000. Energy explor exploit 3 (3): 173–193
2. Chierici GL, Pizzi G, Ciucci GM (1967) Water drive gas reservoirs: uncertainty in reserves evaluation from past history. J Petrol Tech (Feb): 237–244 and (July): 965–967; Trans AIME 240
3. Cullender MH, Smith RV (1956) Practical solution of gas-flow equations for wells and pipelines with large temperature gradients. Trans AIME, 207: 281–287
4. Dake LP (1978) Fundamentals of reservoir engineering. Elsevier, Amsterdam
5. Dietz DN (1965) Determination of average reservoir pressure from build-up surveys. J Petrol Tech (Aug): 955–959; Trans AIME 234
6. Havlena D, Odeh AS (1963) The material balance as an equation of a straight line, part I. J Petrol Tech (Aug): 896–900; Part II, Field cases. J Petrol Tech (July 1964): 815–827; Trans AIME 228 and 231
7. Hurst W (1934) Unsteady flow of fluids in oil reservoirs. Physics (Jan): 5
8. Matthews CS, Brons F, Hazebroek P (1954) A method for the determination of average pressure in a bounded reservoir. Trans AIME, 201: 182–191
9. Merle HA, Kentie CJP, van Opstal GHC, Schneider GMG (1976) The Bachaquero study – a composite analysis of the behaviour of a compaction drive/solution gas drive reservoir. J Petrol Tech (Sept): 1107–1115; Trans AIME 261
10. Moody LF (1944) Friction factors for pipe flow. Trans ASME, p 671
11. Muskat M (1937) The flow of homogeneous fluids through porous media, J.W. Edwards, Ann Arbor
12. Muskat M (1949) Physical principles of oil production. McGraw-Hill, New York
13. Pirson SJ (1958) Oil reservoir engineering. McGraw-Hill, New York
14. Schilthuis R (1936) Active oil and reservoir energy. Trans AIME, 118: 33–52
15. Soave G (1972) Equilibrium constants from a modified Redlich-Kwong equation of state. Chem Eng Sci 27: 1197–1203
16. van Everdingen AF, Hurst W (1949) The application of the Laplace transformation to flow problems in reservoirs. J Petrol Tech (Dec): 305–324B; Trans AIME 186
17. Young KL (1967) Effects of assumptions used to calculate bottom-hole pressures in gas wells. J Petrol Tech (April): 547–550; Trans AIME 240
18. Phase Behavior (1981) SPE Reprint Series No 15, Society of Petroleum Engineers, Dallas
19. Pedersen KS, Fredenslund Aa, Thomassen P (1989) Properties of oils and natural gases. Contributions in petroleum geology and engineering, vol 5. Gulf Publ, Houston

EXERCISES

Exercise 10.1

In a dry gas reservoir (99.8% methane) not in communication with an aquifer (i.e. of the "closed" type), the initial gas in place is:

$$G = 10 \text{ billion } (10 \times 10^9) \text{ m}^3 \text{ at standard conditions.}$$

The initial reservoir pressure is:

$$p_i = 200 \text{ kg/cm}^2 \text{ abs.}$$

and the reservoir temperature is:

$$T_R = 71.1 \,^{\circ}\text{C} \ .$$

Fifteen wells have been drilled and completed for production. The flow equation for an average well, which for simplicity we will assume to be applicable to all the wells, is:

$$q_{sc}(\text{sm}^3/\text{day}) = 150 \, (\bar{p}^2 - p_{wf}^2)^{0.9}$$

where

$$\bar{p} = \text{average reservoir pressure (kg/cm}^2 \text{ abs.)}$$

and

$$p_{wf} = \text{bottom hole flowing pressure (kg/cm}^2 \text{ abs.)}$$

Therefore:

$$p_{wf} = \left[\bar{p}^2 - \left(\frac{q_{sc}}{150} \right)^{1.111} \right]^{0.5} \ .$$

An economic study has estimated that profitability could be maximised by drilling six further wells, making a total of 21.

In order to avoid sanding, the following production constraints must be applied in all wells:

maximum production rate: 250 000 sm^3/day

maximum drawdown $(\bar{p} - p_{wf})$:   15 kg/cm^2

The average downtime per well as a result of closures for various operations (well testing, recompletions and workovers) is estimated to be 10% of the total time (36.5 d/yr).

The objective is to maintain a steady production rate of 1.2 billion sm^3/yr from the reservoir for as long as possible.

Devise a program for the optimum development of this reservoir.

Table E10/1.1. *Gas properties*

p MPa	kg/cm^2	z dimensionless	p/z kg/cm^2
5.516	56.25	0.9536	58.98
8.618	87.88	0.9357	93.92
10.34	105.4	0.9257	113.9
12.07	123.1	0.9190	133.9
13.79	140.6	0.9144	153.8
17.24	175.8	0.9130	192.6
20.68	210.9	0.9199	229.2

Table E10/1.2. *Predicted reservoir behaviour*

Time	Production		Number of wells	Average production per well	Average reservoir pressure	$\bar{p} - p_{wf}$ at end of period	Comments
	In time interval (ΔG_p)	Cumulative (G_p)		(q_g)	($\bar{p}$) at end of period		
(years)	($sm^3 \times 10^9$)	($sm^3 \times 10^9$)		(sm^3/day)	(kg/cm^2 abs)	(kg/cm^2)	
0	zero	zero	15	zero	200.00	zero	
0.5	0.60	0.60	15	243 500	187.95	10.09	
1	0.60	1.20	15	243 500	175.90	10.82	
1.5	0.60	1.80	15	243 500	163.96	11.67	
2	0.60	2.40	15	243 500	152.04	12.66	
2.5	0.60	3.00	15	243 500	140.18	13.85	
3	0.60	3.60	15	243 500	128.53	15.26	$(\bar{p} - p_{wf})$ limit reached
3.5	0.60	4.20	16	228 300	116.94	15.75	One more well drilled
4	0.60	4.80	18	203 000	105.38	15.44	Two more wells drilled
4.5	0.60	5.40	21	174 000	93.89	14.68	Three more wells drilled
5	0.55	5.95	21	160 000	83.19	15.32	Field production rate reduced
5.5	0.48	6.43	21	140 000	73.65	15.09	Further reduction in rate
6	0.40	6.83	21	116 000	65.70	13.76	Further reduction in rate

Solution

The material balance equation for a closed dry gas reservoir is:

$$GB_{gi} = (G - G_p)B_g .$$ (10.6a)

We can restate this as:

$$G\frac{z_i}{p_i} = (G - G_p)\frac{\bar{z}}{\bar{p}}$$

so that:

$$\frac{\bar{p}}{\bar{z}} = \left(1 - \frac{G_p}{G}\right)\frac{p_i}{z_i} .$$ (10.6b)

The gas is almost pure methane at 71.1 °C. Therefore we will be able to use the data already calculated in Ex. 7.1, and summarised in Table E10.1.1.

From these, we can interpolate for $p = 200$ kg/cm^2:

$$\frac{p_i}{z_i} = \frac{200}{z(200)} = 218.90 \text{ kg/cm}^2 \text{ abs.}$$

so that:

$$\frac{\bar{p}}{z} = 218.90[1 - (G_p/G)] .$$ (10/1.1)

We can use Eq. (10/1.1) to calculate the average reservoir pressure for different values of $G_p(t)$.

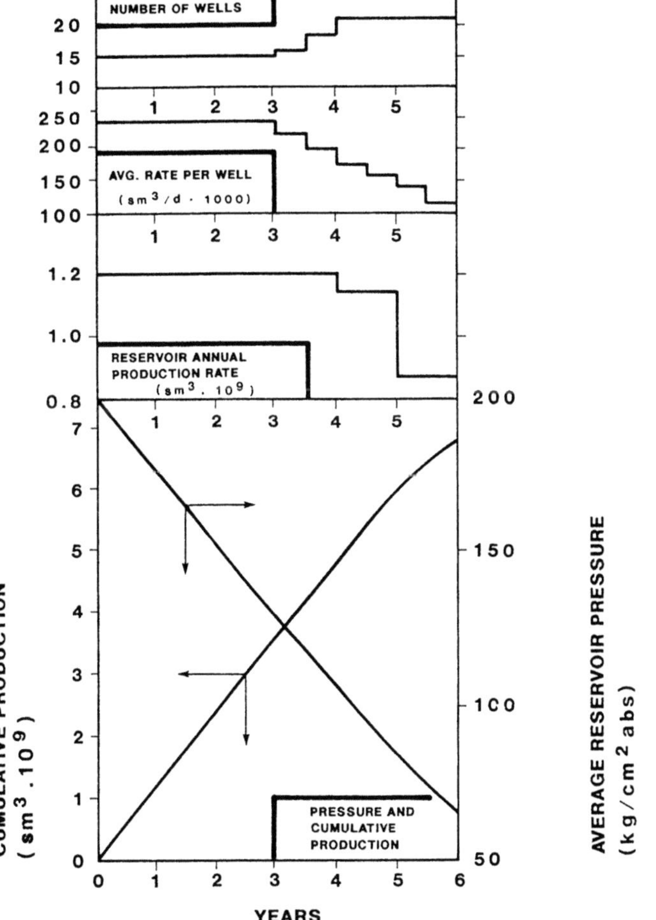

Fig. E10/1.1

$G_p(t)$ itself is the sum of the increments of gas production during each time interval ($\Delta t = 0.5$ yr), taking into account the number of wells and any constraints on individual wells.

The results are listed in Table E10/1.2 and plotted in Fig. E10/1.1.

The calculations have been interrupted at the end of the sixth year, by which time approximately 70% of the gas initially in place had been produced, the production rate had fallen to 0.8 billion sm^3/yr with the recommended maximum number of wells, and the static reservoir pressure was 65.7 kg/cm^2 abs while bottom-hole flowing pressures were about 50 kg/cm^2.

Production from the reservoir can of course be continued. However, at the end of this sixth year, the installation of gas compression facilities should be given serious consideration, since the wellhead pressure is by now very close to the minimum required to deliver the gas to the pipeline (operating range 50–70 kg/cm^2).

A long period of gradually declining productivity will follow. The field will have to be shut down when operating revenue no longer covers costs.

Exercise 10.2

The reservoir described in Ex.10.1 is in contact with an aquifer of unknown size and properties.

There are doubts about the validity of the reserves estimated by the volumetric method. If the reservoir has been on stream for 6 yr, and its production/pressure behaviour is that listed in Table E10/1.2, use the material balance equation to estimate the initial volume, G, of gas in place.

Because none of the wells have intercepted the aquifer or cut water, we have no idea about the extent of water influx, so some assumptions will have to be made about how much of the reservoir might have been swept.

Four scenarios will be investigated, each with a different fraction of the reservoir volume swept by encroached water at the end of year 6, as follows:

  a) 5% of the reservoir volume
  b) 10% of the reservoir volume
  c) 20% of the reservoir volume
  d) 30% of the reservoir volume.

Solution

From Eq. (10.15):

$$\frac{\bar{p}}{\bar{z}} = \frac{p_i}{z_i} \frac{1 - (G_p/G)}{1 - (W_e/V_{g,o})}, \tag{10.15}$$

where:

W_e = volume of water that has invaded the reservoir

$V_{g,o}$ = reservoir pore volume initially occupied by gas

so that: $W_e/V_{g,o}$ = fraction of the hydrocarbon-bearing pore volume that has been invaded by water from the aquifer.

Rearranging, we can write:

$$G = \frac{G_p}{1 - (\bar{p}/\bar{z})(z_i/p_i)[1 - (W_e/V_{g,o})]} . \tag{10/2.1}$$

In the present case we have:

$p_i = 200$ kg/cm^2 abs ,

$z_i = 0.9137$,

giving:

$p_i/z_i = 218.89$ kg/cm^2 abs

and, in the sixth year of production:

$$G_p = 6.83 \times 10^9 \text{ sm}^3 \, ,$$

$$\bar{p} = 65.70 \text{ kg/cm}^2 \text{ abs} \, ,$$

$$\bar{z} = 0.9468 \, ,$$

so that:

$$\bar{p}/\bar{z} = 69.39 \text{ kg/cm}^2 \text{ abs} \, .$$

Equation (10/2.1) therefore becomes:

$$G = \frac{6.83 \times 10^9}{0.683 + 0.317 \left(W_e/V_{g,o}\right)} \, .$$

From this we obtain:

$W_e/V_{g,o}$	$G\,(\text{sm}^3 \times 10^9)$
0.05	9.77
0.10	9.56
0.20	9.15
0.30	8.78

Obviously, for $W_e = 0$, $G = 10 \times 10^9 \text{ sm}^3$.

Exercise 10.3

An oil reservoir, whose upper section is in contact with a gas cap, and lower section with an aquifer, has the following properties:

– depth	$D = 900 \text{ m}$
– initial pressure = bubble point pressure	$p_i = 96.9 \text{ kg/cm}^2$
– temperature	$T_R = 45\,°\text{C}$
– average initial water saturation	$\bar{S}_{wi} = 0.15$
– initial volume of oil in place (std conditions)	$N = 1 \times 10^9 \text{ m}^3$
– $\dfrac{\text{initial volume of gas in gas cap (res conditions)}}{\text{initial volume of oil in place}}$	$m = 0.4$

The reservoir water has the following characteristics:

– average volume factor	$\bar{B}_w = 1.015$
– average isothermal compressibility	$\bar{c}_w = 3.3 \times 10^{-5} \text{ cm}^2/\text{kg}$
The pore compressibility is	$c_p = 1.1 \times 10^{-4} \text{ cm}^2/\text{kg}$

In the first 2 years, the cumulative volumes of fluid produced were:

$$N_p = 50 \times 10^6 \text{ m}^3 \text{ of oil at standard conditions} \, ,$$

$$W_p = 5 \times 10^6 \text{ m}^3 \text{ of water at standard conditions} \, ,$$

with an *average* produced gas/oil ratio:

$$R_p = 48 \text{ m}^3/\text{m}^3 \, .$$

The reservoir pressure had fallen at the end of the second year to:

$$p_2 = 90.2 \text{ kg/cm}^2 \, .$$

	p (kg/cm^2)	B_o (dimensionless)	R_s (m^3/m^3)	B_g (dimensionless)
– at initial conditions	96.9	1.155	32.8	9.969×10^{-3}
– at end of 2nd year	90.2	1.144	30.3	1.075×10^{-2}

Calculate the drive indices (Sect. 10.5.2) and the volume of water that has entered the reservoir from the aquifer by the end of the second year of production.

Solution

Referring to Sect. 10.5.2, we will first calculate the value of the denominator in Eq. (10.60) for the end of year 2.

We have:

$$[\text{Den}] = N_p[B_o + (R_p - R_s)B_g] + W_p B_w$$
$$= 50 \times 10^6 [1.144 + (48 - 30.3)1.075 \times 10^{-2}] + 1.015 \times 5 \times 10^6 = 7.1789 \times 10^7 \ .$$

Next we calculate the three drive indices that can be derived from the available data. We have:

$$\text{DDI} = \frac{N(B_o - B_{oi}) + N(R_{si} - R_s)B_g}{[\text{Den}]}$$

$$= \frac{1 \times 10^9 (1.144 - 1.155) + 1 \times 10^9 (32.8 - 30.3)1.075 \times 10^{-2}}{7.1789 \times 10^7} = 0.2211 \ .$$

$$\text{SDI} = mNB_{oi} \frac{(B_g/B_{gi}) - 1}{[\text{Den}]}$$

$$= \frac{0.4 \times 1 \times 10^9 \times 1.155[(1.075/0.9969) - 1]}{7.1789 \times 10^7} = 0.5042 \ .$$

$$\text{CDI} = \frac{(1 + m)NB_{oi}[((\bar{S}_{wi}c_w + c_p)/(1 - \bar{S}_{wi})]\Delta p}{[\text{Den}]}$$

$$= \frac{1.4 \times 1 \times 10^9 \times 1.155[(0.15 \times 3.3 + 11)/(1 - 0.15)] \times 10^{-5}(96.9 - 90.2)}{7.1789 \times 10^7} = 0.0204 \ .$$

We can now get WDI from Eq. (10.61):

$$\text{WDI} = 1 - \text{DDI} - \text{SDI} - \text{CDI}$$
$$= 1 - 0.2211 - 0.5042 - 0.0204$$
$$= 0.2543$$

Therefore, from Eq. (10.60c):

$$W_e = \text{WDI} [\text{Den}] = 0.2543 \times 7.1789 \times 10^7.$$
$$= 18.26 \times 10^6 \ \text{m}^3 \text{ of water influx into the reservoir in the first two years of production.}$$

$$\diamond \ \diamond \ \diamond$$

Exercise 10.4

Suppose the reservoir of Ex. 10.3 is not in contact with an aquifer. The water drive index WDI is therefore zero (WDI = 0), and there will be no water production ($W_p = 0$).

For the same average producing gas/oil ratio ($R_p = 48$), what quantity of oil will have been produced if the reservoir pressure is the same as in the previous example after 2 years ($p_2 = 90.2$ kg/cm^2)?

Solution

Since WDI = 0, we have:

$$DDI + SDI + CDI = 1 \ .$$

The denominator of Eq. (10.60) reduces to:

$$[Den] = N_p[B_o + (R_p - R_s)B_g]$$

with $W_p = 0$ in this case.

Therefore:

$$[Den] = N_p[1.144 + (48 - 30.3) \, 1.075 \times 10^{-2}] = 1.33428 \, N_p$$

with N_p as yet unknown.

We will next calculate the numerators of the ratios that constitute the various drive indices:

$$[Num \ DDI] = N(B_o - B_{oi}) + N(R_{si} - R_s)B_g$$

$$= 1 \times 10^9(1.144 - 1.155) + 1 \times 10^9(32.8 - 30.3) \, 1.075 \times 10^{-2} = 1.5875 \times 10^7 \ ,$$

$$[Num \ SDI] = mNB_{oi}\left(\frac{B_g}{B_{gi}} - 1\right)$$

$$= 0.4 \times 1 \times 10^9 \times 1.155 \left(\frac{1.075}{0.9969} - 1\right) = 3.6194 \times 10^7 \ ,$$

$$[Num \ CDI] = (1 + m)NB_{oi}\frac{\bar{S}_{wi}c_w + c_p}{1 - \bar{S}_{wi}}\Delta p$$

$$= 1.4 \times 1 \times 10^9 \times 1.155 \, \frac{0.15 \times 3.3 + 11}{1 - 0.15} \times 10^{-5}(96.9 - 90.2) = 0.1465 \times 10^7 \ .$$

We now have the equality:

$$1 = \frac{1.5875 \times 10^7}{1.33428 \, N_p} + \frac{3.6194 \times 10^7}{1.33428 \, N_p} + \frac{0.1465 \times 10^7}{1.33428 \, N_p} \ ,$$

from which:

$$N_p = 40.12 \times 10^6 \text{ m}^3 \ .$$

This represents the quantity of oil that could be produced, in the absence of water drive, for the same reservoir pressure depletion as in Ex. 10.3.

An approximate calculation could have been made as follows.

The volume of fluids extracted in Exercise 10.3, *expressed at mean reservoir pressure*, was:

$$\text{oil:} \quad 50 \times 10^6 \, \frac{B_{oi} + B_o}{2} = 57.475 \times 10^6 \text{ m}^3$$

$$\text{water:} \ 5 \times 10^6 \, \bar{B}_w \qquad = 5.075 \times 10^6 \text{ m}^3$$

$$\overline{\qquad\qquad\qquad\quad 62.550 \times 10^6 \text{ m}^3}$$

The volume of water that entered the reservoir from the aquifer was:

$$W_e = 18.26 \times 10^6 \text{ m}^3$$

Therefore, the *net* volume of fluid leaving the reservoir was:

$$(62.55 - 18.26) \times 10^6 = 44.29 \times 10^6 \text{ m}^3$$

In order for there to be the same drop in pressure, the same volume of fluid – *measured at reservoir conditions* – must be removed.

Under the conditions specified in Exercise 10.3, this is equivalent to removing:

$$N_p = \frac{44.29 \times 10^6}{\dfrac{B_{oi} + B_o}{2}} = 38.53 \times 10^6 \text{ m}^3 \text{ of oil at standard conditions.}$$

There is a difference of -4% between this estimate of N_p and the previously calculated value. This can be attributed to the fact that, in this simplified approach, the reservoir volume of free gas $(R_p - R_s)$ produced by an amount ΔN_p of oil equal to the difference between this estimate and the previous one has not been taken into account.

Author Index

Subject Index

Springer-Verlag and the Environment

We at Springer-Verlag firmly believe that an international science publisher has a special obligation to the environment, and our corporate policies consistently reflect this conviction.

We also expect our business partners – paper mills, printers, packaging manufacturers, etc. – to commit themselves to using environmentally friendly materials and production processes.

The paper in this book is made from low- or no-chlorine pulp and is acid free, in conformance with international standards for paper permanency.